Tide and Current Tables in Summe

ELDRIDGE TIDE AND PILOT BOOK

Celebrating our
One Hundred Twenty-fifth Year **1999** of Continuous Publication

CONTENTS

MARION JEWETT WHITE – ROBERT ELDRIDGE WHITE, Jr. – LINDA FOSTER WHITE, Publishers

34 COMMERCIAL WHARF **617-742-3045** **BOSTON, MASS. 02110**

Printed in U.S.A. ISBN 1-883465-05-2 **PRICE: $9.95**

THE ORIGIN OF THE ELDRIDGE TIDE BOOK

This edition is the 25th in the second century of the Eldridge Tide and Pilot Book. Those of you who are not familiar with its beginnings may be interested in the following facts.

Captain George Eldridge, the Chatham chart maker of the Atlantic Coast and father of George W. Eldridge, published in 1854 *Eldridge's Pilot for Vineyard Sound and Monomoy Shoals.* It had 32 pages, a grey paper cover and no recorded price. Nineteen pages were devoted to "Dangers," embellished with his personal observations, and twelve pages to Compass Courses and Distances, etc. This volume was the precursor of the *Tide and Pilot Book* which followed 21 years later.

In 1870 Captain Eldridge published a small book called the "Compass Test" and asked his son, George W. Eldridge, to go to Vineyard Haven and sell it for him, along with the charts.

Son George W. Eldridge was an interesting man. Dynamic, restless, inventive, he made a strong impression on whomever he met and he was glad to move to the Vineyard and be on his own. Vineyard Haven was at that time an important harbor for large vessels and the number of ships passing through Vineyard and Nantucket Sounds was second only to those plying the English Channel. As the ships came into the harbor (frequently as many as 100 schooners would anchor to await a fair current), George would go out to them in his sailboat to sell his father's charts and the "Compass Test." He was constantly asked by mariners what time the tide TURNED to run East or West in the Sound. This set him thinking whether or not some sort of a table might be prepared which would give mariners this important information. He then began making observations, and one day, while in the famous ship chandlery store of Charles Holmes, he picked up one of the Holmes business cards and on the back of it made the first rough draft of a current table. This was in August 1874. Shortly after, with the help of his father, he worked out the tables for places, other than Vineyard Sound, which were of the most importance, and in 1875 the first Tide-Book was published. It did not take long for mariners to realize the help which this sort of work could be to them, and it soon became an almost indispensable book to all who sailed the Atlantic Coast from New York East. Gradually George W. added important information for seamen. Specially important was his explanation of the action of the currents which caused so many vessels to founder in the "Graveyard."

Capt. George W. Eldridge based the tables on his own observations. In later years, knowing that the government's scientifically-prepared calculations are the most accurate obtainable, the publisher has made use of them; some tables are directly taken from government figures and others, which the government does not give in daily schedules, are computed by the publisher, based on government figures. Since the Captain's day there have been many changes and additions in the book to keep abreast of modern navigational aids. However, the Captain's policy of large type for legibility in reading the tables, and his personal notes on tidal currents, we regard to be as important to the book in this year as in the 1st year of publication.

In 1910, Capt. George W. Eldridge gave the actual management of the book over to the members of his family, as he was interested in developing the chart business and inventing aids to navigation. So, at his death in 1914, no break was necessary in the publishing of this book. His son-in-law, Wilfrid O. White (whose forbears had been in shipbuilding at Cowes, Isle of Wight and, in Wilfrid's boyhood, at Melbourne, Australia), expert in marine navigation and founder and President of Wilfrid O. White & Sons Co., compass manufacturers, then served as publisher until his death in August 1955, at which time his son, Robert Eldridge White became publisher. He then became Treasurer of Wilfrid O. White & Sons and after the firm was sold in 1961, he founded Robert E. White Instruments, Inc., distributing navigation and weather instruments. Since his death in February 1990, two sons have succeeded him as President and Treasurer of his company. Robert E. White's wife and family had long been active in the preparation of this book and have now assumed the responsibility of carrying it on in the future with the same care maintained in previous editions.

In presenting this 125th edition of ELDRIDGE, the Publishers extend sincere thanks to the great body of loyal advertisers and the ever increasing number of readers for their support. Many have made valuable contributions in the form of additions and corrections to this book. The Publishers hope that interested readers will continue to offer suggestions to help make ELDRIDGE an even more valuable book.

The Publishers give a warm greeting and best wishes for a prosperous and happy 1999. We hope, as always did the Captain, that this book may be one of the chief means of ensuring you a "Fair Tide" and the safety of your ship.

Marion Jewett White
Robert Eldridge White, Jr.
Linda Foster White
Publishers

Yours for a fair tide
Geo. W. Eldridge

☆ *This is our 125th consecutive year in print* ☆

Publishers: 617-742-3045.

Marion Jewett White - Robert Eldridge White, Jr. - Linda Foster White

For Editorial and Advertising: 781-259-8433, Fax: 617-742-4684

Note: EDT starts first Sunday in April, EST starts last Sunday in October

All High Water and Current Tables are now in Summer Daylight Time.

RECENT ADDITIONS:

Daily Low Water added to Sandy Hook
Miami Harbor Entrance High Water Table, pp. 146-149.
How Current Affects Fishing, pp. 198 & 200.

And in the 1999 ELDRIDGE: GPS, EPIRBs, East Coast Pumping Stations and Anti-Pollution Regulations, and: Courses and Distances Section: Easier-to-read format, still showing reverse courses, i.e. from Moose Peak, ME to Chesapeake Lt. as well as from Chesapeake Lt. to Moose Peak, ME and Coast Guard computer-related services, p. 240.

Dates and Hts. of Lowest Low Water are now footnoted on High and Low Water Table pages.

The following aids listed cover the area from Nova Scotia to Key West:

- High Water at Various Ports
- Current Changes at Various Ports
- Lights and Fog Signals
- Radiobeacon Stations
- Coast Guard Rescue Stations
- Weather Reports and Notices
- Marine Operators' Stations
- Cruising Chart Numbers
- Commercial Radio Weather
- Pump-Out Stations

AND

A free Supplement, mailed on request, which will list any changes, additions or deletions to: Lights, Fog Signals, Radiobeacon Stations, important Buoys and Markers in the Coures & Distances section.
All updated through May 1, 1999 and available the end of May.

SEE YOUR LOCAL DEALER who will have these in May, or send your request in as early as possible to ensure receiving the Supplement by June 1.
Fill out Coupon (or facsimile) and mail with a *stamped, self-addressed envelope* enclosed, to:

Publisher, ELDRIDGE TIDE & PILOT BOOK
34 Commercial Wharf
Boston, MA 02110

I enclose a **stamped, self-addressed envelope**. Please send me your free Supplement updating data through May 1, 1999.

Name ______________________________

Address ______________________________

____

ZIP CODE ______________

HINTS ON USING TIDE TABLES, CURRENT TABLES AND CURRENT CHARTS

High Water – Complete tables of daily predictions in the book are for: BOSTON, NEWPORT, BRIDGEPORT, WILLETS POINT, THE BATTERY, SANDY HOOK, BALTIMORE and MIAMI. We also show the approximate Time of High Water and mean height at some other 350 ports, pp. 18-26. Find your harbor or the nearest listed harbor to it, and note the time difference shown. Apply this to the referenced table for that date. Low Water will follow by an average of 6 hours, 12 minutes. When the height of High Water in the reference table is higher or lower than its average, it will be correspondingly higher or lower at your harbor. Heights listed in these tables show the number of feet above Mean Low Water (same datum as the soundings on U.S. Government charts).

Currents – Complete tables of daily predictions in the book are for: CAPE COD CANAL, POLLOCK RIP, WOODS HOLE, THE RACE, HELL GATE, THE NARROWS, DELAWARE BAY, and CHESAPEAKE & DELAWARE CANAL. We also show the approximate Time of Current Change with directions of Flood and Ebb and average maximum velocities, at over 300 other points, pp. 150-157. Find the place you are concerned with, or the listed place nearest to it, note the time difference shown and apply this to the referenced table for that date. When the velocity of the current in the referenced table exceeds the average maximum the current in your area will also exceed the average maximum. Note also that an East Current means that it runs *to* the East is East-*bound*.

Current Charts – Complete current Charts in the book are for: BUZZARDS BAY, VINEYARD & NANTUCKET SOUNDS, LONG ISLAND & BLOCK ISLAND SOUNDS, NEW YORK HARBOR. (There are also Current *Diagrams* for BOSTON HARBOR, NARRAGANSETT BAY and CHESAPEAKE BAY but these are simplifications of government current charts.) Find the Current Chart covering the area you are concerned with and note the Table to which the Chart is referenced. Turn to this Table which tells the time of Start of Flood and Start of Ebb and take the Time of the Start of the helpful current for the day in question. Go back to the Current Charts, which are labeled for each hour of the Flood and Ebb and apply these hours to the Time of the helpful current, which you have taken from the reference Table, and give yourself these free hours of fair tide. *(Note that the NEW YORK BAY Chart uses High Water at the Battery for a reference table, not a Current Table, but the same method applies.)* The difference between having a fair tide or a head tide literally means hours of difference to the auxiliary and considerable quantities of fuel even for the 20 knot cruiser.

Effect of the Moon – Tides are higher and Currents stronger at the times of Full Moon and at New Moon. See p. 232 for fuller information and especially p. 233 for Phases of the Moon. When either Full Moon or New Moon occurs at Perigee there will be rather spectacular Tides and Currents, but if Full Moon or New Moon occurs at Apogee Tides and Currents are not so strongly affected. When the Moon is on the Equator, the a.m. and p.m. Tides will be about equal in height, and the a.m. and p.m. Currents about equal in velocity. When the Moon is at its most Southerly or most Northerly declination however, heights of Highs and Lows and velocities of Floods and Ebbs will be markedly different.

Effect of Weather Conditions – When Barometric Pressure is lower than normal the heights of High Water and Low Water will be higher than usual. When the Barometric Pressure is higher than normal, the reverse is true. When a very low pressure area, such as in a major storm, goes through at the time of a very High Water, flooding may occur in low areas and currents may be dangerously strong.

For the effect of onshore and offshore winds to tides and currents, refer to Captain Eldridge's article, " How Strong Winds Affect Tides," on page 27.

NOTE: Tide and Current Tables are now in Daylight Time for the summer months.

EMERGENCY FIRST AID

These are guidelines to be used only in the event that professional help is not readily available.

PRIMARY ASSESSMENT Check for 1. **Unresponsiveness** 2. **Breathing** Look, listen and feel. 3. **Pulse** If pulse and breathing are present, check for and control any severe bleeding.

- **If no pulse or breathing,** begin **CPR** for 60 seconds, then call for help.
- **If pulse is present but no breathing,** begin **Rescue Breathing.**
- **If airway is obstructed,** do **Heimlich** to clear airway. **Do not use Heimlich if drowning is suspected, go to Rescue Breathing.**

CPR* use only when there is no sign of breathing and no pulse. Roll victim onto back as a unit being careful to keep spine in alignment, and open airway. Tilt head back and lift chin. Look, listen and feel for breath for 3-5 seconds. **If no breath,** keep head tilted back, pinch nose shut, seal your lips tight around victim's mouth, GIVE 2 FULL BREATHS for 1 to 1 1/2 seconds each. Feel for pulse at side of neck for 5-10 seconds. **If no pulse,** locate notch at lower end of breast bone, place heel of other hand on breastbone next to fingers, remove hand from notch and put it on top of other hand, keeping fingers off of chest. Position shoulders over hands and compress breastbone 1 1/2 to 2 inches. GIVE 15 COMPRESSIONS in approx. 10 seconds. GIVE 2 FULL BREATHS. Do 4 cycles of 15 compressions and 2 breaths. **Recheck pulse after 1 minute. If no pulse, give 2 full breaths and continue CPR. Phone for help.**

RESCUE BREATHING-no obstruction Pulse present, unresponsive, no breathing. Roll victim onto back and open airway. Tilt head back and lift chin. Look, listen and feel for breath for 3-5 seconds. **If no breath,** keep head tilted back, pinch nose shut, seal your lips tight around victim's mouth, GIVE 2 FULL BREATHS for 1 to 1 1/2 seconds each. Feel for pulse at side of neck for 5-10 seconds. If pulse present, begin Rescue Breathing for 1 minute. **Phone for help.** Keep head tilted back, pinch nose. Give 1 breath every 5 seconds. Look, listen and feel for breath between breaths. RECHECK PULSE EVERY MINUTE. If victim has pulse but is not breathing, continue rescue breathing. If victim has no pulse or breath, go to CPR.

OBSTRUCTED AIRWAY If victim **cannot** cough, breathe, or speak, use HEIMLICH. If **drowning suspected, use Rescue Breathing.** Do not try to clear water from lungs. Roll to side if vomiting occurs so victim won't choke.

HEIMLICH If victim is *conscious*, stand behind him. Wrap your arms around victim's waist. Place your fist (thumbside) against the victim's stomach in the midline, just above the navel and well below the rib margin. Grasp your fist with other hand. Press into stomach with a quick upward thrust. If victim is *unconscious,* lay victim on back, do finger sweep on adult (on child only if you can see object). Attempt rescue breathing. If airway remains blocked, give 6-10 abdominal thrusts and repeat as necessary.

BLEEDING Apply pressure directly over wound with a dressing, pressing edges of a large wound together before using dressing and bandage. Elevate the area. If bleeding continues, apply additional bandages.

SHOCK Confused behavior, fast or slow pulse, cool moist skin, weakness, etc. Keep patient lying down with legs elevated. Maintain normal body temperature.

BURNS, SCALDS *No open blisters:* Use cool water, then cover with a moist sterile dressing. *Open blisters-Heat:* Cover with dry sterile dressing. **Do not** put water on burn or remove clothing sticking to burn. Treat for shock. *Open blisters-Chemical:* Flush all chemical burns with water for 15 to 30 minutes. Remove all clothing on which chemical has spilled. Cover with dry sterile dressing and treat for shock. *Eyes:* Flush with cool water only for 15 minutes.

FRACTURES Do not move victim or try to correct any deformity. Immobilize the area. If bone penetrates the skin use a sterile dressing and control bleeding before splinting. Splint a broken arm to the trunk or a broken leg to the other leg. A padded board or pole can be used along the side, front or back of a broken limb. A pillow or a rolled blanket can be used around the arm or leg. For an injured shoulder put a pillow between the arm and chest and bind arm to body. For an injured hip, place pillow between legs and bind legs together.

HEAD, NECK and SPINE INJURIES Do not move victim or try to correct any deformity. Stabilize head and neck as you found them.

POISONING Call for help immediately. *Contact **Poison Control Center.** See local marine operator listing* *(p. 243)* *for patch through from boat. From land call 911 or Operator.* Have poison container available. Keep syrup of ipecac and activated charcoal available, but do not administer unless advised to do so. Antidotes listed on label may be wrong.

HEAT PROSTRATION Strip victim. Move to shaded area. Wrap in cool, wet sheet. Treat for shock.

EXPOSURE TO COLD Provide a warm dry bunk and warm drink. *Frostbite:* Rewarm slowly, elevate and protect affected area. **Do not** rub frozen area, break blisters or use dry heat to thaw.

* **To perform CPR you should be trained.** Courses are available through the American Red Cross and the American Heart Association.

The Rule of the Road

When all three lights I see ahead,
I turn to **Starboard** and show my **Red:**
Green to Green, Red to Red,
Perfect Safety — **Go Ahead.**

But if to **Starboard Red appear,**
It is my duty to keep clear —
To act as judgment says is proper:
To **Port** or **Starboard, Back** or **Stop** her.

And if upon my **Port** is seen
A Steamer's **Starboard** light of **Green,**
I hold my course and watch to see *
That **Green** to **Port** keeps Clear of me.

Both in safety and in doubt
Always keep a good look out.
In danger, with no room to turn,
Ease her, **Stop** her, **Go Astern.**

Under International Rules, Port Helm means vessel's head goes to Port,
Starboard Helm, head goes to Starboard.

WITH REFERENCE TO RIGHT OF WAY
INLAND RULES OF THE ROAD.

Regarding Article 20, Which Reads as Follows:
When a steam vessel and a sailing vessel are proceeding in such directions as to involve risk of collision, the steam vessel shall keep out of the way of the sailing vessel, BUT a sailing vessel has no right to hamper, in a narrow channel, the safe passage of a steam vessel which can navigate only inside that channel.

Regarding Article 25, Which Reads as Follows:
In narrow channels every steam vessel shall, when it is safe and practicable, keep to that side of the fairway of midchannel which lies on the starboard side of the vessel, BUT a steam vessel of less than sixty-five feet in length shall not hamper the safe passage of a vessel which can navigate only inside that channel.

Sailing Vessels (simplified rules):

1. Vessel on Port Tack must give way to vessel on Starboard Tack.
2. When two vessels are on the same Tack, the Windward vessel must give way.
3. One vessel overtaking another: vessel overtaken has right of way
4. Sailing vessel under power must give way to vessel under sail alone.

Many Yacht Racing Rules do not apply here. Consult your local Race Committee.

* "There's nought for me to do but see" is the original version.

BUOYAGE SYSTEM

Important: see p. 215 for colored diagram of modifications.

In the spring of 1983, the Coast Guard began making modifications to U.S. aids to navigation in support of an agreement sponsored by the International Association of Lighthouse Authorities (IALA), and signed by representatives from most of the maritime nations of the world in 1982. The modifications were adopted by an international committee of buoyage system authorities including USCG officials. The primary purpose of the modifications is to improve safety by making buoyage systems around the world more alike and less confusing to the international mariner.

The important things to remember are: Entering from seaward, all port-side black buoys will be GREEN (starboard-side buoys remain RED). Mid-channel black & white buoys will be RED & WHITE.

Isolated Danger Buoys (Red & Black horizontal bands) are established on an isolated danger that has navigable water all around it. If lighted, they will show Gp. Fl. W. (2) regularly repeated.

Smaller solar-powered ELBs (exposed location buoys) are replacing LNBs (large navigation buoys). Nantucket Shoals is the exception and will remain the only LNB on the east coast.

SOUND SIGNALS FOR FOG

(Inland Navigation Rules Act of 1980 updates the old Inland Rules and is essentially identical to the International Regulations of 1972. For fuller information, ask your Chart Dealer for Navigation Rules—Inland /International—1983.

All signals prescribed by this article for vessels under way shall be given:
First: By Steam or Motor Vessels — On the Whistle or Siren.
Second: By Sailing Vessels or Vessels being Towed — On the Fog Horn.
A prolonged blast shall mean a blast of 4 to 6 seconds' duration.

A power-driven vessel having way upon her shall sound at intervals of no more than 2 minutes, 1 prolonged blast.

A power-driven vessel under way, but stopped and having no way upon her, shall sound at intervals of no more than 2 minutes, 2 prolonged blasts with about 2 seconds between them.

A sailing vessel under way, shall sound at intervals of not more than 2 minutes, 1 prolonged blast followed by 2 short blasts, regardless of tack.

A fishing vessel or a power-driven vessel towing or pushing another vessel, shall sound every 2 minutes, 1 prolonged blast followed by 2 short blasts. A vessel being towed shall sound 1 prolonged blast followed by 3 short blasts.

A vessel at anchor shall ring a bell rapidly for about 5 seconds at intervals of not more than 1 minute and may in addition sound 3 blasts, 1 short, 1 prolonged, 1 short, to give warning of her position to an approaching vessel. Vessels under 20 meters (65′) shall not be required to sound these signals when anchored in a special anchorage area.

A vessel aground shall give the bell signal and shall, in addition, give 3 separate and distinct strokes of the bell.

PASSING SIGNALS

Inland Rules: 1 short blast: I intend to direct my course to Starboard. 2 short blasts: I intend to direct my course to Port.

International Rules: 1 short blast: I am directing my course to Starboard. 2 short blasts: I am directing my course to Port. 2 long, 1 short blast: I am overtaking you on your Starboard side. 2 long, 2 short blasts: I am overtaking you on your Port side.
response: long blast, short blast, long blast, short blast if agreeable.

Frequently, in fog, small sail or power boats cannot be heard or picked up by other vessels' radar. The Coast Guard strongly recommends that, to avoid collisions, all vessels carry Radar Reflectors mounted as high as possible.

NAVIGATION LIGHTS

Inland Navigation Rules Act of 1980 updates the old Inland Rules and is essentially identical to the International Regulations of 1972.

Definition of Lights:

Masthead Light — a white light fixed over the centerline of the vessel showing an unbroken light over an arc of the horizon of 225°, from dead ahead to 22.5° abaft the beam on either side.

Sidelights — a green light on the starboard side and a red light on the port side showing an unbroken light over an arc of the horizon of 112.5°, from dead ahead to 22.5° abaft the beam on either side.

Sternlight — a white light placed as nearly as practicable at the stern showing an unbroken light over an arc of the horizon of 135°, 67.5°, from dead aft to each side of the vessel.

All-round Light — an unbroken light over an arc of the horizon of 360°.

Towing Light — a yellow light with same characteristics as the sternlight.

When under way, in all weathers from sunset to sunrise, every vessel shall carry and exhibit the following lights:

When Under Power Alone or When Under Power and Sail Combined:

Vessels 12 to 50 meters (40-165′): Masthead Light, visible 5 miles (vessels under 20 meters or 65′ may have Masthead Light, visible 3 miles); Sidelights, visible 2 miles; Sternlight, visible 2 miles. When under sail and power combined, a vessel shall show forward where it can best be seen, a conical shape, apex down, in addition to the lights given above.

Vessels under 12 meters (40′): Masthead Light placed as far forward as possible but not necessarily on the centerline, visible 2 miles; Sidelights, visible 1 mile; Sternlight, visible 2 miles; or in lieu of the Masthead and Sternlight, an All-round white Light, visible 2 miles.

Sailing Vessels Under Way (Sail Only):

Vessels 20 to 50 meters (65-165′): Sternlight, visible 2 miles; Sidelights, visible 2 miles.

Vessels 12 to 20 meters (40-65′): Sternlight, visible 2 miles; Sidelights, visible 2 miles. These lights may be combined in one lantern carried at or near the top of the mast where it can best be seen, visible 2 miles.

Vessels 7 to 12 meters (22-40′): Sternlight, visible 2 miles; Sidelights, visible 1 mile, or a combination of these 3 lights in one lantern at or near the top of the mast where it can best be seen, visible 2 miles.

Vessels under 7 meters (22′) and Vessels Under Oars: Sternlight, visible 2 miles; Sidelights, visible 1 mile; or in lieu of these, she shall have ready at hand an electric torch or white lantern to show in time to prevent a collision.

At Anchor: Vessels under 50 meters (165′) must show an All-round white Light where it can best be seen and visible 2 miles.

Vessels under 7 meters (22′) need no light unless they are near a channel, a fairway, an anchorage or area where other vessels navigate.

Vessels Trawling shall show, in addition to the appropriate lights above, 2 All-round Lights in a vertical line, the upper green and the lower white.

Vessels Fishing (other than trawling) shall show, in addition to the appropriate lights above, 2 All-round Lights, the upper red and the lower white.

When Towing or Being Towed:

Towing Vessel: 2 Masthead Lights (if tow less than 200 meters); 3 Masthead Lights in a vertical line forward (if tow exceeds 200 meters); Sidelights; Sternlight; a Yellow Tow Light in vertical line above Sternlight; a Diamond Shape where it can best be seen (if tow exceeds 200 meters).

Vessel being Towed: Sidelights; Sternlight; a Diamond Shape where it can best be seen (if tow exceeds 200 meters).

Note: R. and Y. Flashing Lights now authorized for vessels assigned to Traffic Control, Medical Emergencies, Search and Rescue, Fire-Fighting, Salvage and Disabled Vessels.

SAFETY EQUIPMENT REQUIREMENTS
(Mininum)

Lengths	Bell, Whistle or Horn
<16'	None required, but must have some means of making an efficient sound signal.
16-25'	Mouth, hand or power Whistle audible at least 1/2 mi.
26-39'	Hand or power Whistle audible at least 1 mi.
> 39'	1 Bell; power Whistle audible at least 1 mi.
> 328'	Bell, power Whistle and Gong.

Visual Distress Signals (all with approval no.)

< 16′	By night: 1 electric SOS flashlight.
16-65′	By day only: 1 orange flag; 3 floating or 3 hand-held orange smoke signals (not required on no-power open sailboats less than 26′, or rowboats, or while racing.)
	By day and night: 3 hand-held or 3 pistol or 3 hand-held rocket or 3 pyrotechnic red flares. (All suitable for day use also.)

*** Personal Flotation Devices (approved)**

< 16′	1 Type I, II or III per person
16-65′	1 Type I, II or III per person and 1 Type IV per boat.

---

* Type I - Life Preserver designed to turn an unconscious person in the water from face downward to vertical or slightly backward; to have more than 20 lbs. buoyancy. For offshore cruising.
Type II - Same as Type I but with at least 15-1/2 lbs. buoyancy. For closer, in-shore cruising.
Type III - Same as Type II but with a lesser turning ability to allow for water skiing, etc.
Type IV - Cushion or Ring Buoy to be thrown to a person in the water; not to be worn. Has at least 16-1/2 lbs. buoyancy. For canoes, etc. and as throwable device for all boats.

Portable Fire Extinguishers (approved)

< 26′	1 B-I, if ***no*** fixed extinguisher system in machinery space. (Not required on outboards built so that vapor entrapment cannot occur.)
26-39′	2 B-I or 1 B-II & 1 B-I, if ***no*** fixed exting. system; 1 B-I, ***with*** a fixed exting. system.
40-65′	3 B-I or 1 B-II & 1 B-I, if ***no*** fixed exting. system; 2 B-I or 1 B-II ***with*** a fixed exting. system.

Back-Fire Flame Arrestor

1 approved device per carburetor of all inboard engines installed after April 25, 1940.

At least 2 ventilator ducts fitted with cowls or their equivalent to ventilate efficiently the bilges of every engine and fuel tank compartment of boats built or decked-over after April 25, 1940 and using gasoline or other fuel with a flashpoint less than 110° F.

BOSTON HARBOR CURRENTS

The diagram below shows the direction of the Flood Currents in Boston Harbor at the time of Maximum Flood Current, generally 3½hrs. after Low Water at Boston. Generally speaking the Ebb Currents flow in precisely the opposite direction (note one exception, shown by dotted arrow east of Winthrop), and reach these maximum velocities about 4 hrs. after High Water at Boston.

Basically, the velocities of the Ebb Currents are about the same as those of the Flood Currents. Where the Ebb Current differs by .2 kts., the velocity of the Ebb is shown in parentheses.

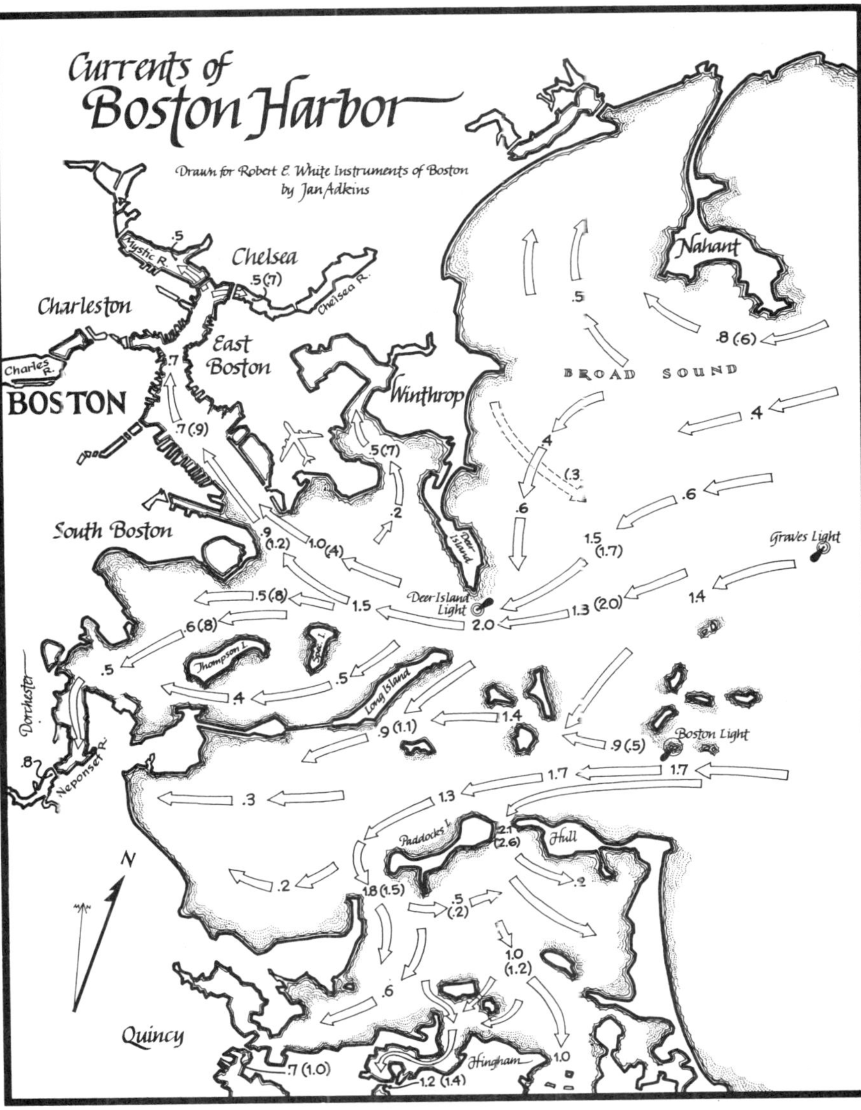

The Velocities shown on this Current Diagram are the maximums normally encountered each month at Full Moon and at New Moon. At other times the velocities will be smaller.

As a rule of thumb, the velocities shown are those found on days when High Water at Boston is 11.0 to 11.5 ft. (see Boston High Water Tables pp. 12-17).When the height of High Water is 10.5′, subtract 10% from the velocities shown; at 10.0′, subtract 20%; at 9.0′, 30%; at 8.0′, 40%; below 7.5′, 50%.

1999 HIGH & LOW WATER
BOSTON, MASS.
42–21N x 71–03W

Standard Time										Standard Time									
DAY OF MONTH	DAY OF WEEK	JANUARY								DAY OF MONTH	DAY OF WEEK	FEBRUARY							
		HIGH				LOW						HIGH				LOW			
		a.m.	Ht.	p.m.	Ht.	a.m.	p.m.					a.m.	Ht.	p.m.	Ht.	a.m.	p.m.		
1	F	10 00	11.9	10 40	10.3	3 45	4 26			1	M	11 29	11.3	...	...	5 13	5 49		
2	S	10 52	11.9	11 32	10.3	4 38	5 17			2	T	12 05	10.1	12 16	11.1	6 01	6 34		
3	S	11 43	11.8	...	...	5 29	6 07			3	W	12 49	10.0	1 01	10.7	6 48	7 17		
4	M	12 23	10.3	12 33	11.5	6 19	6 56			4	T	1 32	9.9	1 46	10.3	7 34	8 00		
5	T	1 13	10.1	1 23	11.0	7 09	7 44			5	F	2 16	9.7	2 32	9.7	8 20	8 44		
6	W	2 02	9.8	2 13	10.4	8 00	8 32			6	S	3 00	9.5	3 20	9.2	9 08	9 29		
7	T	2 51	9.6	3 04	9.8	8 51	9 21			7	S	3 46	9.2	4 10	8.7	9 58	10 17		
8	F	3 42	9.3	3 58	9.3	9 45	10 11			8	M	4 35	9.0	5 04	8.3	10 52	11 08		
9	S	4 34	9.2	4 54	8.8	10 40	11 03			9	T	5 28	8.9	6 02	8.0	11 48	...		
10	S	5 25	9.1	5 50	8.5	11 37	11 55			10	W	6 21	8.9	6 59	8.0	12 02	12 45		
11	M	6 18	9.1	6 47	8.3	...	12 35			11	T	7 15	9.1	7 54	8.1	12 56	1 40		
12	T	7 09	9.2	7 42	8.3	12 48	1 30			12	F	8 07	9.4	8 46	8.4	1 49	2 31		
13	W	7 59	9.4	8 34	8.4	1 38	2 21			13	S	8 56	9.7	9 33	8.8	2 39	3 18		
14	T	8 45	9.6	9 21	8.6	2 27	3 08			14	S	9 42	10.1	10 17	9.2	3 27	4 03		
15	F	9 29	9.9	10 05	8.8	3 13	3 51			15	M	10 26	10.5	10 59	9.6	4 12	4 45		
16	S	10 11	10.2	10 47	9.0	3 56	4 33			16	T	11 09	10.9	11 40	10.1	4 57	5 28		
17	S	10 52	10.4	11 27	9.3	4 39	5 14			17	W	11 52	11.1	...	...	5 42	6 10		
18	M	11 33	10.6	...	...	5 22	5 54			18	T	12 22	10.5	12 37	11.2	6 27	6 54		
19	T	12 08	9.5	12 14	10.8	6 04	6 36			19	F	1 05	10.7	1 24	11.0	7 15	7 40		
20	W	12 49	9.7	12 57	10.8	6 48	7 19			20	S	1 51	10.9	2 14	10.7	8 04	8 28		
21	T	1 31	9.9	1 43	10.7	7 35	8 04			21	S	2 40	10.9	3 08	10.3	8 57	9 19		
22	F	2 17	10.1	2 32	10.5	8 24	8 51			22	M	3 33	10.7	4 06	9.8	9 54	10 14		
23	S	3 05	10.2	3 25	10.1	9 17	9 42			23	T	4 30	10.5	5 08	9.3	10 55	11 14		
24	S	3 57	10.3	4 23	9.8	10 14	10 37			24	W	5 32	10.4	6 15	9.1	11 59	...		
25	M	4 53	10.3	5 24	9.5	11 14	11 35			25	T	6 37	10.3	7 23	9.0	12 17	1 05		
26	T	5 52	10.5	6 29	9.3	...	12 17			26	F	7 42	10.4	8 27	9.2	1 21	2 08		
27	W	6 53	10.6	7 34	9.3	12 35	1 20			27	S	8 43	10.5	9 25	9.5	2 22	3 07		
28	T	7 54	10.8	8 36	9.5	1 36	2 22			28	S	9 39	10.7	10 16	9.8	3 19	3 59		
29	F	8 53	11.1	9 35	9.7	2 35	3 19												
30	S	9 48	11.3	10 29	9.9	3 31	4 13												
31	S	10 40	11.3	11 18	10.0	4 23	5 02												

Dates when Ht. of **Low** Water is below Mean Low with Ht. of lowest given for each period and Date of lowest in ( ):

1st - 6th: -1.7' (2nd)
17th - 23rd: -0.9' (20th)
30th - 31st: -1.2'

1st - 4th: -1.2' (1st)
15th - 23rd: -1.3' (18th)
26th - 28th: -0.6' (28th)

Average Rise and Fall 9 1/2 ft.

When a high tide exceeds av. ht., the *following* low tide will be lower than av. Since there is a high degree of correlation between the height of High Water and the velocities of the Flood and Ebb Currents for that same day, we offer a rough rule of thumb for estimating the current velocities, for ALL the Current Charts and Diagrams in this book. **Rule of Thumb:** Refer to Boston High Water. If the height of High Water is 11.0' or over, use the Current Chart velocities as shown. When the height is 10.5', subtract 10%; at 10.0', subtract 20%; at 9.0', 30%; at 8.0', 40%; below 7.5', 50%.

1999 HIGH & LOW WATER BOSTON, MASS.

42–21N x 71–03W

Standard Time **Daylight Time starts Apr. 4 at 2 a.m.**

DAY OF MONTH	DAY OF WEEK	MARCH						DAY OF MONTH	DAY OF WEEK	APRIL					
		HIGH				LOW				HIGH				LOW	
		a.m.	Ht.	p.m.	Ht.	a.m.	p.m.			a.m.	Ht.	p.m.	Ht.	a.m.	p.m.
1	M	10 29	10.8	11 01	10.0	4 11	4 45	1	T	11 36	10.2	11 54	10.2	5 22	5 41
2	T	11 14	10.8	11 43	10.1	4 58	5 28	2	F	...	..	12 15	10.0	6 02	6 18
3	W	11 57	10.6	...	...	5 43	6 09	3	S	12 30	10.2	12 54	9.8	6 41	6 56
4	T	12 22	10.1	12 38	10.4	6 25	6 48	4	S	1 06	10.1	*2 34	9.5	*8 21	*8 35
5	F	1 01	10.1	1 19	10.1	7 07	7 28	5	M	2 45	9.9	3 15	9.2	9 03	9 16
6	S	1 40	9.9	2 01	9.6	7 50	8 08	6	T	3 26	9.7	4 00	8.8	9 47	10 00
7	S	2 20	9.7	2 45	9.2	8 34	8 51	7	W	4 11	9.4	4 48	8.5	10 34	10 48
8	M	3 03	9.4	3 32	8.7	9 20	9 36	8	T	5 00	9.2	5 41	8.3	11 25	11 41
9	T	3 51	9.2	4 24	8.3	10 10	10 25	9	F	5 54	9.0	6 37	8.2	...	12 19
10	W	4 40	9.0	5 18	8.1	11 03	11 19	10	S	6 49	9.1	7 32	8.5	12 37	1 15
11	T	5 35	8.9	6 16	8.0	...	12 01	11	S	7 46	9.3	8 26	8.9	1 34	2 10
12	F	6 31	9.0	7 13	8.1	12 15	12 57	12	M	8 42	9.7	9 17	9.5	2 30	3 03
13	S	7 27	9.2	8 07	8.5	1 11	1 51	13	T	9 36	10.2	10 06	10.2	3 24	3 53
14	S	8 20	9.7	8 56	9.0	2 05	2 41	14	W	10 27	10.7	10 53	10.9	4 16	4 41
15	M	9 09	10.2	9 42	9.6	2 56	3 29	15	T	11 17	11.1	11 40	11.5	5 06	5 29
16	T	9 57	10.7	10 26	10.2	3 44	4 14	16	F	...	..	12 07	11.4	5 56	6 16
17	W	10 43	11.1	11 10	10.8	4 32	4 59	17	S	12 27	12.0	12 57	11.4	6 45	7 04
18	T	11 30	11.4	11 54	11.3	5 19	5 43	18	S	1 15	12.2	1 49	11.3	7 36	7 54
19	F	...	...	12 17	11.4	6 06	6 29	19	M	2 05	12.1	2 42	10.9	8 28	8 45
20	S	12 40	11.5	1 06	11.3	6 55	7 16	20	T	2 57	11.3	3 38	10.4	9 22	9 39
21	S	1 27	11.6	1 57	10.9	7 46	8 06	21	W	3 53	11.3	4 37	9.9	10 19	10 37
22	M	2 17	11.4	2 52	10.4	8 39	8 58	22	T	4 53	10.7	5 41	9.5	11 19	11 39
23	T	3 12	11.1	3 51	9.8	9 36	9 55	23	F	5 58	10.2	6 47	9.3	...	12 23
24	W	4 11	10.7	4 55	9.3	10 37	10 56	24	S	7 05	9.9	7 52	9.3	12 45	1 27
25	T	5 15	10.3	6 03	9.1	11 42	...	25	S	8 11	9.7	8 53	9.4	1 51	2 29
26	F	6 22	10.0	7 11	9.1	12 01	12 49	26	M	9 12	9.7	9 46	9.7	2 53	3 25
27	S	7 29	10.0	8 14	9.3	1 07	1 52	27	T	10 06	9.7	10 31	9.9	3 49	4 13
28	S	8 31	10.1	9 10	9.5	2 10	2 50	28	W	10 52	9.8	11 11	10.1	4 37	4 55
29	M	9 26	10.2	9 57	9.8	3 06	3 39	29	T	11 34	9.8	11 48	10.2	5 20	5 34
30	T	10 14	10.3	10 39	10.0	3 56	4 23	30	F	...	--	12 14	9.7	6 00	6 11
31	W	10 57	10.3	11 17	10.2	4 41	5 03								

***Daylight Time starts**

Dates when Ht. of **Low** Water is below Mean Low with Ht. of lowest given for each period and Date of lowest in ( ):

1st - 4th: -0.7' (1st - 2nd))
16th - 24th: -1.6' (20th)

14th - 22nd: -2.0' (18th)

Average Rise and Fall 9 1/2 ft.

When a high tide exceeds av. ht., the *following* low tide will be lower than av. Since there is a high degree of correlation between the height of High Water and the velocities of the Flood and Ebb Currents for that same day, we offer a rough rule of thumb for estimating the current velocities, for ALL the Current Charts and Diagrams in this book. **Rule of Thumb:** Refer to Boston High Water. If the height of High Water is 11.0' or over, use the Current Chart velocities as shown. When the height is 10.5', subtract 10%; at 10.0', subtract 20%; at 9.0', 30%; at 8.0', 40%; below 7.5', 50%.

1999 HIGH & LOW WATER
BOSTON, MASS.

42–21N x 71–03W

Daylight Time

DAY OF MONTH	DAY OF WEEK	MAY HIGH a.m.	Ht.	p.m.	Ht.	LOW a.m.	p.m.
1	S	12 24	10.3	**12 52**	9.6	6 38	**6 49**
2	S	1 00	10.3	**1 30**	9.5	7 16	**7 27**
3	M	1 36	10.2	**2 09**	9.3	7 55	**8 06**
4	T	2 15	10.1	**2 50**	9.1	8 36	**8 47**
5	W	2 55	9.9	**3 33**	8.9	9 18	**9 30**
6	T	3 39	9.6	**4 19**	8.7	10 03	**10 17**
7	F	4 26	9.5	**5 08**	8.6	10 51	**11 08**
8	S	5 17	9.3	**6 00**	8.7	11 43	...
9	S	6 13	9.4	**6 55**	9.0	12 03	**12 37**
10	M	7 09	9.5	**7 48**	9.5	1 00	**1 31**
11	T	8 07	9.8	**8 40**	10.1	1 58	**2 25**
12	W	9 03	10.2	**9 31**	10.8	2 54	**3 18**
13	T	9 58	10.6	**10 22**	11.5	3 49	**4 10**
14	F	10 52	10.9	**11 12**	12.0	4 42	**5 00**
15	S	11 45	11.1	...	...	5 34	**5 51**
16	S	12 02	12.4	**12 38**	11.2	6 26	**6 41**
17	M	12 53	12.5	**1 32**	11.1	7 18	**7 33**
18	T	1 45	12.3	**2 26**	10.8	8 10	**8 26**
19	W	2 39	11.9	**3 23**	10.4	9 04	**9 21**
20	T	3 35	11.4	**4 21**	10.0	10 00	**10 19**
21	F	4 35	10.7	**5 22**	9.7	10 58	**11 20**
22	S	5 37	10.2	**6 24**	9.5	11 58	...
23	S	6 41	9.7	**7 24**	9.5	12 23	**12 58**
24	M	7 44	9.4	**8 21**	9.6	1 27	**1 57**
25	T	8 43	9.3	**9 12**	9.7	2 28	**2 50**
26	W	9 37	9.2	**9 58**	9.9	3 23	**3 38**
27	T	10 25	9.2	**10 39**	10.0	4 11	**4 21**
28	F	11 08	9.3	**11 18**	10.2	4 55	**5 02**
29	S	11 49	9.3	**11 55**	10.2	5 35	**5 41**
30	S	...	...	**12 28**	9.3	6 14	**6 20**
31	M	12 32	10.3	**1 07**	9.2	6 52	**6 59**

Daylight Time

DAY OF MONTH	DAY OF WEEK	JUNE HIGH a.m.	Ht.	p.m.	Ht.	LOW a.m.	p.m.
1	T	1 10	10.2	**1 46**	9.2	7 31	**7 39**
2	W	1 49	10.2	**2 26**	9.1	8 11	**8 21**
3	T	2 29	10.1	**3 08**	9.0	8 53	**9 04**
4	F	3 12	9.9	**3 52**	9.0	9 36	**9 50**
5	S	3 57	9.8	**4 38**	9.1	10 22	**10 40**
6	S	4 47	9.7	**5 28**	9.3	11 11	**11 34**
7	M	5 41	9.7	**6 19**	9.6	...	**12 03**
8	T	6 38	9.7	**7 13**	10.0	12 30	**12 57**
9	W	7 37	9.8	**8 08**	10.6	1 29	**1 52**
10	T	8 35	10.0	**9 01**	11.2	2 27	**2 47**
11	F	9 33	10.3	**9 54**	11.7	3 24	**3 41**
12	S	10 30	10.5	**10 47**	12.1	4 20	**4 35**
13	S	11 26	10.7	**11 41**	12.4	5 14	**5 28**
14	M	...	...	**12 21**	10.8	6 07	**6 21**
15	T	12 34	12.4	**1 15**	10.8	7 00	**7 14**
16	W	1 27	12.2	**2 10**	10.6	7 53	**8 07**
17	T	2 21	11.8	**3 04**	10.4	8 45	**9 01**
18	F	3 16	11.3	**3 59**	10.1	9 38	**9 57**
19	S	4 13	10.6	**4 55**	9.9	10 32	**10 55**
20	S	5 10	10.0	**5 52**	9.7	11 27	**11 54**
21	M	6 09	9.5	**6 47**	9.6	...	**12 22**
22	T	7 09	9.1	**7 41**	9.5	12 55	**1 17**
23	W	8 06	8.9	**8 32**	9.6	1 54	**2 09**
24	T	9 01	8.8	**9 20**	9.7	2 49	**2 59**
25	F	9 52	8.8	**10 04**	9.9	3 40	**3 45**
26	S	10 38	8.8	**10 46**	10.0	4 25	**4 29**
27	S	11 21	8.9	**11 27**	10.1	5 08	**5 11**
28	M	...	...	**12 03**	9.0	5 48	**5 53**
29	T	12 06	10.2	**12 43**	9.1	6 27	**6 33**
30	W	12 45	10.3	**1 22**	9.2	7 07	**7 14**

Dates when Ht. of **Low** Water is below Mean Low with Ht. of lowest given for each period and Date of lowest in ( ):

12th - 21st: -2.2' (17th)

11th - 19th: -2.0' (14th - 15th)

Average Rise and Fall 9 1/2 ft.

When a high tide exceeds av. ht., the *following* low tide will be lower than av. Since there is a high degree of correlation between the height of High Water and the velocities of the Flood and Ebb Currents for that same day, we offer a rough rule of thumb for estimating the current velocities, for ALL the Current Charts and Diagrams in this book. **Rule of Thumb:** Refer to Boston High Water. If the height of High Water is 11.0' or over, use the Current Chart velocities as shown. When the height is 10.5', subtract 10%; at 10.0', subtract 20%; at 9.0', 30%; at 8.0', 40%; below 7.5', 50%.

1999 HIGH & LOW WATER BOSTON, MASS.

42–21N x 71–03W

Daylight Time

Day of Month	Day of Week	JULY HIGH a.m.	Ht.	p.m.	Ht.	LOW a.m.	p.m.
1	T	1 25	10.3	2 01	9.2	7 47	7 56
2	F	2 05	10.3	2 42	9.3	8 27	8 40
3	S	2 47	10.2	3 25	9.4	9 10	9 26
4	S	3 33	10.1	4 09	9.6	9 55	10 15
5	M	4 22	10.0	4 58	9.8	10 42	11 08
6	T	5 15	9.9	5 49	10.1	11 33	...
7	W	6 12	9.8	6 43	10.4	12 05	12 27
8	T	7 11	9.7	7 39	10.8	1 04	1 23
9	F	8 13	9.8	8 37	11.2	2 04	2 21
10	S	9 13	9.9	9 32	11.6	3 03	3 18
11	S	10 13	10.1	10 29	11.9	4 01	4 14
12	M	11 10	10.3	11 24	12.1	4 57	5 09
13	T	...	...	12 05	10.5	5 51	6 03
14	W	12 18	12.1	12 58	10.5	6 43	6 56
15	T	1 10	11.9	1 50	10.5	7 33	7 48
16	F	2 02	11.5	2 41	10.4	8 23	8 40
17	S	2 54	11.0	3 31	10.2	9 12	9 32
18	S	3 46	10.4	4 22	9.9	10 01	10 25
19	M	4 39	9.8	5 13	9.7	10 51	11 20
20	T	5 33	9.3	6 04	9.5	11 42	...
21	W	6 29	8.8	6 56	9.4	12 16	12 34
22	T	7 26	8.6	7 49	9.4	1 13	1 27
23	F	8 22	8.4	8 39	9.5	2 10	2 19
24	S	9 15	8.5	9 28	9.6	3 03	3 09
25	S	10 05	8.6	10 14	9.8	3 52	3 56
26	M	10 51	8.8	10 58	10.1	4 37	4 41
27	T	11 34	9.0	11 39	10.3	5 19	5 25
28	W	...	...	12 14	9.2	6 00	6 07
29	T	12 20	10.4	12 54	9.4	6 40	6 49
30	F	1 00	10.6	1 33	9.6	7 20	7 32
31	S	1 41	10.6	2 14	9.8	8 01	8 16

Daylight Time

Day of Month	Day of Week	AUGUST HIGH a.m.	Ht.	p.m.	Ht.	LOW a.m.	p.m.
1	S	2 24	10.6	2 56	10.0	8 43	9 02
2	M	3 10	10.4	3 41	10.2	9 28	9 52
3	T	4 00	10.2	4 29	10.4	10 16	10 45
4	W	4 53	10.0	5 22	10.5	11 07	11 42
5	T	5 51	9.7	6 18	10.6	...	12 02
6	F	6 53	9.5	7 17	10.8	12 43	1 01
7	S	7 56	9.5	8 17	11.0	1 45	2 01
8	S	8 59	9.6	9 17	11.2	2 46	3 01
9	M	10 01	9.3	10 17	11.5	3 46	3 59
10	T	10 56	10.1	11 11	11.6	4 42	4 54
11	W	11 50	10.3	...	...	5 34	5 47
12	T	12 03	11.6	12 39	10.4	6 24	6 38
13	F	12 53	11.4	1 27	10.5	7 11	7 27
14	S	1 42	11.1	2 13	10.4	7 57	8 15
15	S	2 29	10.7	2 58	10.2	8 42	9 03
16	M	3 16	10.2	3 43	9.9	9 27	9 52
17	T	4 05	9.6	4 30	9.7	10 13	10 43
18	W	4 56	9.1	5 19	9.4	11 01	11 35
19	T	5 49	8.7	6 11	9.2	11 52	...
20	F	6 45	8.4	7 04	9.2	12 31	12 45
21	S	7 42	8.2	7 59	9.2	1 27	1 39
22	S	8 38	8.3	8 51	9.4	2 23	2 32
23	M	9 30	8.5	9 41	9.7	3 15	3 23
24	T	10 17	8.8	10 27	10.0	4 02	4 11
25	W	11 01	9.2	11 10	10.3	4 46	4 56
26	T	11 42	9.6	11 52	10.6	5 28	5 40
27	F	...	...	12 22	9.9	6 09	6 23
28	S	12 34	10.8	1 02	10.3	6 50	7 07
29	S	1 17	10.9	1 43	10.6	7 32	7 52
30	M	2 02	10.9	2 27	10.8	8 16	8 40
31	T	2 49	10.7	3 13	10.9	9 02	9 31

Dates when Ht. of **Low** Water is below Mean Low with Ht. of lowest given for each period and Date of lowest in ( ):

July:
9th - 17th: -1.6' (13th - 14th)
30th - 31st: -0.4' (31st)

August:
1st: -0.4' (1st)
8th - 15th: -1.2' (11th - 12th)
27th - 31st: -0.7' (29th - 30th)

Average Rise and Fall 9 1/2 ft.

When a high tide exceeds av. ht., the *following* low tide will be lower than av. Since there is a high degree of correlation between the height of High Water and the velocities of the Flood and Ebb Currents for that same day, we offer a rough rule of thumb for estimating the current velocities, for ALL the Current Charts and Diagrams in this book. **Rule of Thumb:** Refer to Boston High Water. If the height of High Water is 11.0' or over, use the Current Chart velocities as shown. When the height is 10.5', subtract 10%; at 10.0', subtract 20%; at 9.0', 30%; at 8.0', 40%; below 7.5', 50%.

1999 HIGH & LOW WATER
BOSTON, MASS.
42–21N x 71–03W

Daylight Time **Standard Time starts Oct. 31 at 2 a.m.**

DAY OF MONTH	DAY OF WEEK	SEPTEMBER						DAY OF MONTH	DAY OF WEEK	OCTOBER					
		HIGH				LOW				HIGH				LOW	
		a.m.	Ht.	**p.m.**	Ht.	a.m.	**p.m.**			a.m.	Ht.	**p.m.**	Ht.	a.m.	**p.m.**
1	**W**	3 41	10.4	**4 04**	10.9	9 51	**10 25**	1	**F**	4 21	10.0	**4 39**	10.9	10 25	**11 06**
2	**T**	4 36	10.0	**4 59**	10.8	10 45	**11 23**	2	**S**	5 23	9.6	**5 42**	10.6	11 25	**...**
3	**F**	5 36	9.6	**5 58**	10.7	11 42	**...**	3	**S**	6 29	9.4	**6 48**	10.4	12 09	**12 29**
4	**S**	6 40	9.4	**7 01**	10.6	12 25	**12 43**	4	**M**	7 36	9.4	**7 54**	10.3	1 14	**1 34**
5	**S**	7 45	9.4	**8 05**	10.7	1 29	**1 46**	5	**T**	8 39	9.6	**8 58**	10.4	2 18	**2 38**
6	**M**	8 49	9.5	**9 07**	10.8	2 32	**2 48**	6	**W**	9 37	9.9	**9 55**	10.5	3 16	**3 36**
7	**T**	9 49	9.8	**10 06**	11.0	3 32	**3 47**	7	**T**	10 27	10.2	**10 46**	10.5	4 09	**4 29**
8	**W**	10 43	10.1	**10 59**	11.1	4 27	**4 42**	8	**F**	11 12	10.4	**11 32**	10.5	4 56	**5 17**
9	**T**	11 34	10.3	**11 50**	11.1	5 17	**5 32**	9	**S**	11 54	10.5	**...**	...	5 38	**6 01**
10	**F**	...	...	**12 18**	10.5	6 03	**6 20**	10	**S**	12 15	10.4	**12 32**	10.5	6 19	**6 42**
11	**S**	12 35	10.9	**1 00**	10.5	6 46	**7 05**	11	**M**	12 56	10.2	**1 10**	10.5	6 58	**7 23**
12	**S**	1 19	10.7	**1 41**	10.4	7 28	**7 49**	12	**T**	1 36	9.9	**1 48**	10.3	7 38	**8 05**
13	**M**	2 02	10.3	**2 22**	10.2	8 09	**8 33**	13	**W**	2 17	9.6	**2 28**	10.0	8 18	**8 47**
14	**T**	2 46	9.9	**3 04**	10.0	8 52	**9 18**	14	**T**	3 01	9.2	**3 10**	9.7	9 00	**9 32**
15	**W**	3 31	9.4	**3 48**	9.7	9 35	**10 06**	15	**F**	3 47	8.8	**3 56**	9.4	9 45	**10 20**
16	**T**	4 19	8.9	**4 35**	9.4	10 22	**10 56**	16	**S**	4 36	8.5	**4 46**	9.2	10 34	**11 11**
17	**F**	5 11	8.6	**5 27**	9.2	11 11	**11 49**	17	**S**	5 29	8.3	**5 40**	9.0	11 27	**...**
18	**S**	6 06	8.3	**6 21**	9.0	...	**12 05**	18	**M**	6 24	8.3	**6 36**	9.0	12 05	**12 23**
19	**S**	7 03	8.2	**7 18**	9.1	12 45	**1 01**	19	**T**	7 19	8.5	**7 32**	9.2	1 00	**1 19**
20	**M**	7 59	8.3	**8 13**	9.3	1 41	**1 56**	20	**W**	8 11	8.9	**8 26**	9.5	1 53	**2 14**
21	**T**	8 52	8.6	**9 04**	9.6	2 35	**2 49**	21	**T**	9 00	9.5	**9 17**	10.0	2 44	**3 06**
22	**W**	9 40	9.1	**9 53**	10.0	3 24	**3 39**	22	**F**	9 46	10.1	**10 06**	10.4	3 32	**3 55**
23	**T**	10 24	9.6	**10 38**	10.4	4 10	**4 26**	23	**S**	10 30	10.8	**10 54**	10.8	4 18	**4 44**
24	**F**	11 06	10.1	**11 23**	10.8	4 54	**5 11**	24	**S**	11 15	11.4	**11 42**	11.1	5 04	**5 31**
25	**S**	11 48	10.7	**...**	...	5 36	**5 57**	25	**M**	...	...	**12 01**	11.8	5 50	**6 19**
26	**S**	12 08	11.0	**12 30**	11.1	6 20	**6 43**	26	**T**	12 30	11.2	**12 46**	12.1	6 37	**7 09**
27	**M**	12 53	11.1	**1 13**	11.4	7 04	**7 30**	27	**W**	1 20	11.1	**1 35**	12.1	7 25	**7 59**
28	**T**	1 40	11.1	**1 59**	11.5	7 49	**8 19**	28	**T**	2 12	10.8	**2 27**	11.9	8 16	**8 53**
29	**W**	2 30	10.8	**2 48**	11.5	8 38	**9 11**	29	**F**	3 08	10.5	**3 22**	11.5	9 10	**9 49**
30	**T**	3 23	10.4	**3 41**	11.2	9 29	**10 06**	30	**S**	4 07	10.1	**4 22**	11.0	10 07	**10 49**
								31	**S**	*4 10	9.7	***4 26**	10.5	*10 09*	**10 51**

***Standard Time starts**

Dates when Ht. of **Low** Water is below Mean Low with Ht. of lowest given for each period and Date of lowest in ( ):

9th - 10th: -0.7'
27th - 30th: -1.2' (28th)

1st : -0.4'
8th - 10th: -0.2'
23rd - 31st: -1.8' (26th)

Average Rise and Fall 9 1/2 ft.

When a high tide exceeds av. ht., the *following* low tide will be lower than av. Since there is a high degree of correlation between the height of High Water and the velocities of the Flood and Ebb Currents for that same day, we offer a rough rule of thumb for estimating the current velocities, for ALL the Current Charts and Diagrams in this book. **Rule of Thumb:** Refer to Boston High Water. If the height of High Water is 11.0' or over, use the Current Chart velocities as shown. When the height is 10.5', subtract 10%; at 10.0', subtract 20%; at 9.0', 30%; at 8.0', 40%; below 7.5', 50%.

1999 HIGH & LOW WATER
BOSTON, MASS.
42–21N x 71–03W

Standard Time

DAY OF MONTH	DAY OF WEEK	NOVEMBER HIGH a.m.	Ht.	p.m.	Ht.	LOW a.m.	p.m.
1	M	5 15	9.5	5 33	10.2	11 14	11 55
2	T	6 21	9.6	6 40	10.0	...	12 20
3	W	7 23	9.8	7 43	9.9	12 58	1 24
4	T	8 17	10.0	8 39	9.9	1 55	2 22
5	F	9 06	10.2	9 29	9.9	2 46	3 13
6	S	9 48	10.4	10 13	9.9	3 31	3 59
7	S	10 27	10.5	10 54	9.8	4 12	4 40
8	M	11 04	10.5	11 33	9.7	4 51	5 20
9	T	11 42	10.4	...	...	5 30	5 59
10	W	12 12	9.5	12 18	10.3	6 08	6 39
11	T	12 52	9.3	12 57	10.1	6 48	7 19
12	F	1 34	9.0	1 38	9.8	7 30	8 02
13	S	2 17	8.8	2 23	9.6	8 14	8 47
14	S	3 04	8.6	3 10	9.3	9 01	9 35
15	M	3 54	8.5	4 02	9.1	9 52	10 26
16	T	4 45	8.6	4 56	9.1	10 46	11 19
17	W	5 38	8.8	5 52	9.2	11 42	...
18	T	6 30	9.2	6 48	9.5	12 12	12 38
19	F	7 20	9.8	7 42	9.8	1 04	1 33
20	S	8 09	10.5	8 35	10.2	1 55	2 26
21	S	8 57	11.2	9 27	10.6	2 45	3 17
22	M	9 45	11.8	10 18	10.9	3 34	4 08
23	T	10 34	12.2	11 10	11.0	4 23	4 59
24	W	11 23	12.5	...	...	5 13	5 50
25	T	12 02	11.0	12 15	12.4	6 04	6 42
26	F	12 56	10.8	1 08	12.1	6 56	7 35
27	S	1 52	10.5	2 04	11.6	7 51	8 31
28	S	2 50	10.2	3 04	11.0	8 49	9 28
29	M	3 51	9.9	4 07	10.4	9 50	10 28
30	T	4 54	9.7	5 12	9.9	10 54	11 29

Standard Time

DAY OF MONTH	DAY OF WEEK	DECEMBER HIGH a.m.	Ht.	p.m.	Ht.	LOW a.m.	p.m.
1	W	5 57	9.7	6 17	9.6	...	12 01
2	T	6 55	9.7	7 19	9.4	12 30	1 03
3	F	7 50	9.9	8 16	9.3	1 26	2 02
4	S	8 38	10.0	9 06	9.3	2 17	2 53
5	S	9 21	10.2	9 51	9.3	3 02	3 38
6	M	10 01	10.3	10 33	9.3	3 44	4 19
7	T	10 39	10.3	11 12	9.2	4 24	4 58
8	W	11 16	10.3	11 51	9.2	5 04	5 37
9	T	11 55	10.2	...	...	5 43	6 15
10	F	12 30	9.1	12 32	10.1	6 22	6 55
11	S	1 09	9.0	1 12	10.0	7 03	7 35
12	S	1 51	8.9	1 54	9.8	7 46	8 18
13	M	2 34	8.8	2 39	9.5	8 31	9 02
14	T	3 19	8.8	3 27	9.4	9 20	9 50
15	W	4 07	8.9	4 19	9.3	10 12	10 40
16	T	4 58	9.2	5 15	9.2	11 07	11 32
17	F	5 50	9.6	6 12	9.4	...	12 04
18	S	6 42	10.1	7 10	9.6	12 26	1 02
19	S	7 35	10.7	8 07	9.9	1 21	1 58
20	M	8 28	11.3	9 03	10.2	2 15	2 53
21	T	9 20	11.9	9 58	10.5	3 08	3 47
22	W	10 13	12.3	10 52	10.7	4 01	4 40
23	T	11 05	12.4	11 46	10.8	4 53	5 32
24	F	11 58	12.4	...	...	5 46	6 24
25	S	12 40	10.7	12 52	12.1	6 39	7 17
26	S	1 34	10.5	1 47	11.5	7 33	8 10
27	M	2 29	10.3	2 43	10.9	8 29	9 04
28	T	3 26	10.0	3 42	10.2	9 27	9 59
29	W	4 23	9.8	4 43	9.6	10 27	10 56
30	T	5 22	9.6	5 45	9.1	11 30	11 53
31	F	6 19	9.5	6 46	8.8	...	12 32

Dates when Ht. of **Low** Water is below Mean Low with Ht. of lowest given for each period and Date of lowest in ( ):

20th - 29th: -2.2' (24th)

19th - 28th: -2.2' (23rd - 24th)

Average Rise and Fall 9 1/2 ft.

When a high tide exceeds av. ht., the *following* low tide will be lower than av. Since there is a high degree of correlation between the height of High Water and the velocities of the Flood and Ebb Currents for that same day, we offer a rough rule of thumb for estimating the current velocities, for ALL the Current Charts and Diagrams in this book. **Rule of Thumb:** Refer to Boston High Water. If the height of High Water is 11.0' or over, use the Current Chart velocities as shown. When the height is 10.5', subtract 10%; at 10.0', subtract 20%; at 9.0', 30%; at 8.0', 40%; below 7.5', 50%.

TIME OF HIGH WATER

Time figures shown are the *average* differences throughout the year. Rise in feet is mean range. (Low Water times are given *only* when they vary more than 20 min. from High Water times.)

	H.M.			Rise in feet
NOVA SCOTIA, Outer Coast				
Guysborough	3 15	*before*	BOSTON	3.8
Whitehaven	3 35	"	"	4.0
Liscomb River	3 30	"	"	4.2
Sheet Harbor	3 30	"	"	4.2
Ship Harbor	3 30	"	"	4.2
Jeddore Harbor	3 30	"	"	4.3
Halifax	3 23	"	"	4.4
Sable Island, North	3 30	"	"	2.6
Sable Island, South	3 25	"	"	3.9
Chester	3 20	"	"	4.4
Mahone	3 20	"	"	4.5
Lunenburg	3 20	"	"	4.2
Riverport, LeHave R.	3 10	"	"	4.5
Liverpool	3 00	"	"	4.3
Lockeport	3 00	"	"	4.6
Shelbourne	2 50	"	"	4.8
NOVA SCOTIA & NEW BRUNSWICK, Bay of Fundy				
Lower E. Pubnico	1 30	*before*	BOSTON	8.7
Yarmouth	0 40	"	"	11.5
Annapolis	0 40	*after*	"	22.6
Minas Basin				
Parrsboro	1 20	"	"	34.4
Burntcoat Harbor	1 30	"	"	22.6
Amherst Point, Cumberland Basin	1 05	"	"	35.6
Petitcodiac River				
Grindstone Is.	0 50	*after*	"	31.1
Hopewell Cape	0 45	"	"	33.2
Saint John	0 30	"	"	20.8
Indiantown, Saint John R.	2 00	"	"	1.2
L'Etang Harbor	0 30	*after*	"	18.4

NOTE: All figures here are based on the assumption that your watch is set for Atlantic Time and are to be compared *directly* with times shown for Boston High Water. (We have taken into account the time zone difference so please eliminate this in your figuring.)

For time of HIGH & LOW WATER at BOSTON, see Pages 12-17.

Extra HIGH & LOW TIDES at time of full & new moon, see Page 233.

REVERSING FALLS....SAINT JOHN, N.B.

The most turbulence in the gorge occurs on days when the tides are largest. On largest tides the outward fall is between 15 and 16 1/2 feet and is accompanied by a greater turbulence than the inward fall which is between 11 and 12 1/2 feet. The most spectacular effect is at low water and high water at Saint John. Slack water lasts only about 10 minutes and occurs about 2 hr. 25 min. after high water and 3 hr. 50 min. after low water at Saint John. For complete tidal information of Canadian ports see Tide Tables of the Atlantic Coast of Canada. (Purchase tables from nautical dealers in Canadian ports or from the Queen's Printer, Department of Public Printing, Ottawa.

TIME OF HIGH WATER

Time figures shown are the *average* differences throughout the year. Rise in feet is mean range. (Low Water times are given *only* when they vary more than 20 min. from High Water times.)

U.S. ATLANTIC COAST, from Maine southward

	H.M.			Rise in feet
MAINE				
Eastport	0 25	*before*	BOSTON	18.2
Lubec	0 30	"	"	17.5
West Quoddy Head	0 40	"	"	15.8
Cutler, Little River	0 40	"	"	13.4
Shoppee Pt. Englishman Bay	0 35	"	"	12.0
Roque Is. Harbor	0 40	"	"	12.5
Jonesport	0 30	"	"	11.5
Milbridge	0 35	"	"	11.2
Garden Pt. Gouldsboro Bay	0 40	"	"	10.8
Winter Harbor	0 35	"	"	10.1
Bar Harbor	0 35	"	"	10.5
Southwest Harbor, Mt. Desert Island	0 35	"	"	10.2
Penobscot Bay				
Sedgewick, Eggemoggin Reach	0 25	*before*	BOSTON	10.2
Isle Au Haut	0 40	"	"	9.3
Head Harbor	0 40	"	"	9.0
Stonington, Deer Isle	0 35	"	"	9.8
Matinicus Harbor	0 35	"	"	9.2
Vinalhaven	0 30	"	"	9.4
Pulpit Harbor	0 25	"	"	9.8
Castine	0 20	"	"	9.8
Hampden, Penobscot River	0 15	"	"	12.8
Bangor, Penobscot River	0 15	"	"	13.0
Belfast	0 20	"	"	10.0
Camden	0 25	"	"	9.5
Rockland	0 30	"	"	9.8
MAINE, Outer Coast				
Tenants Harbor	0 25	*before*	BOSTON	9.2
Monhegan Island	0 30	"	"	8.8
Thomaston, St. George River	0 20	"	"	9.5
New Harbor	0 25	"	"	9.0
Friendship Harbor	0 30	"	"	9.0
Waldoboro, Medomak River	0 30	"	"	9.5
East Boothbay, Damariscotta River	0 15	"	"	9.8
Boothbay Harbor	0 20	"	"	8.7
Sheepscot River				
Wiscasset	same as		BOSTON	9.5
Five Islands	0 20	*before*	"	9.0
Kennebec River				
Phippsburg	0 05	*after*	BOSTON	8.0
Bath	0 45	"	"	6.5
Richmond	2 35	"	"	8.9
Gardiner ... **high 3 25** *after*, low	4 10	"	"	5.0
Hallowell ... **high 3 40** *after*, low	4 45	"	"	4.3
Augusta ... **high 3 45** *after*, low	5 15	"	"	4.0
Casco Bay				
Small Point Harbor	0 25	*before*	BOSTON	8.8
Cundy Harbor	0 15	"	"	8.9
Harpswell Harbor	0 20	"	"	9.0
Potts Harbor	0 20	"	"	8.8
Great Chebeague Island	0 20	"	"	9.0
Portland	0 15	"	"	9.0

BOSTON Tables, pp. 12-17

When a high tide exceeds av. ht., the *following* low tide will be lower than av.

TIME OF HIGH WATER

Time figures shown are the *average* differences throughout the year. Rise in feet is mean range. (Low Water times are given *only* when they vary more than 20 min. from High Water times.)

	H.M.			Rise in feet
MAINE, Outer Coast, Cont.				
Richmond Island	0 15	*before*	BOSTON	8.9
Wood Island Harbor	0 15	"	"	9.0
Cape Porpoise	same as		"	8.8
Kennebunkport	same as		"	8.6
York Harbor	0 10	*before*	"	8.6
NEW HAMPSHIRE				
Portsmouth	0 05	*after*	BOSTON	8.1
Gosport Harbor, Isles of Shoals	0 15	*before*	"	8.8
Hampton	same as		"	8.3
MASSACHUSETTS, Outer Coast				
Newburyport	0 30	*after*	BOSTON	7.8
Annisquam	same as		"	8.7
Rockport	0 05	*after*	"	8.6
Gloucester	same as		"	8.7
Manchester	same as		"	8.8
Salem	same as		"	9.0
Marblehead	0 05	*before*	"	9.1
Broad Sound				
Nahant	same as		"	9.0
Lynn Harbor	0 10	*after*	"	9.2
Neponset	0 05	*before*	"	9.5
Weymouth, Back River	0 05	*after*	"	9.5
Hingham	same as		"	9.5
Cohasset Harbor	same as		"	8.8
Scituate	same as		"	9.0
Duxbury	same as		"	
Cape Cod Bay				
Gurnet Point	same as		"	9.2
Plymouth	same as		"	9.5
Cape Cod Canal East	same as		"	8.7
Wellfleet	0 15	*after*	"	10.0
Provincetown	0 15	"	"	9.1
Race Point	0 05	*before*	"	9.0
Cape Cod				
Nauset Harbor .... **high 0 30** *after*, low	0 55	*after*	"	6.0
Chatham, Stage Harbor	0 55	"	"	4.0
Chatham, Outside	0 30	"	"	6.7
Monomoy Point	0 35	"	"	3.7
Meeting House Pond	2 00	"	"	4.1
Nantucket Sound				
Dennisport .... **high 1 00** *after*, low	0 35	*after*	BOSTON	3.4
South Yarmouth, Bass River Bridge	1 45	"	"	2.8
Hyannisport .... **high 1 00** *after*, low	0 30	"	"	3.1
Cotuit .... **high 1 15** *after*, low	0 45	"	"	2.5
Succonnesset	0 50	"	"	1.9
Falmouth Inner Harbor	same as		"	1.5
Nantucket Island				
Siasconset	0 15	*after*	BOSTON	1.2
Great Point .... **high 0 40** *after*, low	0 25	"	"	3.1
Nantucket	1 00	"	"	3.0
Tuckernuck Island .... **high 0 45** *after*, low	0 25	"	"	2.6
Muskeget Island, North side	0 20	"	"	2.0

BOSTON Tables, pp. 12-17

When a high tide exceeds av. ht., the *following* low tide will be lower than av.

TIME OF HIGH WATER

Time figures shown are the *average* differences throughout the year. Rise in feet is mean range. (Low Water times are given *only* when they vary more than 20 min. from High Water times.)

		H.M.			Rise in feet
MASSACHUSETTS, Outer Coast, Cont.					
Martha's Vineyard					
Cape Pogue	**high 0 45** *after*, low	same as		BOSTON	2.2
Edgartown	**high 1 00** *after*, low	0 15	*after*	"	2.0
Oak Bluffs	**high 0 30** *after*, low	0 15	*before*	"	1.7
East Chop	**high 0 25** *after*, low	0 15	"	"	1.7
Vineyard Haven	**high 0 25** *after*, low	0 05	"	"	1.7
West Chop	**high 0 15** *after*, low	0 30	"	"	1.4
Lake Tashmoo (inside)		2 30	*before*	"	2.0
Cedar Tree Neck	**high 0 15** *after*, low	1 35	*after*	NEWPORT	2.3
Menemsha Bight	**high 0 05** *after*, low	0 40	"	"	2.7
Gay Head	**high** same as, low	0 50	"	"	3.0
Squibnocket Point	**high 0 40** *before*, low	0 05	"	"	2.9
Wasque Point	**high 2 05** *after*, low	3 25	"	"	1.1
Nomans Island	**high 0 15** *before*, low	0 25	"	"	3.0
Vineyard Sound North Side					
Little Hbr., Woods Hole	**high 0 35** *after*, low	2 25	*after*	NEWPORT	1.3
Oceanographic Inst.	**high 0 27** *after*, low	2 00	"	"	2.0
Tarpaulin Cove	**high 0 15** *after*, low	1 30	"	"	1.9
Quick's Hole, S. side	**high 0 10** *before*, low	0 15	"	"	2.5
Buzzards Bay					
Cuttyhunk, Pond Entrance		same as		NEWPORT	3.4
Penikese Island		0 15	*before*	"	3.4
W. Falmouth Harbor		0 25	*after*	"	4.0
Monument Beach		0 25	"	"	4.0
Pocasset Harbor		0 25	"	"	4.0
Wareham		0 25	"	"	4.1
Bird Island		0 10	"	"	4.2
Marion		0 10	"	"	4.0
Mattapoisett		0 15	"	"	3.9
New Bedford		0 10	"	"	3.7
South Dartmouth		0 30	"	"	3.7
Dumpling Rocks		same as		"	3.8
Westport Harbor	**high 0 10** *after*, low	0 40	"	"	3.0
RHODE ISLAND & MASS, Narragansett Bay					
Sakonnet		0 10	*before*	NEWPORT	3.1
Tiverton		0 20	*after*	"	3.8
Beavertail		same as		"	3.5
Prudence Island		0 10	*after*	"	3.9
Bristol Point		0 20	"	"	4.1
Fall River		0 30	"	"	4.4
Taunton	**high 1 10** *after*, low	2 25	*after*	"	2.8
Warren		0 15	"	"	4.6
Providence		0 10	"	"	4.6
Pawtucket		0 20	"	"	4.5
East Greenwich		0 15	"	"	4.0
Wickford		0 10	"	"	3.8
Narragansett Pier		0 10	*before*	"	3.2
RHODE ISLAND, Outer Coast					
Pt. Judith Harbor	**high 0 05** *before*, low	0 20	*after*	NEWPORT	3.1
Great Salt Pond		same as		"	2.6
Watch Hill	**high 0 45** *after*, low	1 20	*after*	"	2.5

BOSTON Tables, pp. 12-17
NEWPORT Tables, pp. 64-69

When a high tide exceeds av. ht., the *following* low tide will be lower than av.

TIME OF HIGH WATER

Time figures shown are the *average* differences throughout the year. Rise in feet is mean range. (Low Water times are given *only* when they vary more than 20 min. from High Water times.)

	H.M.			Rise in feet
CONNECTICUT, L.I. Sound				
Stonington	2 25	*before*	BRIDGEPORT	2.8
Noank	2 15	"	"	2.3
Thames River				
New London	1 50	*before*	BRIDGEPORT	2.5
Norwich	1 40	"	"	3.0
Connecticut River				
Saybrook Breakwater	0 50	*before*	BRIDGEPORT	3.5
Essex	0 10	"	"	3.0
Madison	0 20	"	"	4.9
Falkner Island	0 15	"	"	5.5
Branford Harbor	0 10	"	"	6.0
New Haven Harbor Entrance	0 10	"	"	6.2
Milford Harbor	0 10	"	"	6.5
Stratford, Housatonic R. **high 0 30** *after*, low	1 00	*after*	"	5.6
Black Rock Harbor Entrance	0 05	*before*	"	7.0
South Norwalk	0 10	*after*	"	7.0
Stamford	0 05	"	"	7.2
Cos Cob Harbor	0 10	"	"	7.2
Greenwich	same as		"	7.4
Great Capt. Island	same as		"	7.3
NEW YORK, Long Island Sound, North Side				
Port Chester	0 10	*before*	WILLETS	7.2
Rye Beach	0 30	"	"	7.2
Mamaroneck	0 10	"	"	7.3
New Rochelle	0 25	"	"	7.2
City Island	0 05	"	"	7.2
Throgs Neck	same as		"	7.0
East River				
Whitestone	same as		WILLETS	7.1
College Point	0 20	*after*	"	6.5
Hunts Point	0 10	"	"	6.9
North Brother Island	0 10	"	"	6.6
Port Morris	0 15	"	"	6.4
Pot Cove, Astoria	2 20	*after*	BATTERY	5.3
Spuyten Duyvil Bridge, Harlem River	1 00	"	"	3.6
NEW YORK, Long Island Sound, South Side				
Port Washington	0 05	*after*	WILLETS	7.3
Execution Rocks	0 05	*before*	"	7.3
Glen Cove	0 10	"	"	7.3
Oyster Bay Harbor	0 10	*after*	BRIDGEPORT	7.2
Cold Spring Harbor	0 10	"	"	7.5
Eaton's Neck Point	0 05	"	"	7.1
Lloyd Harbor Entrance	same as		"	7.5
Northport	0 05	"	"	7.3
Stratford Shoal	0 05	*before*	"	6.6
Port Jefferson Harbor Entrance	same as		"	6.6
Mattituck Inlet	0 05	*after*	"	5.2
Horton Point	0 25	*before*	"	4.0
Shelter Island Sound				
Orient Point	1 15	*before*	BRIDGEPORT	2.5
Greenport **high 0 45** *before*, low	1 00	*before*	"	2.4
Southold	0 10	"	"	2.3
New Suffolk	0 35	*after*	"	2.6
Sag Harbor	0 50	*before*	"	2.5

BRIDGEPORT Tables, pp. 84-89
BATTERY Tables, pp. 108-113
WILLETS PT. Tables, pp. 90-95

When a high tide exceeds av. ht., the *following* low tide will be lower than av.

TIME OF HIGH WATER

Time figures shown are the *average* differences throughout the year. Rise in feet is mean range. (Low Water times are given *only* when they vary more than 20 min. from High Water times.)

	H.M.			Rise in feet
NEW YORK, Long Island Sound, South Side, Cont.				
South Jamesport, Peconic Bay	0 40	*after*	BRIDGEPORT	2.7
Shinnecock Canal	0 40	"	"	2.4
Three Mile Harbor Ent. **high 1 30** *before*, low	1 50	*before*	"	2.4
Promised Land, Napeague Bay	2 00	"	"	2.3
Montauk, North Side	3 10	"	"	2.0
Montauk (Fort Pond Bay)	2 15	"	"	2.1
Shinnecock Inlet	0 50	*before*	SANDY HOOK	2.9
Moriches Inlet	1 00	"	"	2.9
Fire Island Breakwater	0 40	"	"	4.1
Great South Bay				
Bellport	3 45	*after*	SANDY HOOK	0.8
Patchogue	3 25	"	"	0.7
Bay Shore	2 25	"	"	0.6
Babylon	2 10	"	"	0.6
Jones Inlet (Point Lookout)	0 20	*before*	"	3.6
E. Rockaway, Hempstead Bay	0 40	*after*	"	4.0
Bellmore, Hempstead Bay	1 30	"	"	2.0
Freeport, Baldwin Bay	0 40	"	"	3.0
Rockaway Inlet, Jamaica Bay	0 05	*before*	"	5.0
NEW YORK & NEW JERSEY				
New York Harbor				
Coney Island	0 05	*before*	SANDY HOOK	4.8
Fort Hamilton, The Narrows	same as		"	4.8
Bayonne, NJ	0 20	*before*	BATTERY	4.5
Hudson River				
Yonkers	1 10	*after*	BATTERY	3.7
Dobbs Ferry	1 30	"	"	3.4
Tarrytown	1 50	"	"	3.2
West Point	3 15	"	"	2.8
Poughkeepsie	4 30	"	"	3.0
Kingston Point	5 15	"	"	3.7
Catskill	6 35	"	"	4.2
Kill Van Kull				
Constable Hook	0 30	*before*	BATTERY	2.4
New Brighton	0 10	"	"	4.5
Port Richmond	0 05	"	"	4.5
Newark, Passaic R.	0 20	*after*	"	5.0
Bellville, Passaic R. **high 0 05** *after*, low	0 45	"	"	3.0
Passaic, Greg. Ave. Br. **high 0 55** *after*, low	2 00	"	"	2.7
Kearney Pt., Hackensack R. **high 0 15** *after*, low	0 35	"	"	2.8
Secaucus, Hackensack R.	1 15	"	"	2.8
Hackensack, Hackensack R.	1 35	"	"	5.2
Arthur Kill				
Elizabethport	0 25	*after*	SANDY HOOK	4.9
Tottenville	0 05	"	"	5.2
Perth Amboy	0 15	"	"	5.2
New Brunswick, Raritan River	0 45	"	"	5.8
Port Monmouth	same as		"	4.8
Red Bank	1 50	*after*	"	3.0
NEW JERSEY, Outer Coast				
Sea Bright	0 35	*before*	SANDY HOOK	4.5
Asbury Park	0 35	"	"	4.3
Shark R. Inlet **high 0 20** *before*, low	0 35	"	"	4.3

BRIDGEPORT Tables, pp. 84-89
BATTERY Tables, pp. 108-113
SANDY HOOK Tables, pp. 120-125

When a high tide exceeds av. ht., the *following* low tide will be lower than av.

TIME OF HIGH WATER

Time figures shown are the *average* differences throughout the year. Rise in feet is mean range. (Low Water times are given *only* when they vary more than 20 min. from High Water times.)

	H.M.			Rise in feet
NEW JERSEY, Outer Coast, Cont.				
Manasquan Inlet	0 10	*before*	SANDY HOOK	4.0
Manasquan River RR Bridge	0 15	*after*	SANDY HOOK	3.5
Mantoloking, Barnegat Bay	5 35	"	"	0.5
Tom's River (town)	4 35	"	"	0.6
Barnegat Inlet	0 20	*before*	"	3.1
Manahawkin Bridge	2 35	*after*	"	1.5
Beach Haven, Little Egg Harbor	1 50	"	"	2.2
Absecon Creek Entrance	1 05	"	"	3.6
Atlantic City, steel pier	0 25	*before*	"	4.1
Great Egg Harbor Inlet	0 10	*after*	"	3.8
Ocean City, 9th Street Bridge	0 25	"	"	3.7
Great Egg Harbor Bay	0 45	"	"	3.6
Townsend Inlet	0 05	"	"	3.8
Hereford Inlet	same as		"	4.2
Cape May Harbor	0 05	*before*	"	4.5
NEW JERSEY & DELAWARE BAY				
Eastern Shore				
Cape May Point	same as		BATTERY	4.7
Dennis Creek Entrance	1 00	*after*	"	5.6
Mauricetown ... **high 2 00** *after*, low	2 30	"	"	5.9
Millville, Maurice R. ... **high 2 45** *after*, low	3 30	"	"	6.0
Egg Island Point ... **high 0 45** *after*, low	1 10	"	"	5.7
Ben Davis Point	1 20	"	"	6.0
Cohansey River Entrance	1 30	"	"	6.0
Bridgeton ... **high 3 25** *after*, low	2 45	"	"	6.5
Brandywine Shoal Light	1 10	"	"	5.0
Fourteen Ft. Bk. Lt. ... **high 0 30** *after*, low	1 00	"	"	5.2
Ship John Shoal Light	1 30	"	"	5.7
Western Shore				
Delaware Breakwater Harbor	0 10	*after*	BATTERY	4.2
Cape Henlopen	0 05	"	"	4.1
St. Jones River Ent. ... **high 1 05** *after*, low	1 45	"	"	4.8
Delaware River				
Liston Point	2 05	*after*	BATTERY	5.8
Salem, Salem River	3 20	"	"	5.6
Reedy Point	3 00	"	"	5.5
C&D Summit Bridge ... **high 2 25** *after*, low	2 00	"	"	3.5
Chesapeake City ... **high 2 30** *after*, low	1 55	"	"	2.6
New Castle ... **high 3 30** *after*, low	3 50	"	"	5.6
Christina River Ent. ... **high 3 50** *after*, low	4 15	"	"	5.6
Wilmington ... **high 3 55** *after*, low	4 30	"	"	5.8
DELAWARE, MARYLAND & VIRGINIA				
Indian River Inlet (bridge)	0 35	*after*	SANDY HOOK	2.7
Ocean City (pier)	0 25	*before*	"	3.5
Harbor of Refuge, Chincoteague Bay	0 05	*after*	"	2.5
Chincoteague Channel, south end	0 15	"	"	2.2
Chincoteague Island, USCG Station	0 30	"	"	1.6
Metomkin Inlet	0 35	"	"	3.6
Wachapreague Inlet, inside	0 10	"	"	3.9
Quinby Inlet Entrance	0 05	"	"	4.0
Great Machipongo Inlet, inside	0 35	"	"	3.9
Chesapeake Bay, Eastern Shore				
Cape Charles Harbor	0 35	*after*	BATTERY	2.4
Crisfield, Little Annemessex River	4 40	"	"	2.0
Salisbury, Wicomico River	7 10	"	"	3.0

BATTERY Tables, pp. 108-113
SANDY HOOK Tables, pp. 120-125

When a high tide exceeds av. ht., the *following* low tide will be lower than av.

TIME OF HIGH WATER

Time figures shown are the *average* differences throughout the year. Rise in feet is mean range. (Low Water times are given *only* when they vary more than 20 min. from High Water times.)

	H.M.			Rise in feet
DELAWARE, MARYLAND & VIRGINIA, Cont.				
Chesapeake Bay, Eastern Shore, cont.				
Taylors Island, Little Choptank River	3 05	*before*	BALTIMORE	1.2
Cambridge, Choptank River	2 45	"	"	1.6
Oxford, Tred Avon River	3 00	"	"	1.4
St. Michaels, Miles River	2 15	"	"	1.2
Kent Island Narrows	1 40	"	"	1.2
Worton Creek Entrance	1 10	*after*	"	1.3
Georgetown, Sassafras River	2 05	"	"	2.0
Town Neck Point, Elk River	3 15	"	"	2.1
Chesapeake, Western Shore				
Annapolis, Severn River	1 40	*before*	BALTIMORE	0.9
Drum Point, Pawtuxent River	4 50	"	"	1.2
Solomon's Island, Pawtuxent River	4 45	"	"	1.2
Sunnybank, Little Wicomico River	6 30	*after*	BATTERY	0.8
Glebe Point, Great Wicomico River	4 05	"	"	1.2
Windmill Point, Rappahannock River	2 45	"	"	1.2
Orchard Point, Rappahannock River	3 20	"	"	1.4
New Point Comfort, Mobjack Bay	0 45	"	"	2.3
Tue Marshes Light, York River	0 45	"	"	2.2
Perrin River, York River	1 05	"	"	2.3
Goodwin Neck, York River	1 05	"	"	2.3
Norfolk	1 10	"	"	2.8
Chuckatuck Creek Entrance, James River	1 35	"	"	2.8
Jamestown Island, James River	3 50	"	"	2.0
Windmill Point, James River	6 15	"	"	2.7
Cape Henry	same as		"	2.8
NORTH CAROLINA				
Roanoke Sound Channel	1 15	*after*	BATTERY	0.5
Oregon Inlet	0 25	*before*	"	2.0
Oregon Inlet Channel	0 30	"	"	1.2
Oregon Inlet, USCG Station	0 50	"	"	1.7
Hatteras Inlet	0 50	"	"	2.0
Ocracoke Inlet	0 50	"	"	1.9
Beaufort Inlet Channel Range	0 50	"	"	3.2
Morehead City	0 30	"	"	3.1
Bogue Inlet	0 25	"	"	2.2
New River Inlet	0 40	"	"	3.0
New Topsail Inlet	0 35	"	"	3.0
Masonboro Inlet	3 45	*before*	BOSTON	3.8
Bald Head, Cape Fear River	3 50	"	"	4.3
Wilmington	1 30	"	"	4.2
Lockwoods Folly Inlet	4 00	"	"	4.2
SOUTH CAROLINA				
Little River, 1 mile above mouth	3 35	*before*	BOSTON	5.0
Hog Inlet Pier	4 10	"	"	5.0
Murrell's Inlet	3 40	"	"	4.5
Pawley's Island	4 05	"	"	4.9
North Inlet	3 50	"	"	4.5
Winyah Bay Entrance, south jetty	4 00	"	"	4.6
South Island Plantation	3 00	"	"	3.8
Georgetown, Sampit River	2 05	"	"	3.3
Schooner Creek Entrance, Waccamaw River	1 10	"	"	3.2
North Santee River Inlet	3 50	"	"	4.5
South Santee River	2 55	"	"	4.1
Charleston Harbor Entrance, north jetty	3 50	"	"	5.2
Folly Island River, north	3 15	"	"	5.4

BATTERY Tables, pp. 108-113
BALTIMORE Tables, pp. 140-143
BOSTON Tables, p. 12-17

When a high tide exceeds av. ht., the *following* low tide will be lower than av.

TIME OF HIGH WATER

Time figures shown are the *average* differences throughout the year. Rise in feet is mean range. (Low Water times are given *only* when they vary more than 20 min. from High Water times.)

	H.M.			Rise in feet
SOUTH CAROLINA, Cont.				
North Edisto River, Rockville, Bohicket Creek	3 20	*before*	BOSTON	5.7
South Edisto River Marina, Big Bay Creek	3 15	"	"	5.9
Hutchinson Island, Ashepoo River	2 40	"	"	5.9
Fripp Inlet Bridge, St. Helena Sound	3 20	"	"	6.3
Hilton Head, Port Royal Sound	3 20	"	"	6.6
Port Royal, Battery Creek	2 35	"	"	7.2
Beaufort, Beaufort River	2 05	"	"	7.4
Braddock Point, Hilton Head Island	3 05	"	"	6.7
GEORGIA				
Savannah River Entrance	3 10	*before*	BOSTON	6.9
Tybee Creek Entrance	3 20	"	"	6.8
Wilmington River, north entrance	2 30	"	"	7.6
Isle of Hope, Skidaway River	2 20	"	"	7.8
Egg Islands, Ossabaw Sound	3 05	"	"	7.2
Walburg Creek Entrance, St. Catherines Sound	2 45	"	"	7.1
South Newport River, Sapelo Sound	2 35	"	"	7.4
Blackbeard Island	2 50	"	"	6.9
Blackbeard Creek, Blackbeard Island	2 50	"	"	6.5
Sapelo Island, Doboy Sound	3 10	"	"	6.8
Threemile Cut Entrance Darien River	2 25	"	"	6.6
St. Simons Sound Bar	3 10	"	"	6.5
Frederica River, St. Simons Sound	2 20	"	"	7.2
Brunswick, East River, St. Simons Sound	2 15	"	"	7.3
Jekyll Point, St. Andrews Sound	2 35	"	"	6.6
Cumberland Wharf, Cumberland River	2 30	"	"	6.8
FLORIDA, East Coast				
St. Mary's Entrance, north jetty	2 55	*before*	BOSTON	5.8
Nassau Sound	3 00	"	"	5.4
Amelia City, South Amelia River	2 00	"	"	5.6
St. John's River, south jetty	3 20	"	"	4.9
Mayport	2 55	"	"	4.5
Jacksonville R.R. Bridge	0 50	"	"	1.2
St. Augustine Inlet	3 15	"	"	4.5
Ponce Inlet	0 10	*after*	MIAMI	2.3
Cape Canaveral	0 40	*before*	"	3.5
Port Canaveral	0 35	"	"	3.6
Sebastian Inlet Bridge	0 20	"	"	2.2
St. Lucie .... **high 1 00** *after*, low	1 50	*after*	"	1.0
Fort Pierce Inlet, south jetty	0 10	*before*	"	2.6
Stuart .... **high 2 35** *after*, low	3 35	*after*	"	0.9
Jupiter Inlet, south jetty	same as		"	2.5
North Palm Beach, Lake Worth	0 10	"	"	2.9
Port of Palm Beach, Lake Worth	0 05	"	"	2.7
Lake Worth Pier (ocean)	0 15	*before*	"	2.8
Hillsboro Inlet, C.G. Station	0 10	*after*	"	2.5
Hillsboro Inlet Marina	0 20	"	"	2.5
Lauderdale-by-the-Sea, fishing pier	0 10	*before*	"	2.6
Bahia Mar Yt. Club .... **high 0 20** *after*, low	0 35	*after*	"	2.4
Port Everglades, turning basin	0 10	*before*	"	2.6
North Miami Beach, fishing pier	same as		"	2.5
Miami Marina, Biscayne Bay	0 55	*after*	"	2.2
Dinner Key Marina .... **high 1 15** *after*, low	1 50	"	"	1.9
Key Biscayne Yt. Club .... **high 1 05** *after* low	1 35	"	"	2.0
Ocean Reef Harbor, Key Largo	0 15	"	"	2.3
Tavernier, Hawk Channel	0 30	"	"	2.1
Key West	0 50	*before*	BOSTON	1.3

BOSTON Tables, p. 12-17
MIAMI Tables, p. 146-149

When a high tide exceeds av. ht., the *following* low tide will be lower than av.

HOW STRONG WINDS AFFECT TIDES

ON account of the disturbing effect of strong winds, it is not possible to predict at all times the exact time of high or low tide or the *turning of the current.* When the winds blow strong from the **southward** or **eastward**, the effect is to increase the tides, and make them **later**; on the contrary, when the winds are strong from the **northward** or **westward** the tides are diminished, and occur **earlier**. My extended observations on tides and tidal currents have proven that they are rendered inconstant by **distant winds**, as I have been carried out of my course on several occasions in thick weather by irregular currents while making surveys at various places between Boston and New York.

It will be well to remember that tides do not flow steadily until high water, as is generally supposed, but rise by waves, or swells, always making three waves or swells before reaching the highest point, or high water. In other words, the tide begins to rise steadily until it reaches about one-third of its usual height, when it stands still for a short time, and then ebbs or recedes slightly; it then begins to flow again until it has reached about two-thirds of its height, when it stands still and recedes as before; after this it flows until high water. Oftentimes, an inexperienced observer would suppose it was high water, and that ebb tide had commenced, when, in fact, it was only two-thirds flood. The ebb tide works in the same way. I speak of this particularly, because I have observed in many cases where vessels were ashore, or instances wherein high water was of the utmost importance, that the captain or those interested abandoned their labors or hope, when, if they had waited or held on an hour or so longer, their object would have been accomplished. (The state of the tide sometimes determines a matter of life or death.) I have therefore written the above, knowing that in many instances mariners would be benefited by an observance of these explanations and suggestions.

GEO. W. ELDRIDGE

(Today this theory of 3 waves or swells is under dispute but as it is an interesting observation by the Captain we continue to include it.)

Publisher's Note:

An interesting rule-of-thumb in wide usage round the world today in estimating how much the tide will rise or fall during any part of its 6 hour cycle is:

1/12 the first hour	3/12 the fourth
2/12 the second	2/12 the fifth
3/12 the third	1/12 the sixth

This approximation may be helpful to anyone who has run aground.

See Cape Cod Canal Currents and Tide, pages 30 thru 35

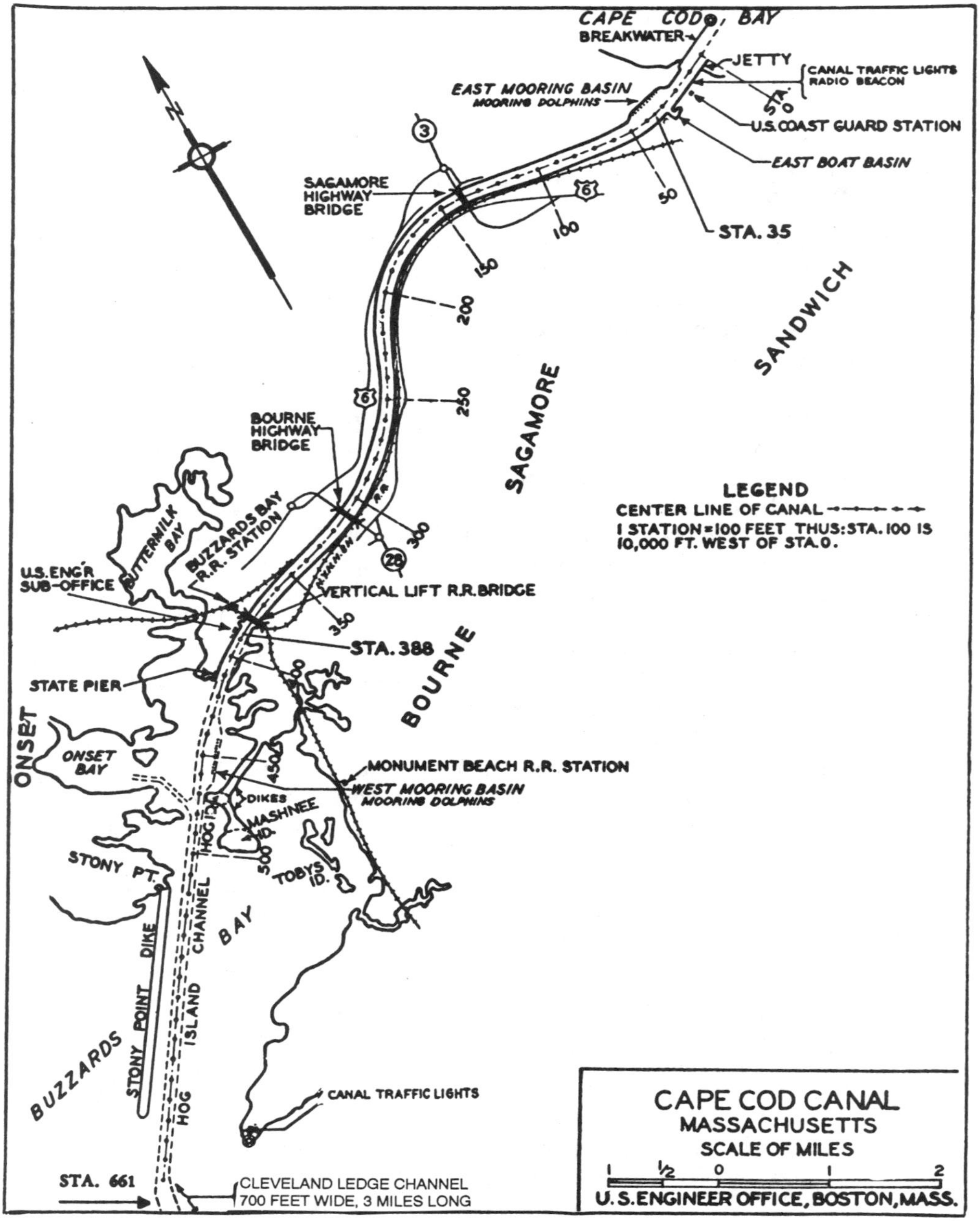

SMALL BOAT BASINS ON EITHER END OF CANAL: On E. end, 13-ft. mean low water, on S. Side of Sandwich, available for mooring small boat traffic; On W. end, channel 15-ft. at mean low water, 100 ft. wide leads from NE side of Hog Is. Ch. abreast of Hog Is. to harbor in Onset Bay. Fuel, supplies and phone service at both locations.

CAPE COD CANAL REGULATIONS

Speed Limit 10 m.p.h. - No excessive wake.
All vessels going with the current have right of way over those going against the current.
Clearance under highway bridges: 135 ft. at mean high water. Buzz. Bay R.R. Bridge maintained in up, or open position, except when lowered for trains or maintenance purposes.

Obtaining Clearance
Vessels over 65' shall not enter the Canal until clearance has been given by radio from the Marine Traffic Controller.
Vessels under 65' may proceed through the Canal without regard to traffic signals.
Vessels of any kind unable to make a through transit of the Canal against a head current of 6 kts. within a time limit of 2-1/2 hrs. are required to obtain helper tug assistance or wait for a fair current prior to receiving clearance from Traffic Controller.
Two-way traffic through Canal (vessels 65 ft. and over) allowed when Dispatcher on duty controlling traffic, considers conditions suitable.

Communications
Direct communiciations are available at all hours by radio or radiotelephone by phoning 508-759-4431. Direct radio communications may be made on VHF radio. Call on Channel 13 to establish contact. Transmissions will then be switched to Channels 12 or 14 as the working channels. Channel 16 is also available but should be limited to emergency situations. Vessels shall maintain a radio guard on Channel 13 during the entire passage.

Traffic Lights
Traffic Lights are at Eastern end at Sandwich (Cape Cod Bay Terminal) and at Western end near Wings Neck (Buzzards Bay entrance).
Entering From EASTERN END: (Lights S. side of ent. to Canal.)

RED LIGHT: Vessels over 65' must stop clear of outer end of channel and wait there until given clearance by radiotel from Marine Traffic Controller.

AMBER LIGHT: Proceed as far as E. Mooring Basin dolphins opp. Sandwich terminal.

GREEN LIGHT: Canal open to WEST-bound traffic.

Entering From WESTERN END: (Lights near Wings Neck at W. Entr. to Hog Is. Ch.)

RED LIGHT: Vessels over 65' must stop clear of channel and wait there until given clearance by radio from the Marine Traffic Controller.

AMBER LIGHT: Proceed as far as the W. Mooring Basin at Hog Is., and wait there for clearance as above.

GREEN LIGHT: Canal open to EAST-bound traffic.

When traffic lights are extinguished: all vessels over 65 ft. cautioned not to enter Canal until clearance given, as above.

Minimum Time Limits of Passage
(Station 35: E. Entr. at Sandwich; Station 388: Admin. Office Buzzards Bay; Station 661: W. Entr. Hog Is. Channel near Wings Neck).

	Station 35 to 388	Station 388 to 661
Against Head Tide	60 minutes	46 minutes
With Fair Tide	30 minutes	23 minutes
Slack Water	45 minutes	35 minutes

1999 CURRENT TABLE
CAPE COD CANAL
At R.R. Bridge

Standard Time

JANUARY

DAY OF MONTH	DAY OF WEEK	CURRENT TURNS EAST Flood Starts a.m.	**p.m.**	Kts.	CURRENT TURNS WEST Ebb Starts a.m.	**p.m.**	Kts.
1	F	2 06	**2 35**	p4.8	8 23	**9 07**	a5.0
2	S	2 58	**3 26**	p4.8	9 12	**9 58**	a5.1
3	S	3 48	**4 15**	p4.8	10 01	**10 48**	a5.1
4	M	4 37	**5 02**	p4.8	10 49	**11 37**	a5.0
5	T	5 25	**5 49**	p4.6	11 37	**...**	4.9
6	W	6 12	**6 37**	p4.4	12 25	**12 26**	p4.7
7	T	7 01	**7 25**	p4.2	1 13	**1 16**	p4.5
8	F	7 52	**8 15**	p3.9	2 03	**2 10**	p4.2
9	S	8 45	**9 08**	p3.7	2 54	**3 07**	4.0
10	S	9 42	**10 03**	p3.6	3 46	**4 07**	a3.9
11	M	10 40	**11 00**	p3.6	4 40	**5 07**	3.8
12	T	11 36	**11 56**	3.6	5 33	**6 05**	a3.9
13	W	...	**12 29**	3.7	6 23	**6 58**	a4.0
14	T	12 47	**1 17**	p3.9	7 10	**7 47**	a4.1
15	F	1 35	**2 02**	p4.1	7 55	**8 33**	a4.3
16	S	2 19	**2 43**	p4.2	8 36	**9 15**	a4.5
17	S	3 01	**3 23**	p4.4	9 15	**9 56**	a4.6
18	M	3 41	**4 02**	p4.5	9 53	**10 36**	a4.7
19	T	4 20	**4 42**	p4.5	10 31	**11 15**	a4.8
20	W	5 00	**5 23**	p4.6	11 11	**11 56**	a4.8
21	T	5 43	**6 06**	p4.5	11 54	**...**	4.7
22	F	6 28	**6 53**	p4.5	12 39	**12 41**	p4.7
23	S	7 18	**7 44**	p4.3	1 26	**1 34**	p4.6
24	S	8 14	**8 40**	p4.2	2 17	**2 34**	4.4
25	M	9 15	**9 42**	4.1	3 14	**3 42**	4.3
26	T	10 20	**10 48**	4.1	4 15	**4 52**	a4.4
27	W	11 26	**11 54**	a4.2	5 17	**6 00**	a4.4
28	T	...	**12 30**	4.3	6 18	**7 04**	a4.6
29	F	12 56	**1 29**	p4.5	7 15	**8 02**	a4.7
30	S	1 54	**2 23**	p4.6	8 09	**8 55**	a4.9
31	S	2 46	**3 13**	p4.7	8 59	**9 44**	a4.9

Standard Time

FEBRUARY

DAY OF MONTH	DAY OF WEEK	CURRENT TURNS EAST Flood Starts a.m.	**p.m.**	Kts.	CURRENT TURNS WEST Ebb Starts a.m.	**p.m.**	Kts.
1	M	3 35	**4 00**	p4.7	9 47	**10 30**	a5.0
2	T	4 20	**4 44**	p4.7	10 33	**11 14**	a4.9
3	W	5 03	**5 27**	p4.5	11 17	**11 57**	a4.8
4	T	5 45	**6 08**	p4.4	...	**12 01**	4.7
5	F	6 28	**6 50**	p4.2	12 39	**12 46**	p4.5
6	S	7 12	**7 34**	3.9	1 22	**1 34**	4.2
7	S	7 59	**8 22**	3.7	2 07	**2 25**	4.0
8	M	8 51	**9 14**	3.5	2 56	**3 22**	a3.9
9	T	9 48	**10 12**	3.4	3 49	**4 24**	a3.8
10	W	10 48	**11 12**	3.4	4 46	**5 25**	a3.8
11	T	11 47	**...**	3.6	5 42	**6 24**	a3.9
12	F	12 11	**12 41**	p3.8	6 35	**7 17**	a4.0
13	S	1 04	**1 30**	p4.0	7 23	**8 04**	a4.2
14	S	1 51	**2 15**	p4.2	8 08	**8 48**	a4.4
15	M	2 35	**2 57**	p4.4	8 49	**9 29**	a4.6
16	T	3 16	**3 38**	p4.6	9 30	**10 09**	a4.8
17	W	3 57	**4 19**	p4.7	10 10	**10 49**	a4.9
18	T	4 38	**5 01**	p4.7	10 52	**11 29**	a5.0
19	F	5 21	**5 45**	p4.7	11 36	**...**	4.9
20	S	6 06	**6 32**	4.6	12 12	**12 24**	p4.8
21	S	6 56	**7 23**	a4.5	12 58	**1 18**	4.7
22	M	7 51	**8 20**	a4.3	1 49	**2 18**	a4.6
23	T	8 52	**9 23**	a4.2	2 47	**3 26**	a4.4
24	W	9 59	**10 31**	a4.1	3 50	**4 38**	a4.4
25	T	11 08	**11 41**	a4.1	4 56	**5 48**	a4.4
26	F	...	**12 15**	4.2	6 01	**6 52**	a4.4
27	S	12 45	**1 15**	p4.4	7 01	**7 49**	a4.6
28	S	1 43	**2 09**	p4.5	7 55	**8 39**	a4.7

The Kts. (knots) columns show the **maximum** predicted velocities of the stronger one of the Flood Currents and the stronger one of the Ebb Currents for each day.
The letter "a" means the velocity shown should occur **after** the **a.m.** Current Change.
The letter "p" means the velocity shown should occur **after** the **p.m.** Current Change (even if next morning).
No "a" or "p" means a.m. and p.m. velocities are the same for that day.
Av. Max. Vel.: Fl. 4.0 Kts., Ebb 4.5 Kts.
Max. Flood & Max. Ebb 3 hrs. after Flood Starts & Ebb Starts, ±20 min.
Average rise and fall: canal east end, 8.7 ft. (time of high water same as Boston); west end, at Monument Beach, 4.0 ft. (time of high water 20 min. after Newport).

See pp. 150-157 for Current Change at other points.

1999 CURRENT TABLE
CAPE COD CANAL
At R.R. Bridge

Standard Time — Daylight Time starts Apr. 4 at 2 a.m.

MARCH

DAY OF MONTH	DAY OF WEEK	CURRENT TURNS: EAST Flood Starts a.m.	p.m.	Kts.	WEST Ebb Starts a.m.	p.m.	Kts.
1	M	2 33	2 58	p4.6	8 45	9 25	a4.8
2	T	3 18	3 41	p4.6	9 31	10 08	a4.8
3	W	4 00	4 22	p4.5	10 15	10 47	a4.8
4	T	4 39	5 01	4.4	10 56	11 25	a4.7
5	F	5 17	5 38	4.3	11 37	...	4.6
6	S	5 55	6 16	a4.2	12 03	12 18	4.4
7	S	6 34	6 56	a4.0	12 41	1 00	a4.3
8	M	7 17	7 40	a3.8	1 21	1 47	a4.1
9	T	8 05	8 30	a3.7	2 06	2 41	a4.0
10	W	9 00	9 28	a3.5	2 58	3 42	a3.8
11	T	10 00	10 30	a3.4	3 57	4 46	a3.7
12	F	11 02	11 33	a3.5	4 58	5 47	a3.8
13	S	...	12 01	3.7	5 56	6 42	a3.9
14	S	12 29	12 55	p4.0	6 49	7 32	a4.2
15	M	1 20	1 43	p4.3	7 37	8 16	4.4
16	T	2 06	2 28	p4.5	8 22	8 58	a4.7
17	W	2 49	3 11	p4.7	9 05	9 39	a4.9
18	T	3 31	3 54	p4.8	9 48	10 20	a5.0
19	F	4 14	4 38	4.8	10 33	11 02	a5.1
20	S	4 59	5 24	4.8	11 20	11 45	5.0
21	S	5 46	6 12	a4.8	...	12 10	4.9
22	M	6 36	7 05	a4.6	12 33	1 05	a4.9
23	T	7 32	8 02	a4.5	1 25	2 06	a4.7
24	W	8 33	9 07	a4.3	2 24	3 13	a4.5
25	T	9 40	10 18	a4.1	3 29	4 24	a4.4
26	F	10 51	11 28	a4.1	4 37	5 32	a4.3
27	S	11 58	...	4.2	5 44	6 35	a4.3
28	S	12 32	12 58	p4.3	6 45	7 29	a4.4
29	M	1 27	1 50	p4.4	7 39	8 18	a4.5
30	T	2 15	2 37	p4.5	8 28	9 01	a4.6
31	W	2 58	3 18	p4.4	9 13	9 41	a4.7

APRIL

DAY OF MONTH	DAY OF WEEK	CURRENT TURNS: EAST Flood Starts a.m.	p.m.	Kts.	WEST Ebb Starts a.m.	p.m.	Kts.
1	T	3 36	3 57	4.4	9 54	10 18	4.6
2	F	4 13	4 33	4.3	10 34	10 53	a4.6
3	S	4 48	5 08	a4.3	11 13	11 28	4.5
4	S	*6 24	*6 45	a4.2	...	*12 52	4.3
5	M	7 02	7 23	a4.1	1 03	1 32	a4.3
6	T	7 43	8 06	a3.9	1 41	2 17	a4.2
7	W	8 28	8 54	a3.8	2 23	3 07	a4.0
8	T	9 19	9 50	a3.6	3 12	4 05	a3.9
9	F	10 17	10 51	a3.6	4 10	5 07	a3.8
10	S	11 19	11 54	a3.6	5 13	6 08	a3.8
11	S	...	12 20	3.8	6 15	7 04	a4.0
12	M	12 53	1 16	p4.0	7 12	7 55	4.2
13	T	1 46	2 08	p4.3	8 04	8 41	4.5
14	W	2 34	2 56	p4.6	8 53	9 25	4.7
15	T	3 20	3 43	p4.7	9 40	10 08	4.9
16	F	4 06	4 29	4.8	10 27	10 51	p5.1
17	S	4 51	5 16	4.9	11 15	11 35	5.1
18	S	5 38	6 04	a4.9	...	12 04	5.0
19	M	6 27	6 54	a4.9	12 21	12 57	a5.1
20	T	7 19	7 48	a4.7	1 10	1 53	a5.0
21	W	8 15	8 47	a4.5	2 04	2 54	a4.8
22	T	9 16	9 52	a4.3	3 03	3 59	a4.6
23	F	10 22	11 01	a4.2	4 09	5 07	a4.4
24	S	11 29	...	4.1	5 17	6 11	a4.3
25	S	12 09	12 34	p4.1	6 23	7 10	a4.3
26	M	1 11	1 33	p4.2	7 23	8 03	a4.3
27	T	2 04	2 25	p4.3	8 18	8 50	a4.4
28	W	2 51	3 10	p4.3	9 06	9 32	4.4
29	T	3 32	3 51	p4.3	9 51	10 11	4.5
30	F	4 10	4 29	a4.3	10 32	10 47	4.5

***Daylight Time starts**

The Kts. (knots) columns show the **maximum** predicted velocities of the stronger one of the Flood Currents and the stronger one of the Ebb Currents for each day.
The letter "a" means the velocity shown should occur **after** the **a.m.** Current Change.
The letter "p" means the velocity shown should occur **after** the **p.m.** Current Change (even if next morning).
No "a" or "p" means a.m. and p.m. velocities are the same for that day.
Av. Max. Vel.: Fl. 4.0 Kts., Ebb 4.5 Kts.
Max. Flood & Max. Ebb 3 hrs. after Flood Starts & Ebb Starts, ±20 min.
Average rise and fall: canal east end, 8.7 ft. (time of high water same as Boston); west end, at Monument Beach, 4.0 ft. (time of high water 20 min. after Newport).

See pp. 150-157 for Current Change at other points.

1999 CURRENT TABLE
CAPE COD CANAL
At R.R. Bridge

MAY (Daylight Time)								JUNE (Daylight Time)							
		CURRENT TURNS								CURRENT TURNS					
		EAST Flood Starts			WEST Ebb Starts					EAST Flood Starts			WEST Ebb Starts		
DAY OF MONTH	DAY OF WEEK	a.m.	**p.m.**	Kts.	a.m.	**p.m.**	Kts.	DAY OF MONTH	DAY OF WEEK	a.m.	**p.m.**	Kts.	a.m.	**p.m.**	Kts.
1	**S**	4 46	**5 05**	a4.3	11 12	**11 22**	p4.5	1	**T**	5 34	**5 52**	a4.2	...	**12 08**	4.2
2	**S**	5 22	**5 40**	a4.3	11 50	**11 56**	p4.5	2	**W**	6 12	**6 31**	a4.2	12 04	**12 48**	a4.4
3	**M**	5 57	**6 17**	a4.2	...	**12 29**	4.2	3	**T**	6 51	**7 13**	a4.1	12 42	**1 30**	a4.3
4	**T**	6 35	**6 55**	a4.1	12 31	**1 09**	a4.4	4	**F**	7 34	**7 58**	a4.1	1 21	**2 15**	a4.3
5	**W**	7 14	**7 37**	a4.0	1 08	**1 52**	a4.3	5	**S**	8 20	**8 48**	a4.0	2 06	**3 04**	a4.2
6	**T**	7 58	**8 24**	a3.9	1 49	**2 40**	a4.1	6	**S**	9 11	**9 43**	a4.0	2 57	**3 56**	a4.1
7	**F**	8 47	**9 17**	a3.8	2 35	**3 33**	a4.0	7	**M**	10 06	**10 41**	a4.0	3 55	**4 50**	a4.1
8	**S**	9 41	**10 16**	a3.8	3 30	**4 31**	a3.9	8	**T**	11 04	**11 41**	a4.1	4 58	**5 45**	a4.2
9	**S**	10 40	**11 17**	a3.8	4 31	**5 29**	a4.0	9	**W**	...	**12 03**	4.2	6 01	**6 39**	p4.4
10	**M**	11 40	**...**	3.9	5 34	**6 24**	a4.1	10	**T**	12 39	**1 02**	p4.3	7 03	**7 32**	p4.6
11	**T**	12 16	**12 38**	p4.1	6 35	**7 16**	p4.3	11	**F**	1 35	**1 58**	p4.5	8 01	**8 23**	p4.8
12	**W**	1 11	**1 33**	p4.3	7 31	**8 05**	p4.6	12	**S**	2 29	**2 52**	4.6	8 57	**9 12**	p5.0
13	**T**	2 03	**2 25**	p4.5	8 25	**8 52**	p4.8	13	**S**	3 22	**3 45**	a4.8	9 51	**10 02**	p5.1
14	**F**	2 53	**3 16**	p4.7	9 17	**9 38**	p5.0	14	**M**	4 13	**4 37**	a4.9	10 45	**10 51**	p5.2
15	**S**	3 42	**4 05**	4.8	10 08	**10 24**	p5.1	15	**T**	5 04	**5 29**	a4.9	11 38	**11 41**	p5.1
16	**S**	4 31	**4 55**	a4.9	10 59	**11 11**	p5.2	16	**W**	5 55	**6 21**	a4.9	...	**12 31**	4.8
17	**M**	5 20	**5 45**	a5.0	11 51	**11 59**	p5.1	17	**T**	6 47	**7 14**	a4.8	12 33	**1 24**	a5.0
18	**T**	6 10	**6 37**	a4.9	...	**12 44**	4.8	18	**F**	7 40	**8 09**	a4.6	1 26	**2 19**	a4.9
19	**W**	7 03	**7 32**	a4.8	12 50	**1 40**	a5.0	19	**S**	8 34	**9 06**	a4.4	2 22	**3 14**	a4.7
20	**T**	7 58	**8 30**	a4.6	1 45	**2 39**	a4.8	20	**S**	9 30	**10 05**	a4.2	3 20	**4 11**	a4.4
21	**F**	8 57	**9 32**	a4.4	2 44	**3 40**	a4.6	21	**M**	10 28	**11 05**	a4.0	4 21	**5 07**	a4.2
22	**S**	9 59	**10 37**	a4.2	3 47	**4 42**	a4.4	22	**T**	11 26	**...**	3.9	5 23	**6 01**	a4.1
23	**S**	11 02	**11 41**	a4.1	4 52	**5 42**	a4.3	23	**W**	12 03	**12 22**	p3.8	6 23	**6 52**	4.0
24	**M**	...	**12 04**	4.0	5 56	**6 39**	a4.2	24	**T**	12 57	**1 15**	3.8	7 19	**7 41**	p4.1
25	**T**	12 41	**1 01**	p4.0	6 56	**7 30**	a4.2	25	**F**	1 47	**2 04**	a3.9	8 11	**8 26**	p4.2
26	**W**	1 34	**1 52**	p4.1	7 50	**8 17**	4.2	26	**S**	2 33	**2 49**	a4.0	8 59	**9 08**	p4.3
27	**T**	2 21	**2 39**	p4.1	8 40	**8 59**	p4.3	27	**S**	3 15	**3 32**	a4.1	9 44	**9 48**	p4.4
28	**F**	3 04	**3 21**	4.1	9 26	**9 39**	p4.4	28	**M**	3 55	**4 12**	a4.1	10 26	**10 27**	p4.5
29	**S**	3 43	**4 00**	a4.2	10 08	**10 17**	p4.5	29	**T**	4 34	**4 51**	a4.2	11 07	**11 04**	p4.5
30	**S**	4 21	**4 37**	a4.2	10 49	**10 53**	p4.5	30	**W**	5 12	**5 29**	a4.2	11 47	**11 41**	p4.5
31	**M**	4 57	**5 14**	a4.2	11 29	**11 29**	p4.5								

The Kts. (knots) columns show the **maximum** predicted velocities of the stronger one of the Flood Currents and the stronger one of the Ebb Currents for each day.
The letter "a" means the velocity shown should occur **after** the **a.m.** Current Change.
The letter "p" means the velocity shown should occur **after** the **p.m.** Current Change (even if next morning).
No "a" or "p" means a.m. and p.m. velocities are the same for that day.
Av. Max. Vel.: Fl. 4.0 Kts., Ebb 4.5 Kts.
Max. Flood & Max. Ebb 3 hrs. after Flood Starts & Ebb Starts, ±20 min.
Average rise and fall: canal east end, 8.7 ft. (time of high water same as Boston); west end, at Monument Beach, 4.0 ft. (time of high water 20 min. after Newport).
See pp. 150-157 for Current Change at other points.

1999 CURRENT TABLE
CAPE COD CANAL
At R.R. Bridge

JULY — Daylight Time

DAY OF MONTH	DAY OF WEEK	CURRENT TURNS EAST Flood Starts a.m.	p.m.	Kts.	CURRENT TURNS WEST Ebb Starts a.m.	p.m.	Kts.
1	T	5 50	**6 08**	a4.3	...	**12 26**	4.2
2	F	6 29	**6 49**	a4.3	12 18	**1 06**	a4.5
3	S	7 10	**7 32**	a4.3	12 58	**1 48**	a4.5
4	S	7 55	**8 20**	a4.2	1 42	**2 33**	a4.4
5	M	8 43	**9 12**	a4.2	2 31	**3 22**	a4.3
6	T	9 36	**10 09**	a4.1	3 27	**4 14**	a4.3
7	W	10 34	**11 10**	a4.1	4 29	**5 10**	4.3
8	T	11 34	...	4.1	5 35	**6 07**	p4.4
9	F	12 11	**12 36**	p4.2	6 40	**7 04**	p4.6
10	S	1 12	**1 36**	4.3	7 43	**7 59**	p4.8
11	S	2 10	**2 34**	a4.5	8 42	**8 53**	p4.9
12	M	3 05	**3 29**	a4.7	9 38	**9 45**	p5.1
13	T	3 58	**4 22**	a4.8	10 32	**10 35**	p5.1
14	W	4 50	**5 13**	a4.9	11 23	**11 25**	p5.1
15	T	5 39	**6 03**	a4.8	...	**12 13**	4.7
16	F	6 28	**6 52**	a4.7	12 15	**1 03**	a5.0
17	S	7 17	**7 42**	a4.6	1 05	**1 52**	a4.9
18	S	8 06	**8 33**	a4.4	1 57	**2 42**	a4.6
19	M	8 57	**9 26**	a4.1	2 50	**3 32**	a4.4
20	T	9 49	**10 21**	a3.9	3 47	**4 25**	a4.2
21	W	10 44	**11 18**	a3.7	4 46	**5 18**	a4.0
22	T	11 40	...	3.6	5 46	**6 11**	p3.9
23	F	12 15	**12 36**	3.6	6 44	**7 03**	p4.0
24	S	1 10	**1 29**	a3.7	7 39	**7 52**	p4.1
25	S	2 00	**2 19**	a3.8	8 30	**8 38**	p4.2
26	M	2 46	**3 04**	a4.0	9 17	**9 21**	p4.4
27	T	3 29	**3 46**	a4.1	10 01	**10 01**	p4.5
28	W	4 09	**4 26**	a4.2	10 42	**10 40**	p4.6
29	T	4 48	**5 05**	a4.3	11 21	**11 18**	p4.6
30	F	5 26	**5 44**	a4.4	12 01	**11 56**	p4.7
31	S	6 06	**6 24**	a4.4	...	**12 38**	4.4

AUGUST — Daylight Time

DAY OF MONTH	DAY OF WEEK	CURRENT TURNS EAST Flood Starts a.m.	p.m.	Kts.	CURRENT TURNS WEST Ebb Starts a.m.	p.m.	Kts.
1	S	6 46	**7 07**	a4.4	12 36	**1 18**	a4.7
2	M	7 30	**7 53**	a4.4	1 20	**2 01**	a4.6
3	T	8 18	**8 44**	a4.3	2 09	**2 49**	a4.5
4	W	9 11	**9 41**	a4.2	3 04	**3 41**	a4.4
5	T	10 09	**10 44**	4.1	4 07	**4 39**	4.3
6	F	11 12	**11 49**	p4.1	5 15	**5 41**	p4.4
7	S	...	**12 17**	4.1	6 24	**6 42**	p4.5
8	S	12 54	**1 21**	a4.3	7 30	**7 42**	p4.7
9	M	1 55	**2 21**	a4.5	8 30	**8 38**	p4.8
10	T	2 52	**3 17**	a4.6	9 26	**9 31**	p5.0
11	W	3 45	**4 08**	a4.7	10 17	**10 21**	p5.0
12	T	4 35	**4 56**	a4.8	11 05	**11 09**	p5.0
13	F	5 21	**5 42**	a4.8	11 51	**11 56**	p4.9
14	S	6 06	**6 26**	a4.7	...	**12 36**	4.6
15	S	6 50	**7 10**	a4.5	12 43	**1 19**	a4.8
16	M	7 34	**7 56**	a4.3	1 30	**2 04**	a4.6
17	T	8 19	**8 43**	a4.0	2 18	**2 49**	a4.3
18	W	9 07	**9 35**	a3.8	3 10	**3 38**	a4.1
19	T	9 59	**10 31**	a3.6	4 07	**4 31**	3.9
20	F	10 56	**11 30**	p3.5	5 07	**5 27**	p3.8
21	S	11 56	...	3.4	6 08	**6 24**	p3.8
22	S	12 29	**12 54**	a3.6	7 06	**7 18**	p4.0
23	M	1 24	**1 48**	a3.7	8 00	**8 07**	p4.2
24	T	2 14	**2 36**	a3.9	8 48	**8 52**	p4.3
25	W	2 59	**3 19**	a4.1	9 32	**9 35**	p4.5
26	T	3 41	**4 00**	a4.3	10 12	**10 14**	p4.7
27	F	4 21	**4 39**	a4.5	10 51	**10 54**	p4.8
28	S	5 00	**5 18**	a4.6	11 29	**11 33**	p4.8
29	S	5 40	**5 58**	a4.6	...	**12 07**	4.7
30	M	6 22	**6 41**	a4.6	12 15	**12 48**	a4.8
31	T	7 06	**7 28**	a4.5	1 00	**1 31**	a4.8

The Kts. (knots) columns show the **maximum** predicted velocities of the stronger one of the Flood Currents and the stronger one of the Ebb Currents for each day.
The letter "a" means the velocity shown should occur **after** the **a.m.** Current Change.
The letter "p" means the velocity shown should occur **after** the **p.m.** Current Change (even if next morning).
No "a" or "p" means a.m. and p.m. velocities are the same for that day.
Av. Max. Vel.: Fl. 4.0 Kts., Ebb 4.5 Kts.
Max. Flood & Max. Ebb 3 hrs. after Flood Starts & Ebb Starts, ±20 min.
Average rise and fall: canal east end, 8.7 ft. (time of high water same as Boston); west end, at Monument Beach, 4.0 ft. (time of high water 20 min. after Newport).
See pp. 150-157 for Current Change at other points.

1999 CURRENT TABLE
CAPE COD CANAL
At R.R. Bridge

Daylight Time **Standard Time starts Oct. 31 at 2 a.m.**

SEPTEMBER

Day of Month	Day of Week	CURRENT TURNS EAST Flood Starts a.m.	p.m.	Kts.	CURRENT TURNS WEST Ebb Starts a.m.	p.m.	Kts.
1	W	7 55	8 20	a4.4	1 50	2 19	a4.6
2	T	8 49	9 18	4.2	2 47	3 13	a4.5
3	F	9 49	10 23	p4.1	3 52	4 14	p4.4
4	S	10 56	11 31	p4.1	5 02	5 20	p4.4
5	S	...	12 05	3.9	6 13	6 26	p4.4
6	M	12 39	1 11	a4.2	7 18	7 28	p4.6
7	T	1 42	2 11	a4.4	8 17	8 26	p4.7
8	W	2 39	3 05	a4.6	9 11	9 18	p4.8
9	T	3 30	3 53	a4.7	9 59	10 07	p4.9
10	F	4 17	4 37	a4.7	10 43	10 52	p4.9
11	S	5 00	5 18	a4.6	11 25	11 36	p4.8
12	S	5 41	5 58	a4.5	...	12 05	4.6
13	M	6 21	6 38	a4.4	12 19	12 44	a4.7
14	T	7 01	7 19	a4.2	1 02	1 24	a4.5
15	W	7 42	8 02	3.9	1 47	2 05	a4.3
16	T	8 26	8 51	3.7	2 35	2 51	4.0
17	F	9 17	9 45	3.5	3 28	3 43	3.8
18	S	10 14	10 45	p3.4	4 28	4 42	p3.7
19	S	11 16	11 47	p3.5	5 31	5 43	p3.8
20	M	...	12 18	3.3	6 31	6 41	p3.9
21	T	12 45	1 15	a3.7	7 26	7 34	p4.1
22	W	1 38	2 04	a3.9	8 14	8 21	p4.3
23	T	2 26	2 49	a4.2	8 58	9 05	p4.6
24	F	3 09	3 30	a4.4	9 39	9 47	p4.8
25	S	3 51	4 11	a4.6	10 18	10 29	p4.9
26	S	4 32	4 52	4.7	10 57	11 11	p5.0
27	M	5 14	5 34	4.7	11 37	11 55	4.9
28	T	5 58	6 19	4.7	...	12 18	4.9
29	W	6 44	7 07	4.6	12 43	1 03	a4.9
30	T	7 35	8 00	p4.5	1 36	1 53	4.7

OCTOBER

Day of Month	Day of Week	CURRENT TURNS EAST Flood Starts a.m.	p.m.	Kts.	CURRENT TURNS WEST Ebb Starts a.m.	p.m.	Kts.
1	F	8 30	9 00	p4.3	2 34	2 49	p4.6
2	S	9 33	10 05	p4.2	3 40	3 53	p4.4
3	S	10 43	11 15	p4.1	4 50	5 02	p4.4
4	M	11 54	...	3.8	6 00	6 11	p4.4
5	T	12 24	1 00	a4.2	7 03	7 14	p4.5
6	W	1 27	1 58	a4.4	8 00	8 11	p4.6
7	T	2 22	2 49	a4.5	8 51	9 03	p4.7
8	F	3 11	3 34	a4.5	9 36	9 50	p4.7
9	S	3 56	4 16	4.5	10 18	10 34	p4.7
10	S	4 36	4 54	a4.5	10 57	11 16	a4.7
11	M	5 15	5 31	4.4	11 34	11 56	a4.6
12	T	5 51	6 08	p4.3	...	12 10	4.5
13	W	6 29	6 46	4.1	12 37	12 46	4.4
14	T	7 08	7 27	p4.0	1 18	1 25	4.2
15	F	7 50	8 13	p3.8	2 04	2 08	4.0
16	S	8 39	9 04	p3.6	2 54	2 57	p3.9
17	S	9 35	10 01	p3.5	3 51	3 55	p3.8
18	M	10 36	11 02	p3.6	4 52	4 58	p3.8
19	T	11 39	...	3.3	5 52	6 00	p3.9
20	W	12 03	12 37	a3.7	6 48	6 56	p4.1
21	T	12 58	1 29	a3.9	7 37	7 47	p4.3
22	F	1 49	2 16	a4.2	8 22	8 34	p4.6
23	S	2 36	3 00	4.4	9 04	9 20	p4.8
24	S	3 21	3 43	p4.7	9 45	10 05	p4.9
25	M	4 05	4 27	p4.8	10 26	10 51	5.0
26	T	4 50	5 12	p4.9	11 08	11 38	a5.1
27	W	5 36	5 59	p4.9	11 52	...	5.1
28	T	6 25	6 49	p4.8	12 29	12 39	p5.0
29	F	7 17	7 44	p4.6	1 23	1 31	p4.8
30	S	8 15	8 43	p4.4	2 23	2 29	p4.7
31	S	*8 18	*8 48	p4.2	*2 27	*2 34	p4.5

***Standard Time starts**

The Kts. (knots) columns show the **maximum** predicted velocities of the stronger one of the Flood Currents and the stronger one of the Ebb Currents for each day.
The letter "a" means the velocity shown should occur **after** the **a.m.** Current Change.
The letter "p" means the velocity shown should occur **after** the **p.m.** Current Change (even if next morning).
No "a" or "p" means a.m. and p.m. velocities are the same for that day.

Av. Max. Vel.: Fl. 4.0 Kts., Ebb 4.5 Kts.

Max. Flood & Max. Ebb 3 hrs. after Flood Starts & Ebb Starts, ±20 min.

Average rise and fall: canal east end, 8.7 ft. (time of high water same as Boston); west end, at Monument Beach, 4.0 ft. (time of high water 20 min. after Newport).

See pp. 150-157 for Current Change at other points.

1999 CURRENT TABLE
CAPE COD CANAL
At R.R. Bridge

Standard Time

NOVEMBER

DAY OF MONTH	DAY OF WEEK	CURRENT TURNS EAST Flood Starts a.m.	p.m.	Kts.	CURRENT TURNS WEST Ebb Starts a.m.	p.m.	Kts.
1	M	9 27	**9 56**	p4.2	3 34	**3 43**	p4.4
2	T	10 37	**11 03**	p4.2	4 40	**4 52**	p4.3
3	W	11 42	...	3.9	5 42	**5 55**	p4.4
4	T	12 05	**12 39**	a4.3	6 37	**6 52**	p4.4
5	F	1 00	**1 28**	a4.3	7 26	**7 44**	p4.5
6	S	1 48	**2 13**	a4.4	8 10	**8 31**	4.5
7	S	2 31	**2 53**	p4.4	8 51	**9 14**	a4.6
8	M	3 11	**3 30**	p4.4	9 28	**9 55**	a4.6
9	T	3 48	**4 06**	p4.3	10 04	**10 35**	a4.6
10	W	4 24	**4 42**	p4.2	10 40	**11 14**	a4.5
11	T	5 01	**5 19**	p4.2	11 15	**11 55**	a4.4
12	F	5 39	**5 59**	p4.0	11 52	...	4.3
13	S	6 20	**6 42**	p3.9	12 37	**12 32**	p4.1
14	S	7 07	**7 29**	p3.8	1 24	**1 18**	p4.0
15	M	7 59	**8 22**	p3.7	2 16	**2 12**	p3.9
16	T	8 56	**9 20**	p3.7	3 13	**3 12**	p3.9
17	W	9 57	**10 19**	p3.8	4 10	**4 15**	p3.9
18	T	10 56	**11 16**	p4.0	5 05	**5 15**	p4.1
19	F	11 51	...	3.9	5 56	**6 11**	p4.3
20	S	12 11	**12 42**	4.2	6 44	**7 04**	p4.5
21	S	1 02	**1 30**	p4.5	7 29	**7 54**	4.7
22	M	1 51	**2 18**	p4.7	8 14	**8 44**	4.9
23	T	2 40	**3 05**	p4.9	8 59	**9 33**	a5.1
24	W	3 28	**3 53**	p5.0	9 44	**10 24**	a5.2
25	T	4 17	**4 43**	p5.0	10 31	**11 16**	a5.2
26	F	5 08	**5 34**	p4.9	11 21	...	5.1
27	S	6 01	**6 28**	p4.7	12 11	**12 14**	p5.0
28	S	6 58	**7 26**	p4.5	1 08	**1 11**	p4.8
29	M	8 00	**8 27**	p4.3	2 09	**2 14**	p4.6
30	T	9 05	**9 31**	p4.2	3 11	**3 20**	p4.4

Standard Time

DECEMBER

DAY OF MONTH	DAY OF WEEK	CURRENT TURNS EAST Flood Starts a.m.	p.m.	Kts.	CURRENT TURNS WEST Ebb Starts a.m.	p.m.	Kts.
1	W	10 11	**10 35**	p4.1	4 13	**4 27**	p4.2
2	T	11 14	**11 36**	p4.1	5 12	**5 30**	p4.2
3	F	...	**12 12**	3.9	6 07	**6 28**	4.2
4	S	12 31	**1 02**	4.1	6 56	**7 21**	a4.3
5	S	1 21	**1 48**	p4.2	7 41	**8 09**	a4.4
6	M	2 05	**2 29**	p4.2	8 22	**8 53**	a4.4
7	T	2 46	**3 07**	p4.2	9 01	**9 35**	a4.5
8	W	3 24	**3 44**	p4.3	9 38	**10 15**	a4.5
9	T	4 01	**4 21**	p4.2	10 14	**10 54**	a4.5
10	F	4 37	**4 57**	p4.2	10 50	**11 33**	a4.5
11	S	5 15	**5 35**	p4.1	11 27	...	4.4
12	S	5 55	**6 16**	p4.1	12 14	**12 05**	p4.3
13	M	6 38	**7 00**	p4.0	12 57	**12 47**	p4.2
14	T	7 26	**7 48**	p3.9	1 43	**1 35**	p4.1
15	W	8 18	**8 41**	p3.9	2 32	**2 30**	p4.0
16	T	9 15	**9 37**	p3.9	3 25	**3 31**	p4.0
17	F	10 14	**10 36**	p4.0	4 20	**4 35**	p4.1
18	S	11 13	**11 35**	p4.1	5 14	**5 37**	p4.3
19	S	...	**12 09**	4.2	6 07	**6 36**	4.4
20	M	12 31	**1 04**	p4.5	6 58	**7 32**	a4.7
21	T	1 26	**1 56**	p4.7	7 47	**8 26**	a4.9
22	W	2 19	**2 47**	p4.9	8 36	**9 19**	a5.1
23	T	3 10	**3 38**	p5.0	9 25	**10 11**	a5.2
24	F	4 02	**4 29**	p5.0	10 14	**11 03**	a5.2
25	S	4 53	**5 20**	p4.9	11 05	**11 55**	a5.2
26	S	5 45	**6 12**	p4.8	11 57	...	5.0
27	M	6 39	**7 05**	p4.6	12 49	**12 52**	p4.8
28	T	7 35	**8 01**	p4.3	1 44	**1 50**	p4.6
29	W	8 34	**9 00**	p4.1	2 41	**2 52**	p4.4
30	T	9 36	**10 00**	p3.9	3 38	**3 56**	p4.2
31	F	10 38	**11 00**	p3.9	4 36	**4 59**	4.0

The Kts. (knots) columns show the **maximum** predicted velocities of the stronger one of the Flood Currents and the stronger one of the Ebb Currents for each day.
The letter "a" means the velocity shown should occur **after** the **a.m.** Current Change.
The letter "p" means the velocity shown should occur **after** the **p.m.** Current Change (even if next morning).
No "a" or "p" means a.m. and p.m. velocities are the same for that day.
Av. Max. Vel.: Fl. 4.0 Kts., Ebb 4.5 Kts.
Max. Flood & Max. Ebb 3 hrs. after Flood Starts & Ebb Starts, ±20 min.
Average rise and fall: canal east end, 8.7 ft. (time of high water same as Boston); west end, at Monument Beach, 4.0 ft. (time of high water 20 min. after Newport).

See pp. 150-157 for Current Change at other points.

1999 CURRENT TABLE
WOODS HOLE, MASS.

0.1 mile SW of Devil's Ft. Is. – 41-31-12N x 70-41-06W (for South End, +10 Flood, +20 Ebb)

JANUARY

Standard Time

DAY OF MONTH	DAY OF WEEK	CURRENT TURNS			
		SOUTHEAST Flood Starts		NORTHWEST Ebb Starts	
		a.m.	**p.m.**	a.m.	**p.m.**
1	F	2 26	**2 55**	9 18	**10 02**
2	S	3 18	**3 46**	10 07	**10 53**
3	S	4 08	**4 35**	10 56	**11 43**
4	M	4 57	**5 22**	11 44	**...**
5	T	5 45	**6 09**	12 32	**12 32**
6	W	6 32	**6 57**	1 20	**1 21**
7	T	7 21	**7 45**	2 08	**2 11**
8	F	8 12	**8 35**	2 58	**3 05**
9	S	9 05	**9 28**	3 49	**4 02**
10	S	10 02	**10 23**	4 41	**5 02**
11	M	11 00	**11 20**	5 35	**6 02**
12	T	11 56	**...**	6 28	**7 00**
13	W	12 16	**12 49**	7 18	**7 53**
14	T	1 07	**1 37**	8 05	**8 42**
15	F	1 55	**2 22**	8 50	**9 28**
16	S	2 39	**3 03**	9 31	**10 10**
17	S	3 21	**3 43**	10 10	**10 51**
18	M	4 01	**4 22**	10 48	**11 31**
19	T	4 40	**5 02**	11 26	**...**
20	W	5 20	**5 43**	12 10	**12 06**
21	T	6 03	**6 26**	12 51	**12 49**
22	F	6 48	**7 13**	1 34	**1 36**
23	S	7 38	**8 04**	2 21	**2 29**
24	S	8 34	**9 00**	3 12	**3 29**
25	M	9 35	**10 02**	4 09	**4 37**
26	T	10 40	**11 08**	5 10	**5 47**
27	W	11 46	**...**	6 12	**6 55**
28	T	12 14	**12 50**	7 13	**7 59**
29	F	1 16	**1 49**	8 10	**8 57**
30	S	2 14	**2 43**	9 04	**9 50**
31	S	3 06	**3 33**	9 54	**10 39**

FEBRUARY

Standard Time

DAY OF MONTH	DAY OF WEEK	CURRENT TURNS			
		SOUTHEAST Flood Starts		NORTHWEST Ebb Starts	
		a.m.	**p.m.**	a.m.	**p.m.**
1	M	3 55	**4 20**	10 42	**11 25**
2	T	4 40	**5 04**	11 28	**...**
3	W	5 23	**5 47**	12 09	**12 12**
4	T	6 05	**6 28**	12 52	**12 56**
5	F	6 48	**7 10**	1 34	**1 41**
6	S	7 32	**7 54**	2 17	**2 29**
7	S	8 19	**8 42**	3 02	**3 20**
8	M	9 11	**9 34**	3 51	**4 17**
9	T	10 08	**10 32**	4 44	**5 19**
10	W	11 08	**11 32**	5 41	**6 20**
11	T	...	**12 07**	6 37	**7 19**
12	F	12 31	**1 01**	7 30	**8 12**
13	S	1 24	**1 50**	8 18	**8 59**
14	S	2 11	**2 35**	9 03	**9 43**
15	M	2 55	**3 17**	9 44	**10 24**
16	T	3 36	**3 58**	10 25	**11 04**
17	W	4 17	**4 39**	11 05	**11 44**
18	T	4 58	**5 21**	11 47	**...**
19	F	5 41	**6 05**	12 24	**12 31**
20	S	6 26	**6 52**	1 07	**1 19**
21	S	7 16	**7 43**	1 53	**2 13**
22	M	8 11	**8 40**	2 44	**3 13**
23	T	9 12	**9 43**	3 42	**4 21**
24	W	10 19	**10 51**	4 45	**5 33**
25	T	11 28	**...**	5 51	**6 43**
26	F	12 01	**12 35**	6 56	**7 47**
27	S	1 05	**1 35**	7 56	**8 44**
28	S	2 03	**2 29**	8 50	**9 34**

Av. Max. Vel.: Fl. 4.5 Kts., Ebb 3.6 Kts.

(There are times when velocities can exceed 7 Kts.)

The best guide for determining daily velocities at Woods Hole is to refer to velocities at Cape Cod Canal (pp. 30-35). If Flood or Ebb Current on a certain day is stronger or weaker than the Av. Max. Vel. at Cape Cod Canal, the Current for Woods Hole for that day will be similarly stronger or weaker than the Av. Max. Vel. at Woods Hole.

To hold longest fair current from Buzzards Bay headed East through Vd. & Nant. Sds. go through Woods Hole 2 1/2 hrs. after flood starts SE in Woods Hole. (Any earlier means adverse currents in the Sounds.)

See pp. 150-157 for Current Changes at other points.

1999 CURRENT TABLE
WOODS HOLE, MASS.

0.1 mile SW of Devil's Ft. Is. – 41-31-12N x 70-41-06W (for South End, +10 Flood, +20 Ebb)

Standard Time **Daylight Time starts Apr. 4 at 2 a.m.**

MARCH						APRIL					
DAY OF MONTH	DAY OF WEEK	CURRENT TURNS				DAY OF MONTH	DAY OF WEEK	CURRENT TURNS			
		SOUTHEAST Flood Starts		NORTHWEST Ebb Starts				SOUTHEAST Flood Starts		NORTHWEST Ebb Starts	
		a.m.	**p.m.**	a.m.	**p.m.**			a.m.	**p.m.**	a.m.	**p.m.**
1	**M**	2 53	**3 18**	9 40	**10 20**	1	**T**	3 56	**4 17**	10 49	**11 13**
2	**T**	3 38	**4 01**	10 26	**11 03**	2	**F**	4 33	**4 53**	11 29	**11 48**
3	**W**	4 20	**4 42**	11 10	**11 42**	3	**S**	5 08	**5 28**	...	**12 08**
4	**T**	4 59	**5 21**	11 51	**...**	4	**S**	*6 44	***7 05**	12 23	***1 47**
5	**F**	5 37	**5 58**	12 20	**12 32**	5	**M**	7 22	**7 43**	1 58	**2 27**
6	**S**	6 15	**6 36**	12 58	**1 13**	6	**T**	8 03	**8 26**	2 36	**3 12**
7	**S**	6 54	**7 16**	1 36	**1 55**	7	**W**	8 48	**9 14**	3 18	**4 02**
8	**M**	7 37	**8 00**	2 16	**2 42**	8	**T**	9 39	**10 10**	4 07	**5 00**
9	**T**	8 25	**8 50**	3 01	**3 36**	9	**F**	10 37	**11 11**	5 05	**6 02**
10	**W**	9 20	**9 48**	3 53	**4 37**	10	**S**	11 39	**...**	6 08	**7 03**
11	**T**	10 20	**10 50**	4 52	**5 41**	11	**S**	12 14	**12 40**	7 10	**7 59**
12	**F**	11 22	**11 43**	5 53	**6 42**	12	**M**	1 13	**1 36**	8 07	**8 50**
13	**S**	...	**12 21**	6 51	**7 37**	13	**T**	2 06	**2 28**	8 59	**9 36**
14	**S**	12 49	**1 15**	7 44	**8 27**	14	**W**	2 54	**3 16**	9 48	**10 20**
15	**M**	1 40	**2 03**	8 32	**9 11**	15	**T**	3 40	**4 03**	10 35	**11 03**
16	**T**	2 26	**2 48**	9 17	**9 53**	16	**F**	4 26	**4 49**	11 22	**11 46**
17	**W**	3 09	**3 31**	10 00	**10 34**	17	**S**	5 11	**5 36**	...	**12 10**
18	**T**	3 51	**4 14**	10 43	**11 15**	18	**S**	5 58	**6 24**	12 30	**12 59**
19	**F**	4 34	**4 58**	11 28	**11 57**	19	**M**	6 47	**7 14**	1 16	**1 52**
20	**S**	5 19	**5 44**	...	**12 15**	20	**T**	7 39	**8 08**	2 05	**2 48**
21	**S**	6 06	**6 32**	12 40	**1 05**	21	**W**	8 35	**9 07**	2 59	**3 49**
22	**M**	6 56	**7 15**	1 28	**2 00**	22	**T**	9 36	**10 12**	3 58	**4 54**
23	**T**	7 52	**8 22**	2 20	**3 01**	23	**F**	10 42	**11 21**	5 04	**6 02**
24	**W**	8 53	**9 27**	3 19	**4 08**	24	**S**	11 49	**...**	6 12	**7 06**
25	**T**	10 00	**10 38**	4 24	**5 19**	25	**S**	12 29	**12 54**	7 18	**8 05**
26	**F**	11 11	**11 48**	5 32	**6 27**	26	**M**	1 31	**1 53**	8 18	**8 58**
27	**S**	...	**12 18**	6 39	**7 30**	27	**T**	2 24	**2 45**	9 13	**9 45**
28	**S**	12 52	**1 18**	7 40	**8 24**	28	**W**	3 11	**3 30**	10 01	**10 27**
29	**M**	1 47	**2 10**	8 34	**9 13**	29	**T**	3 52	**4 11**	10 46	**11 06**
30	**T**	2 35	**2 57**	9 23	**9 56**	30	**F**	4 30	**4 49**	11 27	**11 42**
31	**W**	3 18	**3 38**	10 08	**10 36**						

***Daylight Time starts**

Av. Max. Vel.: Fl. 4.5 Kts., Ebb 3.6 Kts.

(There are times when velocities can exceed 7 Kts.)

The best guide for determining daily velocities at Woods Hole is to refer to velocities at Cape Cod Canal (pp. 30-35). If Flood or Ebb Current on a certain day is stronger or weaker than the Av. Max. Vel. at Cape Cod Canal, the Current for Woods Hole for that day will be similarly stronger or weaker than the Av. Max. Vel. at Woods Hole.

CAUTION: Going *from* Buzzards Bay *into* Vineyard Sound, whether through Woods Hole, or Robinsons Hole or Quicks Hole, *Red* Buoys must be kept on the LEFT or PORT hand, *Black* or *Green* Buoys kept on the RIGHT or STARBOARD hand. You are considered to be proceeding seaward and should thus follow the rules for LEAVING a harbor.

1999 CURRENT TABLE
WOODS HOLE, MASS.

0.1 mile SW of Devil's Ft. Is. – 41-31-12N x 70-41-06W (for South End, +10 Flood, +20 Ebb)

MAY (Daylight Time)						JUNE (Daylight Time)					
DAY OF MONTH	DAY OF WEEK	CURRENT TURNS				DAY OF MONTH	DAY OF WEEK	CURRENT TURNS			
		SOUTHEAST Flood Starts		NORTHWEST Ebb Starts				SOUTHEAST Flood Starts		NORTHWEST Ebb Starts	
		a.m.	**p.m.**	a.m.	**p.m.**			a.m.	**p.m.**	a.m.	**p.m.**
1	S	5 06	**5 25**	...	**12 07**	1	T	5 54	**6 12**	12 24	**1 03**
2	S	5 42	**6 00**	12 17	**12 45**	2	W	6 32	**6 51**	12 59	**1 43**
3	M	6 17	**6 37**	12 51	**1 24**	3	T	7 11	**7 33**	1 37	**2 25**
4	T	6 55	**7 15**	1 26	**2 04**	4	F	7 54	**8 18**	2 16	**3 10**
5	W	7 34	**7 57**	2 03	**2 47**	5	S	8 40	**9 08**	3 01	**3 59**
6	T	8 18	**8 44**	2 42	**3 35**	6	S	9 31	**10 03**	3 52	**4 51**
7	F	9 07	**9 37**	3 30	**4 28**	7	M	10 26	**11 01**	4 50	**5 45**
8	S	10 01	**10 36**	4 25	**5 26**	8	T	11 24	**...**	5 53	**6 40**
9	S	11 00	**11 37**	5 26	**6 24**	9	W	12 01	**12 23**	6 56	**7 34**
10	M	...	**12 01**	6 29	**7 19**	10	T	12 59	**1 22**	7 58	**8 27**
11	T	12 36	**12 58**	7 30	**8 11**	11	F	1 55	**2 18**	8 56	**9 18**
12	W	1 31	**1 53**	8 26	**9 00**	12	S	2 49	**3 12**	9 52	**10 07**
13	T	2 23	**2 45**	9 20	**9 47**	13	S	3 42	**4 05**	10 46	**10 57**
14	F	3 13	**3 36**	10 12	**10 33**	14	M	4 33	**4 57**	11 40	**11 46**
15	S	4 02	**4 25**	11 03	**11 19**	15	T	5 24	**5 49**	...	**12 33**
16	S	4 51	**5 15**	11 54	**...**	16	W	6 15	**6 41**	12 36	**1 26**
17	M	5 40	**6 05**	12 06	**12 46**	17	T	7 07	**7 34**	1 28	**2 19**
18	T	6 30	**7 57**	12 54	**1 39**	18	F	8 00	**8 29**	2 21	**3 14**
19	W	7 23	**7 52**	1 45	**2 35**	19	S	8 54	**9 26**	3 17	**4 09**
20	T	8 18	**8 50**	2 40	**3 34**	20	S	9 50	**10 25**	4 15	**5 06**
21	F	9 17	**9 52**	3 39	**4 35**	21	M	10 48	**11 25**	5 16	**6 02**
22	S	10 19	**10 57**	4 42	**5 37**	22	T	11 46	**...**	6 18	**6 56**
23	S	11 22	**...**	5 47	**6 37**	23	W	12 23	**12 42**	7 18	**7 47**
24	M	12 01	**12 24**	6 51	**7 34**	24	T	1 17	**1 35**	8 14	**8 36**
25	T	1 01	**1 21**	7 51	**8 25**	25	F	2 07	**2 24**	9 06	**9 21**
26	W	1 54	**2 12**	8 45	**9 12**	26	S	2 53	**3 09**	9 54	**10 03**
27	T	2 41	**2 59**	9 35	**9 54**	27	S	3 35	**3 52**	10 39	**10 43**
28	F	3 24	**3 41**	10 21	**10 34**	28	M	4 15	**4 32**	11 21	**11 22**
29	S	4 03	**4 20**	11 03	**11 12**	29	T	4 54	**5 11**	...	**12 02**
30	S	4 41	**4 57**	11 44	**11 48**	30	W	5 32	**5 49**	12 01	**12 42**
31	M	5 17	**5 34**	...	**12 24**						

Av. Max. Vel.: Fl. 4.5 Kts., Ebb 3.6 Kts.
(There are times when velocities can exceed 7 Kts.)

The best guide for determining daily velocities at Woods Hole is to refer to velocities at Cape Cod Canal (pp. 30-35). If Flood or Ebb Current on a certain day is stronger or weaker than the Av. Max. Vel. at Cape Cod Canal, the Current for Woods Hole for that day will be similarly stronger or weaker than the Av. Max. Vel. at Woods Hole.

To hold longest fair current from Buzzards Bay headed East through Vd. & Nant. Sds. go through Woods Hole 2 1/2 hrs. after flood starts SE in Woods Hole. (Any earlier means adverse currents in the Sounds.)

See pp. 150-157 for Current Changes at other points.

1999 CURRENT TABLE
WOODS HOLE, MASS.

0.1 mile SW of Devil's Ft. Is. – 41-31-12N x 70-41-06W (for South End, +10 Flood, +20 Ebb)

JULY (Daylight Time)						AUGUST (Daylight Time)					
DAY OF MONTH	DAY OF WEEK	CURRENT TURNS				DAY OF MONTH	DAY OF WEEK	CURRENT TURNS			
		SOUTHEAST Flood Starts		NORTHWEST Ebb Starts				SOUTHEAST Flood Starts		NORTHWEST Ebb Starts	
		a.m.	**p.m.**	a.m.	**p.m.**			a.m.	**p.m.**	a.m.	**p.m.**
1	T	6 10	**6 28**	12 36	**1 21**	1	S	7 06	**7 27**	1 31	**2 13**
2	F	6 49	**7 09**	1 13	**2 01**	2	M	7 50	**8 13**	2 15	**2 56**
3	S	7 30	**7 52**	1 53	**2 43**	3	T	8 38	**9 04**	3 04	**3 44**
4	S	8 15	**8 40**	2 37	**3 28**	4	W	9 31	**10 01**	3 59	**4 36**
5	M	9 03	**9 32**	3 26	**4 17**	5	T	10 29	**11 04**	5 02	**5 34**
6	T	9 56	**10 29**	4 22	**5 09**	6	F	11 32	**...**	6 10	**6 36**
7	W	10 54	**11 30**	5 24	**6 05**	7	S	12 09	**12 37**	7 19	**7 37**
8	T	11 54	**...**	6 30	**7 02**	8	S	1 14	**1 41**	8 25	**8 37**
9	F	12 31	**12 56**	7 35	**7 59**	9	M	2 15	**2 41**	9 25	**9 33**
10	S	1 32	**1 56**	8 38	**8 54**	10	T	3 12	**3 37**	10 21	**10 26**
11	S	2 30	**2 54**	9 37	**9 48**	11	W	4 05	**4 28**	11 12	**11 16**
12	M	3 25	**3 49**	10 33	**10 40**	12	T	4 55	**5 16**	...	**12 01**
13	T	4 18	**4 42**	11 27	**11 30**	13	F	5 41	**6 02**	12 04	**12 46**
14	W	5 10	**5 33**	...	**12 18**	14	S	6 26	**6 46**	12 51	**1 31**
15	T	5 59	**6 23**	12 20	**1 08**	15	S	7 10	**7 30**	1 38	**2 14**
16	F	6 48	**7 12**	1 10	**1 58**	16	M	7 54	**8 16**	2 25	**2 59**
17	S	7 37	**8 02**	2 00	**2 47**	17	T	8 39	**9 03**	3 13	**3 44**
18	S	8 26	**8 53**	2 52	**3 37**	18	W	9 27	**9 55**	4 05	**4 33**
19	M	9 17	**9 46**	3 45	**4 27**	19	T	10 19	**10 51**	5 02	**5 26**
20	T	10 09	**10 41**	4 42	**5 20**	20	F	11 16	**11 51**	6 02	**6 22**
21	W	11 04	**11 38**	5 41	**6 13**	21	S	...	**12 16**	7 03	**7 19**
22	T	...	**12 01**	6 41	**7 06**	22	S	12 49	**1 14**	8 01	**8 13**
23	F	12 35	**12 56**	7 39	**7 58**	23	M	1 44	**2 08**	8 55	**9 02**
24	S	1 30	**1 49**	8 34	**8 47**	24	T	2 34	**2 56**	9 43	**9 47**
25	S	2 20	**2 39**	9 25	**9 33**	25	W	3 19	**3 39**	10 27	**10 30**
26	M	3 06	**3 24**	10 12	**10 16**	26	T	4 01	**4 20**	11 07	**11 09**
27	T	3 49	**4 06**	10 56	**10 56**	27	F	4 41	**4 59**	11 46	**11 49**
28	W	4 29	**4 46**	11 37	**11 35**	28	S	5 20	**5 38**	...	**12 24**
29	T	5 08	**5 25**	...	**12 16**	29	S	6 00	**6 18**	12 28	**1 02**
30	F	5 46	**6 04**	12 13	**12 55**	30	M	6 42	**7 01**	1 10	**1 43**
31	S	6 26	**6 44**	12 51	**1 33**	31	T	7 26	**7 48**	1 55	**2 26**

Av. Max. Vel.: Fl. 4.5 Kts., Ebb 3.6 Kts.

(There are times when velocities can exceed 7 Kts.)

The best guide for determining daily velocities at Woods Hole is to refer to velocities at Cape Cod Canal (pp. 30-35). If Flood or Ebb Current on a certain day is stronger or weaker than the Av. Max. Vel. at Cape Cod Canal, the Current for Woods Hole for that day will be similarly stronger or weaker than the Av. Max. Vel. at Woods Hole.

CAUTION: Going *from* Buzzards Bay *into* Vineyard Sound, whether through Woods Hole, or Robinsons Hole or Quicks Hole, *Red* Buoys must be kept on the LEFT or PORT hand, *Black* or *Green* Buoys kept on the RIGHT or STARBOARD hand. You are considered to be proceeding seaward and should thus follow the rules for LEAVING a harbor.

1999 CURRENT TABLE
WOODS HOLE, MASS.

0.1 mile SW of Devil's Ft. Is. – 41-31-12N x 70-41-06W (for South End, +10 Flood, +20 Ebb)

Daylight Time **Standard Time starts Oct. 31 at 2 a.m.**

SEPTEMBER

DAY OF MONTH	DAY OF WEEK	SOUTHEAST Flood Starts a.m.	SOUTHEAST Flood Starts p.m.	NORTHWEST Ebb Starts a.m.	NORTHWEST Ebb Starts p.m.
1	W	8 15	8 40	2 45	3 14
2	T	9 09	9 38	3 42	4 08
3	F	10 09	10 43	4 47	5 09
4	S	11 16	11 51	5 57	6 15
5	S	...	12 25	7 08	7 21
6	M	12 59	1 31	8 13	8 23
7	T	2 02	2 31	9 12	9 21
8	W	2 59	3 25	10 06	10 13
9	T	3 50	4 13	10 54	11 02
10	F	4 37	4 57	11 38	11 47
11	S	5 20	5 38	...	12 20
12	S	6 01	6 18	12 31	1 00
13	M	6 41	6 58	1 14	1 39
14	T	7 21	7 39	1 57	2 19
15	W	8 02	8 22	2 42	3 00
16	T	8 46	9 11	3 30	3 46
17	F	9 37	10 05	4 23	4 38
18	S	10 34	11 05	5 23	5 37
19	S	11 36	...	6 26	6 38
20	M	12 07	12 38	7 26	7 36
21	T	1 05	1 35	8 21	8 29
22	W	1 58	2 24	9 09	9 16
23	T	2 46	3 09	9 53	10 00
24	F	3 29	3 50	10 34	10 42
25	S	4 11	4 31	11 13	11 24
26	S	4 52	5 12	11 52	...
27	M	5 34	5 54	12 06	12 32
28	T	6 18	6 39	12 50	1 13
29	W	7 04	7 27	1 38	1 58
30	T	7 55	8 20	2 31	2 48

OCTOBER

DAY OF MONTH	DAY OF WEEK	SOUTHEAST Flood Starts a.m.	SOUTHEAST Flood Starts p.m.	NORTHWEST Ebb Starts a.m.	NORTHWEST Ebb Starts p.m.
1	F	8 50	9 20	3 29	3 44
2	S	9 53	10 25	4 35	4 48
3	S	11 03	11 35	5 45	5 57
4	M	...	12 14	6 55	7 06
5	T	12 44	1 20	7 58	8 09
6	W	1 47	2 18	8 55	9 06
7	T	2 42	3 09	9 46	9 58
8	F	3 31	3 54	10 31	10 45
9	S	4 16	4 36	11 13	11 29
10	S	4 56	5 14	11 52	...
11	M	5 35	5 51	12 11	12 29
12	T	6 11	6 28	12 51	1 05
13	W	6 49	7 06	1 32	1 41
14	T	7 28	7 47	2 13	2 20
15	F	8 10	8 33	2 59	3 03
16	S	8 59	9 24	3 49	3 52
17	S	9 55	10 21	4 46	4 50
18	M	10 56	11 22	5 47	5 53
19	T	11 59	...	6 47	6 55
20	W	12 23	12 57	7 43	7 51
21	T	1 18	1 49	8 32	8 42
22	F	2 09	2 36	9 17	9 29
23	S	2 56	3 20	9 59	10 15
24	S	3 41	4 03	10 40	11 00
25	M	4 25	4 47	11 21	11 46
26	T	5 10	5 32	...	12 03
27	W	5 56	6 19	12 33	12 47
28	T	6 45	7 09	1 24	1 34
29	F	7 37	8 04	2 18	2 26
30	S	8 35	9 03	3 18	3 24
31	S	*8 38	*9 08	*3 22	*3 29

***Standard Time starts**

Av. Max. Vel.: Fl. 4.5 Kts., Ebb 3.6 Kts.

(There are times when velocities can exceed 7 Kts.)

The best guide for determining daily velocities at Woods Hole is to refer to velocities at Cape Cod Canal (pp. 30-35). If Flood or Ebb Current on a certain day is stronger or weaker than the Av. Max. Vel. at Cape Cod Canal, the Current for Woods Hole for that day will be similarly stronger or weaker than the Av. Max. Vel. at Woods Hole.

To hold longest fair current from Buzzards Bay headed East through Vd. & Nant. Sds. go through Woods Hole 2 1/2 hrs. after flood starts SE in Woods Hole. (Any earlier means adverse currents in the Sounds.)

See pp. 150-157 for Current Changes at other points.

1999 CURRENT TABLE
WOODS HOLE, MASS.

0.1 mile SW of Devil's Ft. Is. – 41-31-12N x 70-41-06W (for South End, +10 Flood, +20 Ebb)

Standard Time

NOVEMBER

DAY OF MONTH	DAY OF WEEK	CURRENT TURNS			
		SOUTHEAST Flood Starts		NORTHWEST Ebb Starts	
		a.m.	**p.m.**	a.m.	**p.m.**
1	M	9 47	**10 16**	4 29	**4 38**
2	T	10 57	**11 23**	5 35	**5 47**
3	W	...	**12 02**	6 37	**6 50**
4	T	12 25	**12 59**	7 32	**7 47**
5	F	1 20	**1 48**	8 21	**8 39**
6	S	2 08	**2 33**	9 05	**9 26**
7	S	2 51	**3 13**	9 46	**10 09**
8	M	3 31	**3 50**	10 23	**10 50**
9	T	4 08	**4 26**	10 59	**11 30**
10	W	4 44	**5 02**	11 35	**...**
11	T	521	**5 39**	12 09	**12 10**
12	F	5 59	**6 19**	12 50	**12 47**
13	S	6 40	**7 02**	1 32	**1 27**
14	S	7 27	**7 49**	2 19	**2 13**
15	M	8 19	**8 42**	3 11	**3 07**
16	T	9 16	**9 40**	4 08	**4 07**
17	W	10 17	**10 39**	5 05	**5 10**
18	T	11 16	**11 36**	6 00	**6 10**
19	F	...	**12 11**	6 51	**7 06**
20	S	12 31	**1 02**	7 39	**7 59**
21	S	1 22	**1 50**	8 24	**8 49**
22	M	2 11	**2 38**	9 09	**9 39**
23	T	3 00	**3 25**	9 51	**10 28**
24	W	3 48	**4 13**	10 39	**11 19**
25	T	4 37	**5 03**	11 26	**...**
26	F	5 28	**5 54**	12 11	**12 16**
27	S	6 21	**6 48**	1 06	**1 09**
28	S	7 18	**7 46**	2 03	**2 06**
29	M	8 20	**8 47**	3 04	**3 09**
30	T	9 25	**9 51**	4 06	**4 15**

Standard Time

DECEMBER

DAY OF MONTH	DAY OF WEEK	CURRENT TURNS			
		SOUTHEAST Flood Starts		NORTHWEST Ebb Starts	
		a.m.	**p.m.**	a.m.	**p.m.**
1	W	10 31	**10 55**	5 08	**5 22**
2	T	11 34	**11 56**	6 07	**6 25**
3	F	...	**12 32**	7 02	**7 23**
4	S	12 51	**1 22**	7 51	**8 16**
5	S	1 41	**2 08**	8 36	**9 04**
6	M	2 25	**2 49**	9 17	**9 48**
7	T	3 06	**3 27**	9 56	**10 30**
8	W	3 44	**4 04**	10 33	**11 10**
9	T	4 21	**4 41**	11 09	**11 49**
10	F	4 57	**5 17**	11 45	**...**
11	S	5 35	**5 55**	12 28	**12 22**
12	S	6 15	**6 36**	1 09	**1 00**
13	M	6 58	**7 20**	1 52	**1 42**
14	T	7 46	**8 08**	2 38	**2 30**
15	W	8 38	**9 01**	3 27	**3 25**
16	T	9 35	**9 57**	4 20	**4 26**
17	F	10 34	**10 56**	5 15	**5 30**
18	S	11 33	**11 55**	6 09	**6 32**
19	S	...	**12 29**	7 02	**7 31**
20	M	12 51	**1 24**	7 53	**8 27**
21	T	1 46	**2 16**	8 42	**9 21**
22	W	2 39	**3 07**	9 31	**10 14**
23	T	3 30	**3 58**	10 20	**11 06**
24	F	4 22	**4 49**	11 09	**11 58**
25	S	5 13	**5 40**	...	**12 01**
26	S	6 05	**6 32**	12 50	**12 52**
27	M	6 59	**7 25**	1 44	**1 47**
28	T	7 55	**8 21**	2 39	**2 45**
29	W	8 54	**9 20**	3 36	**3 47**
30	T	9 56	**10 20**	4 33	**4 51**
31	F	10 58	**11 20**	5 31	**5 54**

Av. Max. Vel.: Fl. 4.5 Kts., Ebb 3.6 Kts.
(There are times when velocities can exceed 7 Kts.)

The best guide for determining daily velocities at Woods Hole is to refer to velocities at Cape Cod Canal (pp. 30-35). If Flood or Ebb Current on a certain day is stronger or weaker than the Av. Max. Vel. at Cape Cod Canal, the Current for Woods Hole for that day will be similarly stronger or weaker than the Av. Max. Vel. at Woods Hole.

CAUTION: Going *from* Buzzards Bay *into* Vineyard Sound, whether through Woods Hole, or Robinsons Hole or Quicks Hole, *Red* Buoys must be kept on the LEFT or PORT hand, *Black* or *Green* Buoys kept on the RIGHT or STARBOARD hand. You are considered to be proceeding seaward and should thus follow the rules for LEAVING a harbor.

My dear Captain and Mr. Mate,

As I cannot talk with you, I will do the next best thing. I will write you a letter.

Do you know, Captain and Mr. Mate of a place on the Atlantic Coast that is called "The Graveyard"? I propose to tell you something about it, and do what I can to keep vessels out of it. "The Graveyard" so called, is that part of the coast which lies between Sow and Pigs Rocks and Naushon Island. This place has been called "The Graveyard" for many years, because many a good craft has laid her bones there, and many a captain has lost his reputation there also. If a vessel gets into this graveyard, there must be a cause for it. Did it ever occur to you that seldom does a vessel go ashore on Gay Head, or on the south side of the Sound? but that hundreds of them have been piled up in "The Graveyard," or on the north side of the Sound? I will explain why this is so: if you are bound into Vineyard Sound in thick weather, you will probably refer to the "Gay Head and Cross Rip" table in this book, to see when the tide turns in or out. You will notice at the head of each table that it says, "This table shows the time that the current turns Easterly and Westerly, off Gay Head in ship channel." That means off Gay Head when it bears about South. Now, as a rule, captains figure on the current, after they leave the Lightship, as running Easterly into the Sound, when as a matter of fact the first of the flood between the Lightship and Gay Head runs nearly North; and the current does not begin to run to the eastward until you are well into the Sound, as shewn by the chart on the opposite page. Vessels bound into Vineyard Sound from the Westward will have the current of ebb on the starboard bow. (see arrows on the hulls in the chart on the opposite page)

I have explained this matter, and I leave the rest to your judgment and careful consideration; and thus you will undoubtedly keep your vessel out of "The Graveyard".

Yours for a fair tide,

Geo. W. Eldridge.

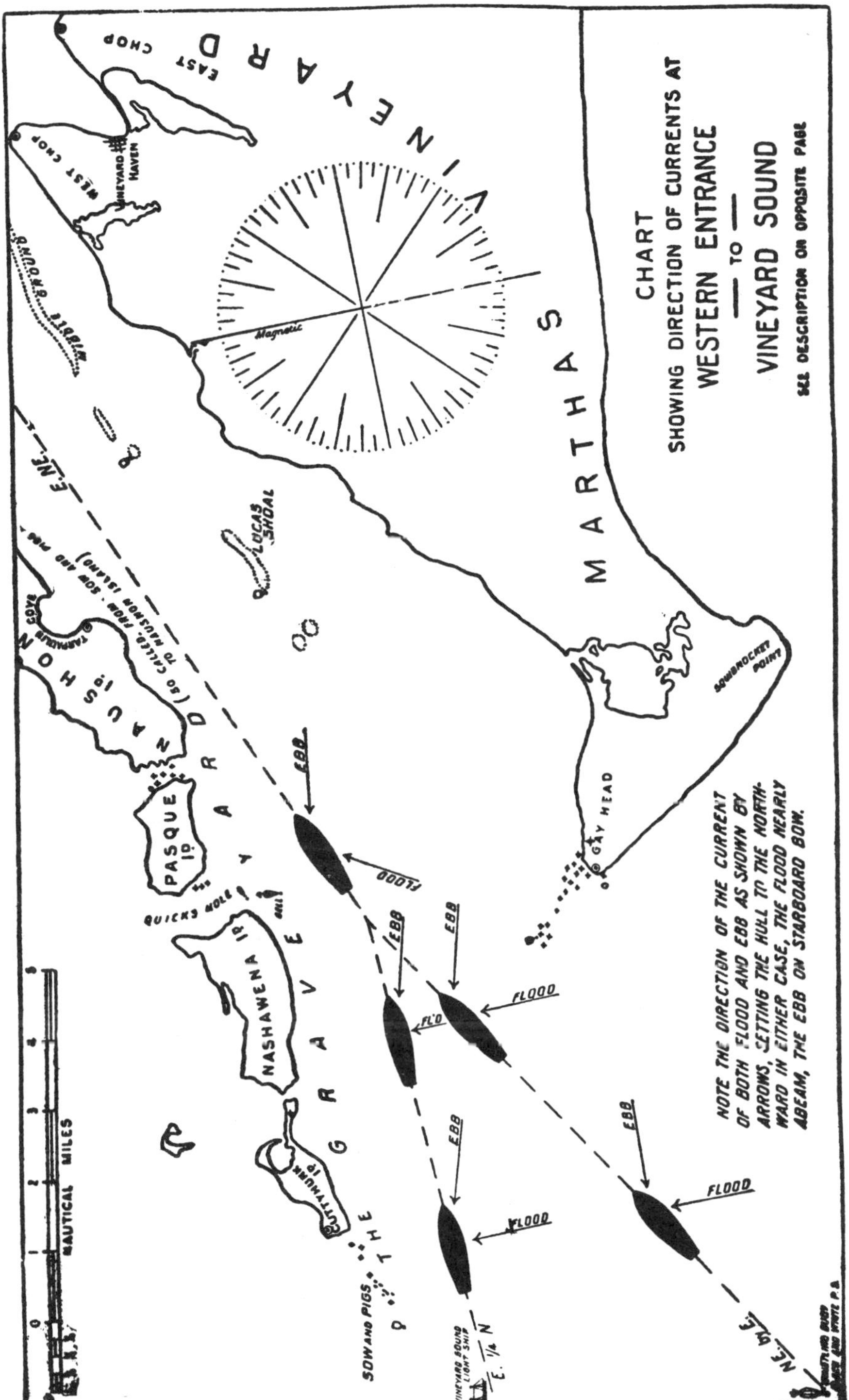

This lightship, shown on Capt. Eldridge's chart, on the western edge, was replaced many years ago by a buoy.

1999 CURRENT TABLE

POLLOCK RIP CHANNEL, MASS.

At Butler Hole – 41-33-00 N x 69-59-00W

JANUARY

Standard Time

DAY OF MONTH	DAY OF WEEK	CURRENT TURNS: NORTHEAST Flood Starts a.m.	p.m.	Kts.	SOUTHWEST Ebb Starts a.m.	p.m.	Kts.
1	F	1 51	2 12	p2.3	8 13	8 55	a1.9
2	S	2 46	3 04	p2.4	9 05	9 47	a2.0
3	S	3 38	3 54	p2.3	9 55	10 38	a2.0
4	M	4 29	4 43	p2.3	10 46	11 28	a1.9
5	T	5 19	5 33	p2.2	11 36	...	1.8
6	W	6 09	6 23	p2.1	12 17	12 28	p1.8
7	T	7 00	7 14	p2.0	1 07	1 21	a1.7
8	F	7 53	8 08	p1.9	1 58	2 17	a1.6
9	S	8 47	9 03	p1.9	2 50	3 14	a1.6
10	S	9 42	10 00	1.8	3 43	4 11	a1.5
11	M	10 36	10 56	1.8	4 35	5 08	a1.5
12	T	11 28	11 50	a1.9	5 27	6 02	a1.6
13	W	...	12 18	2.0	6 17	6 53	a1.6
14	T	12 41	1 03	p2.1	7 04	7 40	a1.7
15	F	1 28	1 45	p2.1	7 48	8 24	a1.8
16	S	2 11	2 25	p2.2	8 29	9 05	a1.8
17	S	2 52	3 03	p2.2	9 09	9 44	a1.9
18	M	3 30	3 40	p2.3	9 48	10 24	a2.0
19	T	4 08	4 19	p2.3	10 28	11 03	2.0
20	W	4 48	4 59	p2.4	11 09	11 45	a2.1
21	T	5 30	5 43	p2.3	11 53	...	2.1
22	F	6 14	6 30	p2.3	12 29	12 42	p2.1
23	S	7 03	7 22	p2.1	1 16	1 35	2.0
24	S	7 57	8 20	2.0	2 08	2 33	a1.9
25	M	8 55	9 22	1.9	3 04	3 36	a1.8
26	T	9 58	10 29	a1.9	4 04	4 42	a1.7
27	W	11 03	11 37	a1.9	5 06	5 48	a1.7
28	T	...	12 07	2.0	6 07	6 52	a1.7
29	F	12 43	1 08	p2.2	7 06	7 51	a1.8
30	S	1 44	2 04	p2.3	8 02	8 45	a1.8
31	S	2 39	2 55	p2.3	8 54	9 36	a1.9

FEBRUARY

Standard Time

DAY OF MONTH	DAY OF WEEK	CURRENT TURNS: NORTHEAST Flood Starts a.m.	p.m.	Kts.	SOUTHWEST Ebb Starts a.m.	p.m.	Kts.
1	M	3 29	3 44	p2.3	9 44	10 23	a1.9
2	T	4 15	4 29	p2.3	10 31	11 08	a1.9
3	W	5 00	5 13	p2.2	11 17	11 52	a1.9
4	T	5 43	5 57	p2.1	...	12 04	1.8
5	F	6 27	6 42	p2.0	12 36	12 51	1.7
6	S	7 13	7 30	p1.9	1 21	1 41	a1.7
7	S	8 02	8 21	1.8	2 09	2 34	a1.6
8	M	8 54	9 16	1.7	2 59	3 30	a1.5
9	T	9 48	10 13	1.7	3 52	4 27	a1.5
10	W	10 43	11 10	a1.8	4 45	5 23	a1.5
11	T	11 36	...	1.9	5 38	6 17	a1.5
12	F	12 05	12 26	p2.0	6 29	7 07	a1.6
13	S	12 55	1 12	p2.1	7 16	7 52	a1.7
14	S	1 41	1 55	p2.2	8 00	8 35	1.8
15	M	2 23	2 35	p2.3	8 42	9 15	1.9
16	T	3 02	3 15	p2.4	9 22	9 55	2.0
17	W	3 41	3 55	p2.4	10 03	10 35	a2.2
18	T	4 21	4 36	p2.4	10 45	11 16	a2.2
19	F	5 02	5 21	p2.4	11 30	...	2.2
20	S	5 47	6 08	p2.3	12 01	12 18	p2.2
21	S	6 36	7 01	a2.2	12 48	1 12	a2.1
22	M	7 30	7 59	a2.1	1 40	2 11	a2.0
23	T	8 30	9 05	a1.9	2 38	3 16	a1.8
24	W	9 37	10 16	a1.9	3 42	4 25	a1.7
25	T	10 46	11 28	a1.9	4 48	5 34	a1.6
26	F	11 55	...	2.0	5 54	6 40	a1.6
27	S	12 36	12 58	p2.2	6 56	7 39	a1.7
28	S	1 36	1 54	p2.3	7 52	8 32	a1.8

The Kts. (knots) columns show the **maximum** predicted velocities of the stronger one of the Flood Currents and the stronger one of the Ebb Currents for each day.
The letter "a" means the velocity shown should occur **after** the **a.m.** Current Change. The letter "p" means the velocity shown should occur **after** the **p.m.** Current Change (even if next morning). No "a" or "p" means a.m. and p.m. velocities are the same for that day.
Av. Max. Vel.: Fl. 2.0 Kts., Ebb 1.8 Kts.
Max. Flood 3 hrs. 20 min. after Flood Starts, ±15 min.
Max. Ebb 2 hrs. 45 min. after Ebb Starts, ±15 min.
Gay Head (1 1/2 mi. NW of): av. max velocity, flood 2.0 kts., ebb 2.0 kts. Time of Fl. and Ebb 1 hr. 35 min. after Pollock Rip. Cross Rip: av. max. velocity, flood 1.3 kts., ebb 0.9 kts. Time of Fl. and Ebb 1 hr. 50 min. after Pollock Rip.
See pp. 150-157 for Current Change at other points.

1999 CURRENT TABLE
POLLOCK RIP CHANNEL, MASS.

At Butler Hole – 41-33-00 N x 69-59-00W

Standard Time **Daylight Time starts April 4 at 2 a.m.**

MARCH

DAY OF MONTH	DAY OF WEEK	NORTHEAST Flood Starts a.m.	NORTHEAST Flood Starts **p.m.**	NORTHEAST Kts.	SOUTHWEST Ebb Starts a.m.	SOUTHWEST Ebb Starts **p.m.**	SOUTHWEST Kts.
1	M	2 28	**2 44**	p2.3	8 43	**9 19**	1.8
2	T	3 14	**3 29**	p2.3	9 30	**10 02**	a1.9
3	W	3 56	**4 11**	p2.3	10 14	**10 43**	a1.9
4	T	4 35	**4 50**	p2.2	10 56	**11 22**	1.8
5	F	5 13	**5 30**	p2.1	11 38	...	1.8
6	S	5 52	**6 10**	a2.0	12 02	**12 22**	a1.8
7	S	6 33	**6 54**	a1.9	12 44	**1 08**	a1.7
8	M	7 18	**7 42**	a1.8	1 29	**1 57**	a1.6
9	T	8 07	**8 35**	a1.8	2 17	**2 50**	a1.6
10	W	9 00	**9 32**	a1.7	3 10	**3 47**	a1.5
11	T	9 56	**10 31**	a1.7	4 05	**4 44**	a1.5
12	F	10 52	**11 27**	a1.8	5 00	**5 39**	a1.5
13	S	11 46	...	1.9	5 53	**6 30**	1.5
14	S	12 20	**12 36**	p2.0	6 43	**7 18**	1.7
15	M	1 07	**1 22**	p2.2	7 29	**8 02**	1.8
16	T	1 51	**2 05**	p2.3	8 13	**8 43**	2.0
17	W	2 32	**2 48**	p2.4	8 55	**9 24**	2.1
18	T	3 12	**3 30**	p2.4	9 38	**10 05**	2.2
19	F	3 53	**4 14**	2.4	10 22	**10 48**	a2.3
20	S	4 36	**5 00**	2.4	11 08	**11 33**	a2.3
21	S	5 22	**5 49**	a2.4	11 58	...	2.2
22	M	6 12	**6 43**	a2.3	12 22	**12 53**	a2.1
23	T	7 08	**7 44**	a2.1	1 17	**1 54**	a1.9
24	W	8 10	**8 52**	a2.0	2 17	**3 00**	a1.8
25	T	9 19	**10 05**	a1.9	3 24	**4 10**	a1.6
26	F	10 31	**11 18**	a1.9	4 33	**5 20**	1.5
27	S	11 41	...	2.1	5 40	**6 24**	p1.6
28	S	12 24	**12 43**	p2.2	6 42	**7 21**	p1.7
29	M	1 21	**1 38**	p2.3	7 38	**8 11**	p1.8
30	T	2 10	**2 26**	p2.3	8 27	**8 56**	1.8
31	W	2 53	**3 09**	p2.2	9 12	**9 37**	1.8

APRIL

DAY OF MONTH	DAY OF WEEK	NORTHEAST Flood Starts a.m.	NORTHEAST Flood Starts **p.m.**	NORTHEAST Kts.	SOUTHWEST Ebb Starts a.m.	SOUTHWEST Ebb Starts **p.m.**	SOUTHWEST Kts.
1	T	3 32	**3 48**	p2.2	9 54	**10 15**	1.8
2	F	4 08	**4 25**	2.1	10 33	**10 52**	1.8
3	S	4 43	**5 02**	a2.1	11 13	**11 29**	1.8
4	S	*6 19	***6 41**	a2.0	...	***12 54**	1.7
5	M	6 57	**7 22**	a2.0	1 09	**1 37**	1.7
6	T	7 39	**8 08**	a1.9	1 52	**2 24**	a1.7
7	W	8 26	**8 59**	a1.8	2 39	**3 15**	a1.6
8	T	9 17	**9 55**	a1.8	3 31	**4 09**	1.5
9	F	10 12	**10 52**	a1.8	4 26	**5 05**	1.5
10	S	11 09	**11 49**	a1.8	5 22	**5 59**	1.5
11	S	...	**12 05**	1.9	6 16	**6 51**	1.6
12	M	12 42	**12 58**	p2.0	7 08	**7 40**	1.7
13	T	1 31	**1 47**	p2.1	7 57	**8 26**	p1.9
14	W	2 17	**2 34**	p2.3	8 44	**9 10**	2.0
15	T	3 00	**3 20**	p2.3	9 29	**9 53**	p2.2
16	F	3 43	**4 06**	2.4	10 14	**10 36**	2.2
17	S	4 27	**4 52**	2.4	11 01	**11 21**	p2.3
18	S	5 12	**5 41**	a2.5	11 50	...	2.2
19	M	6 00	**6 33**	a2.4	12 09	**12 42**	a2.2
20	T	6 53	**7 30**	a2.3	1 01	**1 39**	a2.1
21	W	7 50	**8 32**	a2.1	1 57	**2 40**	a1.9
22	T	8 53	**9 41**	a2.0	2 59	**3 46**	a1.7
23	F	10 02	**10 52**	a2.0	4 07	**4 53**	a1.6
24	S	11 13	...	2.0	5 15	**5 59**	1.5
25	S	12 01	**12 20**	p2.1	6 22	**7 00**	p1.6
26	M	1 03	**1 21**	p2.2	7 23	**7 55**	p1.7
27	T	1 57	**2 15**	p2.2	8 17	**8 43**	p1.8
28	W	2 45	**3 02**	p2.2	9 06	**9 27**	1.8
29	T	3 26	**3 44**	a2.2	9 50	**10 07**	1.8
30	F	4 04	**4 23**	a2.2	10 31	**10 44**	p1.8

***Daylight Time starts**

The Kts. (knots) columns show the **maximum** predicted velocities of the stronger one of the Flood Currents and the stronger one of the Ebb Currents for each day.
The letter "a" means the velocity shown should occur **after** the **a.m.** Current Change. The letter "p" means the velocity shown should occur **after** the **p.m.** Current Change (even if next morning). No "a" or "p" means a.m. and p.m. velocities are the same for that day.

Av. Max. Vel.: Fl. 2.0 Kts., Ebb 1.8 Kts.
Max. Flood 3 hrs. 20 min. after Flood Starts, ±15 min.
Max. Ebb 2 hrs. 45 min. after Ebb Starts, ±15 min.

Gay Head (1 1/2 mi. NW of): av. max velocity, flood 2.0 kts., ebb 2.0 kts. Time of Fl. and Ebb 1 hr. 35 min. after Pollock Rip. Cross Rip: av. max. velocity, flood 1.3 kts., ebb 0.9 kts. Time of Fl. and Ebb 1 hr. 50 min. after Pollock Rip.

See pp. 150-157 for Current Change at other points.

1999 CURRENT TABLE
POLLOCK RIP CHANNEL, MASS.

At Butler Hole – 41-33-00 N x 69-59-00W

Daylight Time

MAY

DAY OF MONTH	DAY OF WEEK	CURRENT TURNS — NORTHEAST Flood Starts a.m.	p.m.	Kts.	SOUTHWEST Ebb Starts a.m.	p.m.	Kts.
1	S	4 39	**4 59**	a2.1	11 10	**11 21**	p1.8
2	S	5 13	**5 36**	a2.1	11 48	**11 58**	p1.8
3	M	5 48	**6 14**	a2.1	...	**12 28**	1.7
4	T	6 26	**6 54**	a2.0	12 37	**1 10**	a1.8
5	W	7 06	**7 39**	a2.0	1 19	**1 55**	a1.7
6	T	7 51	**8 28**	a1.9	2 05	**2 43**	a1.7
7	F	8 40	**9 20**	a1.9	2 55	**3 34**	1.6
8	S	9 33	**10 15**	a1.9	3 48	**4 27**	1.6
9	S	10 28	**11 10**	a1.9	4 44	**5 20**	1.6
10	M	11 24	...	1.9	5 39	**6 12**	p1.7
11	T	12 04	**12 20**	p2.0	6 33	**7 02**	p1.8
12	W	12 55	**1 13**	p2.1	7 25	**7 51**	p1.9
13	T	1 43	**2 04**	p2.2	8 15	**8 37**	p2.1
14	F	2 30	**2 54**	a2.3	9 05	**9 24**	p2.2
15	S	3 17	**3 44**	a2.4	9 54	**10 10**	p2.2
16	S	4 04	**4 34**	a2.4	10 44	**10 58**	p2.2
17	M	4 53	**5 25**	a2.5	11 35	**11 49**	2.1
18	T	5 43	**6 20**	a2.4	...	**12 29**	2.0
19	W	6 37	**7 18**	a2.3	12 42	**1 26**	a2.0
20	T	7 35	**8 19**	a2.2	1 40	**2 25**	a1.9
21	F	8 37	**9 25**	a2.1	2 42	**3 28**	a1.7
22	S	9 42	**10 31**	a2.0	3 47	**4 31**	1.6
23	S	10 48	**11 35**	a2.0	4 53	**5 32**	p1.6
24	M	11 52	...	2.1	5 56	**6 29**	p1.6
25	T	12 34	**12 51**	p2.1	6 56	**7 22**	p1.7
26	W	1 26	**1 45**	2.1	7 50	**8 10**	p1.8
27	T	2 13	**2 32**	2.1	8 39	**8 54**	p1.8
28	F	2 56	**3 16**	a2.2	9 24	**9 35**	p1.8
29	S	3 34	**3 56**	a2.2	10 06	**10 14**	p1.8
30	S	4 10	**4 33**	a2.1	10 45	**10 52**	p1.8
31	M	4 46	**5 11**	a2.1	11 24	**11 30**	p1.8

Daylight Time

JUNE

DAY OF MONTH	DAY OF WEEK	CURRENT TURNS — NORTHEAST Flood Starts a.m.	p.m.	Kts.	SOUTHWEST Ebb Starts a.m.	p.m.	Kts.
1	T	5 21	**5 49**	a2.1	...	**12 04**	1.7
2	W	5 58	**6 29**	a2.1	12 09	**12 44**	a1.8
3	T	6 37	**7 11**	a2.1	12 50	**1 27**	a1.8
4	F	7 20	**7 57**	a2.1	1 34	**2 12**	a1.8
5	S	8 07	**8 47**	a2.0	2 22	**3 00**	1.7
6	S	8 57	**9 38**	a2.0	3 13	**3 50**	1.7
7	M	9 51	**10 32**	a2.0	4 07	**4 42**	1.7
8	T	10 47	**11 26**	a2.0	5 03	**5 35**	p1.8
9	W	11 45	...	2.0	6 00	**6 27**	p1.8
10	T	12 20	**12 42**	2.0	6 56	**7 18**	p1.9
11	F	1 13	**1 38**	2.1	7 50	**8 09**	p2.0
12	S	2 05	**2 32**	a2.2	8 44	**8 59**	p2.1
13	S	2 56	**3 26**	a2.3	9 37	**9 50**	p2.1
14	M	3 47	**4 19**	a2.4	10 30	**10 41**	p2.1
15	T	4 38	**5 13**	a2.4	11 23	**11 33**	p2.0
16	W	5 30	**6 08**	a2.4	...	**12 16**	1.9
17	T	6 23	**7 04**	a2.3	12 27	**1 11**	a2.0
18	F	7 19	**8 01**	a2.2	1 23	**2 07**	a1.8
19	S	8 17	**9 01**	a2.1	2 22	**3 04**	1.7
20	S	9 17	**10 01**	a2.0	3 22	**4 01**	1.6
21	M	10 18	**11 00**	a2.0	4 24	**4 58**	p1.6
22	T	11 18	**11 57**	a2.0	5 25	**5 53**	p1.6
23	W	...	**12 16**	2.0	6 23	**6 46**	p1.7
24	T	12 49	**1 10**	2.0	7 18	**7 35**	p1.7
25	F	1 38	**2 00**	a2.1	8 09	**8 21**	p1.7
26	S	2 23	**2 45**	a2.1	8 55	**9 04**	p1.7
27	S	3 04	**3 28**	a2.1	9 39	**9 45**	p1.8
28	M	3 42	**4 07**	a2.2	10 20	**10 24**	p1.8
29	T	4 19	**4 46**	a2.2	10 59	**11 03**	p1.8
30	W	4 55	**5 24**	a2.2	11 38	**11 42**	p1.9

The Kts. (knots) columns show the **maximum** predicted velocities of the stronger one of the Flood Currents and the stronger one of the Ebb Currents for each day.
The letter "a" means the velocity shown should occur **after** the **a.m.** Current Change. The letter "p" means the velocity shown should occur **after** the **p.m.** Current Change (even if next morning). No "a" or "p" means a.m. and p.m. velocities are the same for that day.
Av. Max. Vel.: Fl. 2.0 Kts., Ebb 1.8 Kts.
Max. Flood 3 hrs. 20 min. after Flood Starts, ±15 min.
Max. Ebb 2 hrs. 45 min. after Ebb Starts, ±15 min.
Gay Head (1 1/2 mi. NW of): av. max velocity, flood 2.0 kts., ebb 2.0 kts. Time of Fl. and Ebb 1 hr. 35 min. after Pollock Rip. Cross Rip: av. max. velocity, flood 1.3 kts., ebb 0.9 kts. Time of Fl. and Ebb 1 hr. 50 min. after Pollock Rip.
See pp. 150-157 for Current Change at other points.

1999 CURRENT TABLE
POLLOCK RIP CHANNEL, MASS.

At Butler Hole – 41-33-00 N x 69-59-00W

JULY

Daylight Time

DAY OF MONTH	DAY OF WEEK	NORTHEAST Flood Starts a.m.	NORTHEAST Flood Starts **p.m.**	NORTHEAST Kts.	SOUTHWEST Ebb Starts a.m.	SOUTHWEST Ebb Starts **p.m.**	SOUTHWEST Kts.
1	T	5 32	**6 02**	a2.2	...	**12 18**	1.8
2	F	6 11	**6 43**	a2.2	12 22	**12 59**	a1.9
3	S	6 52	**7 26**	a2.2	1 05	**1 41**	a1.9
4	S	7 37	**8 13**	a2.2	1 51	**2 27**	a1.9
5	M	8 26	**9 03**	a2.1	2 40	**3 15**	a1.9
6	T	9 19	**9 56**	a2.0	3 34	**4 07**	1.8
7	W	10 15	**10 52**	a2.0	4 31	**5 00**	1.8
8	T	11 15	**11 50**	p2.0	5 31	**5 56**	p1.8
9	F	...	**12 17**	1.9	6 31	**6 52**	p1.9
10	S	12 48	**1 18**	a2.1	7 31	**7 47**	p1.9
11	S	1 45	**2 17**	a2.2	8 29	**8 42**	p1.9
12	M	2 41	**3 14**	a2.3	9 25	**9 35**	p2.0
13	T	3 34	**4 09**	a2.3	10 19	**10 27**	p2.0
14	W	4 27	**5 02**	a2.4	11 11	**11 19**	p2.0
15	T	5 18	**5 53**	a2.4	...	**12 02**	1.8
16	F	6 09	**6 45**	a2.3	12 11	**12 52**	a1.9
17	S	7 00	**7 37**	a2.2	1 04	**1 43**	a1.8
18	S	7 52	**8 29**	a2.1	1 58	**2 35**	1.7
19	M	8 46	**9 24**	a2.0	2 54	**3 27**	1.6
20	T	9 42	**10 19**	a1.9	3 51	**4 20**	p1.6
21	W	10 39	**11 15**	1.8	4 49	**5 14**	p1.6
22	T	11 37	...	1.8	5 47	**6 07**	p1.6
23	F	12 09	**12 33**	a1.9	6 43	**6 58**	p1.6
24	S	1 00	**1 25**	a2.0	7 36	**7 47**	p1.6
25	S	1 48	**2 14**	a2.1	8 25	**8 33**	p1.7
26	M	2 32	**2 59**	a2.1	9 10	**9 16**	p1.8
27	T	3 13	**3 40**	a2.2	9 52	**9 57**	p1.8
28	W	3 51	**4 19**	a2.2	10 32	**10 36**	p1.9
29	T	4 29	**4 56**	a2.2	11 10	**11 15**	p2.0
30	F	5 06	**5 34**	a2.3	11 49	**11 54**	p2.0
31	S	5 44	**6 13**	a2.3	...	**12 28**	1.9

AUGUST

Daylight Time

DAY OF MONTH	DAY OF WEEK	NORTHEAST Flood Starts a.m.	NORTHEAST Flood Starts **p.m.**	NORTHEAST Kts.	SOUTHWEST Ebb Starts a.m.	SOUTHWEST Ebb Starts **p.m.**	SOUTHWEST Kts.
1	S	6 25	**6 54**	a2.3	12 36	**1 09**	2.0
2	M	7 09	**7 40**	a2.2	1 21	**1 54**	2.0
3	T	7 58	**8 29**	a2.2	2 11	**2 42**	a2.0
4	W	8 51	**9 24**	2.0	3 05	**3 34**	1.9
5	T	9 49	**10 23**	1.9	4 04	**4 31**	1.8
6	F	10 53	**11 26**	p1.9	5 07	**5 31**	1.7
7	S	11 59	...	1.8	6 12	**6 32**	p1.7
8	S	12 30	**1 06**	a2.0	7 16	**7 32**	p1.8
9	M	1 32	**2 08**	a2.1	8 17	**8 30**	p1.8
10	T	2 30	**3 06**	a2.2	9 14	**9 24**	p1.9
11	W	3 25	**3 59**	a2.3	10 07	**10 16**	p1.9
12	T	4 16	**4 48**	a2.3	10 56	**11 06**	p1.9
13	F	5 05	**5 35**	a2.3	11 43	**11 54**	p1.9
14	S	5 51	**6 21**	a2.3	...	**12 29**	1.8
15	S	6 37	**7 06**	a2.2	12 42	**1 14**	1.8
16	M	7 24	**7 53**	a2.0	1 31	**2 01**	1.7
17	T	8 12	**8 43**	a1.9	2 22	**2 49**	1.6
18	W	9 04	**9 35**	1.8	3 16	**3 40**	1.5
19	T	9 59	**10 30**	p1.8	4 12	**4 33**	p1.5
20	F	10 57	**11 26**	p1.8	5 09	**5 28**	p1.5
21	S	11 55	...	1.7	6 06	**6 22**	p1.5
22	S	12 20	**12 51**	a1.9	7 00	**7 13**	p1.6
23	M	1 11	**1 42**	a2.0	7 51	**8 02**	p1.7
24	T	1 58	**2 28**	a2.1	8 37	**8 46**	p1.8
25	W	2 42	**3 10**	a2.2	9 20	**9 28**	p1.9
26	T	3 22	**3 48**	a2.2	10 00	**10 08**	p2.0
27	F	4 00	**4 26**	a2.3	10 39	**10 47**	p2.1
28	S	4 38	**5 03**	a2.3	11 17	**11 27**	p2.1
29	S	5 18	**5 42**	a2.4	11 56	...	2.1
30	M	5 59	**6 23**	a2.3	12 09	**12 38**	a2.2
31	T	6 44	**7 09**	a2.3	12 55	**1 22**	2.1

The Kts. (knots) columns show the **maximum** predicted velocities of the stronger one of the Flood Currents and the stronger one of the Ebb Currents for each day.
The letter "a" means the velocity shown should occur **after** the **a.m.** Current Change. The letter "p" means the velocity shown should occur **after** the **p.m.** Current Change (even if next morning). No "a" or "p" means a.m. and p.m. velocities are the same for that day.
Av. Max. Vel.: Fl. 2.0 Kts., Ebb 1.8 Kts.
Max. Flood 3 hrs. 20 min. after Flood Starts, ±15 min.
Max. Ebb 2 hrs. 45 min. after Ebb Starts, ±15 min.
Gay Head (1 1/2 mi. NW of): av. max velocity, flood 2.0 kts., ebb 2.0 kts. Time of Fl. and Ebb 1 hr. 35 min. after Pollock Rip. Cross Rip: av. max. velocity, flood 1.3 kts., ebb 0.9 kts. Time of Fl. and Ebb 1 hr. 50 min. after Pollock Rip.
See pp. 150-157 for Current Change at other points.

1999 CURRENT TABLE

POLLOCK RIP CHANNEL, MASS.

At Butler Hole – 41-33-00 N x 69-59-00W

Daylight Time — **Standard Time starts Oct. 31 at 2 a.m.**

SEPTEMBER

DAY OF MONTH	DAY OF WEEK	CURRENT TURNS NORTHEAST Flood Starts a.m.	**p.m.**	Kts.	CURRENT TURNS SOUTHWEST Ebb Starts a.m.	**p.m.**	Kts.
1	**W**	7 33	**8 00**	a2.2	1 45	**2 12**	2.0
2	**T**	8 28	**8 56**	2.0	2 41	**3 07**	1.9
3	**F**	9 30	**9 59**	p1.9	3 43	**4 07**	1.7
4	**S**	10 38	**11 07**	p1.9	4 49	**5 12**	1.6
5	**S**	11 49	**...**	1.7	5 58	**6 18**	p1.6
6	**M**	12 16	**12 58**	a2.0	7 04	**7 21**	p1.7
7	**T**	1 22	**2 01**	a2.1	8 05	**8 20**	p1.8
8	**W**	2 21	**2 57**	a2.2	9 01	**9 14**	p1.9
9	**T**	3 15	**3 46**	a2.3	9 51	**10 04**	p1.9
10	**F**	4 03	**4 31**	a2.3	10 37	**10 50**	p1.9
11	**S**	4 48	**5 13**	a2.3	11 20	**11 35**	p1.9
12	**S**	5 30	**5 53**	a2.2	...	**12 01**	1.8
13	**M**	6 12	**6 33**	a2.1	12 19	**12 43**	1.8
14	**T**	6 54	**7 16**	1.9	1 04	**1 26**	1.7
15	**W**	7 39	**8 01**	1.8	1 51	**2 11**	1.6
16	**T**	8 27	**8 51**	p1.8	2 41	**3 00**	1.5
17	**F**	9 21	**9 45**	p1.7	3 35	**3 53**	p1.5
18	**S**	10 18	**10 41**	p1.8	4 31	**4 49**	1.4
19	**S**	11 17	**11 38**	p1.8	5 28	**5 45**	p1.5
20	**M**	...	**12 14**	1.7	6 23	**6 39**	p1.5
21	**T**	12 32	**1 07**	a1.9	7 14	**7 29**	p1.6
22	**W**	1 22	**1 54**	a2.1	8 02	**8 15**	p1.8
23	**T**	2 07	**2 36**	a2.2	8 45	**8 58**	p1.9
24	**F**	2 50	**3 15**	a2.2	9 26	**9 39**	p2.0
25	**S**	3 30	**3 54**	a2.3	10 05	**10 20**	p2.1
26	**S**	4 10	**4 32**	a2.4	10 44	**11 01**	p2.2
27	**M**	4 51	**5 12**	2.4	11 24	**11 45**	2.2
28	**T**	5 35	**5 55**	p2.4	...	**12 07**	2.2
29	**W**	6 22	**6 43**	p2.3	12 33	**12 54**	a2.2
30	**T**	7 13	**7 35**	p2.2	1 25	**1 46**	2.0

OCTOBER

DAY OF MONTH	DAY OF WEEK	CURRENT TURNS NORTHEAST Flood Starts a.m.	**p.m.**	Kts.	CURRENT TURNS SOUTHWEST Ebb Starts a.m.	**p.m.**	Kts.
1	**F**	8 11	**8 35**	p2.0	2 22	**2 44**	a1.9
2	**S**	9 16	**9 41**	p1.9	3 26	**3 48**	1.7
3	**S**	10 27	**10 53**	p1.9	4 35	**4 57**	p1.6
4	**M**	11 40	**...**	1.7	5 44	**6 05**	p1.6
5	**T**	12 04	**12 49**	a2.0	6 49	**7 10**	1.6
6	**W**	1 09	**1 49**	a2.2	7 49	**8 08**	1.7
7	**T**	2 08	**2 42**	a2.3	8 42	**9 01**	1.8
8	**F**	3 00	**3 28**	a2.3	9 30	**9 48**	p1.9
9	**S**	3 46	**4 09**	a2.3	10 13	**10 33**	a1.9
10	**S**	4 28	**4 48**	2.2	10 53	**11 15**	1.8
11	**M**	5 08	**5 25**	2.1	11 32	**11 56**	a1.8
12	**T**	5 46	**6 02**	p2.0	...	**12 11**	1.8
13	**W**	6 25	**6 41**	p1.9	12 38	**12 52**	1.7
14	**T**	7 07	**7 23**	p1.9	1 22	**1 35**	1.6
15	**F**	7 54	**8 10**	p1.8	2 09	**2 23**	p1.6
16	**S**	8 45	**9 02**	p1.8	3 00	**3 15**	1.5
17	**S**	9 41	**9 57**	p1.8	3 54	**4 10**	1.4
18	**M**	10 39	**10 54**	p1.8	4 49	**5 06**	p1.5
19	**T**	11 35	**11 50**	p1.9	5 43	**6 01**	1.5
20	**W**	...	**12 28**	1.7	6 35	**6 53**	1.6
21	**T**	12 42	**1 16**	a2.0	7 23	**7 41**	p1.8
22	**F**	1 30	**2 00**	a2.1	8 08	**8 26**	1.9
23	**S**	2 16	**2 42**	2.2	8 50	**9 10**	2.0
24	**S**	3 00	**3 22**	2.3	9 31	**9 54**	2.1
25	**M**	3 43	**4 03**	p2.4	10 13	**10 38**	2.2
26	**T**	4 27	**4 46**	p2.5	10 56	**11 25**	a2.3
27	**W**	5 14	**5 32**	p2.4	11 41	**...**	2.2
28	**T**	6 03	**6 22**	p2.4	12 14	**12 30**	2.1
29	**F**	6 58	**7 16**	p2.2	1 08	**1 25**	2.0
30	**S**	7 58	**8 17**	p2.1	2 07	**2 25**	1.8
31	**S**	*8 04	***8 25**	p2.0	*2 11	***2 31**	a1.7

***Standard Time starts**

The Kts. (knots) columns show the **maximum** predicted velocities of the stronger one of the Flood Currents and the stronger one of the Ebb Currents for each day.

The letter "a" means the velocity shown should occur **after** the **a.m.** Current Change. The letter "p" means the velocity shown should occur **after** the **p.m.** Current Change (even if next morning). No "a" or "p" means a.m. and p.m. velocities are the same for that day.

Av. Max. Vel.: Fl. 2.0 Kts., Ebb 1.8 Kts.

Max. Flood 3 hrs. 20 min. after Flood Starts, ±15 min.

Max. Ebb 2 hrs. 45 min. after Ebb Starts, ±15 min.

Gay Head (1 1/2 mi. NW of): av. max velocity, flood 2.0 kts., ebb 2.0 kts. Time of Fl. and Ebb 1 hr. 35 min. after Pollock Rip. Cross Rip: av. max. velocity, flood 1.3 kts., ebb 0.9 kts. Time of Fl. and Ebb 1 hr. 50 min. after Pollock Rip.

See pp. 150-157 for Current Change at other points.

1999 CURRENT TABLE

POLLOCK RIP CHANNEL, MASS.

At Butler Hole – 41-33-00 N x 69-59-00W

Standard Time

NOVEMBER

DAY OF MONTH	DAY OF WEEK	CURRENT TURNS NORTHEAST Flood Starts a.m.	**p.m.**	Kts.	CURRENT TURNS SOUTHWEST Ebb Starts a.m.	**p.m.**	Kts.
1	M	9 15	**9 36**	p2.0	3 18	**3 40**	a1.6
2	T	10 26	**10 46**	p2.1	4 25	**4 49**	1.5
3	W	11 32	**11 51**	p2.2	5 29	**5 53**	1.6
4	T	...	**12 30**	2.1	6 26	**6 51**	1.7
5	F	12 48	**1 20**	2.2	7 18	**7 43**	1.8
6	S	1 39	**2 05**	2.2	8 04	**8 30**	1.8
7	S	2 25	**2 45**	2.2	8 47	**9 13**	1.8
8	M	3 06	**3 22**	p2.2	9 26	**9 54**	a1.8
9	T	3 44	**3 58**	p2.1	10 04	**10 33**	a1.8
10	W	4 22	**4 34**	p2.1	10 42	**11 14**	a1.8
11	T	5 00	**5 11**	p2.0	11 21	**11 55**	a1.7
12	F	5 40	**5 51**	p2.0	...	**12 03**	1.7
13	S	6 24	**6 35**	p1.9	12 40	**12 49**	1.6
14	S	7 12	**7 23**	p1.9	1 27	**1 38**	p1.6
15	M	8 04	**8 15**	p1.9	2 17	**2 31**	1.5
16	T	8 59	**9 10**	p1.9	3 10	**3 26**	1.5
17	W	9 53	**10 06**	p1.9	4 02	**4 21**	1.6
18	T	10 46	**11 00**	p2.0	4 54	**5 15**	a1.7
19	F	11 36	**11 52**	p2.1	5 43	**6 06**	a1.8
20	S	...	**12 23**	2.1	6 30	**6 55**	1.9
21	S	12 42	**1 09**	p2.2	7 16	**7 43**	2.0
22	M	1 31	**1 54**	p2.3	8 00	**8 31**	2.1
23	T	2 19	**2 39**	p2.4	8 45	**9 19**	a2.2
24	W	3 07	**3 26**	p2.5	9 32	**10 08**	a2.2
25	T	3 57	**4 14**	p2.5	10 20	**11 00**	a2.2
26	F	4 49	**5 06**	p2.4	11 12	**11 55**	a2.1
27	S	5 45	**6 02**	p2.3	...	**12 07**	2.0
28	S	6 45	**7 02**	p2.1	12 53	**1 08**	1.8
29	M	7 49	**8 07**	p2.0	1 54	**2 12**	1.7
30	T	8 56	**9 14**	p2.0	2 57	**3 19**	1.6

Standard Time

DECEMBER

DAY OF MONTH	DAY OF WEEK	CURRENT TURNS NORTHEAST Flood Starts a.m.	**p.m.**	Kts.	CURRENT TURNS SOUTHWEST Ebb Starts a.m.	**p.m.**	Kts.
1	W	10 02	**10 21**	p2.0	4 00	**4 26**	a1.6
2	T	11 05	**11 25**	p2.1	5 01	**5 29**	1.6
3	F	...	**12 02**	2.0	5 57	**6 27**	a1.7
4	S	12 22	**12 53**	2.1	6 49	**7 20**	a1.8
5	S	1 14	**1 38**	p2.2	7 36	**8 07**	a1.8
6	M	2 00	**2 20**	p2.2	8 19	**8 51**	a1.8
7	T	2 42	**2 57**	p2.2	8 59	**9 32**	a1.8
8	W	3 21	**3 33**	p2.2	9 38	**10 11**	a1.8
9	T	3 59	**4 09**	p2.1	10 16	**10 50**	a1.8
10	F	4 36	**4 45**	p2.1	10 54	**11 30**	a1.8
11	S	5 15	**5 23**	p2.1	11 35	**...**	1.8
12	S	5 56	**6 04**	p2.0	12 12	**12 18**	1.7
13	M	6 40	**6 49**	p2.0	12 55	**1 04**	1.7
14	T	7 28	**7 37**	p2.0	1 42	**1 54**	1.7
15	W	8 18	**8 29**	p1.9	2 30	**2 46**	1.7
16	T	9 10	**9 24**	p1.9	3 21	**3 41**	a1.7
17	F	10 03	**10 20**	p1.9	4 12	**4 37**	1.7
18	S	10 56	**11 16**	p2.0	5 03	**5 32**	a1.8
19	S	11 49	**...**	2.0	5 54	**6 27**	a1.9
20	M	12 12	**12 40**	p2.2	6 45	**7 20**	a2.0
21	T	1 06	**1 30**	p2.3	7 34	**8 12**	a2.1
22	W	1 59	**2 20**	p2.4	8 24	**9 03**	a2.1
23	T	2 52	**3 10**	p2.5	9 13	**9 55**	a2.2
24	F	3 44	**4 01**	p2.5	10 04	**10 47**	a2.2
25	S	4 37	**4 54**	p2.4	10 57	**11 41**	a2.1
26	S	5 32	**5 48**	p2.3	11 52	**...**	2.0
27	M	6 28	**6 45**	p2.2	12 36	**12 49**	p1.9
28	T	7 27	**7 44**	p2.1	1 32	**1 50**	1.7
29	W	8 28	**8 47**	p2.0	2 30	**2 53**	a1.7
30	T	9 30	**9 50**	p1.9	3 29	**3 56**	a1.6
31	F	10 31	**10 52**	p1.9	4 27	**4 59**	a1.6

The Kts. (knots) columns show the **maximum** predicted velocities of the stronger one of the Flood Currents and the stronger one of the Ebb Currents for each day.
The letter "a" means the velocity shown should occur **after** the **a.m.** Current Change. The letter "p" means the velocity shown should occur **after** the **p.m.** Current Change (even if next morning). No "a" or "p" means a.m. and p.m. velocities are the same for that day.
Av. Max. Vel.: Fl. 2.0 Kts., Ebb 1.8 Kts.
Max. Flood 3 hrs. 20 min. after Flood Starts, ±15 min.
Max. Ebb 2 hrs. 45 min. after Ebb Starts, ±15 min.
Gay Head (1 1/2 mi. NW of): av. max velocity, flood 2.0 kts., ebb 2.0 kts. Time of Fl. and Ebb 1 hr. 35 min. after Pollock Rip. Cross Rip: av. max. velocity, flood 1.3 kts., ebb 0.9 kts. Time of Fl. and Ebb 1 hr. 50 min. after Pollock Rip.
See pp. 150-157 for Current Change at other points.

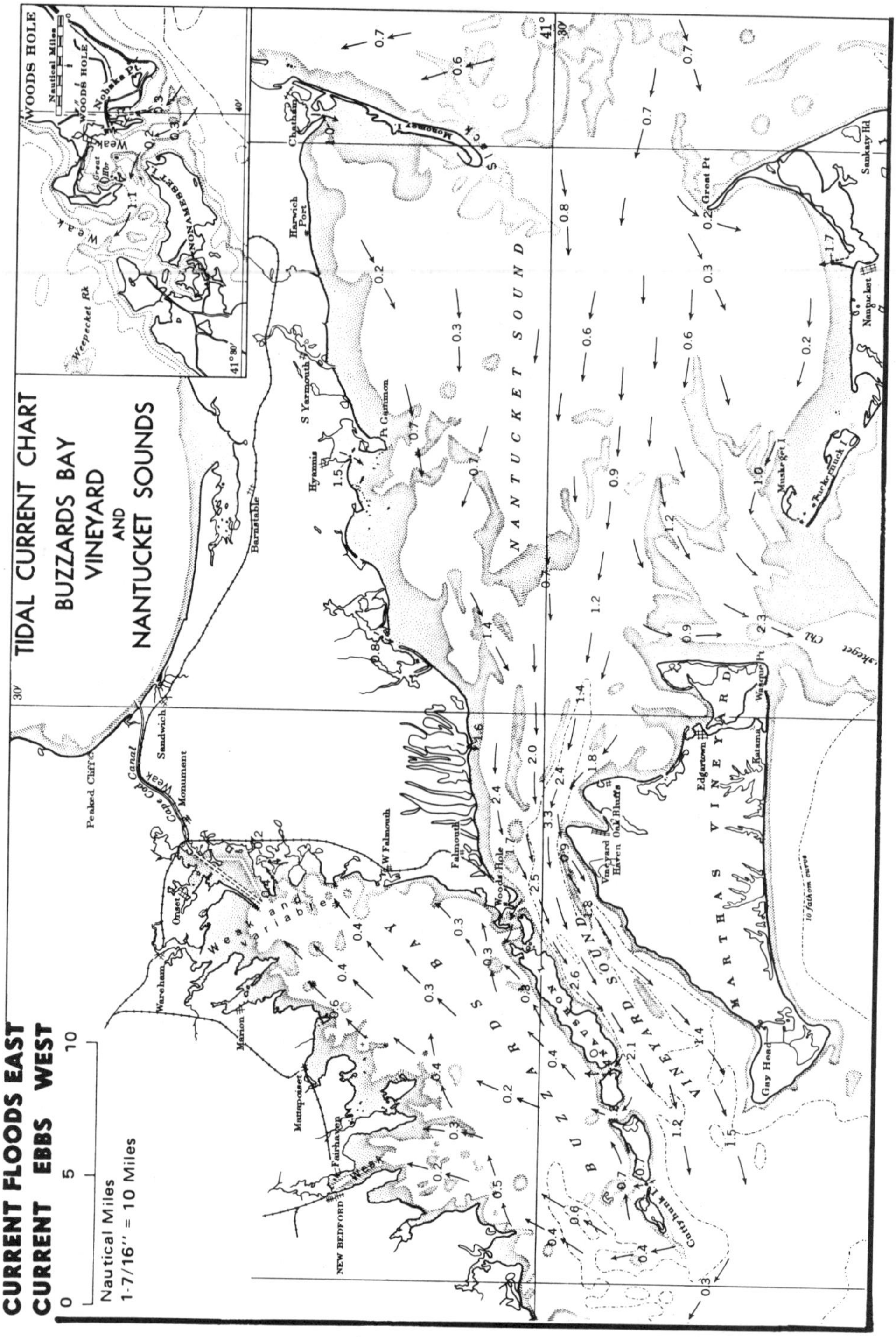

FLOOD STARTS AT POLLOCK RIP CHANNEL
OR: 4 HOURS **AFTER** HIGH WATER AT BOSTON

Velocities shown are at Spring Tides. **See Note at bottom of Boston Tables: Rule-of-Thumb for Current Velocities.** *(Pollock Rip Ch. is SE of Monomoy Pt.)*

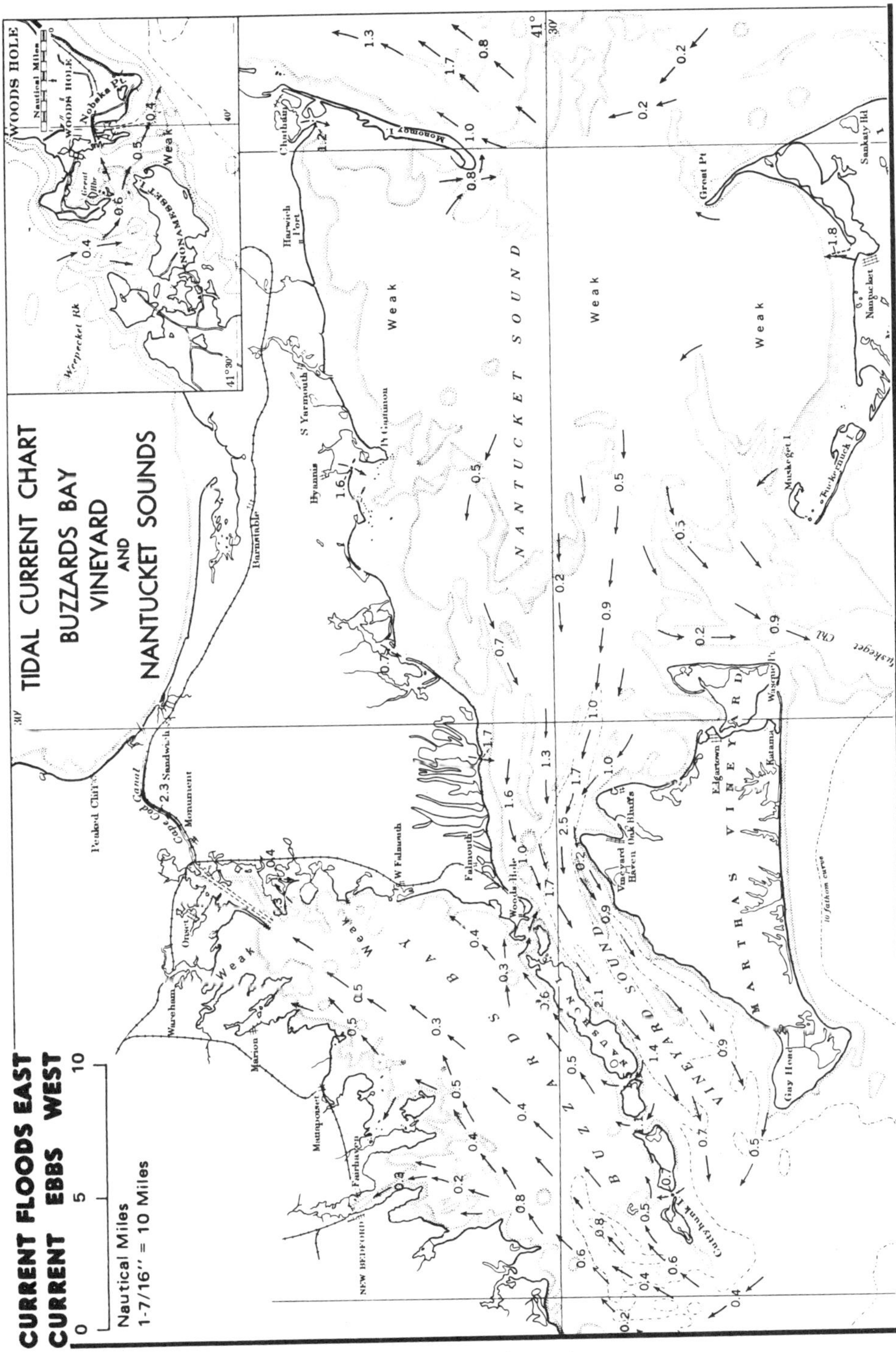

1 HOUR **AFTER** FLOOD STARTS AT POLLOCK RIP CHANNEL
OR: 5 HOURS **AFTER** HIGH WATER AT BOSTON

Velocities shown are at Spring Tides. **See Note at bottom of Boston Tables: Rule-of-Thumb for Current Velocities.** *(Pollock Rip Ch. is SE of Monomoy Pt.)*

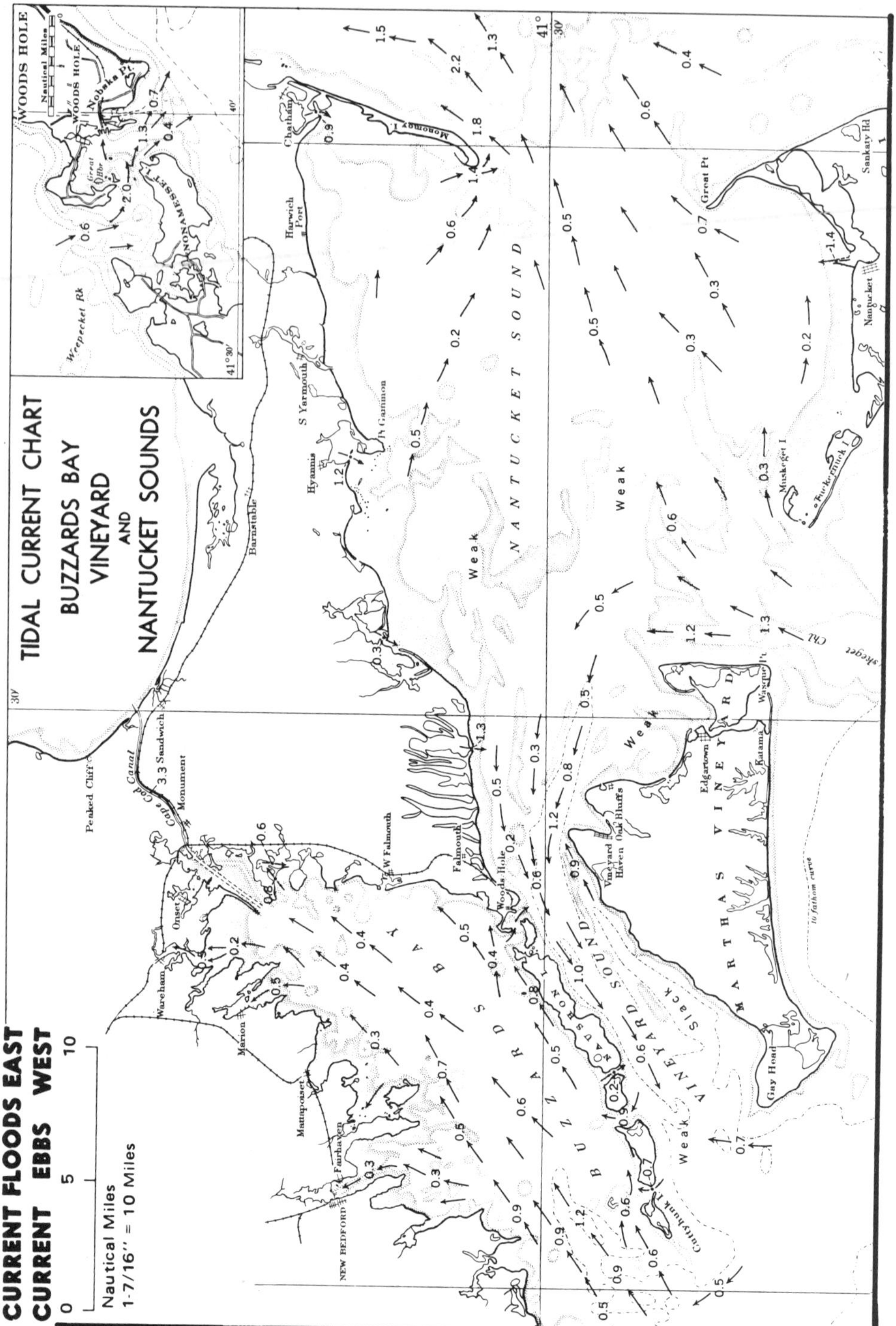

2 HOURS **AFTER** FLOOD STARTS AT POLLOCK RIP CHANNEL
OR: LOW WATER AT BOSTON

Velocities shown are at Spring Tides. **See Note at bottom of Boston Tables: Rule-of-Thumb for Current Velocities.** *(Pollock Rip Ch. is SE of Monomoy Pt.)*

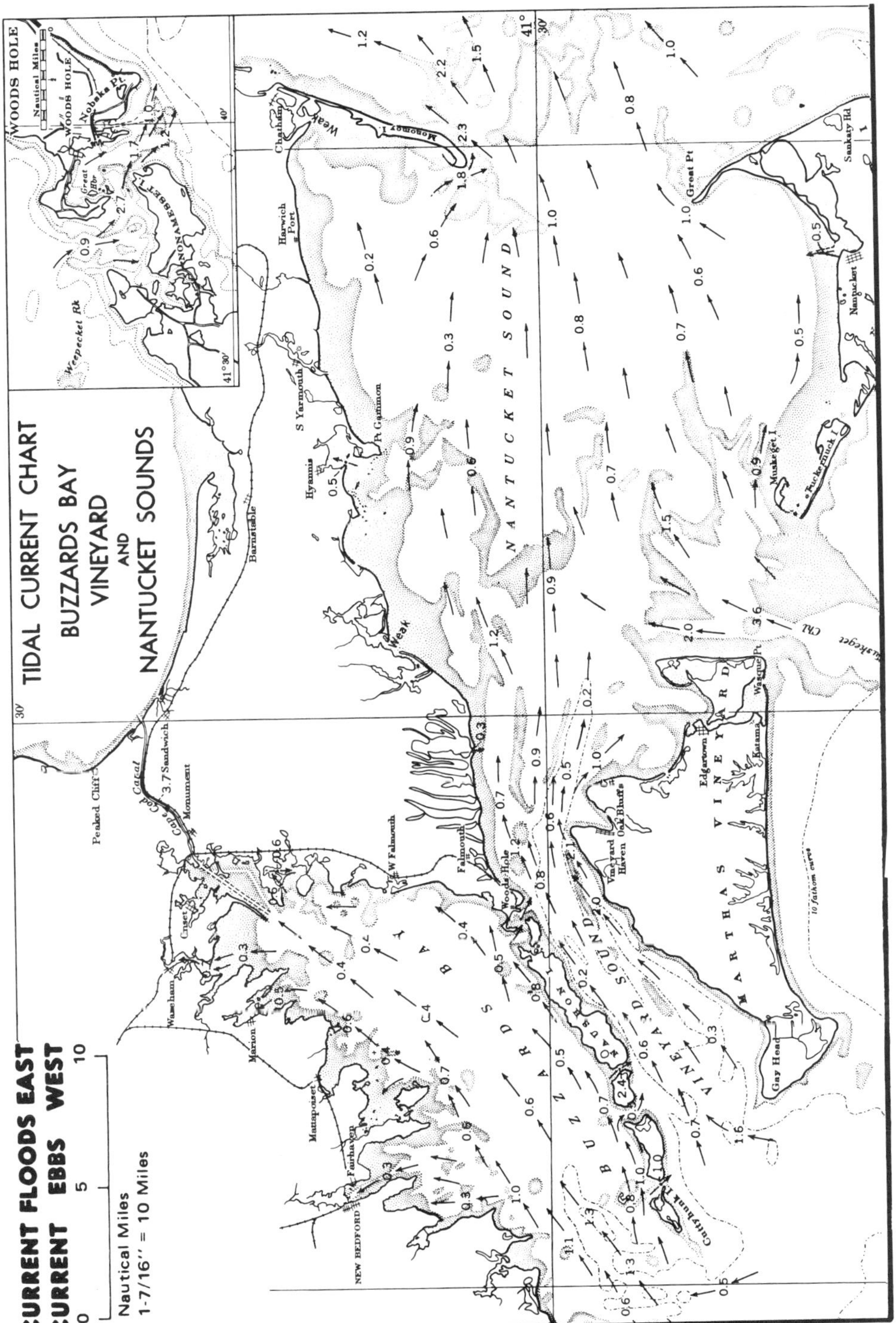

3 HOURS **AFTER** FLOOD STARTS AT POLLOCK RIP CHANNEL
OR: 1 HOUR **AFTER** LOW WATER AT BOSTON

Velocities shown are at Spring Tides. **See Note at bottom of Boston Tables: Rule-of-Thumb for Current Velocities.** *(Pollock Rip Ch. is SE of Monomoy Pt.)*

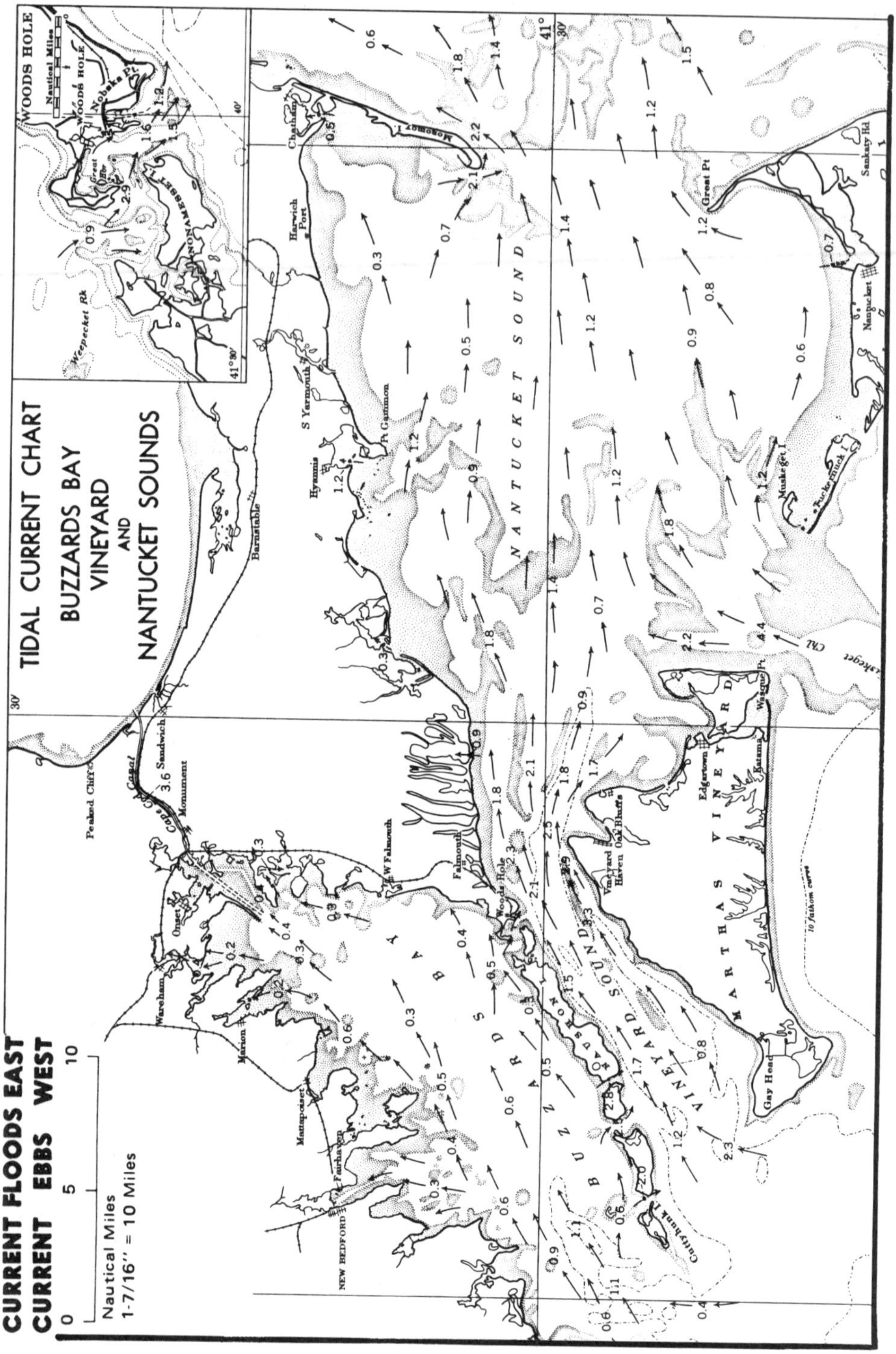

4 HOURS **AFTER** FLOOD STARTS AT POLLOCK RIP CHANNEL
OR: 2 HOURS **AFTER** LOW WATER AT BOSTON

Velocities shown are at Spring Tides. **See Note at bottom of Boston Tables: Rule-of-Thumb for Current Velocities.** *(Pollock Rip Ch. is SE of Monomoy Pt.)*

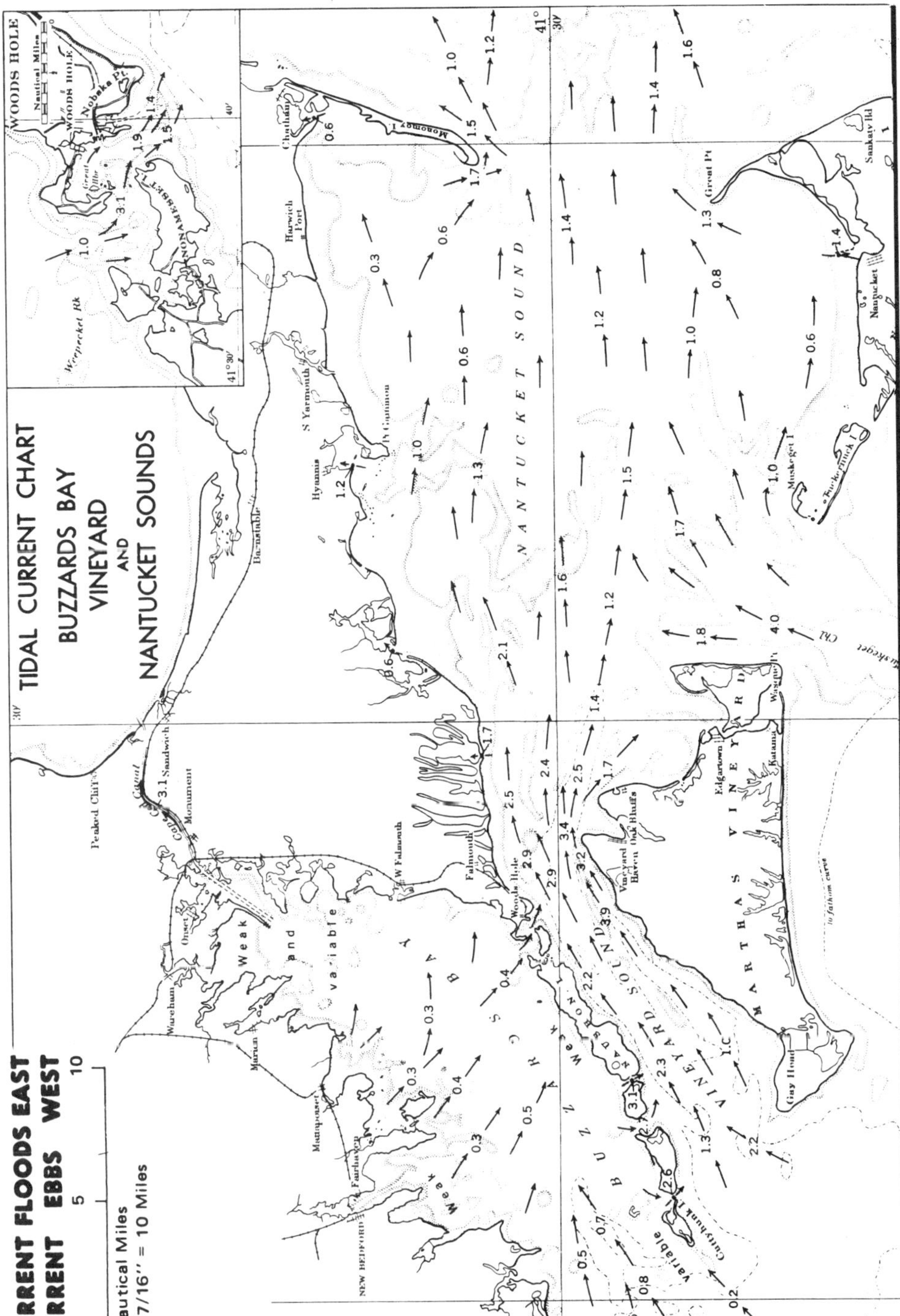

5 HOURS **AFTER** FLOOD STARTS AT POLLOCK RIP CHANNEL
OR: 3 HOURS **AFTER** LOW WATER AT BOSTON

Velocities shown are at Spring Tides. **See Note at bottom of Boston Tables: Rule-of-Thumb for Current Velocities.** *(Pollock Rip Ch. is SE of Monomoy Pt.)*

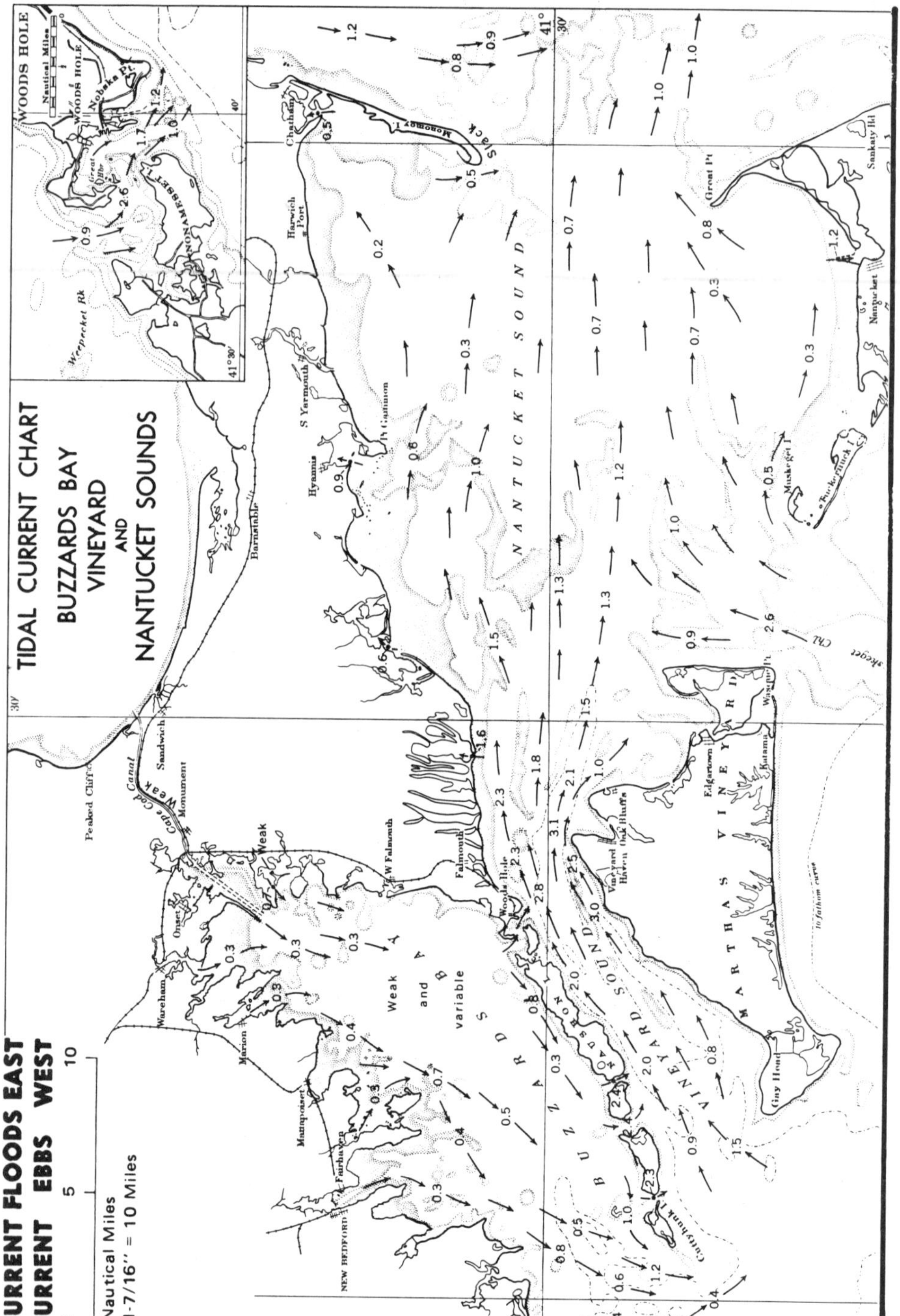

EBB STARTS AT POLLOCK RIP CHANNEL
OR: 4 HOURS **AFTER** LOW WATER AT BOSTON

Velocities shown are at Spring Tides. **See Note at bottom of Boston Tables: Rule-of-Thumb for Current Velocities.** *(Pollock Rip Ch. is SE of Monomoy Pt.)*

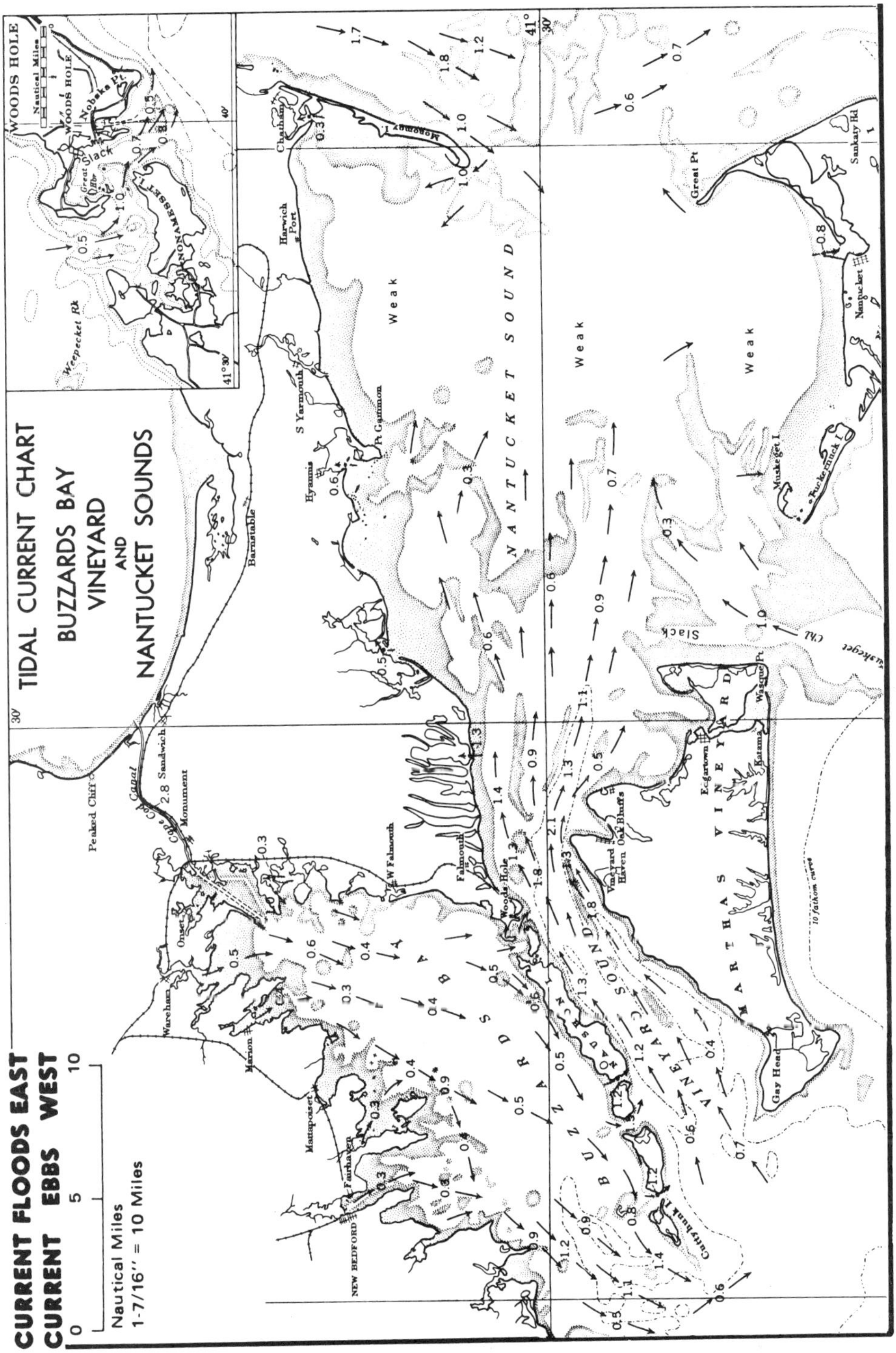

1 HOUR **AFTER** EBB STARTS AT POLLOCK RIP CHANNEL
OR: 5 HOURS **AFTER** LOW WATER AT BOSTON

Velocities shown are at Spring Tides. **See Note at bottom of Boston Tables: Rule-of-Thumb for Current Velocities.** *(Pollock Rip Ch. is SE of Monomoy Pt.)*

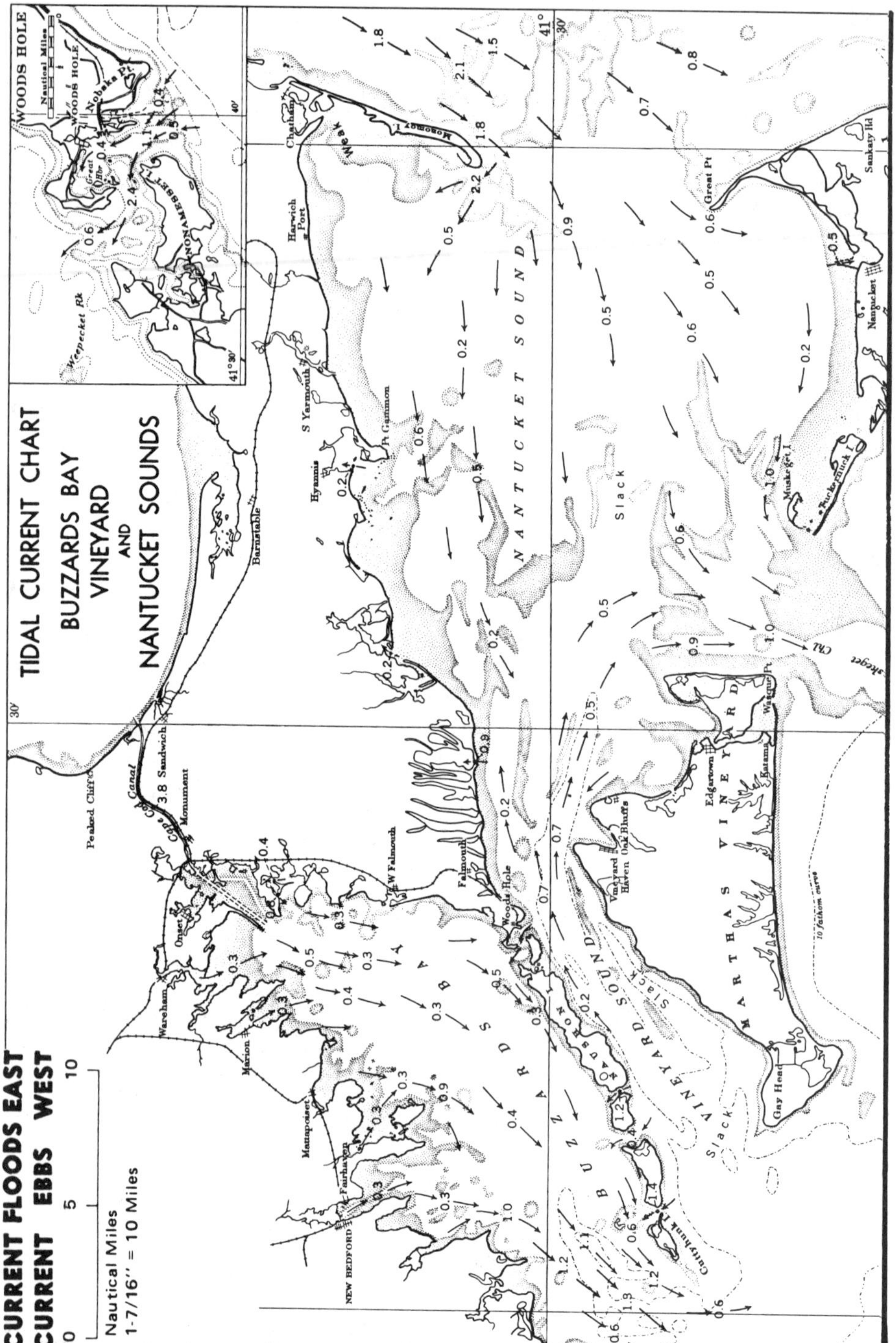

2 HOURS **AFTER** EBB STARTS AT POLLOCK RIP CHANNEL
OR: HIGH WATER AT BOSTON

Velocities shown are at Spring Tides. **See Note at bottom of Boston Tables: Rule-of-Thumb for Current Velocities.** *(Pollock Rip Ch. is SE of Monomoy Pt.)*

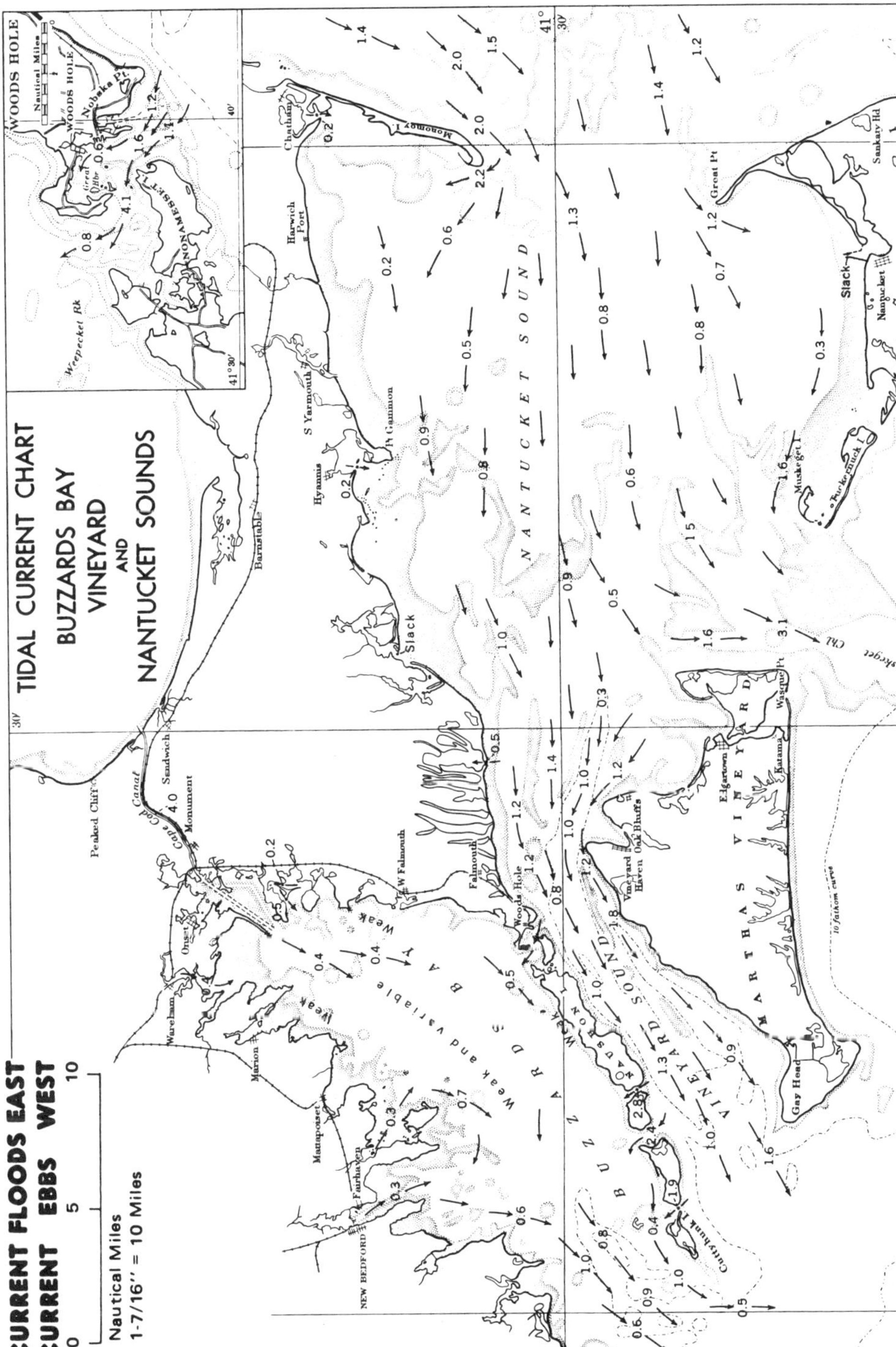

3 HOURS **AFTER** EBB STARTS AT POLLOCK RIP CHANNEL OR: 1 HOUR **AFTER** HIGH WATER AT BOSTON

Velocities shown are at Spring Tides. **See Note at bottom of Boston Tables: Rule-of-Thumb for Current Velocities.** *(Pollock Rip Ch. is SE of Monomoy Pt.)*

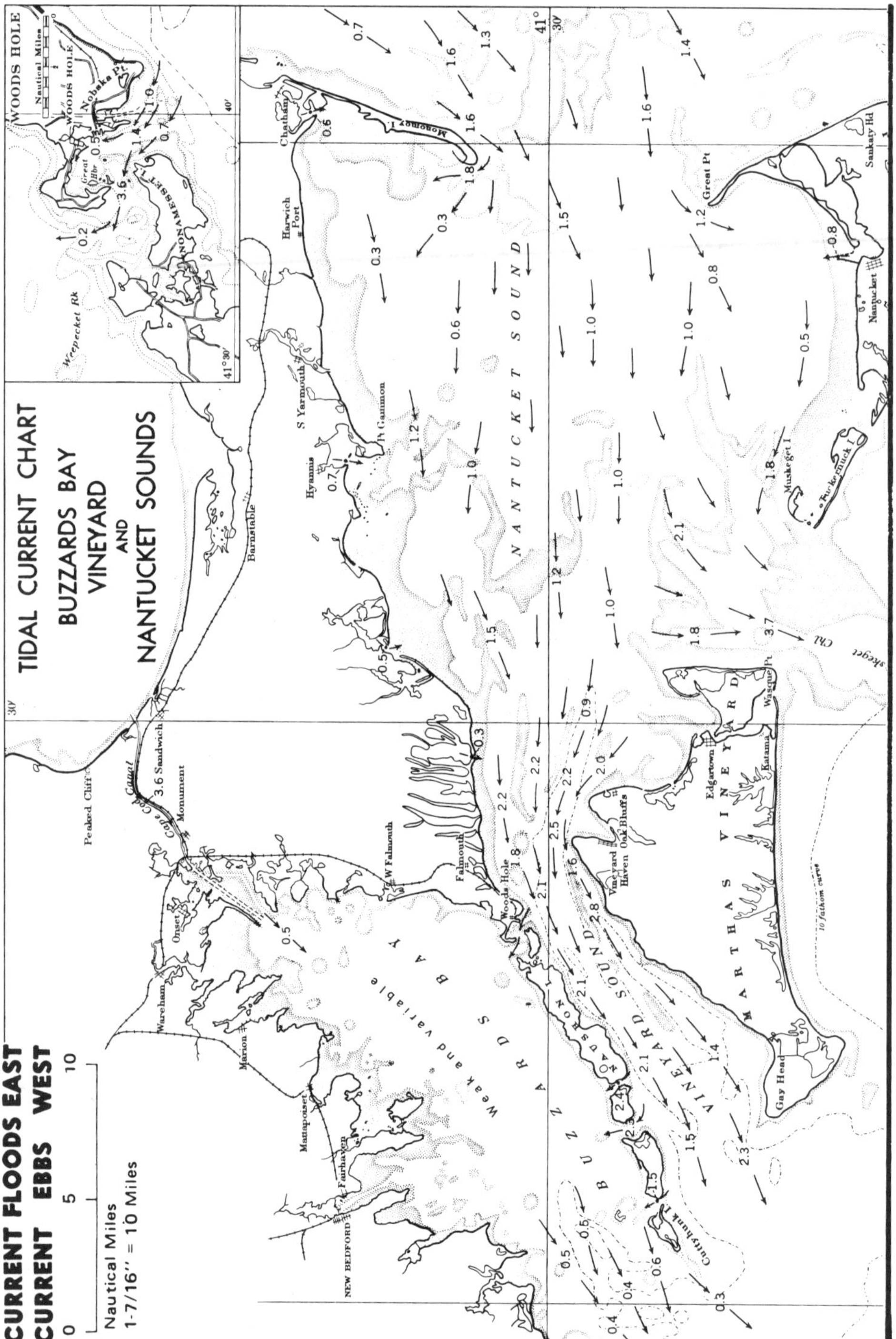

4 HOURS **AFTER** EBB STARTS AT POLLOCK RIP CHANNEL
OR: 2 HOURS **AFTER** HIGH WATER AT BOSTON

Velocities shown are at Spring Tides. **See Note at bottom of Boston Tables: Rule-of-Thumb for Current Velocities.** *(Pollock Rip Ch. is SE of Monomoy Pt.)*

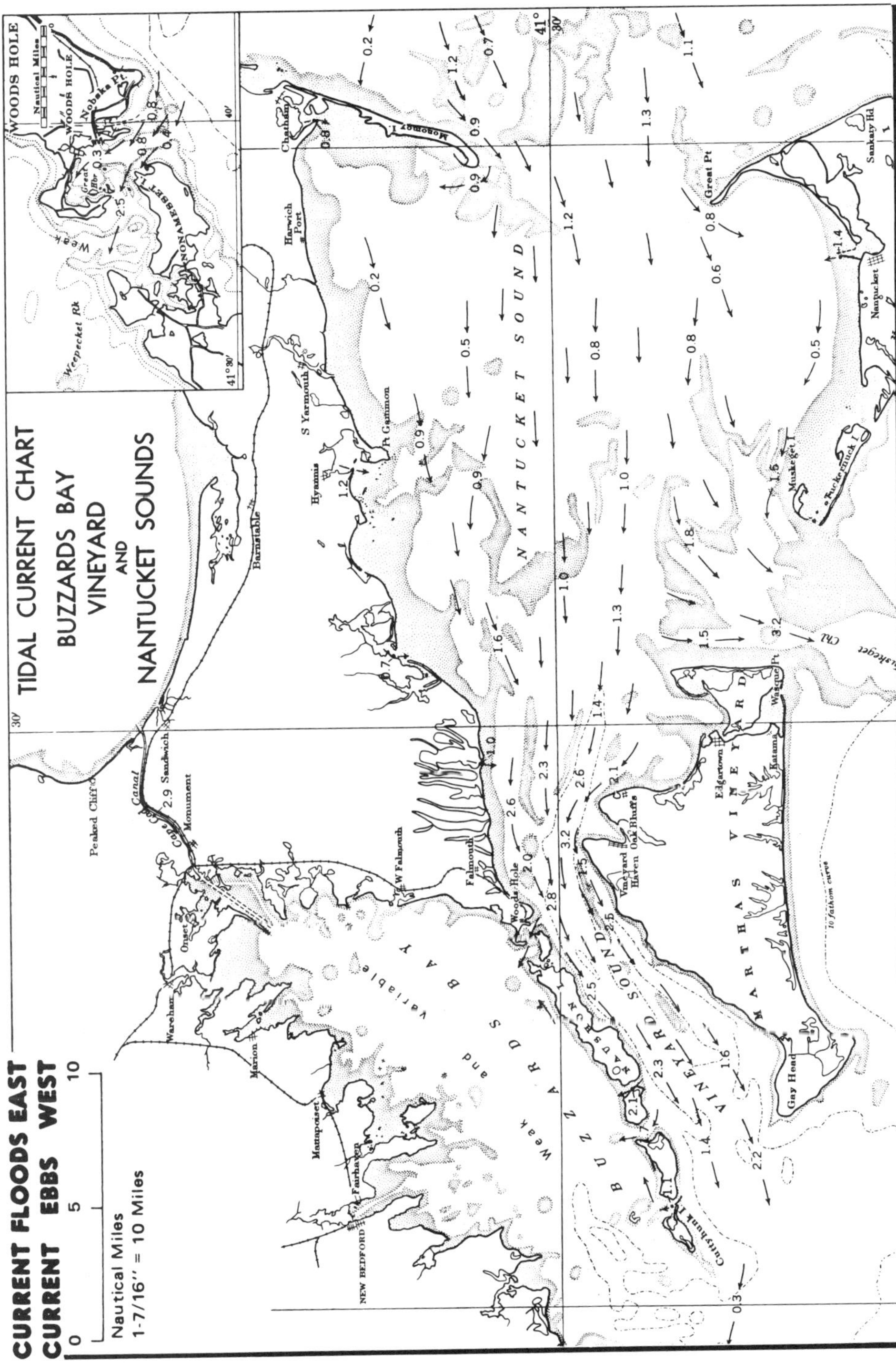

5 HOURS **AFTER** EBB STARTS AT POLLOCK RIP CHANNEL
OR: 3 HOURS **AFTER** HIGH WATER AT BOSTON

Velocities shown are at Spring Tides. **See Note at bottom of Boston Tables: Rule-of-Thumb for Current Velocities.** *(Pollock Rip Ch. is SE of Monomoy Pt.)*

ADJUSTING YOUR COMPASS

Nothing can equal the expert services of a professional Compass Adjuster, but in the event you are in a situation where these are not available, the following information should be helpful.

Most modern compasses are regularly fitted with a magnetic corrector system attached to or inside the bottom of the compass, or inside the bottom of the cylinder, if binnacle-mounted. Such "B.I.C.'s" (Built-in-Correctors) are exceedingly easy to use and are capable of removing virtually all the deviations of a well-located compass.

B.I.C.'s generally consist of two horizontal shafts, one running athwartship and one running fore and aft. At each end of each shaft is a magnet. The magnets on opposite ends of the same shaft have their polarities reversed from each other and when these magnets (or magnetic poles of these magnets) are all horizontal they are equidistant from the directive magnets of the compass and have zero effect on the compass. When a shaft is rotated to any degree, however, the magnetic poles are no longer in neutral and the upper ends of the magnets on the shaft will now be closer to the compass directive magnets and create a magnetic field which will deflect the compass. The usual B.I.C. will remove about 15° of error.

The shaft which runs thwartship corrects on North and South headings and has zero effect on East and West. The fore and aft shaft corrects on East and West headings and has zero effect on North and South. B.I.C.'s in binnacles generally have a locking device to prevent accidental rotation of the shafts. B.I.C.'s within the compass generally do not. Turning a shaft 90° from neutral creates maximum effect in one direction; another 90° turn brings it back to neutral; a third turn of 90° creates maximum effect in the opposite direction; a fourth turn of 90° brings it back to neutral.

Pick a quiet day and a swinging area with calm conditions. Have someone else at the tiller or wheel and the engine just ticking over. Using your hand-bearing compass at the "Magic spot", get the boat headed with 5 or 6° of any cardinal point, say North. Give your helmsman an exact mark to steer by such as a point of land, a distant roof or anything he can aim for as he will be steering by that mark and not by the compass. While steady on the mark, sight forward against the mast and get the exact course from your hand-bearing compass, say 4° Magnetic. While he continues to hold the boat on the mark, slowly and delicately rotate the thwartship shaft by trial and error until the steering compass reads 4°. When rotating the shaft from the neutral position, you will get a large effect with only a few degrees of rotation and as you approach the maximum correction you will have only a very small effect from continued rotation. Now, swing around to either East or West and following the same procedure with the fore and aft shaft, make your steering compass agree exactly with the course your hand-bearing compass has established. Your compass *adjusting* is now in all probability completed.

It is necessary now to check for deviation all around. Compare the steering compass heading to your hand-bearing compass heading at least every 45° on all the cardinal and intercardinal points *including* the headings already checked. If the steering compass reads too little, the deviation is Easterly on that heading and the number of degrees must always be *subtracted* when steering that magnetic course. If the steering compass reads too much, the deviation is Westerly on that heading and the number of degrees must always be *added* when steering that magnetic course. If there is as much as a 3° difference between the deviation on one heading and a heading 45° away, you should check every 15° between those headings. (When all through, it's a good idea to tape the ends of the shafts so they won't shift position.) Remember also, when plotting your deviation on a graph or otherwise, to connect the points with a smooth curve and the curve will give the best indication of the deviations on unchecked intermediate headings.

If you find, on the 8 headings you have checked, a preponderance of Easterly deviations over Westerly deviations or vice versa, add up the total of Easterly, subtract the total of Westerly and divide by 8. The result tells you the amount by which your compass is misaligned. Let's suppose it works out to 1° Easterly. This means your lubbers line is off to port 1°. If you now shim the compass 1° clockwise, all your Easterly deviations will be reduced by 1°, your Westerly deviations increased by 1° and your 0 deviations will become 1° Westerly. You will now find that the total of your Easterly deviations is the very same as your Westerly total, proving that you have eliminated misalignment error.

NARRAGANSETT BAY CURRENTS

This Current Diagram shows average maximum currents with a normal range (3.5 ft.) of tides at Newport. (See pages 64-69.)

Maximum Ebb Currents, 3 hours after High Water at Newport, are shown by double-headed arrows and velocities are underlined.

Maximum Flood Currents, 2½ hours before High Water at Newport, are shown by single-headed arrows and velocities are not underlined.

When height shown is 3.0 ft., subtract 30% from all velocities shown. When height is 4.0 ft., add 20%; when 4.5 ft., add 40%; when 5.0 ft., add 60%.

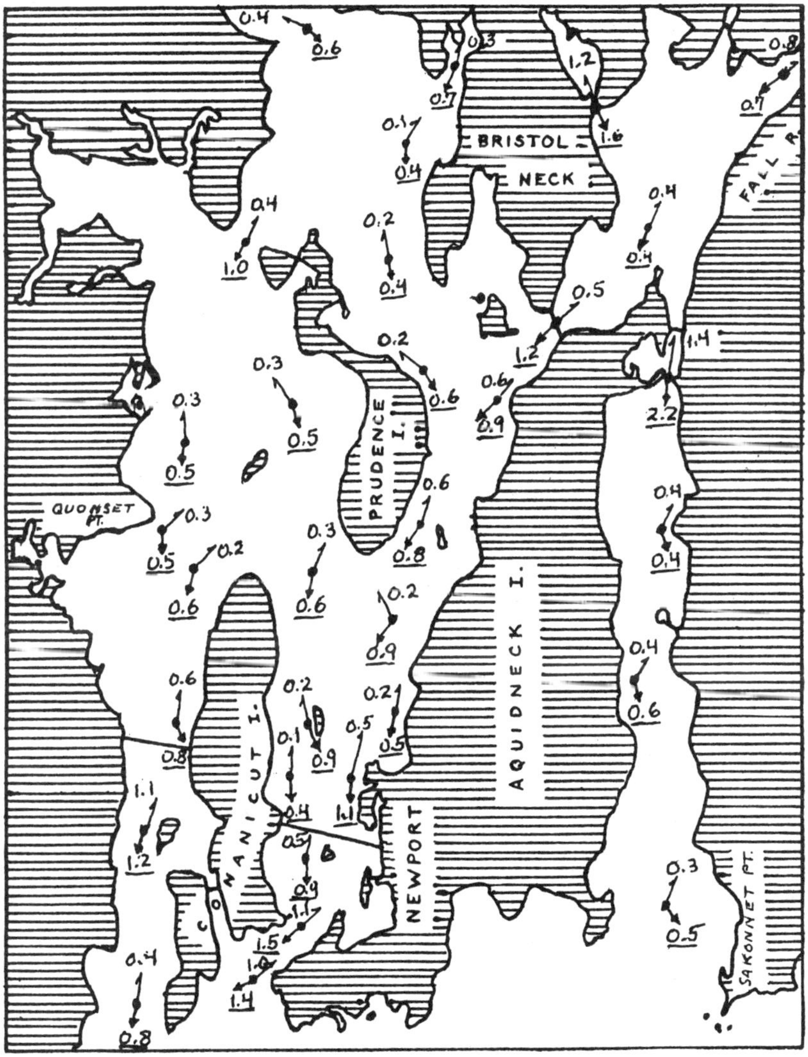

1999 HIGH & LOW WATER NEWPORT, R.I.

41–30N x 71–20W

Standard Time

DAY OF MONTH	DAY OF WEEK	JANUARY					
		HIGH				LOW	
		a.m.	Ht.	p.m.	Ht.	a.m.	p.m.
1	F	6 42	4.8	7 06	3.9	12 01	12 55
2	S	7 32	4.8	7 56	4.0	12 54	1 43
3	S	8 21	4.6	8 45	3.9	1 43	2 29
4	M	9 09	4.4	9 34	3.8	2 30	3 12
5	T	9 57	4.1	10 23	3.6	3 16	3 54
6	W	10 45	3.7	11 13	3.4	4 01	4 34
7	T	11 35	3.4	...	...	4 46	5 13
8	F	12 05	3.2	12 26	3.1	5 33	5 55
9	S	12 59	3.1	1 20	2.9	6 26	6 41
10	S	1 53	3.1	2 13	2.7	7 28	7 33
11	M	2 47	3.1	3 08	2.7	8 41	8 30
12	T	3 39	3.2	4 00	2.8	9 49	9 26
13	W	4 29	3.4	4 49	2.9	10 39	10 17
14	T	5 15	3.6	5 36	3.0	11 19	11 03
15	F	5 59	3.8	6 20	3.2	11 57	11 46
16	S	6 42	3.9	7 03	3.3	...	12 34
17	S	7 24	4.0	7 46	3.4	12 28	1 11
18	M	8 06	4.1	8 28	3.5	1 10	1 49
19	T	8 48	4.1	9 12	3.6	1 52	2 28
20	W	9 32	4.0	9 57	3.6	2 36	3 09
21	T	10 18	3.8	10 46	3.6	3 22	3 52
22	F	11 08	3.6	11 39	3.6	4 12	4 39
23	S	...	...	12 03	3.4	5 06	5 30
24	S	12 37	3.6	1 03	3.3	6 07	6 28
25	M	1 38	3.7	2 06	3.2	7 17	7 32
26	T	2 41	3.8	3 09	3.2	8 36	8 43
27	W	3 43	4.0	4 10	3.3	9 55	9 54
28	T	4 41	4.2	5 07	3.5	11 03	10 59
29	F	5 36	4.3	6 00	3.7	11 59	11 56
30	S	6 27	4.4	6 50	3.8	...	12 47
31	S	7 15	4.4	7 37	3.9	12 47	1 30

Standard Time

DAY OF MONTH	DAY OF WEEK	FEBRUARY					
		HIGH				LOW	
		a.m.	Ht.	p.m.	Ht.	a.m.	p.m.
1	M	8 02	4.4	8 23	3.9	1 33	2 09
2	T	8 47	4.2	9 08	3.8	2 16	2 46
3	W	9 30	3.9	9 53	3.7	2 55	3 19
4	T	10 14	3.6	10 38	3.5	3 33	3 51
5	F	10 58	3.3	11 24	3.3	4 09	4 24
6	S	11 44	3.0	...	...	4 47	5 00
7	S	12 13	3.1	12 34	2.8	5 28	5 40
8	M	1 06	3.0	1 28	2.6	6 17	6 28
9	T	2 03	2.9	2 27	2.5	7 15	7 24
10	W	2 58	3.0	3 23	2.6	8 25	8 28
11	T	3 53	3.1	4 17	2.7	9 38	9 32
12	F	4 43	3.4	5 07	3.0	10 37	10 31
13	S	5 31	3.6	5 53	3.2	11 24	11 22
14	S	6 16	3.8	6 38	3.5	...	12 06
15	M	6 59	4.0	7 21	3.7	12 09	12 46
16	T	7 43	4.2	8 04	3.9	12 54	1 26
17	W	8 26	4.2	8 48	4.0	1 39	2 07
18	T	9 11	4.1	9 34	4.1	2 24	2 48
19	F	9 58	4.0	10 23	4.0	3 11	3 32
20	S	10 48	3.7	11 16	3.9	4 00	4 18
21	S	11 43	3.5	...	...	4 53	5 09
22	M	12 13	3.8	12 42	3.3	5 53	6 06
23	T	1 16	3.7	1 47	3.1	7 03	7 13
24	W	2 21	3.7	2 52	3.1	8 27	8 32
25	T	3 25	3.8	3 54	3.2	9 52	9 53
26	F	4 25	3.9	4 51	3.4	10 59	11 01
27	S	5 20	4.0	5 44	3.6	11 51	11 56
28	S	6 10	4.1	6 31	3.8	...	12 34

Dates when Ht. of **Low** Water is below Mean Low with Ht. of lowest given for each period and Date of lowest in ( ):

January:
1st - 5th: -0.7' (1st - 3rd)
17th - 31st: -0.5' (19th - 20th)
: -0.7' (31st)

February:
1st - 4th: -0.7' (1st)
15th - 22nd: -0.8' (18th)
27th: -0.4'

Average Rise and Fall 3.5 ft.

When a high tide exceeds av. ht., the *following* low tide will be lower than av.

1999 HIGH & LOW WATER
NEWPORT, R.I.
41–30N x 71–20W

Standard Time **Daylight Time starts Apr. 4 at 2 a.m.**

DAY OF MONTH	DAY OF WEEK	MARCH						DAY OF MONTH	DAY OF WEEK	APRIL					
		HIGH				LOW				HIGH				LOW	
		a.m.	Ht.	p.m.	Ht.	a.m.	p.m.			a.m.	Ht.	p.m.	Ht.	a.m.	p.m.
1	M	6 56	4.1	7 16	3.9	12 42	1 11	1	T	7 55	3.8	8 12	4.0	1 37	1 38
2	T	7 40	4.1	7 59	4.0	1 22	1 44	2	F	8 34	3.6	8 51	3.9	2 06	2 06
3	W	8 22	4.0	8 40	3.9	1 58	2 14	3	S	9 13	3.5	9 30	3.8	2 35	2 35
4	T	9 02	3.8	9 21	3.8	2 32	2 42	4	S	*10 53	3.2	*11 11	3.5	*4 05	*4 06
5	F	9 43	3.5	10 03	3.6	3 03	3 11	5	M	11 35	3.0	11 54	3.3	4 38	4 40
6	S	10 24	3.3	10 45	3.4	3 35	3 42	6	T	...	...	12 20	2.8	5 15	5 19
7	S	11 07	3.0	11 31	3.2	4 09	4 16	7	W	12 41	3.1	1 12	2.6	5 57	6 05
8	M	11 54	2.8	...	...	4 46	4 55	8	T	1 35	3.0	2 10	2.6	6 47	6 59
9	T	12 22	3.0	12 48	2.6	5 30	5 40	9	F	2 35	2.9	3 11	2.7	7 46	8 03
10	W	1 16	2.9	1 46	2.5	6 23	6 35	10	S	3 34	3.0	4 08	2.9	8 52	9 16
11	T	2 16	2.9	2 46	2.6	7 27	7 40	11	S	4 32	3.2	5 02	3.2	10 00	10 27
12	F	3 14	3.0	3 43	2.7	8 39	8 51	12	M	5 26	3.5	5 52	3.6	11 01	11 30
13	S	4 09	3.2	4 35	3.0	9 47	9 58	13	T	6 16	3.8	6 40	4.1	11 54	...
14	S	4 59	3.5	5 23	3.4	10 43	10 57	14	W	7 05	4.0	7 27	4.4	12 26	12 44
15	M	5 47	3.8	6 09	3.7	11 32	11 48	15	T	7 52	4.2	8 14	4.7	1 17	1 31
16	T	6 33	4.0	6 54	4.1	...	12 16	16	F	8 40	4.3	9 02	4.9	2 07	2 18
17	W	7 18	4.2	7 39	4.3	12 37	12 59	17	S	9 29	4.2	9 51	4.9	2 56	3 04
18	T	8 03	4.3	8 24	4.5	1 24	1 43	18	S	10 19	4.1	10 42	4.7	3 46	3 52
19	F	8 50	4.2	9 12	4.5	2 11	2 26	19	M	11 12	3.9	11 36	4.5	4 37	4 43
20	S	9 38	4.0	10 02	4.4	2 59	3 12	20	T	...	...	12 08	3.6	5 31	5 37
21	S	10 29	3.8	10 55	4.2	3 49	3 59	21	W	12 34	4.2	1 08	3.4	6 31	6 39
22	M	11 25	3.5	11 53	4.0	4 42	4 51	22	T	1 36	3.9	2 12	3.3	7 42	7 55
23	T	...	...	12 25	3.3	5 42	5 50	23	F	2 40	3.6	3 15	3.3	9 04	9 29
24	W	12 56	3.8	1 30	3.2	6 53	7 02	24	S	3 43	3.5	4 15	3.4	10 21	10 52
25	T	2 02	3.7	2 35	3.2	8 20	8 31	25	S	4 42	3.5	5 10	3.6	11 20	11 52
26	F	3 06	3.6	3 37	3.3	9 44	9 57	26	M	5 35	3.5	5 59	3.8	...	12 04
27	S	4 06	3.7	4 33	3.5	10 46	11 02	27	T	6 22	3.5	6 43	3.9	12 37	12 39
28	S	5 00	3.7	5 24	3.7	11 34	11 51	28	W	7 06	3.6	7 25	4.1	1 14	1 08
29	M	5 49	3.8	6 09	3.9	...	12 12	29	T	7 47	3.6	8 05	4.1	1 45	1 34
30	T	6 33	3.8	6 52	4.0	12 32	12 43	30	F	8 27	3.6	8 44	4.1	2 13	2 02
31	W	7 15	3.8	7 32	4.1	1 06	1 11								

***Daylight Time starts**

Dates when Ht. of **Low** Water is below Mean Low with Ht. of lowest given for each period and Date of lowest in ( ):

March	April
1st - 5th: -0.5' (1st - 3rd)	1st - 3rd: -0.3'
15th - 23rd: -0.9' (18th - 20th)	13th - 20th: -1.0' (16th - 17th)
30th - 31st: -0.3'	

Average Rise and Fall 3.5 ft.

When a high tide exceeds av. ht., the *following* low tide will be lower than av.

1999 HIGH & LOW WATER
NEWPORT, R.I.
41–30N x 71–20W

DAY OF MONTH	DAY OF WEEK	MAY HIGH a.m.	Ht.	p.m.	Ht.	LOW a.m.	p.m.	DAY OF MONTH	DAY OF WEEK	JUNE HIGH a.m.	Ht.	p.m.	Ht.	LOW a.m.	p.m.
		Daylight Time								Daylight Time					
1	S	9 06	3.5	**9 23**	4.0	2 40	**2 31**	1	T	10 03	3.3	**10 17**	3.9	3 18	**3 11**
2	S	9 46	3.4	**10 02**	3.9	3 09	**3 02**	2	W	10 45	3.2	**10 58**	3.7	3 52	**3 49**
3	M	10 26	3.2	**10 42**	3.7	3 40	**3 36**	3	T	11 30	3.1	**11 42**	3.6	4 30	**4 31**
4	T	11 09	3.1	**11 23**	3.5	4 13	**4 12**	4	F	...	...	**12 17**	3.1	5 11	**5 18**
5	W	11 54	2.9	...	...	4 50	**4 52**	5	S	12 31	3.4	**1 09**	3.1	5 56	**6 11**
6	T	12 09	3.3	**12 43**	2.8	5 32	**5 38**	6	S	1 24	3.3	**2 04**	3.3	6 47	**7 12**
7	F	1 00	3.2	**1 38**	2.8	6 19	**6 32**	7	M	2 22	3.3	**3 01**	3.5	7 44	**8 20**
8	S	1 57	3.1	**2 36**	2.9	7 14	**7 35**	8	T	3 22	3.4	**3 57**	3.8	8 45	**9 31**
9	S	2 57	3.1	**3 34**	3.2	8 15	**8 46**	9	W	4 22	3.5	**4 54**	4.2	9 47	**10 40**
10	M	3 55	3.3	**4 29**	3.5	9 19	**9 57**	10	T	5 17	3.7	**5 46**	4.6	10 48	**11 44**
11	T	4 52	3.5	**5 21**	3.9	10 21	**11 04**	11	F	6 12	3.9	**6 39**	4.9	11 46	...
12	W	5 45	3.7	**6 12**	4.4	11 19	...	12	S	7 06	4.1	**7 31**	5.1	12 42	**12 42**
13	T	6 37	4.0	**7 01**	4.7	12 03	**12 13**	13	S	7 58	4.2	**8 22**	5.2	1 37	**1 36**
14	F	7 28	4.1	**7 51**	5.0	12 58	**1 04**	14	M	8 50	4.2	**9 14**	5.1	2 29	**2 29**
15	S	8 18	4.2	**8 41**	5.1	1 51	**1 55**	15	T	9 43	4.2	**10 05**	4.9	3 21	**3 22**
16	S	9 09	4.2	**9 31**	5.1	2 42	**2 45**	16	W	10 36	4.1	**10 58**	4.6	4 12	**4 15**
17	M	10 01	4.1	**10 23**	4.9	3 33	**3 36**	17	T	11 29	4.0	**11 51**	4.3	5 02	**5 09**
18	T	10 54	4.0	**11 17**	4.6	4 25	**4 28**	18	F	...	...	**12 24**	3.8	5 54	**6 08**
19	W	11 50	3.8	...	...	5 19	**5 24**	19	S	12 46	3.9	**1 20**	3.7	6 46	**7 12**
20	T	12 14	4.2	**12 49**	3.6	6 17	**6 27**	20	S	1 42	3.6	**2 17**	3.6	7 41	**8 27**
21	F	1 13	3.9	**1 49**	3.5	7 20	**7 41**	21	M	2 38	3.4	**3 12**	3.6	8 37	**9 46**
22	S	2 14	3.6	**2 49**	3.5	8 30	**9 09**	22	T	3 33	3.2	**4 05**	3.7	9 31	**10 51**
23	S	3 14	3.4	**3 47**	3.5	9 40	**10 29**	23	W	4 26	3.1	**4 55**	3.7	10 20	**11 41**
24	M	4 10	3.3	**4 40**	3.7	10 37	**11 29**	24	T	5 16	3.2	**5 42**	3.9	11 02	...
25	T	5 03	3.3	**5 29**	3.8	11 21	...	25	F	6 03	3.2	**6 26**	4.0	12 19	**-A-**
26	W	5 51	3.3	**6 14**	3.9	12 14	**12 01**	26	S	6 48	3.3	**7 09**	4.1	12 50	**12 19**
27	T	6 35	3.4	**6 56**	4.1	12 50	**12 26**	27	S	7 32	3.4	**7 51**	4.2	1 20	**12 56**
28	F	7 17	3.4	**7 37**	4.1	1 19	**12 56**	28	M	8 15	3.5	**8 32**	4.2	1 51	**1 34**
29	S	7 59	3.5	**8 17**	4.2	1 47	**1 27**	29	T	8 57	3.5	**9 12**	4.2	2 23	**2 12**
30	S	8 40	3.5	**8 56**	4.1	2 15	**2 00**	30	W	9 39	3.5	**9 53**	4.1	2 57	**2 51**
31	M	9 21	3.4	**9 36**	4.0	2 45	**2 35**								

A also at 11:41 a.m.

Dates when Ht. of **Low** Water is below Mean Low with Ht. of lowest given for each period and Date of lowest in ( ):

12th - 19th: -0.9' (16th)

10th - 17th: -0.7' (13th - 15th)

Average Rise and Fall 3.5 ft.

When a high tide exceeds av. ht., the *following* low tide will be lower than av.

1999 HIGH & LOW WATER NEWPORT, R.I.

41–30N x 71–20W

Daylight Time | Daylight Time

DAY OF MONTH	DAY OF WEEK	JULY						DAY OF MONTH	DAY OF WEEK	AUGUST					
		HIGH				LOW				HIGH				LOW	
		a.m.	Ht.	p.m.	Ht.	a.m.	p.m.			a.m.	Ht.	p.m.	Ht.	a.m.	p.m.
1	T	10 21	3.5	10 35	4.0	3 33	3 32	1	S	11 24	3.9	11 41	3.9	4 28	4 47
2	F	11 05	3.5	11 18	3.8	4 10	4 15	2	M	...	...	12 13	3.9	5 11	5 38
3	S	11 51	3.5	...	...	4 50	5 02	3	T	12 32	3.7	1 07	4.0	5 58	6 35
4	S	12 05	3.7	12 40	3.5	5 34	5 54	4	W	1 29	3.5	2 06	4.0	6 51	7 39
5	M	12 57	3.5	1 34	3.6	6 22	6 52	5	T	2 30	3.4	3 07	4.1	7 51	8 52
6	T	1 53	3.5	2 31	3.8	7 16	7 58	6	F	3 34	3.4	4 09	4.3	8 57	10 10
7	W	2 53	3.4	3 30	4.0	8 15	9 09	7	S	4 37	3.5	5 08	4.5	10 08	11 23
8	T	3 54	3.5	4 28	4.3	9 18	10 21	8	S	5 36	3.7	6 05	4.7	11 17	...
9	F	4 55	3.6	5 26	4.6	10 23	11 29	9	M	6 33	4.0	6 59	4.8	12 24	12 20
10	S	5 52	3.8	6 20	4.9	11 26	...	10	T	7 24	4.2	7 49	4.9	1 17	1 17
11	S	6 47	4.0	7 14	5.0	12 30	12 27	11	W	8 14	4.3	8 37	4.8	2 04	2 08
12	M	7 41	4.2	8 05	5.1	1 26	1 23	12	T	9 02	4.4	9 24	4.7	2 47	2 56
13	T	8 33	4.3	8 56	5.0	2 18	2 17	13	F	9 49	4.4	10 10	4.4	3 27	3 40
14	W	9 23	4.3	9 46	4.9	3 06	3 09	14	S	10 35	4.3	10 55	4.1	4 05	4 22
15	T	10 14	4.3	10 35	4.6	3 53	3 59	15	S	11 22	4.1	11 40	3.8	4 40	5 03
16	F	11 04	4.1	11 25	4.2	4 37	4 49	16	M	...	...	12 09	3.9	5 14	5 43
17	S	11 55	4.0	...	...	5 20	5 38	17	T	12 27	3.5	12 58	3.7	5 49	6 26
18	S	12 15	3.9	12 46	3.8	6 02	6 30	18	W	1 17	3.2	1 50	3.5	6 28	7 14
19	M	1 06	3.5	1 39	3.7	6 43	7 25	19	T	2 11	3.0	2 45	3.4	7 13	8 13
20	T	1 58	3.3	2 33	3.6	7 27	8 29	20	F	3 08	2.9	3 41	3.4	8 07	9 26
21	W	2 53	3.1	3 27	3.6	8 15	9 44	21	S	4 05	2.9	4 35	3.5	9 09	10 40
22	T	3 47	3.0	4 19	3.6	9 08	10 51	22	S	5 00	3.0	5 25	3.7	10 13	11 33
23	F	4 40	3.0	5 09	3.7	10 03	11 39	23	M	5 50	3.2	6 12	3.9	11 12	...
24	S	5 31	3.1	5 56	3.9	10 56	...	24	T	6 37	3.5	6 57	4.1	12 14	12 04
25	S	6 19	3.2	6 41	4.0	12 16	-A-	25	W	7 21	3.7	7 40	4.3	12 52	12 50
26	M	7 05	3.4	7 24	4.2	12 50	12 29	26	T	8 03	3.9	8 22	4.4	1 28	1 34
27	T	7 49	3.6	8 07	4.3	1 24	1 12	27	F	8 45	4.1	9 04	4.4	2 06	2 17
28	W	8 31	3.7	8 48	4.3	1 58	1 53	28	S	9 28	4.3	9 47	4.3	2 44	3 00
29	T	9 13	3.8	9 29	4.3	2 34	2 34	29	S	10 12	4.3	10 32	4.2	3 23	3 45
30	F	9 55	3.8	10 11	4.2	3 10	3 16	30	M	10 58	4.3	11 19	4.0	4 04	4 33
31	S	10 38	3.9	10 55	4.1	3 48	4 00	31	T	11 48	4.3	...	...	4 48	5 24

A also at 11:44 a.m.

Dates when Ht. of **Low** Water is below Mean Low with Ht. of lowest given for each period and Date of lowest in ( ):

10th - 16th: -0.5' (12th - 14th)

10th - 13th: -0.3'
28th - 30th: -0.3'

Average Rise and Fall 3.5 ft.

When a high tide exceeds av. ht., the *following* low tide will be lower than av.

1999 HIGH & LOW WATER NEWPORT, R.I.

41–30N x 71–20W

Daylight Time **Standard Time starts Oct. 31 at 2 a.m.**

DAY OF MONTH	DAY OF WEEK	SEPTEMBER						DAY OF MONTH	DAY OF WEEK	OCTOBER					
		HIGH				LOW				HIGH				LOW	
		a.m.	Ht.	**p.m.**	Ht.	a.m.	**p.m.**			a.m.	Ht.	**p.m.**	Ht.	a.m.	**p.m.**
1	W	12 11	3.8	**12 43**	4.2	5 36	**6 20**	1	F	12 53	3.5	**1 25**	4.2	6 14	**7 16**
2	T	1 09	3.5	**1 44**	4.2	6 30	**7 26**	2	S	1 57	3.4	**2 30**	4.0	7 21	**8 38**
3	F	2 13	3.4	**2 48**	4.1	7 32	**8 43**	3	S	3 04	3.4	**3 35**	4.0	8 42	**10 05**
4	S	3 18	3.4	**3 52**	4.2	8 45	**10 08**	4	M	4 08	3.5	**4 37**	4.0	10 11	**11 13**
5	S	4 23	3.5	**4 53**	4.3	10 05	**11 21**	5	T	5 06	3.7	**5 33**	4.1	11 24	**...**
6	M	5 22	3.7	**5 50**	4.4	11 19	**...**	6	W	5 59	4.0	**6 23**	4.2	12 05	**12 21**
7	T	6 17	4.0	**6 42**	4.5	12 18	**12 21**	7	T	6 47	4.2	**7 09**	4.2	12 46	**1 06**
8	W	7 07	4.2	**7 31**	4.6	1 05	**1 13**	8	F	7 31	4.3	**7 52**	4.2	1 22	**1 46**
9	T	7 55	4.4	**8 17**	4.5	1 46	**1 58**	9	S	8 14	4.4	**8 35**	4.1	1 53	**2 20**
10	F	8 39	4.5	**9 00**	4.4	2 22	**2 39**	10	S	8 54	4.4	**9 14**	3.9	2 23	**2 52**
11	S	9 22	4.4	**9 42**	4.2	2 56	**3 17**	11	M	9 35	4.3	**9 54**	3.7	2 51	**3 23**
12	S	10 05	4.3	**10 24**	3.9	3 28	**3 52**	12	T	10 15	4.1	**10 34**	3.5	3 21	**3 53**
13	M	10 47	4.1	**11 06**	3.6	3 58	**4 26**	13	W	10 57	3.9	**11 17**	3.2	3 52	**4 26**
14	T	11 31	3.9	**11 50**	3.3	4 30	**5 01**	14	T	11 40	3.6	**...**	...	4 25	**5 02**
15	W	...	...	**12 18**	3.6	5 03	**5 39**	15	F	12 03	3.0	**12 28**	3.4	5 03	**5 44**
16	T	12 38	3.1	**1 08**	3.4	5 41	**6 22**	16	S	12 56	2.8	**1 21**	3.2	5 47	**6 33**
17	F	1 31	2.9	**2 03**	3.3	6 25	**7 14**	17	S	1 54	2.7	**2 19**	3.1	6 39	**7 31**
18	S	2 30	2.8	**3 01**	3.3	7 18	**8 19**	18	M	2 54	2.8	**3 18**	3.2	7 42	**8 38**
19	S	3 30	2.8	**3 58**	3.3	8 22	**9 33**	19	T	3 52	2.9	**4 14**	3.3	8 53	**9 44**
20	M	4 27	3.0	**4 51**	3.5	9 32	**10 39**	20	W	4 45	3.2	**5 06**	3.6	10 04	**10 42**
21	T	5 19	3.2	**5 40**	3.7	10 39	**11 30**	21	T	5 34	3.6	**5 54**	3.8	11 06	**11 33**
22	W	6 07	3.5	**6 27**	4.0	11 36	**...**	22	F	6 20	4.0	**6 41**	4.0	...	**12 01**
23	T	6 51	3.9	**7 11**	4.2	12 13	**12 26**	23	S	7 05	4.4	**7 27**	4.2	12 20	**12 51**
24	F	7 34	4.2	**7 54**	4.3	12 55	**1 13**	24	S	7 50	4.7	**8 13**	4.3	1 05	**1 39**
25	S	8 17	4.4	**8 38**	4.4	1 35	**1 58**	25	M	8 37	4.9	**9 00**	4.3	1 50	**2 27**
26	S	9 01	4.6	**9 23**	4.4	2 16	**2 44**	26	T	9 24	4.9	**9 49**	4.2	2 35	**3 16**
27	M	9 47	4.7	**10 09**	4.2	2 58	**3 30**	27	W	10 14	4.9	**10 41**	4.0	3 22	**4 06**
28	T	10 35	4.7	**10 59**	4.0	3 42	**4 19**	28	T	11 07	4.7	**11 36**	3.8	4 11	**4 59**
29	W	11 26	4.5	**11 53**	3.8	4 28	**5 11**	29	F	...	...	**12 04**	4.4	5 04	**5 58**
30	T	...	...	**12 23**	4.4	5 18	**6 08**	30	S	12 36	3.6	**1 06**	4.1	6 03	**7 05**
								31	S	1 40	3.4	***1 10**	3.9	*6 13	***7 25**

***Standard Time starts**

Dates when Ht. of **Low** Water is below Mean Low with Ht. of lowest given for each period and Date of lowest in ( ):

8th - 10th: -0.2'
24th - 28th: -0.5' (25th - 28th)
22nd - 29th: -0.8' (26th)

Average Rise and Fall 3.5 ft.

When a high tide exceeds av. ht., the *following* low tide will be lower than av.

1999 HIGH & LOW WATER
NEWPORT, R.I.
41–30N x 71–20W

NOVEMBER (Standard Time)

DAY OF MONTH	DAY OF WEEK	HIGH a.m.	Ht.	HIGH p.m.	Ht.	LOW a.m.	LOW p.m.
1	M	1 45	3.4	**2 14**	3.8	7 40	**8 47**
2	T	2 48	3.5	**3 15**	3.7	9 11	**9 52**
3	W	3 45	3.7	**4 10**	3.7	10 20	**10 42**
4	T	4 37	3.9	**4 59**	3.8	11 12	**11 21**
5	F	5 24	4.1	**5 45**	3.8	11 54	**11 54**
6	S	6 07	4.2	**6 27**	3.8	...	**12 29**
7	S	6 48	4.3	**7 07**	3.7	12 22	**1 00**
8	M	7 28	4.3	**7 47**	3.6	12 50	**1 28**
9	T	8 08	4.2	**8 27**	3.5	1 18	**1 5**
10	W	8 47	4.0	**9 07**	3.3	1 49	**2 27**
11	T	9 27	3.8	**9 49**	3.1	2 21	**2 59**
12	F	10 09	3.6	**10 34**	3.0	2 56	**3 35**
13	S	10 54	3.4	**11 24**	2.8	3 34	**4 15**
14	S	11 44	3.2	**...**	...	4 18	**5 01**
15	M	12 18	2.7	**12 38**	3.1	5 08	**5 53**
16	T	1 16	2.8	**1 36**	3.1	6 08	**6 53**
17	W	2 13	3.0	**2 34**	3.2	7 16	**7 55**
18	T	3 08	3.3	**3 29**	3.4	8 27	**8 57**
19	F	4 00	3.7	**4 21**	3.6	9 34	**9 54**
20	S	4 50	4.1	**5 12**	3.8	10 34	**10 47**
21	S	5 38	4.5	**6 01**	4.0	11 29	**11 37**
22	M	6 27	4.8	**6 50**	4.1	...	**12 22**
23	T	7 15	5.0	**7 40**	4.2	12 27	**1 12**
24	W	8 05	5.0	**8 31**	4.1	1 16	**2 03**
25	T	8 56	4.9	**9 24**	4.0	2 06	**2 55**
26	F	9 50	4.7	**10 19**	3.8	2 58	**3 48**
27	S	10 46	4.4	**11 17**	3.6	3 52	**4 45**
28	S	11 44	4.1	**...**	...	4 52	**5 47**
29	M	12 18	3.5	**12 45**	3.8	6 00	**6 57**
30	T	1 20	3.4	**1 47**	3.5	7 24	**8 11**

DECEMBER (Standard Time)

DAY OF MONTH	DAY OF WEEK	HIGH a.m.	Ht.	HIGH p.m.	Ht.	LOW a.m.	LOW p.m.
1	W	2 21	3.5	**2 46**	3.4	8 52	**9 17**
2	T	3 18	3.6	**3 41**	3.3	10 02	**10 08**
3	F	4 10	3.7	**4 31**	3.3	10 55	**10 49**
4	S	4 57	3.8	**5 17**	3.4	11 37	**11 22**
5	S	5 41	4.0	**6 00**	3.4	-A-	**12 11**
6	M	6 22	4.0	**6 41**	3.4	...	**12 39**
7	T	7 03	4.1	**7 22**	3.4	12 21	**1 07**
8	W	7 42	4.0	**8 03**	3.4	12 51	**1 35**
9	T	8 23	4.0	**8 45**	3.3	1 24	**2 05**
10	F	9 02	3.8	**9 25**	3.1	1 58	**2 38**
11	S	9 43	3.6	**10 08**	3.0	2 34	**3 13**
12	S	10 26	3.5	**10 54**	2.9	3 13	**3 51**
13	M	11 11	3.3	**11 43**	2.9	3 56	**4 33**
14	T	...	...	**12 01**	3.1	4 44	**5 21**
15	W	12 37	2.9	**12 57**	3.1	5 40	**6 15**
16	T	1 34	3.1	**1 55**	3.1	6 44	**7 14**
17	F	2 31	3.3	**2 54**	3.2	7 53	**8 16**
18	S	3 27	3.7	**3 51**	3.4	9 04	**9 18**
19	S	4 22	4.1	**4 45**	3.6	10 10	**10 18**
20	M	5 14	4.4	**5 39**	3.8	11 11	**11 15**
21	T	6 06	4.7	**6 31**	4.0	...	**12 07**
22	W	6 57	4.9	**7 23**	4.1	12 09	**1 00**
23	T	7 48	5.0	**8 14**	4.1	1 02	**1 52**
24	F	8 40	4.9	**9 06**	4.1	1 55	**2 42**
25	S	9 32	4.6	**10 00**	3.9	2 47	**3 33**
26	S	10 25	4.3	**10 54**	3.8	3 41	**4 24**
27	M	11 20	4.0	**11 51**	3.6	4 37	**5 17**
28	T	...	...	**12 16**	3.6	5 37	**6 13**
29	W	12 49	3.5	**1 13**	3.3	6 47	**7 14**
30	T	1 47	3.4	**2 11**	3.1	8 09	**8 18**
31	F	2 44	3.4	**3 06**	3.0	9 28	**9 18**

A also at 11:51 p.m.

Dates when Ht. of **Low** Water is below Mean Low with Ht. of lowest given for each period and Date of lowest in ( ):

20th - 27th: -0.9' (23rd - 25th)   19th - 27th: -1.0' (23rd)

Average Rise and Fall 3.5 ft.

When a high tide exceeds av. ht., the *following* low tide will be lower than av.

1999 CURRENT TABLE
THE RACE, L.I. SOUND

(Near Valiant Rock – 41-14-12N x 72-03-36W)

Standard Time

JANUARY

DAY OF MONTH	DAY OF WEEK	NORTHWEST Flood Starts a.m.	NORTHWEST Flood Starts **p.m.**	Kts.	SOUTHEAST Ebb Starts a.m.	SOUTHEAST Ebb Starts **p.m.**	Kts.
1	F	4 13	**4 57**	a3.6	10 27	**11 04**	a4.2
2	S	5 05	**5 46**	a3.6	11 17	**11 54**	a4.2
3	S	5 55	**6 33**	a3.5	...	**12 06**	4.0
4	M	6 45	**7 20**	a3.3	12 42	**12 54**	p3.8
5	T	7 36	**8 07**	3.0	1 31	**1 42**	p3.5
6	W	8 27	**8 54**	2.7	2 19	**2 31**	p3.2
7	T	9 20	**9 43**	p2.5	3 09	**3 22**	2.8
8	F	10 16	**10 34**	p2.3	3 59	**4 16**	a2.6
9	S	11 14	**11 27**	p2.1	4 52	**5 13**	a2.5
10	S	...	**12 13**	1.8	5 46	**6 12**	a2.4
11	M	12 21	**1 10**	a2.1	6 40	**7 10**	a2.5
12	T	1 14	**2 03**	a2.1	7 31	**8 04**	a2.6
13	W	2 05	**2 52**	a2.2	8 20	**8 53**	a2.8
14	T	2 52	**3 37**	a2.4	9 05	**9 39**	a3.0
15	F	3 36	**4 18**	a2.6	9 47	**10 21**	a3.1
16	S	4 18	**4 57**	2.7	10 27	**11 01**	a3.3
17	S	4 58	**5 35**	2.9	11 06	**11 40**	a3.4
18	M	5 38	**6 12**	3.0	11 45	**...**	3.5
19	T	6 19	**6 51**	3.1	12 18	**12 25**	p3.6
20	W	7 03	**7 32**	3.1	12 58	**1 07**	p3.5
21	T	7 50	**8 16**	p3.1	1 41	**1 53**	p3.4
22	F	8 42	**9 04**	p3.1	2 27	**2 43**	3.2
23	S	9 39	**9 58**	p3.0	3 19	**3 40**	a3.2
24	S	10 42	**10 57**	p2.9	4 16	**4 44**	a3.2
25	M	11 49	**...**	2.5	5 18	**5 52**	a3.2
26	T	12 01	**12 56**	a2.9	6 23	**7 00**	a3.3
27	W	1 06	**2 00**	a2.9	7 27	**8 05**	a3.4
28	T	2 09	**2 59**	a3.1	8 27	**9 05**	a3.6
29	F	3 08	**3 53**	a3.2	9 23	**10 00**	a3.8
30	S	4 02	**4 43**	a3.3	10 16	**10 50**	a3.9
31	S	4 53	**5 30**	3.3	11 04	**11 37**	a3.9

Standard Time

FEBRUARY

DAY OF MONTH	DAY OF WEEK	NORTHWEST Flood Starts a.m.	NORTHWEST Flood Starts **p.m.**	Kts.	SOUTHEAST Ebb Starts a.m.	SOUTHEAST Ebb Starts **p.m.**	Kts.
1	M	5 41	**6 14**	a3.3	11 50	**...**	3.8
2	T	6 27	**6 56**	a3.2	12 22	**12 34**	p3.7
3	W	7 12	**7 37**	3.0	1 05	**1 17**	p3.4
4	T	7 57	**8 18**	p2.8	1 47	**2 00**	3.1
5	F	8 43	**9 00**	2.5	2 30	**2 44**	a2.9
6	S	9 32	**9 46**	p2.3	3 14	**3 32**	a2.7
7	S	10 26	**10 36**	p2.1	4 01	**4 24**	a2.5
8	M	11 24	**11 30**	p2.0	4 54	**5 23**	a2.4
9	T	...	**12 24**	1.7	5 50	**6 24**	a2.3
10	W	12 28	**1 22**	a2.0	6 47	**7 24**	a2.4
11	T	1 25	**2 16**	a2.1	7 41	**8 18**	a2.6
12	F	2 18	**3 04**	2.2	8 32	**9 07**	a2.8
13	S	3 07	**3 48**	2.5	9 18	**9 52**	a3.1
14	S	3 52	**4 28**	2.8	10 02	**10 33**	a3.3
15	M	4 35	**5 07**	p3.1	10 43	**11 13**	a3.6
16	T	5 17	**5 46**	p3.3	11 24	**11 52**	a3.7
17	W	6 00	**6 25**	p3.5	...	**12 06**	3.8
18	T	6 44	**7 07**	p3.5	12 33	**12 50**	p3.8
19	F	7 32	**7 52**	p3.5	1 17	**1 36**	a3.7
20	S	8 24	**8 41**	p3.3	2 04	**2 27**	a3.6
21	S	9 20	**9 35**	p3.1	2 55	**3 23**	a3.5
22	M	10 23	**10 36**	p2.9	3 53	**4 26**	a3.3
23	T	11 30	**11 44**	p2.7	4 57	**5 35**	a3.2
24	W	...	**12 39**	2.4	6 05	**6 46**	a3.1
25	T	12 53	**1 45**	a2.7	7 12	**7 52**	a3.2
26	F	1 59	**2 45**	a2.8	8 15	**8 52**	a3.3
27	S	2 58	**3 38**	2.9	9 12	**9 45**	a3.5
28	S	3 52	**4 26**	3.1	10 03	**10 33**	a3.6

The Kts. (knots) columns show the **maximum** predicted velocities of the stronger one of the Flood Currents and the stronger one of the Ebb Currents for each day. The letter "a" means the velocity shown should occur **after** the **a.m.** Current Change. The letter "p" means the velocity shown should occur **after** the **p.m.** Current Change (even if next morning). No "a" or "p" means a.m. and p.m. velocities are the same for that day.

Av. Max. Vel.: Fl. 2.9 Kts., Ebb 3.5 Kts.

Max. Flood 2 hrs. 45 min. after Flood Starts, ±15 min.

Max. Ebb 3 hrs. 25 min. after Ebb Starts, ±15 min.

Use THE RACE tables with current charts on pages 78 through 83.

See pp. 150-157 for Current Change at other points.

1999 CURRENT TABLE
THE RACE, L.I. SOUND

(Near Valiant Rock – 41-14-12N x 72-03-36W)

Standard Time — **Daylight Time starts Apr. 4 at 2 a.m.**

MARCH

DAY OF MONTH	DAY OF WEEK	CURRENT TURNS NORTHWEST Flood Starts a.m.	**p.m.**	Kts.	SOUTHEAST Ebb Starts a.m.	**p.m.**	Kts.
1	M	4 41	**5 09**	3.1	10 49	**11 17**	a3.6
2	T	5 25	**5 50**	3.1	11 32	**11 57**	a3.6
3	W	6 07	**6 28**	3.0	...	**12 13**	3.5
4	T	6 47	**7 05**	2.9	12 36	**12 52**	3.3
5	F	7 27	**7 41**	p2.8	1 13	**1 30**	a3.2
6	S	8 08	**8 20**	2.5	1 50	**2 10**	a3.0
7	S	8 52	**9 01**	2.3	2 30	**2 52**	a2.8
8	M	9 40	**9 49**	p2.1	3 13	**3 41**	a2.6
9	T	10 35	**10 43**	p1.9	4 02	**4 36**	a2.4
10	W	11 35	**11 44**	p1.9	4 58	**5 38**	a2.3
11	T	...	**12 37**	1.8	5 59	**6 42**	a2.3
12	F	12 46	**1 34**	a2.0	7 00	**7 40**	a2.5
13	S	1 44	**2 26**	2.2	7 56	**8 31**	a2.7
14	S	2 36	**3 12**	p2.6	8 46	**9 18**	a3.0
15	M	3 24	**3 55**	p3.0	9 33	**10 01**	3.3
16	T	4 10	**4 36**	p3.3	10 18	**10 43**	3.6
17	W	4 55	**5 17**	p3.6	11 02	**11 25**	p3.9
18	T	5 40	**5 59**	p3.7	11 46	**...**	3.9
19	F	6 26	**6 43**	p3.8	12 08	**12 32**	a4.1
20	S	7 14	**7 29**	p3.7	12 53	**1 20**	a4.1
21	S	8 06	**8 20**	3.4	1 42	**2 12**	a4.0
22	M	9 03	**9 17**	3.1	2 35	**3 09**	a3.7
23	T	10 05	**10 21**	2.8	3 34	**4 13**	a3.4
24	W	11 12	**11 31**	p2.6	4 39	**5 22**	a3.2
25	T	...	**12 21**	2.4	5 48	**6 32**	a3.0
26	F	12 42	**1 26**	2.5	6 57	**7 37**	a3.0
27	S	1 48	**2 25**	2.6	8 00	**8 35**	a3.1
28	S	2 46	**3 17**	p2.8	8 56	**9 26**	a3.2
29	M	3 38	**4 03**	p2.9	9 46	**10 11**	3.3
30	T	4 24	**4 44**	p3.0	10 30	**10 52**	3.3
31	W	5 06	**5 22**	p3.0	11 11	**11 29**	p3.4

APRIL

DAY OF MONTH	DAY OF WEEK	CURRENT TURNS NORTHWEST Flood Starts a.m.	**p.m.**	Kts.	SOUTHEAST Ebb Starts a.m.	**p.m.**	Kts.
1	T	5 45	**5 58**	2.9	11 49	**...**	3.2
2	F	6 22	**6 32**	2.8	12 05	**12 26**	a3.3
3	S	6 59	**7 07**	2.7	12 39	**1 02**	a3.2
4	S	*8 37	***8 44**	2.5	1 14	***2 40**	a3.1
5	M	9 18	**9 24**	2.3	2 50	**3 20**	a2.9
6	T	10 03	**10 10**	2.1	3 31	**4 05**	a2.7
7	W	10 54	**11 03**	2.0	4 17	**4 58**	a2.5
8	T	11 50	**...**	1.9	5 12	**5 57**	a2.3
9	F	12 03	**12 51**	1.9	6 13	**7 00**	a2.3
10	S	1 07	**1 49**	p2.1	7 17	**7 59**	a2.5
11	S	2 09	**2 43**	p2.4	8 17	**8 52**	2.7
12	M	3 04	**3 32**	p2.7	9 12	**9 41**	p3.1
13	T	3 56	**4 18**	p3.1	10 03	**10 27**	p3.6
14	W	4 44	**5 03**	p3.5	10 51	**11 12**	p4.0
15	T	5 32	**5 47**	p3.8	11 39	**11 58**	p4.2
16	F	6 19	**6 33**	p3.9	...	**12 26**	3.9
17	S	7 08	**7 20**	p3.9	12 44	**1 14**	a4.4
18	S	7 58	**8 10**	3.7	1 32	**2 05**	a4.3
19	M	8 50	**9 03**	a3.5	2 22	**2 58**	a4.1
20	T	9 47	**10 02**	a3.2	3 17	**3 56**	a3.8
21	W	10 48	**11 07**	a2.8	4 16	**4 59**	a3.5
22	T	11 53	**...**	2.6	5 21	**6 06**	a3.2
23	F	12 17	**12 58**	2.5	6 29	**7 13**	a3.0
24	S	1 26	**2 01**	p2.5	7 36	**8 15**	a2.9
25	S	2 31	**2 58**	p2.6	8 38	**9 11**	2.9
26	M	3 28	**3 49**	p2.7	9 33	**10 00**	p3.1
27	T	4 18	**4 34**	p2.8	10 23	**10 43**	p3.2
28	W	5 03	**5 14**	p2.8	11 07	**11 23**	p3.3
29	T	5 43	**5 52**	p2.8	11 47	**11 59**	p3.3
30	F	6 21	**6 27**	p2.8	...	**12 25**	2.9

***Daylight Time starts**

The Kts. (knots) columns show the **maximum** predicted velocities of the stronger one of the Flood Currents and the stronger one of the Ebb Currents for each day. The letter "a" means the velocity shown should occur **after** the **a.m.** Current Change. The letter "p" means the velocity shown should occur **after** the **p.m.** Current Change (even if next morning). No "a" or "p" means a.m. and p.m. velocities are the same for that day.

Av. Max. Vel.: Fl. 2.9 Kts., Ebb 3.5 Kts.

Max. Flood 2 hrs. 45 min. after Flood Starts, ±15 min.

Max. Ebb 3 hrs. 25 min. after Ebb Starts, ±15 min.

Use THE RACE tables with current charts on pages 78 through 83.

See pp. 150-157 for Current Change at other points.

1999 CURRENT TABLE
THE RACE, L.I. SOUND

(Near Valiant Rock – 41-14-12N x 72-03-36W)

MAY (Daylight Time)		CURRENT TURNS						JUNE (Daylight Time)		CURRENT TURNS					
		NORTHWEST Flood Starts			SOUTHEAST Ebb Starts					NORTHWEST Flood Starts			SOUTHEAST Ebb Starts		
DAY OF MONTH	DAY OF WEEK	a.m.	**p.m.**	Kts.	a.m.	**p.m.**	Kts.	DAY OF MONTH	DAY OF WEEK	a.m.	**p.m.**	Kts.	a.m.	**p.m.**	Kts.
1	S	6 57	**7 02**	2.7	12 34	**1 02**	a3.3	1	T	7 47	**7 48**	2.5	1 16	**1 52**	a3.1
2	S	7 33	**7 37**	2.6	1 08	**1 38**	a3.2	2	W	8 25	**8 29**	a2.5	1 53	**2 31**	a3.0
3	M	8 10	**8 14**	2.5	1 42	**2 15**	a3.1	3	T	9 06	**9 13**	a2.4	2 32	**3 13**	a2.9
4	T	8 50	**8 54**	2.4	2 18	**2 54**	a2.9	4	F	9 50	**10 04**	a2.4	3 15	**3 59**	a2.8
5	W	9 32	**9 39**	a2.3	2 58	**3 38**	a2.8	5	S	10 37	**11 00**	a2.4	4 04	**4 50**	a2.7
6	T	10 19	**10 31**	2.1	3 42	**4 27**	a2.6	6	S	11 29	**...**	2.4	4 59	**5 45**	a2.7
7	F	11 12	**11 29**	a2.1	4 34	**5 22**	a2.5	7	M	12 01	**12 24**	p2.6	6 00	**6 42**	p2.8
8	S	...	**12 07**	2.1	5 33	**6 20**	a2.5	8	T	1 03	**1 20**	p2.8	7 04	**7 39**	p3.1
9	S	12 32	**1 04**	p2.3	6 36	**7 19**	a2.6	9	W	2 04	**2 16**	p3.0	8 06	**8 34**	p3.5
10	M	1 34	**2 00**	p2.6	7 38	**8 14**	p2.9	10	T	3 03	**3 11**	p3.3	9 06	**9 28**	p3.8
11	T	2 33	**2 52**	p2.9	8 38	**9 06**	p3.3	11	F	3 58	**4 05**	p3.6	10 03	**10 21**	p4.1
12	W	3 28	**3 42**	p3.3	9 33	**9 56**	p3.8	12	S	4 52	**4 58**	p3.7	10 57	**11 13**	p4.3
13	T	4 20	**4 32**	p3.6	10 26	**10 45**	p4.1	13	S	5 44	**5 51**	p3.8	11 50	**...**	3.7
14	F	5 10	**5 20**	p3.8	11 17	**11 33**	p4.4	14	M	6 35	**6 44**	p3.8	12 05	**12 42**	a4.4
15	S	6 00	**6 10**	p3.9	...	**12 07**	3.9	15	T	7 26	**7 37**	3.6	12 57	**1 35**	a4.3
16	S	6 51	**7 00**	p3.9	12 22	**12 58**	a4.5	16	W	8 18	**8 32**	a3.4	1 49	**2 28**	a4.1
17	M	7 42	**7 53**	3.7	1 13	**1 50**	a4.4	17	T	9 10	**9 29**	a3.2	2 42	**3 23**	a3.8
18	T	8 35	**8 48**	a3.5	2 05	**2 44**	a4.2	18	F	10 04	**10 29**	a3.0	3 37	**4 18**	a3.5
19	W	9 30	**9 47**	a3.2	3 00	**3 42**	a3.9	19	S	10 58	**11 30**	a2.7	4 34	**5 15**	a3.2
20	T	10 28	**10 51**	a2.9	3 58	**4 42**	a3.5	20	S	11 54	**...**	2.5	5 33	**6 13**	a2.8
21	F	11 28	**11 57**	a2.7	5 00	**5 44**	a3.2	21	M	12 32	**12 50**	p2.4	6 33	**7 09**	p2.7
22	S	...	**12 29**	2.5	6 04	**6 47**	a2.9	22	T	1 32	**1 44**	p2.3	7 33	**8 02**	p2.7
23	S	1 03	**1 29**	p2.5	7 08	**7 45**	a2.8	23	W	2 29	**2 36**	p2.3	8 29	**8 52**	p2.8
24	M	2 06	**2 24**	p2.5	8 09	**8 39**	p2.8	24	T	3 20	**3 24**	p2.4	9 21	**9 38**	p2.9
25	T	3 02	**3 14**	p2.5	9 04	**9 28**	p2.9	25	F	4 07	**4 09**	p2.5	10 09	**10 21**	p3.0
26	W	3 52	**4 00**	p2.6	9 54	**10 12**	p3.1	26	S	4 51	**4 51**	p2.5	10 53	**11 01**	p3.1
27	T	4 37	**4 42**	p2.6	10 39	**10 52**	p3.2	27	S	5 31	**5 31**	p2.6	11 35	**11 40**	p3.2
28	F	5 18	**5 21**	p2.7	11 21	**11 30**	p3.2	28	M	6 10	**6 09**	p2.6	...	**12 14**	2.6
29	S	5 57	**5 58**	p2.7	...	**12 01**	2.7	29	T	6 47	**6 47**	p2.7	12 17	**12 52**	a3.2
30	S	6 34	**6 34**	p2.7	12 06	**12 38**	a3.2	30	W	7 24	**7 26**	p2.7	12 54	**1 30**	a3.2
31	M	7 10	**7 11**	2.6	12 41	**1 15**	a3.2								

The Kts. (knots) columns show the **maximum** predicted velocities of the stronger one of the Flood Currents and the stronger one of the Ebb Currents for each day. The letter "a" means the velocity shown should occur **after** the **a.m.** Current Change. The letter "p" means the velocity shown should occur **after** the **p.m.** Current Change (even if next morning). No "a" or "p" means a.m. and p.m. velocities are the same for that day.

Av. Max. Vel.: Fl. 2.9 Kts., Ebb 3.5 Kts.

Max. Flood 2 hrs. 45 min. after Flood Starts, ±15 min.

Max. Ebb 3 hrs. 25 min. after Ebb Starts, ±15 min.

Use THE RACE tables with current charts on pages 78 through 83.

See pp. 150-157 for Current Change at other points.

1999 CURRENT TABLE
THE RACE, L.I. SOUND

(Near Valiant Rock – 41-14-12N x 72-03-36W)

Daylight Time

JULY

DAY OF MONTH	DAY OF WEEK	CURRENT TURNS NORTHWEST Flood Starts a.m.	**p.m.**	Kts.	CURRENT TURNS SOUTHEAST Ebb Starts a.m.	**p.m.**	Kts.
1	T	8 01	**8 07**	a2.7	1 31	**2 08**	a3.2
2	F	8 40	**8 51**	a2.7	2 10	**2 48**	a3.1
3	S	9 21	**9 40**	a2.7	2 52	**3 31**	a3.1
4	S	10 06	**10 34**	a2.7	3 39	**4 19**	a3.0
5	M	10 56	**11 33**	a2.7	4 32	**5 12**	a2.9
6	T	11 50	**...**	2.8	5 32	**6 09**	p3.0
7	W	12 36	**12 48**	p2.9	6 36	**7 09**	p3.2
8	T	1 39	**1 48**	p3.0	7 41	**8 09**	p3.5
9	F	2 41	**2 48**	p3.2	8 44	**9 07**	p3.7
10	S	3 40	**3 46**	p3.4	9 44	**10 04**	p4.0
11	S	4 36	**4 43**	p3.6	10 41	**10 58**	p4.1
12	M	5 29	**5 37**	p3.6	11 35	**11 51**	p4.2
13	T	6 20	**6 30**	p3.6	...	**12 27**	3.6
14	W	7 09	**7 22**	3.5	12 42	**1 18**	a4.2
15	T	7 58	**8 14**	a3.4	1 32	**2 08**	a4.0
16	F	8 46	**9 06**	a3.2	2 22	**2 58**	a3.7
17	S	9 34	**10 00**	a3.0	3 12	**3 48**	a3.4
18	S	10 23	**10 55**	a2.7	4 03	**4 39**	a3.1
19	M	11 14	**11 53**	a2.5	4 57	**5 32**	2.7
20	T	...	**12 07**	2.3	5 53	**6 26**	2.5
21	W	12 52	**1 01**	p2.2	6 52	**7 20**	p2.5
22	T	1 50	**1 55**	p2.1	7 50	**8 13**	p2.6
23	F	2 44	**2 47**	p2.2	8 45	**9 03**	p2.7
24	S	3 35	**3 36**	p2.3	9 37	**9 50**	p2.8
25	S	4 21	**4 22**	p2.4	10 24	**10 33**	p3.0
26	M	5 04	**5 05**	p2.6	11 07	**11 14**	p3.2
27	T	5 43	**5 45**	p2.7	11 47	**11 53**	p3.3
28	W	6 21	**6 24**	p2.8	...	**12 26**	2.8
29	T	6 58	**7 04**	p2.9	12 31	**1 03**	a3.4
30	F	7 34	**7 45**	2.9	1 09	**1 41**	a3.4
31	S	8 12	**8 29**	a3.0	1 49	**2 20**	a3.4

Daylight Time

AUGUST

DAY OF MONTH	DAY OF WEEK	CURRENT TURNS NORTHWEST Flood Starts a.m.	**p.m.**	Kts.	CURRENT TURNS SOUTHEAST Ebb Starts a.m.	**p.m.**	Kts.
1	S	8 53	**9 17**	a3.0	2 31	**3 03**	a3.3
2	M	9 37	**10 10**	a3.0	3 18	**3 50**	a3.2
3	T	10 27	**11 09**	a3.0	4 10	**4 43**	p3.1
4	W	11 23	**...**	2.9	5 09	**5 42**	p3.1
5	T	12 13	**12 24**	p2.9	6 14	**6 46**	p3.2
6	F	1 19	**1 29**	p2.9	7 22	**7 50**	p3.3
7	S	2 24	**2 33**	p3.0	8 29	**8 53**	p3.5
8	S	3 25	**3 35**	p3.2	9 31	**9 52**	p3.8
9	M	4 22	**4 32**	p3.3	10 28	**10 47**	p3.9
10	T	5 14	**5 26**	p3.4	11 21	**11 38**	p4.0
11	W	6 03	**6 17**	3.4	...	**12 11**	3.6
12	T	6 49	**7 05**	3.4	12 27	**12 58**	a4.0
13	F	7 34	**7 53**	a3.3	1 13	**1 43**	a3.8
14	S	8 17	**8 39**	a3.1	1 59	**2 28**	a3.6
15	S	9 00	**9 27**	a2.9	2 44	**3 12**	a3.3
16	M	9 44	**10 17**	a2.7	3 30	**3 58**	2.9
17	T	10 31	**11 10**	a2.4	4 18	**4 46**	2.6
18	W	11 21	**...**	2.2	5 11	**5 38**	p2.5
19	T	12 07	**12 16**	p2.0	6 08	**6 34**	p2.4
20	F	1 07	**1 14**	p2.0	7 08	**7 31**	p2.4
21	S	2 05	**2 11**	p2.0	8 08	**8 26**	p2.5
22	S	2 59	**3 04**	p2.2	9 02	**9 17**	p2.7
23	M	3 48	**3 53**	p2.4	9 52	**10 04**	p2.9
24	T	4 32	**4 38**	p2.6	10 36	**10 47**	p3.2
25	W	5 12	**5 20**	p2.8	11 17	**11 28**	p3.4
26	T	5 50	**6 00**	p3.0	11 55	**...**	3.1
27	F	6 27	**6 41**	p3.2	12 07	**12 33**	a3.5
28	S	7 04	**7 23**	a3.3	12 47	**1 12**	a3.6
29	S	7 44	**8 08**	a3.3	1 28	**1 52**	3.6
30	M	8 26	**8 56**	a3.3	2 11	**2 36**	3.5
31	T	9 12	**9 49**	a3.2	2 59	**3 25**	p3.5

The Kts. (knots) columns show the **maximum** predicted velocities of the stronger one of the Flood Currents and the stronger one of the Ebb Currents for each day. The letter "a" means the velocity shown should occur **after** the **a.m.** Current Change. The letter "p" means the velocity shown should occur **after** the **p.m.** Current Change (even if next morning). No "a" or "p" means a.m. and p.m. velocities are the same for that day.

Av. Max. Vel.: Fl. 2.9 Kts., Ebb 3.5 Kts.

Max. Flood 2 hrs. 45 min. after Flood Starts, ±15 min.

Max. Ebb 3 hrs. 25 min. after Ebb Starts, ±15 min.

Use THE RACE tables with current charts on pages 78 through 83.

See pp. 150-157 for Current Change at other points.

1999 CURRENT TABLE
THE RACE, L.I. SOUND

(Near Valiant Rock – 41-14-12N x 72-03-36W)

Daylight Time **Standard Time starts Oct. 31 at 2 a.m.**

SEPTEMBER

DAY OF MONTH	DAY OF WEEK	NORTHWEST Flood Starts a.m.	p.m.	Kts.	SOUTHEAST Ebb Starts a.m.	p.m.	Kts.
1	W	10 03	10 49	a3.1	3 52	4 19	p3.3
2	T	11 02	11 54	a2.9	4 52	5 21	p3.2
3	F	...	12 07	2.8	5 59	6 28	p3.2
4	S	1 03	1 17	p2.7	7 09	7 36	p3.2
5	S	2 09	2 24	p2.8	8 17	8 41	p3.4
6	M	3 11	3 27	p3.0	9 19	9 41	p3.5
7	T	4 07	4 23	p3.1	10 15	10 34	p3.7
8	W	4 57	5 14	p3.3	11 05	11 24	p3.7
9	T	5 43	6 02	3.3	11 51	...	3.6
10	F	6 26	6 46	a3.3	12 10	12 34	a3.7
11	S	7 07	7 29	a3.2	12 53	1 15	a3.6
12	S	7 46	8 11	a3.0	1 34	1 55	a3.4
13	M	8 25	8 53	a2.8	2 15	2 34	3.1
14	T	9 05	9 38	a2.6	2 56	3 15	p2.9
15	W	9 48	10 27	a2.3	3 41	3 59	p2.6
16	T	10 36	11 21	a2.1	4 29	4 49	p2.4
17	F	11 31	...	1.9	5 25	5 45	p2.3
18	S	12 21	12 32	p1.8	6 27	6 46	p2.3
19	S	1 22	1 34	p1.9	7 28	7 46	p2.4
20	M	2 19	2 31	p2.1	8 26	8 41	p2.6
21	T	3 09	3 23	p2.3	9 16	9 31	p2.9
22	W	3 55	4 09	p2.7	10 01	10 17	p3.2
23	T	4 36	4 53	p3.0	10 43	11 00	p3.4
24	F	5 16	5 35	p3.2	11 22	11 41	p3.6
25	S	5 55	6 18	3.4	...	12 02	3.8
26	S	6 34	7 01	a3.6	12 23	12 43	p3.9
27	M	7 16	7 48	a3.6	1 07	1 25	p4.0
28	T	8 01	8 37	a3.6	1 53	2 12	p3.9
29	W	8 50	9 31	a3.4	2 42	3 02	p3.7
30	T	9 44	10 31	a3.1	3 37	3 59	p3.5

OCTOBER

DAY OF MONTH	DAY OF WEEK	NORTHWEST Flood Starts a.m.	p.m.	Kts.	SOUTHEAST Ebb Starts a.m.	p.m.	Kts.
1	F	10 47	11 37	a2.9	4 39	5 03	p3.2
2	S	11 56	...	2.7	5 47	6 13	p3.1
3	S	12 46	1 08	p2.6	6 57	7 22	p3.1
4	M	1 52	2 16	2.6	8 04	8 28	p3.2
5	T	2 53	3 17	p2.8	9 04	9 27	p3.3
6	W	3 47	4 11	p3.0	9 57	10 19	p3.4
7	T	4 36	5 00	a3.1	10 45	11 07	3.4
8	F	5 20	5 44	3.1	11 28	11 50	a3.5
9	S	6 00	6 26	a3.1	...	12 08	3.5
10	S	6 38	7 05	a3.0	12 31	12 46	p3.5
11	M	7 15	7 43	a2.9	1 09	1 22	p3.3
12	T	7 51	8 22	a2.7	1 47	1 58	p3.1
13	W	8 29	9 03	a2.5	2 26	2 35	p2.9
14	T	9 10	9 48	a2.3	3 07	3 16	p2.7
15	F	9 56	10 39	a2.1	3 53	4 03	p2.5
16	S	10 50	11 36	a1.9	4 46	4 58	p2.3
17	S	11 51	...	1.8	5 45	5 59	p2.3
18	M	12 35	12 55	p1.9	6 47	7 02	p2.4
19	T	1 33	1 55	2.0	7 44	8 01	p2.6
20	W	2 25	2 49	2.3	8 36	8 55	p2.8
21	T	3 13	3 39	p2.7	9 23	9 44	p3.1
22	F	3 57	4 25	p3.1	10 07	10 30	3.4
23	S	4 40	5 10	p3.4	10 50	11 16	a3.8
24	S	5 23	5 55	3.6	11 33	...	4.1
25	M	6 06	6 41	a3.8	12 01	12 17	p4.3
26	T	6 51	7 30	a3.8	12 47	1 03	p4.3
27	W	7 40	8 21	a3.7	1 36	1 52	p4.2
28	T	8 32	9 15	a3.5	2 28	2 44	p3.9
29	F	9 29	10 15	a3.2	3 24	3 43	p3.6
30	S	10 33	11 19	a2.9	4 26	4 47	p3.3
31	S	*10 43	*11 25	2.6	*4 33	*4 55	p3.1

***Standard Time starts**

The Kts. (knots) columns show the **maximum** predicted velocities of the stronger one of the Flood Currents and the stronger one of the Ebb Currents for each day. The letter "a" means the velocity shown should occur **after** the **a.m.** Current Change. The letter "p" means the velocity shown should occur **after** the **p.m.** Current Change (even if next morning). No "a" or "p" means a.m. and p.m. velocities are the same for that day.

Av. Max. Vel.: Fl. 2.9 Kts., Ebb 3.5 Kts.

Max. Flood 2 hrs. 45 min. after Flood Starts, ±15 min.

Max. Ebb 3 hrs. 25 min. after Ebb Starts, ±15 min.

Use THE RACE tables with current charts on pages 78 through 83.

See pp. 150-157 for Current Change at other points.

1999 CURRENT TABLE
THE RACE, L.I. SOUND

(Near Valiant Rock – 41-14-12N x 72-03-36W)

Standard Time

NOVEMBER

DAY OF MONTH	DAY OF WEEK	NORTHWEST Flood Starts a.m.	NORTHWEST Flood Starts p.m.	NORTHWEST Flood Starts Kts.	SOUTHEAST Ebb Starts a.m.	SOUTHEAST Ebb Starts p.m.	SOUTHEAST Ebb Starts Kts.
1	M	11 54	...	2.5	5 41	6 04	p3.0
2	T	12 30	1 02	a2.6	6 45	7 09	p3.0
3	W	1 29	2 02	a2.7	7 43	8 07	3.0
4	T	2 22	2 55	a2.8	8 35	8 59	a3.2
5	F	3 10	3 42	a2.9	9 21	9 46	a3.3
6	S	3 53	4 25	a2.9	10 03	10 29	a3.4
7	S	4 33	5 04	a2.9	10 41	11 08	a3.4
8	M	5 10	5 42	a2.9	11 18	11 46	a3.4
9	T	5 46	6 19	a2.8	11 53	...	3.3
10	W	6 21	6 56	a2.7	12 23	12 27	p3.1
11	T	6 58	7 35	a2.5	1 01	1 04	p3.0
12	F	7 38	8 17	a2.3	1 40	1 42	p2.8
13	S	8 23	9 03	a2.2	2 23	2 26	p2.6
14	S	9 14	9 54	2.0	3 12	3 16	p2.4
15	M	10 12	10 48	p2.0	4 06	4 14	p2.4
16	T	11 14	11 44	p2.2	5 03	5 16	p2.4
17	W	...	12 16	2.1	6 00	6 18	p2.6
18	T	12 38	1 14	a2.4	6 54	7 16	p2.8
19	F	1 30	2 07	2.7	7 44	8 10	3.1
20	S	2 19	2 58	a3.1	8 33	9 02	a3.6
21	S	3 07	3 47	3.4	9 20	9 51	a3.9
22	M	3 54	4 35	a3.7	10 07	10 41	a4.2
23	T	4 42	5 24	a3.9	10 55	11 30	a4.4
24	W	5 31	6 14	a3.9	11 44	...	4.4
25	T	6 23	7 05	a3.8	12 21	12 35	p4.3
26	F	7 17	7 59	a3.6	1 14	1 29	p4.0
27	S	8 15	8 56	a3.2	2 10	2 26	p3.7
28	S	9 18	9 56	a2.9	3 10	3 28	p3.4
29	M	10 25	10 58	p2.7	4 13	4 33	p3.1
30	T	11 33	...	2.4	5 16	5 39	p2.9

Standard Time

DECEMBER

DAY OF MONTH	DAY OF WEEK	NORTHWEST Flood Starts a.m.	NORTHWEST Flood Starts p.m.	NORTHWEST Flood Starts Kts.	SOUTHEAST Ebb Starts a.m.	SOUTHEAST Ebb Starts p.m.	SOUTHEAST Ebb Starts Kts.
1	W	12 01	12 39	a2.6	6 18	6 42	2.8
2	T	12 58	1 38	a2.6	7 15	7 41	a2.9
3	F	1 52	2 32	a2.6	8 07	8 34	a3.0
4	S	2 40	3 19	a2.7	8 53	9 22	a3.1
5	S	3 25	4 03	a2.7	9 36	10 06	a3.2
6	M	4 06	4 42	a2.7	10 15	10 46	a3.3
7	T	4 44	5 20	a2.7	10 52	11 24	a3.3
8	W	5 21	5 57	a2.7	11 28	...	3.2
9	T	5 57	6 33	a2.6	12 02	12 03	p3.2
10	F	6 34	7 10	a2.6	12 39	12 39	p3.1
11	S	7 14	7 49	2.4	1 17	1 16	p2.9
12	S	7 56	8 31	p2.4	1 57	1 57	p2.8
13	M	8 44	9 16	p2.3	2 40	2 42	p2.7
14	T	9 37	10 05	p2.3	3 28	3 34	p2.6
15	W	10 36	10 58	p2.4	4 20	4 33	p2.5
16	T	11 37	11 53	p2.6	5 15	5 36	2.6
17	F	...	12 38	2.4	6 12	6 38	a2.8
18	S	12 49	1 37	a2.8	7 08	7 39	a3.2
19	S	1 44	2 33	a3.1	8 02	8 36	a3.6
20	M	2 38	3 26	a3.4	8 55	9 30	a3.9
21	T	3 31	4 18	a3.7	9 47	10 23	a4.2
22	W	4 24	5 08	a3.8	10 38	11 15	a4.4
23	T	5 16	5 59	a3.9	11 29	...	4.4
24	F	6 09	6 50	a3.8	12 07	12 21	p4.3
25	S	7 03	7 41	a3.6	12 59	1 14	p4.1
26	S	7 59	8 34	a3.3	1 53	2 08	p3.8
27	M	8 58	9 29	p3.0	2 48	3 05	p3.4
28	T	10 00	10 25	p2.7	3 45	4 04	3.0
29	W	11 03	11 23	p2.5	4 44	5 06	a2.9
30	T	...	12 07	2.1	5 43	6 08	a2.7
31	F	12 21	1 07	a2.4	6 40	7 08	a2.7

The Kts. (knots) columns show the **maximum** predicted velocities of the stronger one of the Flood Currents and the stronger one of the Ebb Currents for each day. The letter "a" means the velocity shown should occur **after** the **a.m.** Current Change. The letter "p" means the velocity shown should occur **after** the **p.m.** Current Change (even if next morning). No "a" or "p" means a.m. and p.m. velocities are the same for that day.

Av. Max. Vel.: Fl. 2.9 Kts., Ebb 3.5 Kts.

Max. Flood 2 hrs. 45 min. after Flood Starts, ±15 min.

Max. Ebb 3 hrs. 25 min. after Ebb Starts, ±15 min.

Use THE RACE tables with current charts on pages 78 through 83.

See pp. 150-157 for Current Change at other points.

HOLDING A FAIR TIDE BETWEEN EASTERN LONG ISLAND AND NANTUCKET

There is a curious current phenomenon which can be used to advantage by every vessel, and particularly the slower cruiser or auxiliary, in making the passage *either* way between eastern Long Island Sound, on the west, and Buzzards Bay, Vineyard and Nantucket Sounds on the east.

Note in the very simplified diagram below, that in L. Id. Sd., the Ebb Current flows to the *east,* and in Buzzards Bay, Vineyard and Nantucket Sounds the Ebb Current flows *west.* (Off Newport, these opposed Ebb Currents merge and flow *south.*) The reverse is also true: the Flood Current flows *west* through L.I. Sd. and *east* through Buzzards Bay, Vd. and Nan. Sds. (Half arrow indicates Ebb Current, whole, Flood.)

In making a *complete* passage through the area of the diagram, simply ride the favoring Ebb Current toward Newport from either direction and, pick up the favoring Flood Current in leaving the Newport area.

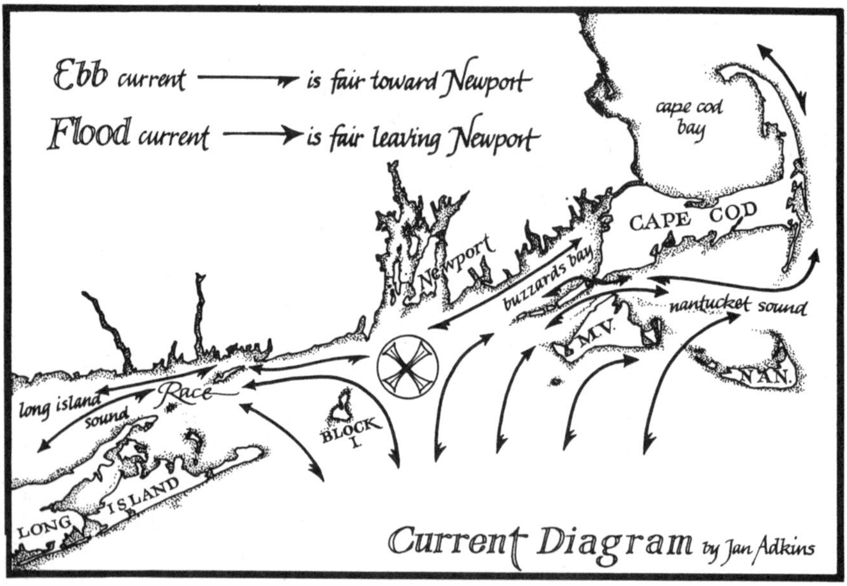

Arrive at "X" at the times shown for "Current Turns Northwest at The Race," tables, pp. 70-75.

The E-W currents between Pt. Judith and Cuttyhunk are only 1/2 to 1 kt., while those to the West of Pt. Judith and to the East of Cuttyhunk are much greater. Bearing this in mind, those making *only a partial trip* through the area may find it better even to buck a slight head tide in the Pt. Judith-Cuttyhunk area so as to pick up the maximum hours of strong favoring currents beyond those points.

For example, if headed for the Cape Cod Canal, refer to the Vineyard & Nantucket Sound Current Charts, pp. 50-61 and arrive just N. of Cuttyhunk as Flood Starts at Pollock Rip, pp. 44-49 to ensure the most favorable currents. If headed for Nantucket, refer to the same Charts and arrive just S. of Cuttyhunk at 3 hours after Flood Starts at Pollock Rip, pp. 44-49. If headed into Long Island Sound, refer to the L. I. Sound Current Charts, pp. 78-83 and arrive at Pt. Judith when Flood Current turns West at The Race, p. 81.

COPING WITH CURRENTS

When going directly with or against a current, our piloting problems are simple: there is no change in course and our speed over the bottom is easily figured, but we tend to guess a bit when the current is at some other angle. Where these currents are strong, as between N.Y. and Nantucket, it will be vital to figure the factors carefully, especially in haze or fog.

The Tables below tell, 1, how many degrees to change your course; 2, by what percent your speed is decreased (bold-face figure), with the current off the bow; 3, or by what percent it is increased, with the current off the stern.

First, estimate your boat's speed through the water. Then refer to the appropriate TIDAL CURRENT CHART (see pp. 50-61 or 78-83) and estimate the current's speed. Put these two in the form of a ratio, for example: boat speed is 8 kts, current 2 kts; ratio is 4 to 1.

Second, using the same CURRENT CHART, estimate the relative direction of the current to the nearest 15°. Example: your desired course is 60°, the current is from the East, or a relative angle of 30° on your stbd. bow.

Third, enter the Tables under the Ratio of 4.0; drop down to the 30° block of numbers (indicated in the left margin). The top figure in the block shows you must change your course 7°, always toward the current, and in this example, to 67°.

The bold-face figure below, 22%, is the amount by which your speed over the bottom will be decreased, i.e. from 8 kts down to 6¼.

Had the current been 30° off your *stern,* instead of your bow, you would apply the 3rd figure, 21%, adding it to your 8 kts, making your true speed about 9.7 kts.

RATIOS OF BOAT SPEED TO CURRENT SPEED

Relative Angle of Current		2	2½	3	3½	4	5	6	7	8	10	12	15	20
0° from	°	0	0	0	0	0	0	0	0	0	0	0	0	0
Bow	−%	**50**	**40**	**33**	**29**	**25**	**20**	**17**	**14**	**12**	**10**	**8.3**	**6.7**	**5.0**
Stern	+%	50	40	33	29	25	20	17	14	12	10	8.3	6.7	5.0
15° from	°	7.0	6.0	5.0	4.0	3.5	3.0	2.5	2.0	1.5	1.5	1.0	1.0	0.5
Bow	−%	**49**	**39**	**33**	**28**	**24**	**20**	**16**	**14**	**12**	**10**	**8.0**	**6.4**	**4.8**
Stern	+%	48	38	32	27	24	19	16	14	12	10	8.0	6.4	4.8
30° from	°	14	11	9.5	8.0	7.0	5.5	4.5	4.0	3.0	2.5	2.0	2.0	1.0
Bow	−%	**46**	**36**	**30**	**26**	**22**	**18**	**15**	**13**	**11**	**8.8**	**7.3**	**5.9**	**4.3**
Stern	+%	40	33	28	24	21	17	14	12	11	8.6	7.1	5.7	4.3
45° from	°	20	16	13	11	10	8.0	7.0	5.5	5.0	4.0	3.0	2.5	1.5
Bow	−%	**42**	**32**	**26**	**22**	**19**	**15**	**12**	**11**	**9.2**	**7.4**	**6.1**	**4.9**	**3.6**
Stern	+%	29	24	21	18	16	13	11	10	8.4	6.8	5.7	4.5	3.4
60° from	°	25	20	16	14	12	9.5	8.0	7.0	6.0	4.5	3.5	3.0	2.0
Bow	−%	**34**	**26**	**21**	**18**	**15**	**11**	**9.3**	**7.8**	**6.7**	**5.4**	**4.4**	**3.5**	**2.6**
Stern	+%	16	14	13	11	10	8.6	7.3	6.4	5.7	4.6	3.8	3.1	2.4
75° from	°	29	23	18	16	14	10	9.0	7.5	6.5	5.0	4.0	3.5	2.5
Bow	−%	**25**	**18**	**14**	**11**	**9.1**	**6.8**	**5.5**	**4.5**	**3.8**	**3.0**	**2.5**	**1.9**	**1.4**
Stern	+%	0.8	2.5	3.7	3.8	3.6	3.6	3.1	2.9	2.6	2.2	1.9	1.5	1.2
90°	°	30	24	19	17	14	11	9.5	8.0	7.0	5.5	4.5	3.5	2.5
Abeam	−%	**13**	**8.6**	**5.4**	**4.1**	**3.0**	**1.8**	**1.4**	**1.0**	**0.7**	**0.4**	**0.3**	**0.2**	**0.1**

Note: In general, while rounding a headland where head current is strong, hug the shore as far as safety will permit or go well out. (Current is usually apt to be strongest between these two points.)

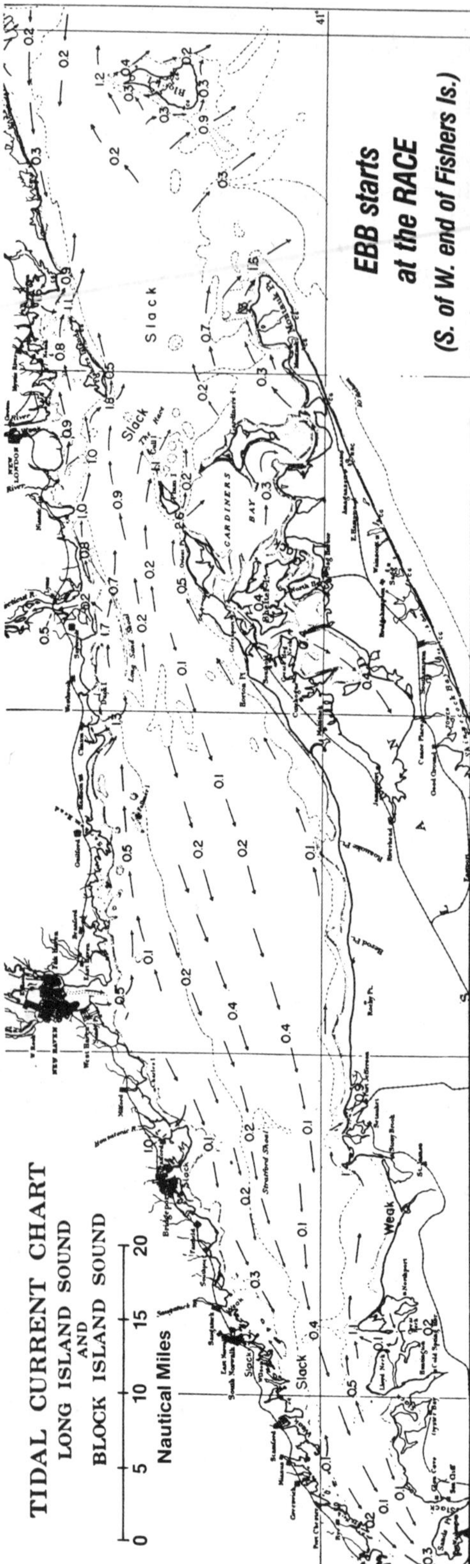

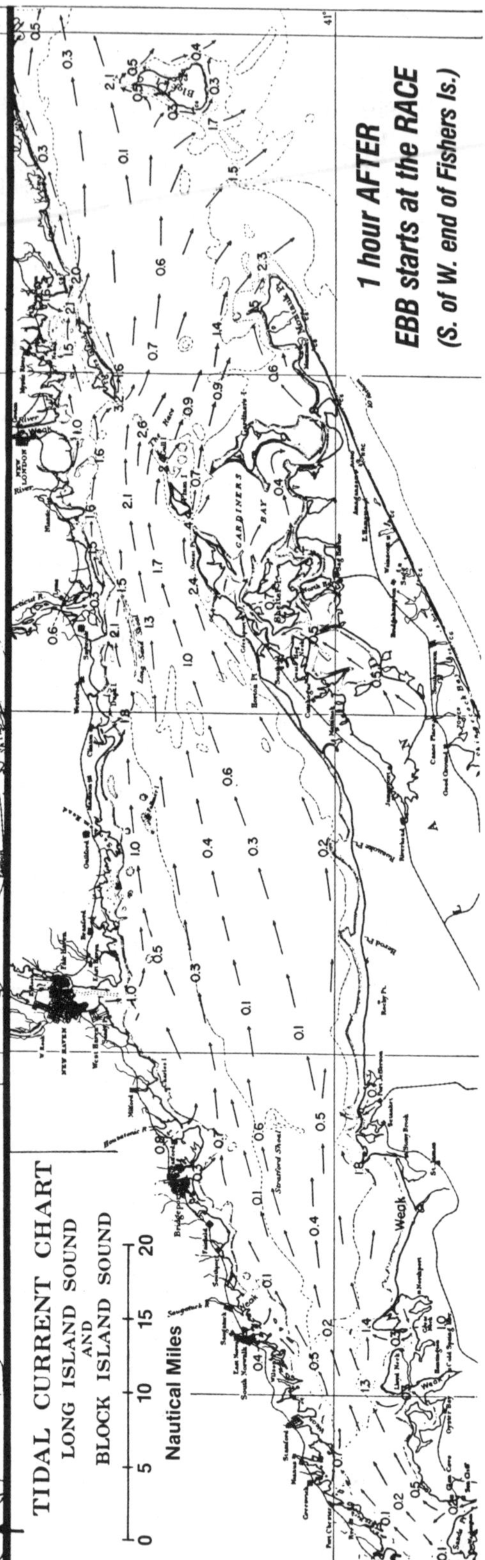

In May 1977 NOS put out 13 new charts for this area, but so awkward to use that, despite some differences (rarely major) in current directions and velocities, we believe the old presentation comprehensibly more useful.

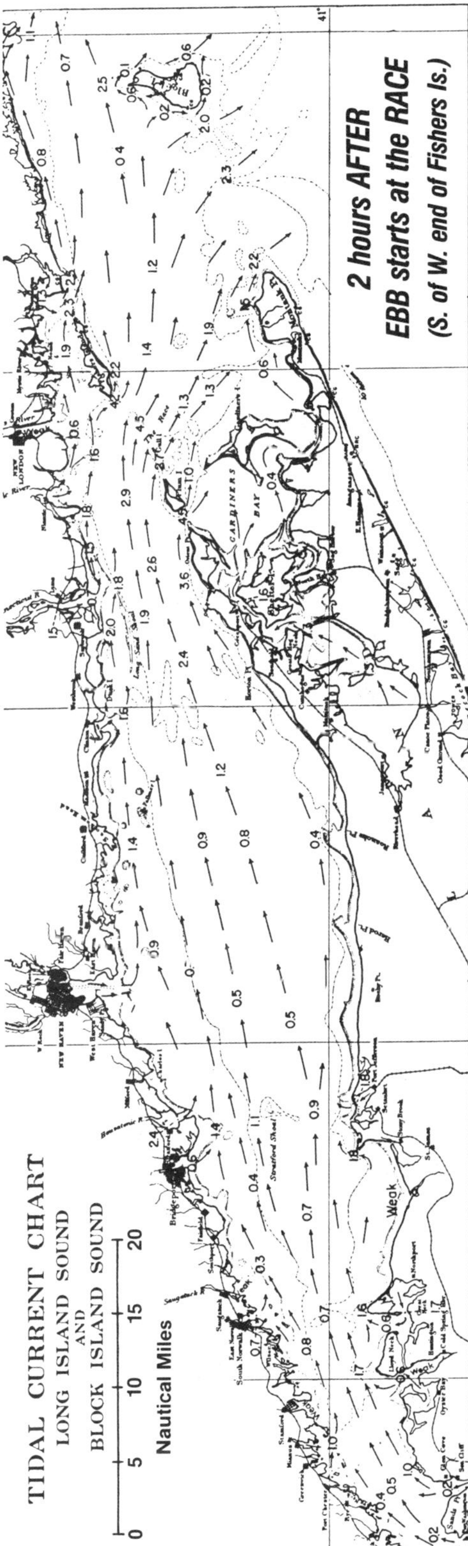

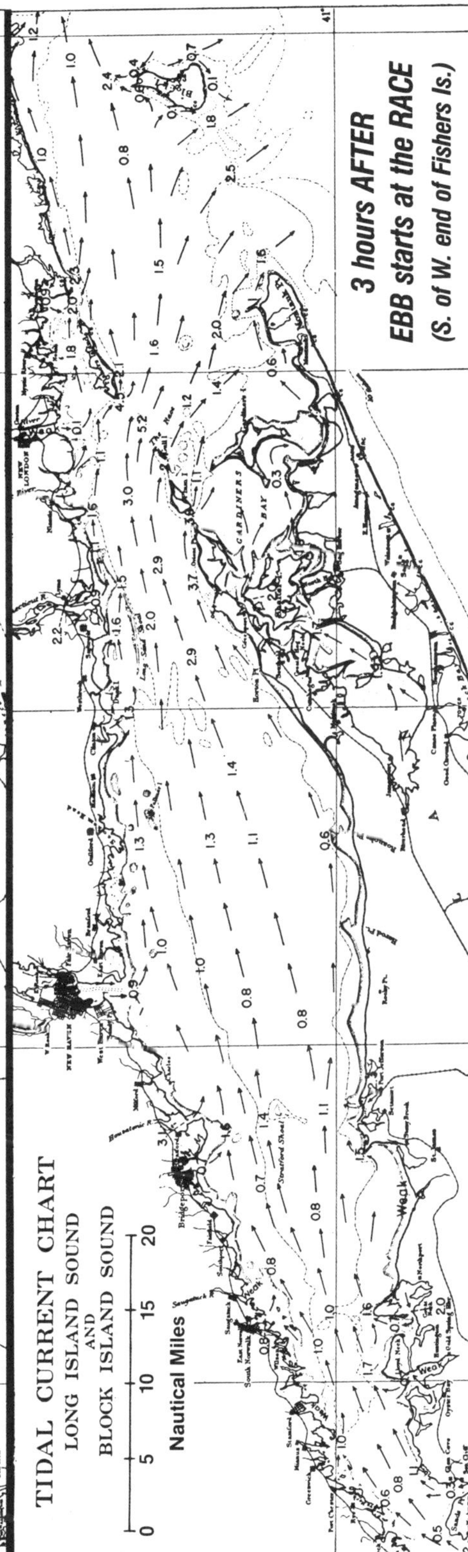

In May 1977 NOS put out 13 new charts for this area, but so awkward to use that, despite some differences (rarely major) in current directions and velocities, we believe the old presentation comprehensibly more useful.

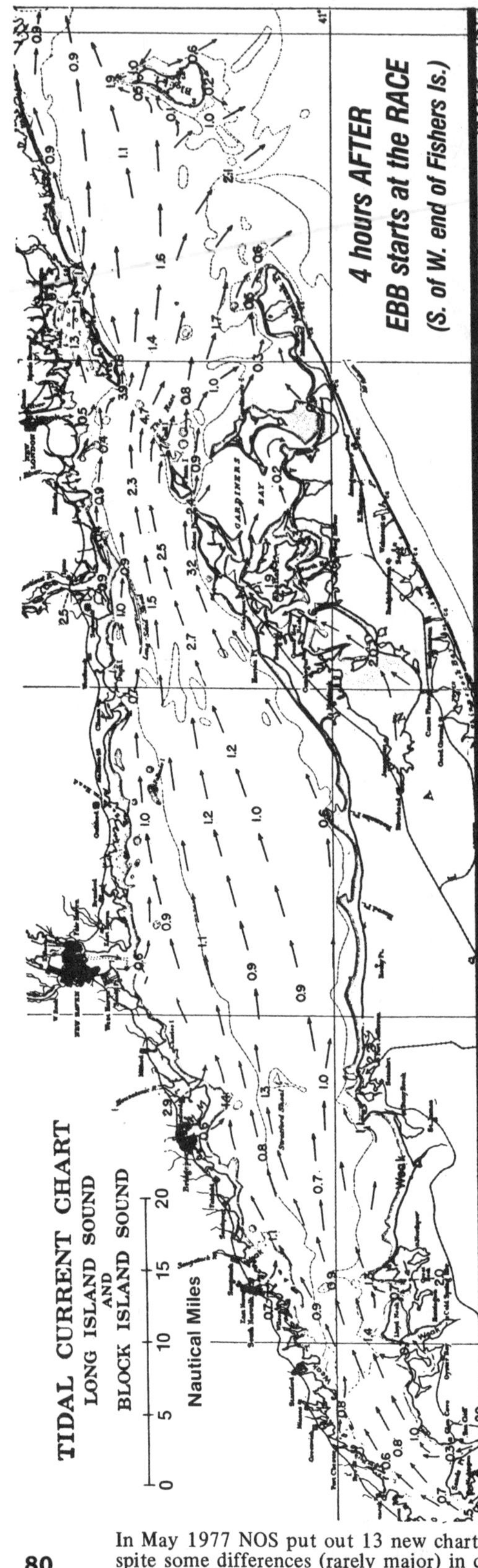

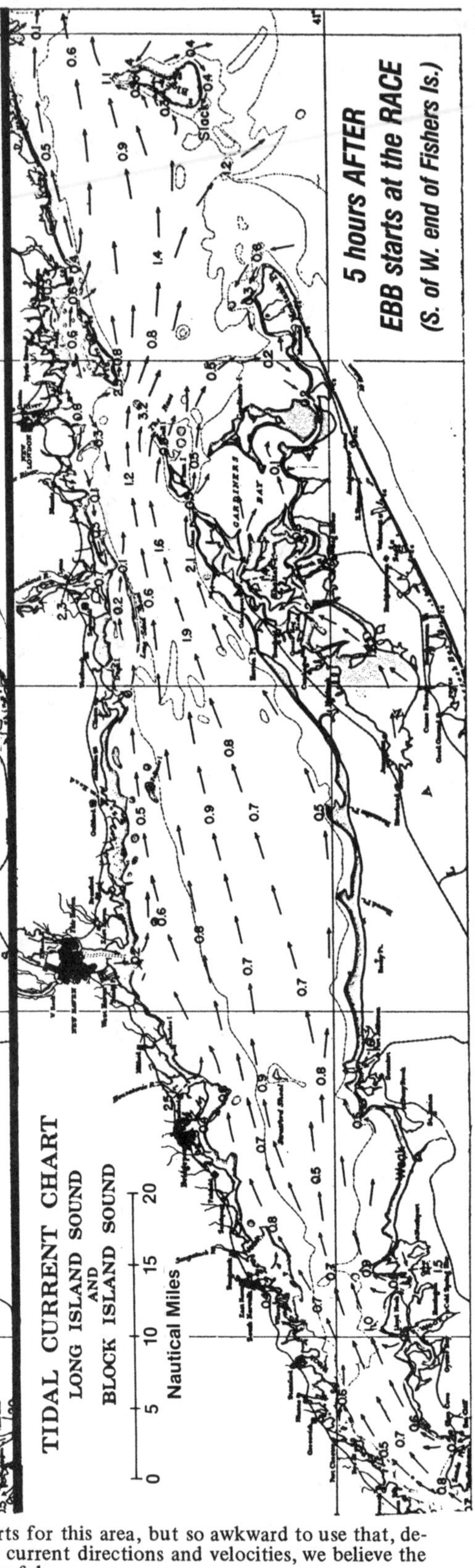

In May 1977 NOS put out 13 new charts for this area, but so awkward to use that, despite some differences (rarely major) in current directions and velocities, we believe the old presentation comprehensibly more useful.

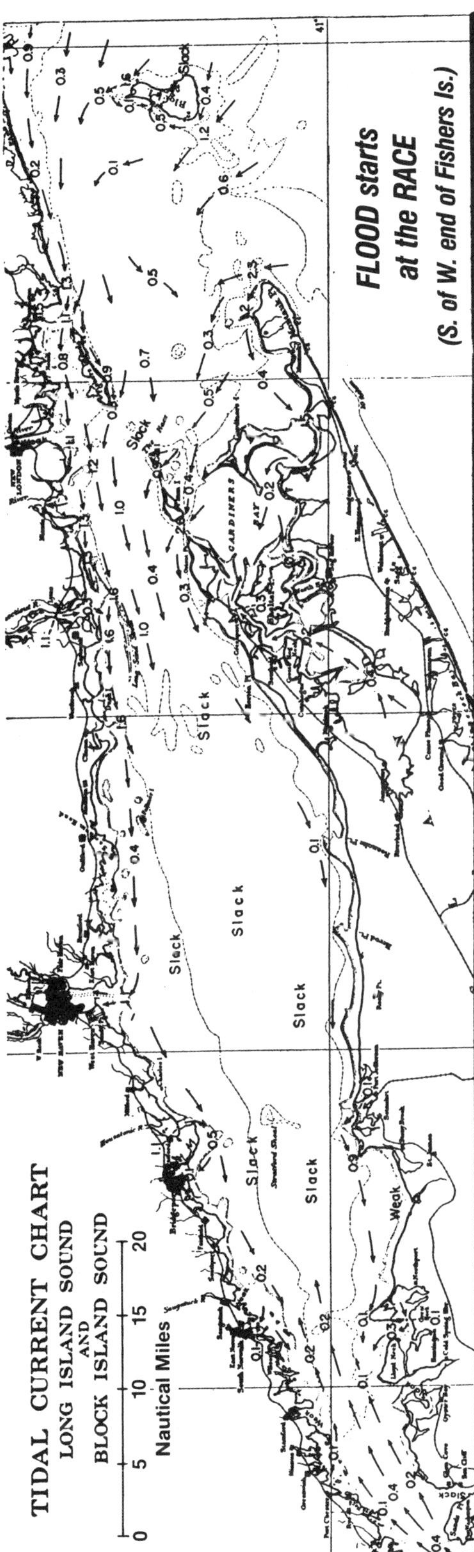

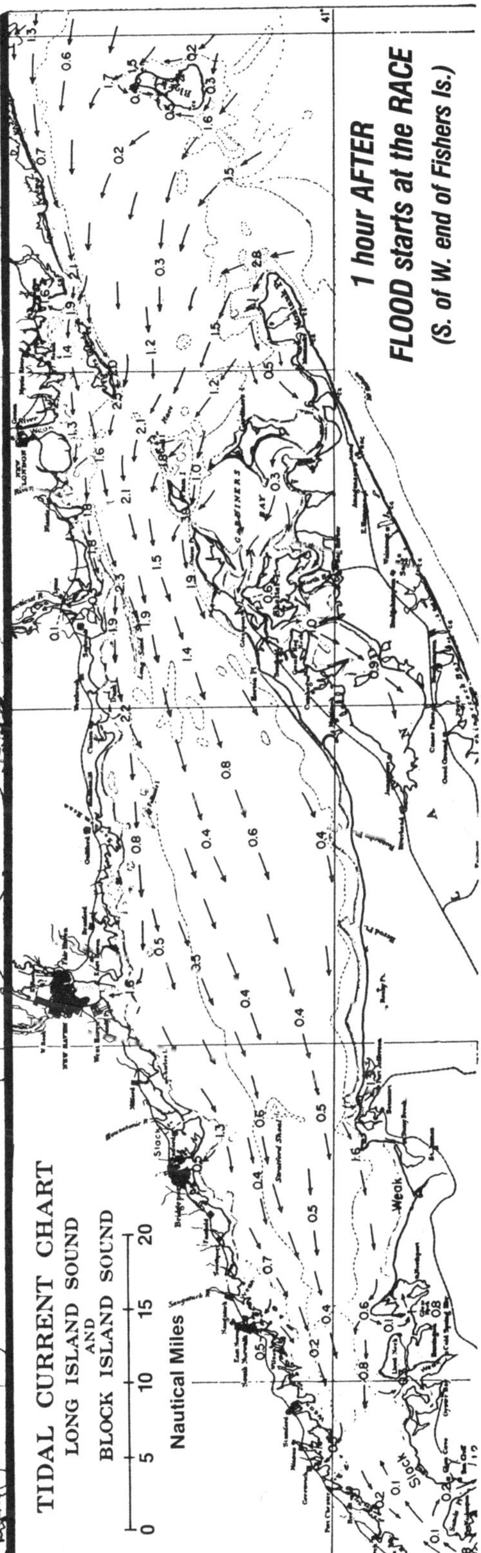

In May 1977 NOS put out 13 new charts for this area, but so awkward to use that, despite some differences (rarely major) in current directions and velocities, we believe the old presentation comprehensibly more useful.

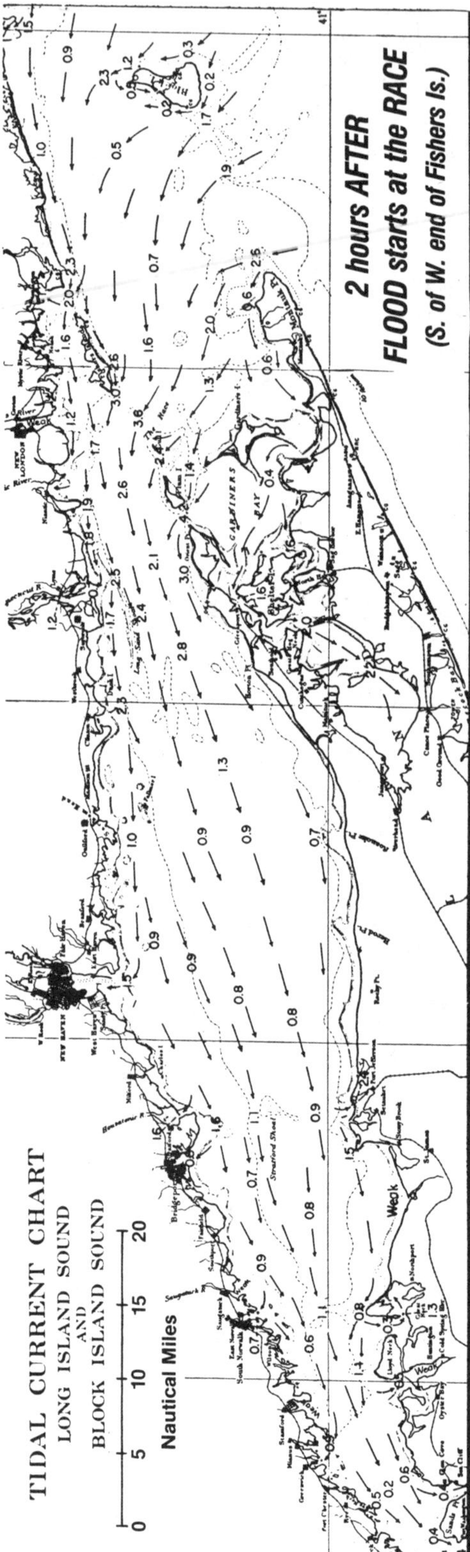

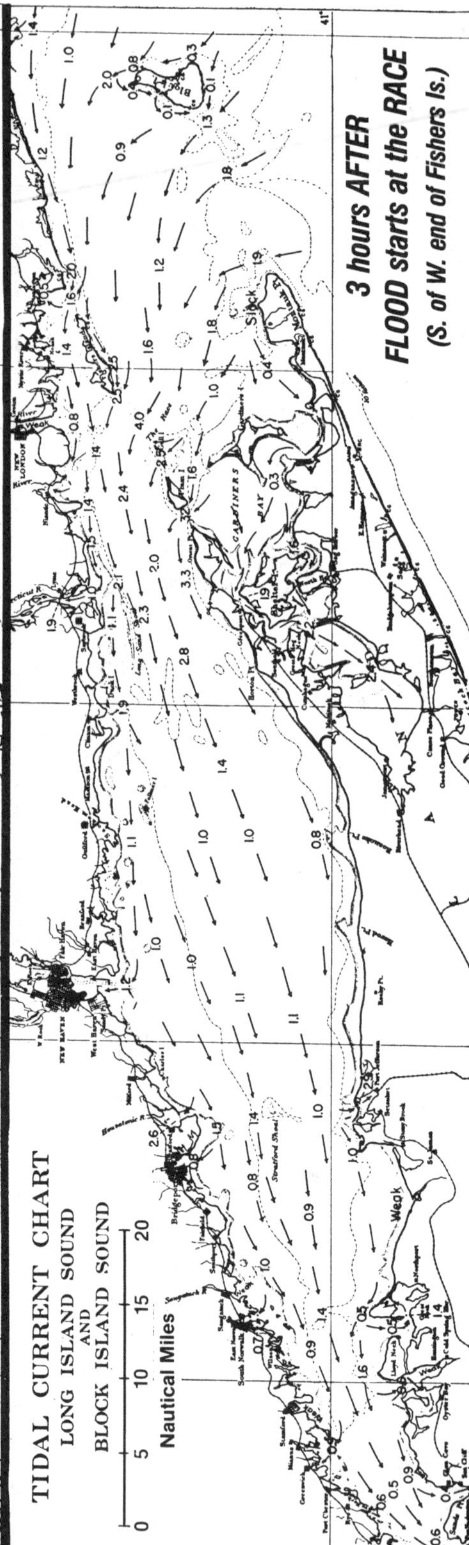

In May 1977 NOS put out 13 new charts for this area, but so awkward to use that, despite some differences (rarely major) in current directions and velocities, we believe the old presentation comprehensibly more useful.

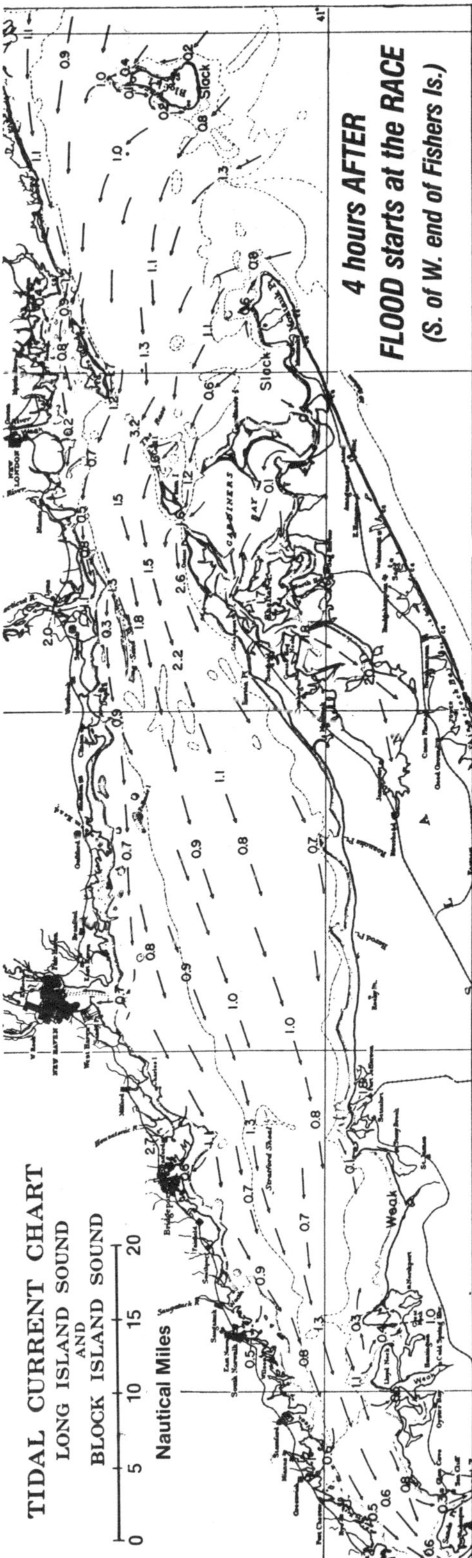

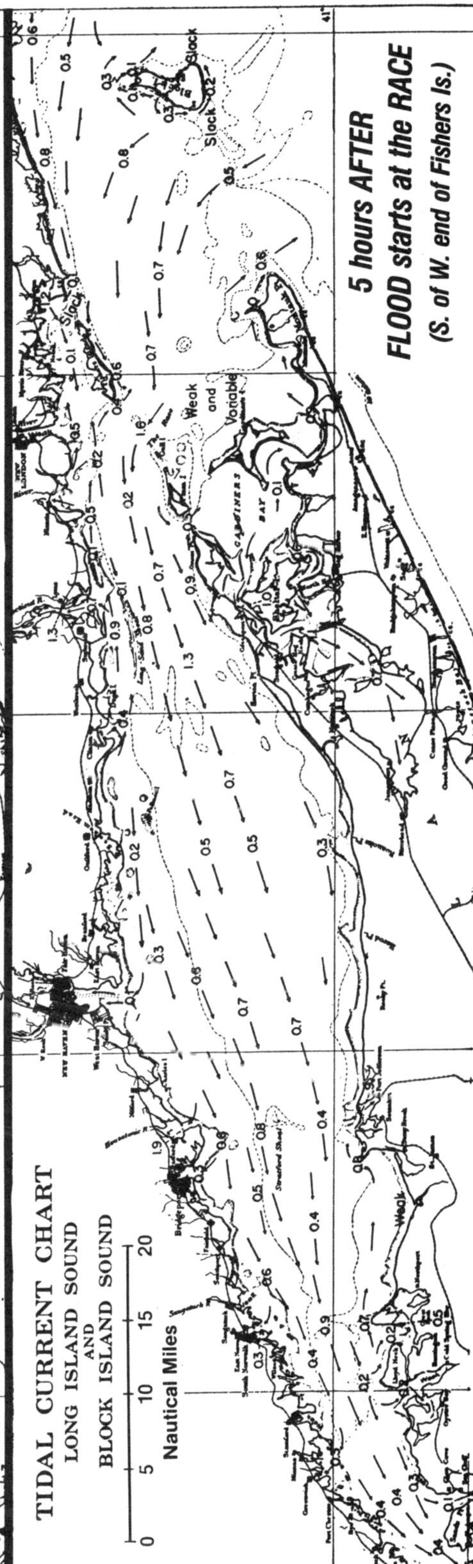

In May 1977 NOS put out 13 new charts for this area, but so awkward to use that, despite some differences (rarely major) in current directions and velocities, we believe the old presentation comprehensibly more useful.

1999 HIGH & LOW WATER BRIDGEPORT, CONN.

41–10N x 73–11W

Standard Time

DAY OF MONTH	DAY OF WEEK	JANUARY HIGH a.m.	Ht.	p.m.	Ht.	LOW a.m.	p.m.
1	F	10 08	7.9	10 42	6.9	3 56	4 36
2	S	11 00	7.9	11 33	6.9	4 49	5 27
3	S	11 51	7.8	...	...	5 40	6 16
4	M	12 23	6.9	12 40	7.5	6 31	7 04
5	T	1 12	6.9	1 30	7.2	7 22	7 52
6	W	2 01	6.7	2 20	6.9	8 13	8 41
7	T	2 51	6.6	3 11	6.5	9 06	9 31
8	F	3 42	6.5	4 03	6.2	10 00	10 22
9	S	4 35	6.4	4 59	5.9	10 55	11 14
10	S	5 27	6.4	5 53	5.8	11 51	...
11	M	6 20	6.4	6 47	5.8	12 06	12 45
12	T	7 10	6.6	7 38	5.9	12 58	1 37
13	W	7 58	6.7	8 27	6.0	1 47	2 25
14	T	8 44	6.9	9 12	6.2	2 34	3 10
15	F	9 26	7.1	9 54	6.4	3 18	3 52
16	S	10 07	7.3	10 35	6.6	4 00	4 32
17	S	10 48	7.5	11 15	6.8	4 40	5 12
18	M	11 28	7.6	11 55	6.9	5 21	5 51
19	T	...	...	12 09	7.6	6 02	6 31
20	W	12 37	7.0	12 52	7.5	6 45	7 13
21	T	1 20	7.1	1 38	7.3	7 31	7 57
22	F	2 07	7.1	2 28	7.0	8 22	8 46
23	S	2 58	7.1	3 23	6.7	9 17	9 40
24	S	3 54	7.0	4 23	6.4	10 18	10 39
25	M	4 55	6.9	5 28	6.1	11 23	11 43
26	T	5 59	6.9	6 35	6.1	...	12 30
27	W	7 04	7.1	7 41	6.2	12 48	1 35
28	T	8 06	7.2	8 42	6.4	1 51	2 36
29	F	9 04	7.4	9 38	6.6	2 50	3 32
30	S	9 58	7.5	10 29	6.8	3 46	4 23
31	S	10 48	7.6	11 17	6.9	4 38	5 11

Standard Time

DAY OF MONTH	DAY OF WEEK	FEBRUARY HIGH a.m.	Ht.	p.m.	Ht.	LOW a.m.	p.m.
1	M	11 36	7.5	...	...	5 26	5 57
2	T	12 02	7.0	12 21	7.3	6 13	6 40
3	W	12 46	6.9	1 05	7.1	6 59	7 23
4	T	1 30	6.8	1 49	6.7	7 44	8 06
5	F	2 13	6.7	2 34	6.4	8 30	8 49
6	S	2 58	6.5	3 20	6.1	9 17	9 35
7	S	3 46	6.4	4 10	5.9	10 08	10 25
8	M	4 36	6.3	5 03	5.7	11 01	11 17
9	T	5 30	6.3	6 00	5.7	11 57	...
10	W	6 23	6.3	6 54	5.8	12 11	12 51
11	T	7 16	6.5	7 47	5.9	1 05	1 44
12	F	8 06	6.8	8 36	6.2	1 56	2 33
13	S	8 53	7.0	9 22	6.5	2 45	3 18
14	S	9 38	7.3	10 05	6.8	3 30	4 01
15	M	10 22	7.5	10 47	7.1	4 14	4 43
16	T	11 05	7.6	11 29	7.3	4 57	5 24
17	W	11 48	7.7	...	...	5 41	6 06
18	T	12 12	7.5	12 33	7.6	6 26	6 49
19	F	12 57	7.5	1 20	7.3	7 13	7 35
20	S	1 45	7.4	2 11	7.0	8 04	8 25
21	S	2 37	7.3	3 07	6.6	9 00	9 20
22	M	3 34	7.1	4 08	6.3	10 01	10 22
23	T	4 37	6.9	5 15	6.1	11 08	11 28
24	W	5 44	6.8	6 24	6.0	...	12 16
25	T	6 52	6.8	7 31	6.2	12 37	1 22
26	F	7 56	7.0	8 31	6.4	1 42	2 24
27	S	8 55	7.2	9 25	6.7	2 42	3 18
28	S	9 47	7.3	10 14	6.9	3 36	4 07

Dates when Ht. of **Low** Water is below Mean Low with Ht. of lowest given for each period and Date of lowest in ( ):

January:
1st - 6th: -1.2' (1st - 2nd)
16th - 23rd: -0.6' (18th - 20th)
27th - 31st: -1.0'

February:
1st - 4th: -0.9' (1st)
14th - 22nd: -0.9' (17th - 18th)
26th - 28th: -0.6' (28th)

Average Rise and Fall 6.7 ft.

When a high tide exceeds av. ht., the *following* low tide will be lower than av.

1999 HIGH & LOW WATER
BRIDGEPORT, CONN.
41–10N x 73–11W

Standard Time **Daylight Time starts Apr. 4 at 2 a.m.**

DAY OF MONTH	DAY OF WEEK	MARCH HIGH a.m.	Ht.	p.m.	Ht.	MARCH LOW a.m.	p.m.	DAY OF MONTH	DAY OF WEEK	APRIL HIGH a.m.	Ht.	p.m.	Ht.	APRIL LOW a.m.	p.m.
1	M	10 35	7.4	10 58	7.1	4 25	4 52	1	T	11 35	7.0	11 49	7.3	5 30	5 45
2	T	11 18	7.3	11 39	7.1	5 10	5 34	2	F	...	...	12 12	6.9	6 08	6 21
3	W	11 59	7.2	...	...	5 53	6 13	3	S	12 25	7.2	12 49	6.7	6 45	6 57
4	T	12 18	7.1	12 39	6.9	6 34	6 52	4	S	1 02	7.1	*2 27	6.5	*8 23	*8 34
5	F	12 57	7.0	1 18	6.7	7 14	7 30	5	M	2 40	6.9	3 07	6.4	9 02	9 14
6	S	1 36	6.8	1 58	6.5	7 55	8 10	6	T	3 20	6.8	3 50	6.3	9 45	9 58
7	S	2 17	6.7	2 41	6.2	8 37	8 52	7	W	4 05	6.7	4 38	6.2	10 32	10 48
8	M	3 00	6.5	3 27	6.0	9 23	9 38	8	T	4 56	6.6	5 31	6.2	11 23	11 43
9	T	3 49	6.4	4 18	5.9	10 13	10 29	9	F	5 52	6.6	6 29	6.3	...	12 19
10	W	4 40	6.3	5 12	5.8	11 08	11 25	10	S	6 49	6.6	7 24	6.5	12 42	1 15
11	T	5 35	6.4	6 09	5.9	...	12 04	11	S	7 47	6.8	8 19	6.9	1 40	2 10
12	F	6 32	6.5	7 05	6.1	12 22	1 00	12	M	8 42	7.1	9 10	7.2	2 36	3 02
13	S	7 27	6.7	7 57	6.5	1 18	1 52	13	T	9 35	7.3	9 59	7.6	3 29	3 51
14	S	8 18	7.0	8 46	6.8	2 10	2 41	14	W	10 26	7.5	10 47	8.0	4 19	4 39
15	M	9 07	7.3	9 32	7.2	3 00	3 27	15	T	11 15	7.7	11 35	8.2	5 08	5 27
16	T	9 54	7.6	10 17	7.5	3 47	4 11	16	F	...	...	12 05	7.7	5 57	6 14
17	W	10 40	7.7	11 02	7.8	4 33	4 55	17	S	12 23	8.3	12 55	7.6	6 47	7 03
18	T	11 26	7.7	11 47	7.9	5 19	5 40	18	S	1 13	8.3	1 46	7.4	7 38	7 55
19	F	...	...	12 14	7.6	6 06	6 26	19	M	2 05	8.1	2 41	7.2	8 32	8 50
20	S	12 34	7.9	1 03	7.4	6 55	7 14	20	T	3 01	7.8	3 39	6.9	9 29	9 50
21	S	1 24	7.8	1 56	7.1	7 48	8 06	21	W	4 01	7.4	4 42	6.7	10 31	10 55
22	M	2 18	7.5	2 53	6.7	8 44	9 04	22	T	5 06	7.1	5 48	6.6	11 35	...
23	T	3 17	7.2	3 55	6.4	9 46	10 08	23	F	6 14	6.9	6 55	6.6	12 03	12 41
24	W	4 21	7.0	5 03	6.2	10 53	11 17	24	S	7 21	6.8	7 57	6.8	1 11	1 44
25	T	5 30	6.8	6 12	6.2	...	12 01	25	S	8 24	6.8	8 53	7.0	2 15	2 42
26	F	6 39	6.8	7 18	6.4	12 26	1 07	26	M	9 20	6.9	9 43	7.2	3 11	3 33
27	S	7 44	6.9	8 16	6.7	1 32	2 06	27	T	10 09	7.0	10 27	7.3	4 01	4 19
28	S	8 41	7.0	9 08	7.0	2 30	2 59	28	W	10 52	7.0	11 07	7.4	4 46	5 00
29	M	9 32	7.1	9 53	7.2	3 22	3 46	29	T	11 32	6.9	11 44	7.4	5 27	5 38
30	T	10 16	7.2	10 34	7.3	4 08	4 28	30	F	...	...	12 09	6.9	6 05	6 15
31	W	10 57	7.1	11 12	7.3	4 50	5 07								

***Daylight Time starts**

Dates when Ht. of **Low** Water is below Mean Low with Ht. of lowest given for each period and Date of lowest in ( ):

March	April
1st - 5th: -0.6' (1st - 3rd)	1st - 3rd: -0.4' (1st - 2nd)
15th - 23rd: -1.2' (19th - 20th)	14th - 21st: -1.3' (17th - 18th)
30th - 31st: -0.4'	30th: -0.3'

Average Rise and Fall 6.7 ft.

When a high tide exceeds av. ht., the *following* low tide will be lower than av.

1999 HIGH & LOW WATER BRIDGEPORT, CONN.

41–10N x 73–11W

Daylight Time

Day of Month	Day of Week	MAY HIGH a.m.	Ht.	p.m.	Ht.	LOW a.m.	p.m.
1	S	12 20	7.4	**12 46**	6.8	6 42	**6 51**
2	S	12 55	7.4	**1 22**	6.7	7 18	**7 27**
3	M	1 31	7.3	**1 59**	6.6	7 55	**8 03**
4	T	2 08	7.2	**2 38**	6.6	8 33	**8 42**
5	W	2 47	7.1	**3 20**	6.5	9 13	**9 25**
6	T	3 31	7.0	**4 06**	6.5	9 57	**10 13**
7	F	4 19	6.9	**4 56**	6.6	10 46	**11 07**
8	S	5 13	6.8	**5 51**	6.7	11 39	**...**
9	S	6 12	6.8	**6 47**	6.9	12 06	**12 34**
10	M	7 10	6.9	**7 42**	7.2	1 05	**1 30**
11	T	8 08	7.0	**8 36**	7.6	2 03	**2 24**
12	W	9 04	7.2	**9 29**	7.9	2 59	**3 17**
13	T	9 59	7.4	**10 20**	8.2	3 53	**4 09**
14	F	10 52	7.5	**11 11**	8.4	4 46	**5 00**
15	S	11 44	7.6	**...**	...	5 37	**5 52**
16	S	12 02	8.5	**12 37**	7.5	6 29	**6 44**
17	M	12 54	8.4	**1 31**	7.4	7 22	**7 38**
18	T	1 48	8.2	**2 26**	7.2	8 16	**8 35**
19	W	2 44	7.9	**3 24**	7.1	9 13	**9 35**
20	T	3 44	7.5	**4 25**	6.9	10 12	**10 38**
21	F	4 46	7.1	**5 27**	6.8	11 13	**11 43**
22	S	5 51	6.9	**6 29**	6.9	...	**12 14**
23	S	6 55	6.7	**7 28**	7.0	12 48	**1 14**
24	M	7 55	6.6	**8 23**	7.1	1 49	**2 09**
25	T	8 50	6.6	**9 11**	7.3	2 44	**3 00**
26	W	9 39	6.7	**9 56**	7.4	3 34	**3 46**
27	T	10 23	6.7	**10 36**	7.4	4 19	**4 28**
28	F	11 04	6.7	**11 14**	7.5	5 00	**5 08**
29	S	11 42	6.7	**11 51**	7.5	5 39	**5 46**
30	S	...	...	**12 19**	6.7	6 16	**6 22**
31	M	12 27	7.5	**12 56**	6.7	6 52	**6 59**

Daylight Time

Day of Month	Day of Week	JUNE HIGH a.m.	Ht.	p.m.	Ht.	LOW a.m.	p.m.
1	T	1 04	7.4	**1 34**	6.8	7 29	**7 37**
2	W	1 41	7.4	**2 12**	6.8	8 06	**8 16**
3	T	2 20	7.3	**2 53**	6.8	8 46	**8 58**
4	F	3 03	7.2	**3 37**	6.9	9 28	**9 45**
5	S	3 49	7.1	**4 26**	6.9	10 14	**10 37**
6	S	4 41	7.0	**5 18**	7.1	11 04	**11 34**
7	M	5 38	6.9	**6 13**	7.2	11 58	**...**
8	T	6 37	6.8	**7 09**	7.5	12 34	**12 54**
9	W	7 39	6.9	**8 07**	7.7	1 34	**1 51**
10	T	8 38	7.0	**9 02**	8.0	2 33	**2 48**
11	F	9 36	7.1	**9 57**	8.2	3 31	**3 44**
12	S	10 32	7.2	**10 51**	8.4	4 26	**4 39**
13	S	11 27	7.3	**11 45**	8.4	5 20	**5 33**
14	M	...	...	**12 21**	7.4	6 13	**6 28**
15	T	12 38	8.4	**1 15**	7.4	7 06	**7 22**
16	W	1 32	8.1	**2 09**	7.3	8 00	**8 18**
17	T	2 27	7.8	**3 04**	7.2	8 53	**9 16**
18	F	3 23	7.5	**4 01**	7.1	9 48	**10 15**
19	S	4 21	7.1	**4 58**	7.0	10 44	**11 15**
20	S	5 21	6.8	**5 55**	7.0	11 41	**...**
21	M	6 20	6.5	**6 51**	7.0	12 16	**12 36**
22	T	7 18	6.4	**7 45**	7.0	1 14	**1 31**
23	W	8 13	6.3	**8 35**	7.1	2 09	**2 22**
24	T	9 04	6.4	**9 21**	7.2	3 00	**3 10**
25	F	9 50	6.4	**10 04**	7.3	3 47	**3 54**
26	S	10 33	6.5	**10 44**	7.4	4 30	**4 36**
27	S	11 13	6.6	**11 23**	7.5	5 10	**5 16**
28	M	11 52	6.7	**...**	...	5 49	**5 55**
29	T	12 01	7.6	**12 30**	6.8	6 26	**6 34**
30	W	12 38	7.6	**1 08**	6.9	7 03	**7 12**

Dates when Ht. of **Low** Water is below Mean Low with Ht. of lowest given for each period and Date of lowest in ( ):

13th - 20th: -1.3' (16th)

11th - 18th: -1.2' (14th)

Average Rise and Fall 6.7 ft.

When a high tide exceeds av. ht., the *following* low tide will be lower than av.

1999 HIGH & LOW WATER
BRIDGEPORT, CONN.
41–10N x 73–11W

Daylight Time

Day of Month	Day of Week	JULY HIGH a.m.	Ht.	p.m.	Ht.	LOW a.m.	p.m.
1	T	1 16	7.6	1 47	7.0	7 41	7 52
2	F	1 56	7.5	2 27	7.1	8 19	8 34
3	S	2 38	7.4	3 10	7.2	9 00	9 21
4	S	3 24	7.2	3 57	7.2	9 45	10 12
5	M	4 15	7.1	4 48	7.3	10 34	11 08
6	T	5 11	6.9	5 44	7.4	11 27	...
7	W	6 11	6.7	6 42	7.5	12 08	12 25
8	T	7 14	6.7	7 42	7.7	1 11	1 26
9	F	8 18	6.7	8 43	7.9	2 13	2 26
10	S	9 18	6.8	9 40	8.1	3 13	3 26
11	S	10 17	7.0	10 37	8.2	4 11	4 24
12	M	11 13	7.2	11 31	8.2	5 06	5 20
13	T	...	...	12 06	7.3	5 59	6 14
14	W	12 24	8.2	12 58	7.4	6 50	7 07
15	T	1 16	8.0	1 49	7.4	7 40	8 00
16	F	2 07	7.7	2 40	7.3	8 30	8 53
17	S	2 59	7.4	3 30	7.2	9 20	9 46
18	S	3 51	7.0	4 22	7.0	10 10	10 41
19	M	4 45	6.6	5 15	6.9	11 02	11 37
20	T	5 40	6.3	6 08	6.9	11 54	...
21	W	6 36	6.2	7 02	6.8	12 33	12 48
22	T	7 31	6.1	7 53	6.9	1 29	1 40
23	F	8 24	6.2	8 43	7.0	2 21	2 31
24	S	9 13	6.3	9 29	7.2	3 10	3 19
25	S	9 59	6.5	10 13	7.4	3 56	4 04
26	M	10 42	6.7	10 54	7.5	4 39	4 47
27	T	11 23	6.9	11 34	7.6	5 19	5 28
28	W	...	...	12 02	7.0	5 58	6 08
29	T	12 13	7.7	12 40	7.2	6 36	6 48
30	F	12 52	7.7	1 20	7.3	7 14	7 28
31	S	1 33	7.7	2 00	7.4	7 53	8 12

Daylight Time

Day of Month	Day of Week	AUGUST HIGH a.m.	Ht.	p.m.	Ht.	LOW a.m.	p.m.
1	S	2 16	7.5	2 44	7.5	8 34	8 58
2	M	3 02	7.3	3 31	7.5	9 19	9 49
3	T	3 53	7.1	4 23	7.5	10 08	10 46
4	W	4 50	6.8	5 20	7.4	11 03	11 48
5	T	5 52	6.6	6 21	7.4	...	12 04
6	F	6 57	6.5	7 25	7.5	12 53	1 08
7	S	8 03	6.5	8 28	7.6	1 58	2 13
8	S	9 06	6.7	9 29	7.8	3 00	3 15
9	M	10 06	7.0	10 30	7.9	3 58	4 14
10	T	11 00	7.2	11 19	8.0	4 53	5 09
11	W	11 50	7.4	...	...	5 43	6 01
12	T	12 10	8.0	12 39	7.5	6 31	6 50
13	F	12 58	7.8	1 25	7.5	7 17	7 38
14	S	1 45	7.5	2 10	7.4	8 02	8 26
15	S	2 31	7.2	2 56	7.2	8 47	9 14
16	M	3 18	6.8	3 42	7.1	9 33	10 03
17	T	4 06	6.5	4 31	6.9	10 20	10 54
18	W	4 57	6.2	5 22	6.8	11 10	11 48
19	T	5 51	6.1	6 15	6.7	...	12 03
20	F	6 46	6.0	7 09	6.7	12 44	12 57
21	S	7 42	6.1	8 02	6.9	1 38	1 51
22	S	8 34	6.3	8 52	7.1	2 30	2 43
23	M	9 23	6.6	9 39	7.3	3 19	3 31
24	T	10 08	6.8	10 23	7.5	4 04	4 16
25	W	10 50	7.1	11 05	7.7	4 46	4 59
26	T	11 31	7.4	11 46	7.8	5 26	5 41
27	F	...	...	12 11	7.6	6 05	6 22
28	S	12 28	7.8	12 51	7.8	6 45	7 05
29	S	1 10	7.8	1 33	7.8	7 25	7 49
30	M	1 55	7.6	2 18	7.8	8 08	8 37
31	T	2 43	7.4	3 07	7.8	8 55	9 30

Dates when Ht. of **Low** Water is below Mean Low with Ht. of lowest given for each period and Date of lowest in ( ):

10th - 17th: -0.9' (13th - 14th)

9th - 14th: - 0.7' (11th - 12th)
27th - 31st: -0.3' (28th - 30th)

Average Rise and Fall 6.7 ft.

When a high tide exceeds av. ht., the *following* low tide will be lower than av.

1999 HIGH & LOW WATER
BRIDGEPORT, CONN.
41–10N x 73–11W

Daylight Time **Standard Time starts Oct. 31 at 2 a.m.**

SEPTEMBER

DAY OF MONTH	DAY OF WEEK	HIGH a.m.	Ht.	HIGH p.m.	Ht.	LOW a.m.	LOW p.m.
1	W	3 35	7.0	4 01	7.6	9 47	10 28
2	T	4 34	6.7	5 01	7.4	10 45	11 31
3	F	5 38	6.5	6 06	7.3	11 50	...
4	S	6 46	6.4	7 13	7.3	12 39	12 58
5	S	7 54	6.5	8 19	7.4	1 46	2 05
6	M	8 57	6.8	9 20	7.6	2 48	3 08
7	T	9 54	7.1	10 16	7.7	3 46	4 05
8	W	10 45	7.4	11 07	7.8	4 38	4 58
9	T	11 34	7.5	11 55	7.7	5 25	5 46
10	F	...	...	12 17	7.6	6 10	6 31
11	S	12 38	7.6	12 59	7.6	6 52	7 15
12	S	1 20	7.3	1 40	7.5	7 33	7 58
13	M	2 02	7.0	2 21	7.3	8 13	8 41
14	T	2 44	6.8	3 03	7.1	8 55	9 25
15	W	3 28	6.5	3 47	6.9	9 38	10 12
16	T	4 15	6.3	4 35	6.8	10 26	11 03
17	F	5 07	6.1	5 28	6.7	11 18	11 57
18	S	6 02	6.1	6 23	6.7	...	12 13
19	S	6 58	6.2	7 19	6.8	12 53	1 10
20	M	7 53	6.4	8 12	7.0	1 47	2 05
21	T	8 44	6.7	9 02	7.2	2 38	2 56
22	W	9 31	7.1	9 49	7.5	3 25	3 44
23	T	10 15	7.4	10 34	7.7	4 09	4 29
24	F	10 58	7.8	11 18	7.8	4 51	5 13
25	S	11 40	8.0	...	...	5 33	5 57
26	S	12 02	7.9	12 23	8.2	6 15	6 42
27	M	12 47	7.8	1 08	8.2	6 58	7 29
28	T	1 35	7.6	1 55	8.2	7 44	8 19
29	W	2 25	7.3	2 46	8.0	8 34	9 13
30	T	3 20	7.0	3 43	7.7	9 29	10 12

OCTOBER

DAY OF MONTH	DAY OF WEEK	HIGH a.m.	Ht.	HIGH p.m.	Ht.	LOW a.m.	LOW p.m.
1	F	4 21	6.7	4 45	7.4	10 31	11 17
2	S	5 27	6.5	5 53	7.2	11 39	...
3	S	6 36	6.5	7 02	7.1	12 25	12 50
4	M	7 43	6.7	8 09	7.2	1 32	1 57
5	T	8 45	7.0	9 09	7.3	2 34	2 59
6	W	9 39	7.3	10 03	7.4	3 29	3 54
7	T	10 28	7.5	10 51	7.5	4 19	4 44
8	F	11 12	7.6	11 35	7.4	5 04	5 29
9	S	11 54	7.7	...	...	5 46	6 11
10	S	12 16	7.3	12 31	7.6	6 25	6 51
11	M	12 55	7.1	1 09	7.5	7 03	7 30
12	T	1 33	6.9	1 47	7.4	7 41	8 09
13	W	2 12	6.7	2 26	7.2	8 19	8 50
14	T	2 53	6.5	3 08	7.0	9 00	9 33
15	F	3 38	6.3	3 53	6.8	9 45	10 21
16	S	4 26	6.3	4 43	6.7	10 35	11 12
17	S	5 19	6.3	5 37	6.7	11 30	...
18	M	6 15	6.4	6 34	6.7	12 06	12 28
19	T	7 10	6.6	7 30	6.9	1 01	1 25
20	W	8 03	6.9	8 24	7.1	1 54	2 19
21	T	8 52	7.3	9 14	7.3	2 43	3 10
22	F	9 39	7.7	10 03	7.5	3 31	3 58
23	S	10 25	8.0	10 51	7.7	4 17	4 46
24	S	11 11	8.3	11 38	7.7	5 02	5 33
25	M	11 57	8.5	...	...	5 47	6 20
26	T	12 26	7.7	12 45	8.5	6 34	7 10
27	W	1 16	7.5	1 35	8.3	7 24	8 02
28	T	2 09	7.3	2 29	8.1	8 17	8 58
29	F	3 06	7.0	3 27	7.7	9 15	9 57
30	S	4 08	6.8	4 30	7.3	10 19	11 02
31	S	*4 14	6.7	*4 38	7.1	*10 27	*11 08

***Standard Time starts**

Dates when Ht. of **Low** Water is below Mean Low with Ht. of lowest given for each period and Date of lowest in ( ):

8th - 11th: -0.4' (9th - 10th)
25th - 29th: -0.7' (26th - 27th)

7th - 11th: -0.3' (8th - 9th)
23rd - 29th: -1.1' (25th - 26th)

Average Rise and Fall 6.7 ft.

When a high tide exceeds av. ht., the *following* low tide will be lower than av.

1999 HIGH & LOW WATER
BRIDGEPORT, CONN.
41–10N x 73–11W

		Standard Time								Standard Time					
DAY OF MONTH	DAY OF WEEK	NOVEMBER						DAY OF MONTH	DAY OF WEEK	DECEMBER					
		HIGH				LOW				HIGH				LOW	
		a.m.	Ht.	p.m.	Ht.	a.m.	p.m.			a.m.	Ht.	p.m.	Ht.	a.m.	p.m.
1	M	5 21	6.7	**5 47**	6.9	11 37	...	1	W	6 00	6.8	**6 28**	6.6	...	**12 21**
2	T	6 26	6.8	**6 53**	6.9	12 13	**12 43**	2	T	6 59	7.0	**7 27**	6.5	12 44	**1 20**
3	W	7 26	7.1	**7 52**	7.0	1 13	**1 44**	3	F	7 52	7.1	**8 20**	6.5	1 38	**2 14**
4	T	8 19	7.3	**8 45**	7.0	2 07	**2 37**	4	S	8 39	7.2	**9 07**	6.5	2 28	**3 02**
5	F	9 06	7.5	**9 31**	7.0	2 56	**3 25**	5	S	9 22	7.3	**9 49**	6.5	3 12	**3 45**
6	S	9 48	7.6	**10 14**	7.0	3 40	**4 09**	6	M	10 02	7.4	**10 29**	6.5	3 53	**4 26**
7	S	10 28	7.6	**10 53**	6.9	4 20	**4 49**	7	T	10 40	7.4	**11 06**	6.5	4 32	**5 03**
8	M	11 05	7.5	**11 30**	6.8	4 58	**5 27**	8	W	11 16	7.3	**11 43**	6.5	5 10	**5 40**
9	T	11 42	7.5	...	...	5 35	**6 04**	9	T	11 53	7.3	...	...	5 46	**6 17**
10	W	12 07	6.7	**12 18**	7.4	6 12	**6 42**	10	F	12 20	6.5	**12 29**	7.2	6 23	**6 53**
11	T	12 45	6.6	**12 55**	7.2	6 49	**7 20**	11	S	12 57	6.5	**1 06**	7.1	7 01	**7 31**
12	F	1 24	6.5	**1 34**	7.1	7 28	**8 00**	12	S	1 37	6.5	**1 47**	7.0	7 42	**8 11**
13	S	2 05	6.4	**2 17**	6.9	8 11	**8 43**	13	M	2 19	6.6	**2 30**	6.9	8 26	**8 54**
14	S	2 51	6.4	**3 03**	6.8	8 58	**9 31**	14	T	3 04	6.6	**3 19**	6.7	9 15	**9 41**
15	M	3 40	6.4	**3 55**	6.7	9 50	**10 22**	15	W	3 54	6.7	**4 12**	6.6	10 09	**10 33**
16	T	4 33	6.5	**4 51**	6.7	10 46	**11 15**	16	T	4 47	6.8	**5 09**	6.5	11 07	**11 28**
17	W	5 27	6.7	**5 48**	6.7	11 45	...	17	F	5 43	7.0	**6 09**	6.5	...	**12 06**
18	T	6 21	7.0	**6 45**	6.8	12 09	**12 42**	18	S	6 39	7.3	**7 09**	6.6	12 24	**1 05**
19	F	7 14	7.4	**7 40**	7.0	1 02	**1 36**	19	S	7 35	7.6	**8 07**	6.8	1 20	**2 03**
20	S	8 05	7.7	**8 33**	7.2	1 54	**2 29**	20	M	8 30	7.9	**9 03**	6.9	2 16	**2 59**
21	S	8 56	8.1	**9 25**	7.3	2 44	**3 20**	21	T	9 24	8.1	**9 58**	7.1	3 11	**3 53**
22	M	9 45	8.3	**10 16**	7.4	3 34	**4 11**	22	W	10 18	8.3	**10 52**	7.2	4 05	**4 46**
23	T	10 35	8.5	**11 08**	7.4	4 23	**5 02**	23	T	11 11	8.3	**11 45**	7.2	4 59	**5 38**
24	W	11 26	8.5	...	...	5 14	**5 53**	24	F	...	...	**12 04**	8.1	5 53	**6 31**
25	T	12 01	7.4	**12 18**	8.3	6 07	**6 47**	25	S	12 39	7.2	**12 58**	7.9	6 48	**7 24**
26	F	12 55	7.2	**1 13**	8.0	7 02	**7 42**	26	S	1 33	7.1	**1 54**	7.5	7 44	**8 19**
27	S	1 51	7.1	**2 11**	7.6	8 01	**8 40**	27	M	2 29	6.9	**2 51**	7.1	8 43	**9 15**
28	S	2 51	6.9	**3 13**	7.2	9 03	**9 41**	28	T	3 27	6.8	**3 51**	6.7	9 44	**10 13**
29	M	3 54	6.8	**4 18**	6.9	10 09	**10 44**	29	W	4 26	6.7	**4 53**	6.3	10 46	**11 11**
30	T	4 58	6.7	**5 24**	6.7	11 16	**11 45**	30	T	5 26	6.6	**5 54**	6.1	11 49	...
								31	F	6 24	6.7	**6 53**	6.0	12 09	**12 48**

Dates when Ht. of **Low** Water is below Mean Low with Ht. of lowest given for each period and Date of lowest in ( ):

6th - 7th: -0.3'
20th - 28th: -1.3' (23rd - 24th)

19th - 27th: -1.4' (22nd - 23rd)

Average Rise and Fall 6.7 ft.

When a high tide exceeds av. ht., the *following* low tide will be lower than av.

1999 HIGH & LOW WATER WILLETS POINT, N.Y.

40–48N x 73–47W

Standard Time

DAY OF MONTH	DAY OF WEEK	JANUARY					
		HIGH				LOW	
		a.m.	Ht.	p.m.	Ht.	a.m.	p.m.
1	F	10 30	8.2	11 13	7.3	4 46	5 30
2	S	11 23	8.2	...	...	5 39	6 20
3	S	12 04	7.4	12 13	8.1	6 30	7 09
4	M	12 54	7.4	1 03	7.8	7 20	7 57
5	T	1 43	7.3	1 54	7.5	8 11	8 44
6	W	2 34	7.2	2 46	7.1	9 03	9 33
7	T	3 26	7.0	3 44	6.8	9 57	10 22
8	F	4 22	6.9	4 46	6.5	10 53	11 14
9	S	5 20	6.8	5 48	6.3	11 51	...
10	S	6 16	6.8	6 46	6.2	12 07	12 48
11	M	7 10	6.8	7 41	6.2	12 59	1 42
12	T	8 01	6.9	8 31	6.3	1 50	2 32
13	W	8 46	7.1	9 17	6.5	2 36	3 18
14	T	9 25	7.2	9 58	6.6	3 19	4 01
15	F	9 53	7.3	10 31	6.7	3 56	4 40
16	S	10 09	7.5	10 50	6.9	4 26	5 13
17	S	10 38	7.8	11 13	7.2	4 54	5 42
18	M	11 14	8.0	11 46	7.4	5 26	6 07
19	T	11 55	8.1	...	...	6 04	6 38
20	W	12 26	7.6	12 39	8.1	6 46	7 16
21	T	1 09	7.8	1 25	7.9	7 32	7 58
22	F	1 56	7.8	2 16	7.6	8 22	8 45
23	S	2 46	7.7	3 10	7.3	9 18	9 38
24	S	3 41	7.6	4 10	6.8	10 26	10 38
25	M	4 42	7.4	5 20	6.5	...	12 06
26	T	5 53	7.3	6 50	6.4	12 01	1 29
27	W	7 19	7.4	8 18	6.5	1 37	2 35
28	T	8 38	7.6	9 22	6.8	2 46	3 34
29	F	9 40	7.8	10 17	7.1	3 46	4 27
30	S	10 33	8.0	11 07	7.3	4 40	5 17
31	S	11 23	8.0	11 53	7.5	5 30	6 04

Standard Time

DAY OF MONTH	DAY OF WEEK	FEBRUARY					
		HIGH				LOW	
		a.m.	Ht.	p.m.	Ht.	a.m.	p.m.
1	M	...	...	12 08	7.9	6 17	6 49
2	T	12 37	7.5	12 51	7.7	7 03	7 31
3	W	1 19	7.4	1 32	7.5	7 47	8 12
4	T	1 58	7.3	2 10	7.1	8 30	8 49
5	F	2 35	7.1	2 48	6.8	9 12	9 21
6	S	3 11	6.9	3 29	6.5	9 56	9 40
7	S	3 51	6.7	4 19	6.2	10 48	10 12
8	M	4 37	6.6	5 31	6.0	11 49	11 00
9	T	5 39	6.5	6 47	6.0	...	12 50
10	W	6 54	6.6	7 45	6.1	12 04	1 47
11	T	7 50	6.7	8 36	6.3	1 32	2 38
12	F	8 32	7.0	9 17	6.6	2 29	3 24
13	S	9 03	7.3	9 49	6.9	3 15	4 05
14	S	9 36	7.6	10 14	7.2	3 56	4 41
15	M	10 14	7.9	10 46	7.5	4 34	5 14
16	T	10 55	8.1	11 23	7.9	5 12	5 45
17	W	11 38	8.2	...	...	5 53	6 20
18	T	12 05	8.1	12 23	8.2	6 36	6 59
19	F	12 49	8.2	1 10	8.0	7 22	7 41
20	S	1 36	8.2	2 01	7.7	8 13	8 29
21	S	2 26	8.0	2 56	7.2	9 12	9 22
22	M	3 22	7.7	3 58	6.8	10 33	10 31
23	T	4 26	7.4	5 20	6.4	...	12 04
24	W	5 52	7.1	7 05	6.4	12 15	1 18
25	T	7 32	7.2	8 16	6.7	1 34	2 22
26	F	8 40	7.5	9 14	7.0	2 39	3 19
27	S	9 37	7.7	10 06	7.3	3 36	4 11
28	S	10 27	7.9	10 53	7.6	4 28	4 59

Dates when Ht. of **Low** Water is below Mean Low with Ht. of lowest given for each period and Date of lowest in ( ):

1st - 5th: -1.4' (1st - 2nd)
16th - 22nd: -0.6' (19th - 21st)
27th - 31st: -1.2' (30th - 31st)

1st - 3rd: -1.1' (1st)
15th - 21st: -0.8' (17th - 18th)
25th - 28th: -0.9' (28th)

Average Rise and Fall 7.1 ft.

When a high tide exceeds av. ht., the *following* low tide will be lower than av.

1999 HIGH & LOW WATER
WILLETS POINT, N.Y.
40–48N x 73–47W

Standard Time **Daylight Time starts Apr. 4 at 2 a.m.**

DAY OF MONTH	DAY OF WEEK	MARCH						DAY OF MONTH	DAY OF WEEK	APRIL					
		HIGH				LOW				HIGH				LOW	
		a.m.	Ht.	p.m.	Ht.	a.m.	p.m.			a.m.	Ht.	p.m.	Ht.	a.m.	p.m.
1	M	11 13	7.9	11 36	7.7	5 16	5 44	1	T	...	...	12 12	7.6	6 19	6 33
2	T	11 55	7.9	...	...	6 01	6 25	2	F	12 20	7.8	12 42	7.4	6 55	6 59
3	W	12 15	7.7	12 33	7.7	6 42	7 03	3	S	12 39	7.6	1 02	7.2	7 23	7 04
4	T	12 50	7.6	1 07	7.4	7 21	7 36	4	S	12 56	7.5	*2 22	7.0	*8 36	*8 22
5	F	1 19	7.5	1 35	7.2	7 55	7 57	5	M	2 25	7.5	2 54	6.9	8 54	8 56
6	S	1 41	7.3	2 00	6.9	8 21	8 05	6	T	3 02	7.4	3 33	6.8	9 29	9 38
7	S	2 08	7.2	2 32	6.7	8 38	8 34	7	W	3 44	7.3	4 19	6.7	10 13	10 26
8	M	2 44	7.0	3 13	6.5	9 10	9 15	8	T	4 32	7.2	5 10	6.6	11 04	11 20
9	T	3 27	6.9	4 01	6.3	9 55	10 02	9	F	5 26	7.1	6 08	6.7	...	12 01
10	W	4 15	6.8	4 55	6.2	10 51	10 57	10	S	6 24	7.1	7 08	6.9	12 20	1 07
11	T	5 09	6.7	5 59	6.2	...	12 09	11	S	7 26	7.2	8 09	7.2	1 28	2 20
12	F	6 09	6.8	7 10	6.4	12 01	1 40	12	M	8 29	7.4	9 06	7.6	2 43	3 24
13	S	7 12	7.0	8 09	6.8	1 14	2 35	13	T	9 28	7.7	9 57	8.1	3 50	4 17
14	S	8 10	7.3	8 53	7.2	2 28	3 21	14	W	10 22	8.0	10 46	8.5	4 47	5 04
15	M	9 01	7.7	9 34	7.6	3 23	4 02	15	T	11 13	8.2	11 33	8.8	5 38	5 50
16	T	9 49	8.0	10 16	8.1	4 11	4 40	16	F	...	...	12 03	8.3	6 28	6 36
17	W	10 34	8.2	10 58	8.4	4 56	5 18	17	S	12 21	8.9	12 53	8.2	7 18	7 23
18	T	11 20	8.3	11 42	8.6	5 40	5 58	18	S	1 09	8.9	1 45	8.0	8 12	8 15
19	F	...	...	12 08	8.3	6 27	6 40	19	M	2 00	8.6	2 41	7.7	9 11	9 13
20	S	12 28	8.6	12 57	8.1	7 16	7 26	20	T	2 55	8.2	3 45	7.3	10 17	10 26
21	S	1 17	8.5	1 49	7.7	8 11	8 17	21	W	3 59	7.8	5 09	7.0	11 27	11 46
22	M	2 09	8.2	2 46	7.2	9 17	9 18	22	T	5 29	7.4	6 33	7.0	...	12 35
23	T	3 07	7.7	3 57	6.8	10 37	10 47	23	F	7 00	7.2	7 41	7.1	12 59	1 40
24	W	4 20	7.3	5 40	6.6	11 53	...	24	S	8 09	7.3	8 41	7.4	2 05	2 39
25	T	6 09	7.1	6 59	6.7	12 13	1 02	25	S	9 08	7.5	9 34	7.7	3 05	3 34
26	F	7 27	7.3	8 03	7.0	1 24	2 04	26	M	10 01	7.6	10 23	7.9	3 59	4 23
27	S	8 29	7.5	8 58	7.4	2 25	2 59	27	T	10 48	7.7	11 07	8.0	4 48	5 09
28	S	9 23	7.7	9 47	7.7	3 21	3 50	28	W	11 32	7.6	11 47	8.0	5 34	5 51
29	M	10 11	7.8	10 32	7.8	4 11	4 37	29	T	...	...	12 12	7.6	6 16	6 29
30	T	10 55	7.8	11 13	7.9	4 57	5 19	30	F	12 23	7.9	12 48	7.4	6 54	7 01
31	W	11 36	7.8	11 50	7.9	5 40	5 58								

***Daylight Time starts**

Dates when Ht. of **Low** Water is below Mean Low with Ht. of lowest given for each period and Date of lowest in ( ):

1st - 4th: -0.9' (1st)
16th - 22nd: -0.1' (19th - 20th)
27th - 31st: -0.5' (29th - 31st)

1st - 2nd: -0.4' (1st)
14th - 21st: -1.2' (17th - 18th)
27th - 30th: -0.4' (29th)

Average Rise and Fall 7.1 ft.

When a high tide exceeds av. ht., the *following* low tide will be lower than av.

1999 HIGH & LOW WATER
WILLETS POINT, N.Y.
40–48N x 73–47W

Daylight Time

DAY OF MONTH	DAY OF WEEK	MAY HIGH a.m.	Ht.	HIGH p.m.	Ht.	LOW a.m.	LOW p.m.
1	S	12 50	7.8	1 18	7.2	7 29	7 22
2	S	1 03	7.7	1 33	7.1	7 56	7 26
3	M	1 20	7.7	1 52	7.1	8 06	7 50
4	T	1 51	7.7	2 23	7.1	8 25	8 27
5	W	2 29	7.7	3 02	7.1	9 00	9 10
6	T	3 12	7.6	3 47	7.1	9 43	9 57
7	F	3 59	7.5	4 37	7.1	10 31	10 50
8	S	4 52	7.4	5 31	7.2	11 25	11 49
9	S	5 51	7.3	6 31	7.3	...	12 23
10	M	6 51	7.3	7 29	7.6	12 54	1 26
11	T	7 54	7.5	8 28	8.0	2 07	2 31
12	W	8 57	7.6	9 25	8.4	3 22	3 34
13	T	9 57	7.8	10 18	8.7	4 26	4 31
14	F	10 53	8.0	11 10	8.9	5 23	5 25
15	S	11 47	8.0	...	...	6 17	6 18
16	S	12 02	9.0	12 42	8.0	7 12	7 12
17	M	12 54	8.9	1 37	7.9	8 07	8 10
18	T	1 48	8.6	2 38	7.6	9 05	9 13
19	W	2 46	8.2	3 46	7.4	10 06	10 21
20	T	3 57	7.8	5 01	7.3	11 08	11 31
21	F	5 23	7.4	6 12	7.3	...	12 11
22	S	6 38	7.3	7 15	7.4	12 38	1 11
23	S	7 43	7.2	8 12	7.5	1 41	2 09
24	M	8 40	7.3	9 05	7.7	2 39	3 02
25	T	9 32	7.3	9 53	7.9	3 32	3 52
26	W	10 21	7.4	10 38	7.9	4 22	4 38
27	T	11 05	7.4	11 19	7.9	5 07	5 20
28	F	11 46	7.3	11 55	7.9	5 49	5 57
29	S	...	...	12 23	7.2	6 28	6 30
30	S	12 21	7.8	12 54	7.1	7 03	6 49
31	M	12 31	7.7	1 09	7.1	7 31	6 58

Daylight Time

DAY OF MONTH	DAY OF WEEK	JUNE HIGH a.m.	Ht.	HIGH p.m.	Ht.	LOW a.m.	LOW p.m.
1	T	12 51	7.8	1 26	7.1	7 45	7 26
2	W	1 23	7.8	1 57	7.2	8 04	8 03
3	T	2 02	7.9	2 36	7.3	8 37	8 46
4	F	2 45	7.8	3 20	7.4	9 18	9 34
5	S	3 33	7.8	4 09	7.5	10 04	10 26
6	S	4 25	7.6	5 01	7.6	10 55	11 23
7	M	5 21	7.5	5 58	7.7	11 50	...
8	T	6 22	7.4	6 57	7.9	12 27	12 50
9	W	7 27	7.3	7 58	8.1	1 41	1 54
10	T	8 31	7.4	8 58	8.4	3 04	3 02
11	F	9 36	7.5	9 57	8.6	4 13	4 09
12	S	10 39	7.7	10 53	8.8	5 13	5 11
13	S	11 38	7.8	11 49	8.8	6 09	6 10
14	M	...	...	12 36	7.8	7 04	7 08
15	T	12 44	8.7	1 33	7.8	7 58	8 05
16	W	1 41	8.5	2 33	7.7	8 52	9 05
17	T	2 42	8.1	3 35	7.6	9 47	10 05
18	F	3 48	7.8	4 38	7.5	10 43	11 07
19	S	4 59	7.4	5 41	7.4	11 40	...
20	S	6 06	7.2	6 41	7.4	12 09	12 37
21	M	7 09	7.0	7 38	7.5	1 09	1 33
22	T	8 06	7.0	8 31	7.6	2 07	2 26
23	W	8 59	7.0	9 21	7.6	3 00	3 16
24	T	9 49	7.0	10 07	7.7	3 51	4 03
25	F	10 35	7.1	10 49	7.7	4 37	4 46
26	S	11 18	7.1	11 26	7.7	5 21	5 25
27	S	11 57	7.1	11 53	7.7	6 01	5 58
28	M	...	...	12 28	7.1	6 36	6 21
29	T	12 03	7.8	12 43	7.1	7 06	6 37
30	W	12 26	7.9	1 01	7.3	7 25	7 07

Dates when Ht. of **Low** Water is below Mean Low with Ht. of lowest given for each period and Date of lowest in ( ):

13th - 20th: -1.3' (16th)

11th - 18th: -1.2' (14th - 15th)

Average Rise and Fall 7.1 ft.

When a high tide exceeds av. ht., the *following* low tide will be lower than av.

1999 HIGH & LOW WATER
WILLETS POINT, N.Y.
40–48N x 73–47W

Daylight Time								Daylight Time							
DAY OF MONTH	DAY OF WEEK	JULY						DAY OF MONTH	DAY OF WEEK	AUGUST					
		HIGH				LOW				HIGH				LOW	
		a.m.	Ht.	p.m.	Ht.	a.m.	p.m.			a.m.	Ht.	p.m.	Ht.	a.m.	p.m.
1	T	1 00	8.0	1 33	7.4	7 45	7 44	1	S	2 03	8.2	2 31	8.2	8 34	8 56
2	F	1 39	8.1	2 11	7.6	8 17	8 27	2	M	2 50	8.1	3 18	8.2	9 17	9 47
3	S	2 23	8.0	2 55	7.7	8 56	9 14	3	T	3 41	7.8	4 10	8.1	10 06	10 44
4	S	3 10	7.9	3 43	7.8	9 40	10 05	4	W	4 37	7.5	5 06	8.0	10 59	11 55
5	M	4 02	7.7	4 34	7.9	10 29	11 01	5	T	5 38	7.1	6 08	7.9	...	12 01
6	T	4 57	7.5	5 30	7.9	11 22	...	6	F	6 49	6.9	7 18	7.8	1 36	1 23
7	W	5 57	7.3	6 30	8.0	12 06	12 21	7	S	8 16	6.9	8 39	7.9	2 55	3 00
8	T	7 03	7.1	7 33	8.1	1 28	1 28	8	S	9 38	7.1	9 56	8.1	3 59	4 09
9	F	8 15	7.1	8 41	8.2	2 58	2 48	9	M	10 41	7.4	10 58	8.3	4 56	5 07
10	S	9 28	7.2	9 46	8.4	4 06	4 05	10	T	11 35	7.7	11 52	8.4	5 48	6 01
11	S	10 37	7.4	10 49	8.5	5 05	5 10	11	W	...	...	12 24	7.9	6 38	6 52
12	M	11 37	7.6	11 48	8.6	6 00	6 07	12	T	12 41	8.4	1 11	8.0	7 25	7 40
13	T	...	...	12 33	7.8	6 53	7 02	13	F	1 28	8.3	1 56	8.0	8 10	8 28
14	W	12 44	8.5	1 26	7.8	7 43	7 56	14	S	2 13	8.0	2 39	7.9	8 54	9 14
15	T	1 38	8.4	2 18	7.8	8 33	8 49	15	S	2 58	7.7	3 22	7.7	9 36	10 02
16	F	2 32	8.1	3 11	7.7	9 23	9 42	16	M	3 43	7.3	4 06	7.4	10 18	10 51
17	S	3 27	7.7	4 05	7.6	10 13	10 37	17	T	4 33	7.0	4 55	7.2	11 00	11 45
18	S	4 25	7.4	5 01	7.4	11 03	11 34	18	W	5 32	6.7	5 54	7.1	11 45	...
19	M	5 26	7.1	5 58	7.3	11 56	...	19	T	6 37	6.5	6 59	7.0	12 42	12 42
20	T	6 27	6.8	6 56	7.3	12 31	12 49	20	F	7 39	6.5	7 59	7.0	1 40	1 42
21	W	7 26	6.7	7 52	7.3	1 29	1 43	21	S	8 36	6.6	8 53	7.2	2 35	2 39
22	T	8 22	6.7	8 44	7.3	2 24	2 35	22	S	9 27	6.8	9 39	7.3	3 26	3 29
23	F	9 15	6.8	9 33	7.4	3 16	3 24	23	M	10 11	7.0	10 16	7.5	4 12	4 14
24	S	10 03	6.9	10 17	7.5	4 04	4 10	24	T	10 48	7.3	10 41	7.8	4 53	4 53
25	S	10 47	7.0	10 55	7.6	4 49	4 51	25	W	11 13	7.5	11 05	8.0	5 30	5 26
26	M	11 26	7.1	11 21	7.7	5 29	5 27	26	T	11 34	7.8	11 39	8.2	6 00	5 58
27	T	11 56	7.2	11 35	7.9	6 06	5 55	27	F	...	...	12 04	8.1	6 27	6 33
28	W	...	...	12 11	7.4	6 36	6 19	28	S	12 17	8.4	12 41	8.4	6 56	7 12
29	T	12 02	8.0	12 34	7.6	6 59	6 50	29	S	12 59	8.4	1 23	8.5	7 32	7 55
30	F	12 38	8.2	1 08	7.9	7 23	7 28	30	M	1 44	8.3	2 07	8.6	8 12	8 42
31	S	1 19	8.3	1 47	8.0	7 55	8 10	31	T	2 32	8.1	2 55	8.5	8 56	9 34

Dates when Ht. of **Low** Water is below Mean Low with Ht. of lowest given for each period and Date of lowest in ( ):

10th - 16th: -1.1' (13th)

9th - 14th: -0.8' (11th - 12th)
28th - 30th: -0.2'

Average Rise and Fall 7.1 ft.

When a high tide exceeds av. ht., the *following* low tide will be lower than av.

1999 HIGH & LOW WATER WILLETS POINT, N.Y.

40–48N x 73–47W

Daylight Time — **Standard Time starts Oct. 31 at 2 a.m.**

SEPTEMBER

Day of Month	Day of Week	High a.m.	Ht.	High p.m.	Ht.	Low a.m.	Low p.m.
1	W	3 24	7.8	3 48	8.3	9 46	10 36
2	T	4 21	7.4	4 46	8.0	10 43	...
3	F	5 27	7.0	5 54	7.7	12 08	12 01
4	S	6 58	6.8	7 27	7.6	1 37	1 48
5	S	8 34	7.0	8 58	7.7	2 46	3 03
6	M	9 39	7.3	10 02	8.0	3 47	4 04
7	T	10 34	7.7	10 57	8.2	4 41	4 59
8	W	11 24	8.0	11 45	8.3	5 32	5 50
9	T	...	...	12 10	8.1	6 18	6 37
10	F	12 30	8.3	12 51	8.2	7 02	7 22
11	S	1 12	8.1	1 30	8.1	7 43	8 04
12	S	1 51	7.9	2 04	7.9	8 21	8 44
13	M	2 26	7.6	2 34	7.7	8 54	9 22
14	T	2 58	7.3	3 02	7.5	9 14	9 57
15	W	3 31	7.0	3 36	7.3	9 30	10 27
16	T	4 11	6.7	4 18	7.1	10 06	11 08
17	F	5 01	6.5	5 07	7.0	10 52	...
18	S	6 09	6.5	6 05	6.9	12 30	-A-
19	S	7 42	6.6	7 14	6.9	1 41	1 01
20	M	8 38	6.8	8 28	7.1	2 37	2 34
21	T	9 22	7.1	9 14	7.4	3 26	3 30
22	W	9 54	7.4	9 53	7.7	4 08	4 15
23	T	10 22	7.8	10 32	8.1	4 45	4 56
24	F	10 56	8.2	11 13	8.3	5 19	5 35
25	S	11 34	8.6	11 56	8.4	5 52	6 15
26	S	...	...	12 16	8.8	6 28	6 57
27	M	12 40	8.5	12 59	8.9	7 08	7 43
28	T	1 27	8.3	1 46	8.9	7 51	8 32
29	W	2 17	8.0	2 35	8.6	8 39	9 30
30	T	3 11	7.7	3 30	8.2	9 33	10 45

A also at 11:49 a.m.

OCTOBER

Day of Month	Day of Week	High a.m.	Ht.	High p.m.	Ht.	Low a.m.	Low p.m.
1	F	4 12	7.2	4 32	7.8	10 43	...
2	S	5 34	6.9	5 57	7.4	12 12	12 27
3	S	7 18	6.9	7 46	7.4	1 26	1 48
4	M	8 29	7.2	8 56	7.6	2 31	2 54
5	T	9 28	7.6	9 54	7.9	3 29	3 52
6	W	10 20	7.9	10 45	8.1	4 22	4 45
7	T	11 06	8.2	11 31	8.1	5 11	5 33
8	F	11 49	8.3	...	...	5 56	6 18
9	S	12 15	8.0	12 29	8.2	6 37	7 01
10	S	12 53	7.9	1 03	8.1	7 15	7 40
11	M	1 28	7.6	1 30	7.9	7 48	8 16
12	T	1 57	7.4	1 49	7.7	8 08	8 44
13	W	2 19	7.2	2 15	7.6	8 15	8 56
14	T	2 47	7.0	2 49	7.4	8 43	9 21
15	F	3 24	6.8	3 30	7.3	9 24	10 01
16	S	4 08	6.7	4 17	7.1	10 11	10 50
17	S	4 59	6.7	5 10	7.0	11 04	11 48
18	M	5 57	6.7	6 09	7.0	...	12 05
19	T	6 59	6.9	7 11	7.1	12 55	1 16
20	W	7 58	7.2	8 13	7.3	2 09	2 33
21	T	8 50	7.6	9 09	7.6	3 06	3 34
22	F	9 36	8.1	10 00	7.9	3 53	4 25
23	S	10 21	8.5	10 48	8.1	4 36	5 12
24	S	11 06	8.8	11 35	8.3	5 19	5 58
25	M	11 51	9.0	...	...	6 02	6 45
26	T	12 23	8.3	12 38	9.1	6 48	7 35
27	W	1 12	8.1	1 27	8.9	7 36	8 30
28	T	2 05	7.9	2 19	8.6	8 29	9 34
29	F	3 03	7.5	3 17	8.1	9 33	10 46
30	S	4 13	7.2	4 27	7.6	11 00	11 59
31	S	*4 48	7.0	*5 13	7.3	*11 23	...

***Standard Time starts**

Dates when Ht. of **Low** Water is below Mean Low with Ht. of lowest given for each period and Date of lowest in ( ):

September:
7th - 11th: -0.6' (9th)
25th - 29th: -0.6' (26th - 27th)

October:
6th - 10th: -0.5' (8th - 9th)
23rd - 29th: -1.0 (25th - 26th)

Average Rise and Fall 7.1 ft.

When a high tide exceeds av. ht., the *following* low tide will be lower than av.

1999 HIGH & LOW WATER WILLETS POINT, N.Y.

40–48N x 73–47W

Standard Time

DAY OF MONTH	DAY OF WEEK	NOVEMBER HIGH a.m.	Ht.	HIGH p.m.	Ht.	LOW a.m.	LOW p.m.
1	M	6 08	7.1	6 36	7.3	12 07	12 34
2	T	7 12	7.4	7 40	7.4	1 09	1 37
3	W	8 08	7.7	8 35	7.6	2 06	2 34
4	T	8 58	7.9	9 25	7.7	2 58	3 25
5	F	9 45	8.1	10 11	7.7	3 46	4 13
6	S	10 27	8.1	10 53	7.6	4 30	4 57
7	S	11 05	8.1	11 32	7.5	5 11	5 39
8	M	11 38	7.9	...	...	5 48	6 17
9	T	12 08	7.3	12 02	7.8	6 17	6 50
10	W	12 33	7.1	12 14	7.6	6 30	7 13
11	T	12 49	7.0	12 40	7.6	6 40	7 21
12	F	1 14	6.9	1 15	7.5	7 12	7 48
13	S	1 50	6.9	1 55	7.4	7 53	8 27
14	S	2 32	6.9	2 41	7.3	8 39	9 12
15	M	3 19	6.9	3 32	7.1	9 31	10 04
16	T	4 12	7.0	4 28	7.1	10 28	11 00
17	W	5 08	7.1	5 27	7.0	11 31	...
18	T	6 06	7.4	6 29	7.1	12 01	12 42
19	F	7 03	7.7	7 31	7.3	1 02	1 54
20	S	7 58	8.1	8 29	7.6	2 02	2 57
21	S	8 50	8.5	9 24	7.8	2 58	3 53
22	M	9 41	8.8	10 17	7.9	3 50	4 46
23	T	10 31	9.0	11 09	7.9	4 42	5 38
24	W	11 22	8.9	...	...	5 34	6 32
25	T	12 02	7.9	12 14	8.7	6 29	7 29
26	F	12 58	7.7	1 09	8.4	7 30	8 29
27	S	2 00	7.4	2 10	7.9	8 39	9 33
28	S	3 15	7.2	3 29	7.4	9 53	10 38
29	M	4 34	7.1	4 59	7.1	11 05	11 41
30	T	5 44	7.2	6 12	7.0	...	12 12

Standard Time

DAY OF MONTH	DAY OF WEEK	DECEMBER HIGH a.m.	Ht.	HIGH p.m.	Ht.	LOW a.m.	LOW p.m.
1	W	6 46	7.3	7 14	7.1	12 41	1 13
2	T	7 41	7.5	8 10	7.1	1 38	2 10
3	F	8 33	7.7	9 01	7.2	2 30	3 02
4	S	9 20	7.8	9 47	7.2	3 19	3 50
5	S	10 03	7.8	10 31	7.2	4 03	4 34
6	M	10 42	7.8	11 10	7.1	4 44	5 15
7	T	11 16	7.7	11 46	7.0	5 21	5 53
8	W	11 39	7.6	...	...	5 51	6 27
9	T	12 15	6.9	-A-	...	6 05	6 51
10	F	12 26	6.9	12 13	7.5	6 18	6 59
11	S	12 48	6.9	12 48	7.5	6 49	7 23
12	S	1 21	7.0	1 28	7.5	7 29	7 59
13	M	2 01	7.1	2 12	7.4	8 13	8 42
14	T	2 46	7.2	3 01	7.3	9 02	9 29
15	W	3 36	7.2	3 55	7.1	9 56	10 21
16	T	4 30	7.4	4 53	7.0	10 56	11 17
17	F	5 27	7.5	5 54	6.9	...	12 03
18	S	6 26	7.7	6 58	7.0	12 18	1 21
19	S	7 26	8.0	8 03	7.1	1 22	2 38
20	M	8 25	8.3	9 06	7.3	2 29	3 41
21	T	9 22	8.5	10 05	7.5	3 33	4 38
22	W	10 17	8.7	11 02	7.6	4 34	5 33
23	T	11 12	8.6	11 58	7.7	5 32	6 26
24	F	...	...	12 07	8.5	6 30	7 20
25	S	12 55	7.6	1 04	8.2	7 29	8 15
26	S	1 54	7.5	2 06	7.8	8 31	9 11
27	M	2 59	7.3	3 17	7.3	9 34	10 09
28	T	4 05	7.2	4 31	7.0	10 39	11 09
29	W	5 11	7.1	5 40	6.7	11 42	...
30	T	6 12	7.1	6 42	6.6	12 07	12 43
31	F	7 09	7.2	7 39	6.6	1 04	1 40

A also at 11:50 a.m. 7.5

Dates when Ht. of **Low** Water is below Mean Low with Ht. of lowest given for each period and Date of lowest in ( ):

November:
4th - 8th: -0.6' (6th)
20th - 28th: -1.4' (24th)

December:
3rd - 7th: -0.5' (5th)
19th - 27th: -1.5' (22nd - 23rd)

Average Rise and Fall 7.1 ft.

When a high tide exceeds av. ht., the *following* low tide will be lower than av.

1999 CURRENT TABLE
HELL GATE, EAST RIVER, N.Y.

(Off Mill Rock – 40-46-42N x 73-56-18W)

Standard Time

JANUARY

DAY OF MONTH	DAY OF WEEK	CURRENT TURNS NORTHEAST Flood Starts a.m.	**p.m.**	Kts.	SOUTHWEST Ebb Starts a.m.	**p.m.**	Kts.
1	F	2 56	**3 33**	3.8	8 57	**9 26**	a5.2
2	S	3 50	**4 25**	a3.9	9 50	**10 17**	a5.2
3	S	4 42	**5 16**	a3.8	10 41	**11 08**	a5.1
4	M	5 33	**6 06**	a3.7	11 31	**11 57**	a5.0
5	T	6 24	**6 56**	a3.6	...	**12 21**	4.9
6	W	7 15	**7 46**	a3.4	12 46	**1 10**	a4.7
7	T	8 08	**8 37**	a3.2	1 35	**2 00**	a4.5
8	F	9 01	**9 29**	a3.0	2 25	**2 51**	4.3
9	S	9 55	**10 21**	2.8	3 16	**3 42**	a4.2
10	S	10 49	**11 13**	2.8	4 08	**4 35**	4.1
11	M	11 41	**...**	2.8	4 59	**5 26**	4.1
12	T	12 02	**12 30**	p2.9	5 49	**6 16**	4.2
13	W	12 48	**1 17**	3.0	6 37	**7 03**	a4.4
14	T	1 32	**2 01**	3.1	7 23	**7 49**	4.5
15	F	2 14	**2 43**	3.3	8 07	**8 32**	a4.7
16	S	2 54	**3 23**	3.4	8 50	**9 13**	a4.9
17	S	3 33	**4 03**	a3.6	9 32	**9 54**	a5.0
18	M	4 13	**4 43**	a3.7	10 14	**10 35**	a5.1
19	T	4 53	**5 23**	a3.7	10 56	**11 17**	a5.1
20	W	5 36	**6 05**	a3.7	11 39	**...**	5.1
21	T	6 21	**6 50**	a3.7	12 01	**12 25**	5.0
22	F	7 10	**7 39**	a3.6	12 48	**1 15**	a5.0
23	S	8 05	**8 33**	a3.5	1 39	**2 09**	a4.9
24	S	9 06	**9 34**	3.3	2 35	**3 08**	a4.8
25	M	10 12	**10 38**	p3.3	3 36	**4 11**	a4.7
26	T	11 21	**11 45**	p3.3	4 41	**5 17**	a4.7
27	W	...	**12 27**	3.3	5 47	**6 22**	a4.7
28	T	12 48	**1 29**	3.4	6 50	**7 23**	a4.8
29	F	1 48	**2 26**	3.6	7 50	**8 19**	a4.9
30	S	2 44	**3 19**	3.7	8 45	**9 12**	a5.0
31	S	3 36	**4 08**	a3.8	9 37	**10 01**	a5.0

Standard Time

FEBRUARY

DAY OF MONTH	DAY OF WEEK	CURRENT TURNS NORTHEAST Flood Starts a.m.	**p.m.**	Kts.	SOUTHWEST Ebb Starts a.m.	**p.m.**	Kts.
1	M	4 25	**4 55**	a3.8	10 26	**10 48**	a5.0
2	T	5 12	**5 40**	a3.8	11 12	**11 33**	4.9
3	W	5 58	**6 25**	a3.7	11 57	**...**	4.8
4	T	6 43	**7 09**	a3.5	12 18	**12 42**	4.7
5	F	7 28	**7 53**	a3.3	1 02	**1 26**	a4.6
6	S	8 15	**8 39**	a3.1	1 47	**2 12**	a4.4
7	S	9 04	**9 28**	a2.9	2 33	**3 00**	a4.3
8	M	9 56	**10 19**	2.8	3 22	**3 51**	a4.1
9	T	10 49	**11 12**	p2.8	4 14	**4 44**	a4.1
10	W	11 43	**...**	2.8	5 07	**5 37**	a4.1
11	T	12 03	**12 35**	2.9	6 00	**6 29**	4.2
12	F	12 52	**1 23**	p3.1	6 51	**7 18**	a4.4
13	S	1 38	**2 09**	p3.3	7 39	**8 04**	4.6
14	S	2 22	**2 52**	3.5	8 25	**8 48**	4.8
15	M	3 05	**3 34**	a3.7	9 10	**9 31**	5.0
16	T	3 47	**4 15**	3.8	9 54	**10 14**	5.1
17	W	4 30	**4 57**	a3.9	10 38	**10 58**	5.2
18	T	5 14	**5 40**	a3.9	11 23	**11 43**	5.2
19	F	6 01	**6 26**	a3.9	...	**12 10**	5.1
20	S	6 51	**7 15**	a3.8	12 30	**1 00**	a5.1
21	S	7 45	**8 09**	a3.6	1 22	**1 54**	a5.0
22	M	8 46	**9 11**	a3.4	2 19	**2 53**	a4.8
23	T	9 53	**10 18**	3.2	3 21	**3 58**	a4.6
24	W	11 03	**11 28**	3.2	4 28	**5 05**	a4.5
25	T	...	**12 11**	3.2	5 36	**6 11**	a4.4
26	F	12 34	**1 14**	a3.4	6 41	**7 12**	4.5
27	S	1 34	**2 09**	3.5	7 40	**8 07**	4.6
28	S	2 29	**3 00**	a3.7	8 34	**8 57**	4.8

The Kts. (knots) columns show the **maximum** predicted velocities of the stronger one of the Flood Currents and the stronger one of the Ebb Currents for each day. The letter "a" means the velocity shown should occur **after** the **a.m.** Current Change. The letter "p" means the velocity shown should occur **after** the **p.m.** Current Change (even if next morning). No "a" or "p" means a.m. and p.m. velocities are the same for that day.

Av. Max. Vel.: Fl. 3.4 Kts., Ebb 4.6 Kts.

Max. Flood & Max. Ebb 3 hrs. after Flood Starts & Ebb Starts, ±10 min.

At **City Island** the Current turns 2 hours before Hell Gate. At **Throg's Neck** the Current turns 1 hour before Hell Gate. At **Whitestone Pt.** the Current turns 35 min. before Hell Gate. At **College Pt.** the Current turns 20 min. before Hell Gate.

1999 CURRENT TABLE
HELL GATE, EAST RIVER, N.Y.
(Off Mill Rock – 40-46-42N x 73-56-18W)

Standard Time **Daylight Time starts Apr. 4 at 2 a.m.**

MARCH

DAY OF MONTH	DAY OF WEEK	CURRENT TURNS NORTHEAST Flood Starts a.m.	p.m.	Kts.	SOUTHWEST Ebb Starts a.m.	p.m.	Kts.
1	M	3 18	**3 46**	a3.8	9 22	**9 43**	4.8
2	T	4 03	**4 29**	a3.8	10 08	**10 26**	4.9
3	W	4 46	**5 09**	a3.8	10 50	**11 08**	4.8
4	T	5 27	**5 49**	a3.7	11 32	**11 48**	4.8
5	F	6 08	**6 29**	a3.6	...	**12 12**	4.7
6	S	6 48	**7 08**	a3.4	12 29	**12 53**	a4.6
7	S	7 30	**7 50**	a3.3	1 10	**1 36**	a4.5
8	M	8 14	**8 35**	a3.1	1 53	**2 21**	a4.4
9	T	9 03	**9 24**	a2.9	2 40	**3 10**	a4.2
10	W	9 56	**10 18**	2.8	3 31	**4 03**	a4.1
11	T	10 53	**11 14**	p2.9	4 26	**4 58**	a4.1
12	F	11 49	**...**	2.9	5 23	**5 53**	4.1
13	S	12 08	**12 42**	p3.1	6 18	**6 45**	4.3
14	S	1 00	**1 31**	3.3	7 10	**7 34**	4.5
15	M	1 48	**2 18**	p3.6	7 59	**8 21**	p4.8
16	T	2 35	**3 02**	3.8	8 46	**9 06**	p5.0
17	W	3 21	**3 46**	a4.0	9 33	**9 52**	p5.2
18	T	4 07	**4 30**	a4.1	10 19	**10 37**	p5.3
19	F	4 54	**5 16**	a4.1	11 06	**11 24**	p5.3
20	S	5 42	**6 03**	a4.0	11 54	**...**	5.1
21	S	6 34	**6 55**	a3.9	12 14	**12 46**	a5.2
22	M	7 29	**7 51**	a3.7	1 07	**1 41**	a5.0
23	T	8 30	**8 54**	3.4	2 05	**2 41**	a4.7
24	W	9 37	**10 03**	3.2	3 08	**3 46**	a4.5
25	T	10 47	**11 14**	p3.2	4 16	**4 54**	a4.3
26	F	11 55	**...**	3.2	5 24	**5 58**	4.2
27	S	12 20	**12 56**	3.3	6 28	**6 58**	4.3
28	S	1 19	**1 49**	3.5	7 26	**7 50**	4.5
29	M	2 11	**2 37**	3.6	8 16	**8 37**	p4.7
30	T	2 57	**3 19**	3.7	9 02	**9 20**	p4.8
31	W	3 39	**3 59**	a3.8	9 44	**10 00**	p4.8

APRIL

DAY OF MONTH	DAY OF WEEK	CURRENT TURNS NORTHEAST Flood Starts a.m.	p.m.	Kts.	SOUTHWEST Ebb Starts a.m.	p.m.	Kts.
1	T	4 19	**4 37**	a3.8	10 24	**10 39**	p4.8
2	F	4 57	**5 15**	a3.7	11 03	**11 17**	p4.8
3	S	5 35	**5 51**	a3.6	11 41	**11 55**	p4.7
4	S	*7 13	***7 29**	a3.5	...	***1 21**	4.5
5	M	7 52	**8 08**	a3.3	1 35	**2 01**	a4.6
6	T	8 34	**8 50**	a3.2	2 16	**2 44**	a4.5
7	W	9 20	**9 37**	3.0	3 01	**3 31**	a4.3
8	T	10 11	**10 30**	2.9	3 50	**4 23**	a4.2
9	F	11 08	**11 29**	2.9	4 45	**5 18**	a4.1
10	S	...	**12 06**	3.0	5 43	**6 14**	a4.2
11	S	12 27	**1 02**	3.1	6 41	**7 09**	p4.4
12	M	1 24	**1 54**	p3.4	7 36	**8 01**	p4.6
13	T	2 17	**2 44**	3.6	8 29	**8 51**	p4.9
14	W	3 08	**3 32**	p3.9	9 19	**9 39**	p5.1
15	T	3 57	**4 19**	4.0	10 08	**10 27**	p5.3
16	F	4 46	**5 07**	4.1	10 57	**11 16**	p5.4
17	S	5 36	**5 55**	4.1	11 46	**...**	5.2
18	S	6 27	**6 46**	4.0	12 06	**12 37**	a5.3
19	M	7 21	**7 40**	a3.9	12 57	**1 30**	a5.2
20	T	8 18	**8 39**	3.6	1 52	**2 26**	a5.0
21	W	9 19	**9 43**	a3.4	2 51	**3 27**	a4.7
22	T	10 25	**10 52**	3.2	3 54	**4 31**	a4.4
23	F	11 32	**...**	3.1	5 00	**5 36**	a4.3
24	S	12 01	**12 37**	3.2	6 06	**6 38**	4.2
25	S	1 04	**1 35**	p3.3	7 08	**7 34**	4.3
26	M	2 00	**2 25**	3.4	8 02	**8 24**	p4.5
27	T	2 49	**3 10**	3.5	8 50	**9 09**	p4.6
28	W	3 33	**3 51**	3.6	9 34	**9 50**	p4.7
29	T	4 13	**4 30**	3.6	10 14	**10 29**	p4.8
30	F	4 52	**5 07**	3.6	10 53	**11 07**	p4.8

***Daylight Time starts**

The Kts. (knots) columns show the **maximum** predicted velocities of the stronger one of the Flood Currents and the stronger one of the Ebb Currents for each day. The letter "a" means the velocity shown should occur **after** the **a.m.** Current Change. The letter "p" means the velocity shown should occur **after** the **p.m.** Current Change (even if next morning). No "a" or "p" means a.m. and p.m. velocities are the same for that day.

Av. Max. Vel.: Fl. 3.4 Kts., Ebb 4.6 Kts.

Max. Flood & Max. Ebb 3 hrs. after Flood Starts & Ebb Starts, ±10 min.

See pp. 150-157 for Current Change at other points.

1999 CURRENT TABLE
HELL GATE, EAST RIVER, N.Y.
(Off Mill Rock – 40-46-42N x 73-56-18W)

Daylight Time

MAY

DAY OF MONTH	DAY OF WEEK	CURRENT TURNS: NORTHEAST Flood Starts a.m.	**p.m.**	Kts.	SOUTHWEST Ebb Starts a.m.	**p.m.**	Kts.
1	S	5 29	5 43	3.6	11 31	11 45	p4.8
2	S	6 06	6 20	a3.6	...	12 09	4.7
3	M	6 44	6 56	a3.5	12 22	12 48	a4.8
4	T	7 23	7 35	a3.4	1 01	1 28	a4.7
5	W	8 04	8 16	3.2	1 42	2 09	a4.6
6	T	8 48	9 02	3.1	2 25	2 55	a4.5
7	F	9 37	9 54	3.0	3 13	3 44	a4.4
8	S	10 31	10 52	3.0	4 06	4 38	a4.3
9	S	11 28	11 53	p3.2	5 04	5 34	a4.4
10	M	...	12 25	3.2	6 03	6 31	4.5
11	T	12 52	1 20	p3.4	7 01	7 26	p4.8
12	W	1 49	2 14	p3.7	7 57	8 20	p5.0
13	T	2 44	3 05	p3.9	8 50	9 12	p5.2
14	F	3 37	3 56	p4.0	9 43	10 04	p5.3
15	S	4 29	4 47	p4.1	10 34	10 55	p5.4
16	S	5 22	5 39	p4.1	11 26	11 47	p5.3
17	M	6 15	6 33	3.9	...	12 19	5.1
18	T	7 10	7 29	3.8	12 41	1 13	a5.2
19	W	8 07	8 28	a3.6	1 36	2 10	a5.0
20	T	9 08	9 31	a3.4	2 35	3 08	a4.7
21	F	10 10	10 36	3.2	3 35	4 09	a4.5
22	S	11 13	11 41	3.1	4 38	5 10	a4.3
23	S	...	12 13	3.1	5 39	6 09	4.2
24	M	12 41	1 08	3.1	6 37	7 02	p4.3
25	T	1 34	1 57	3.2	7 29	7 51	p4.4
26	W	2 22	2 41	3.3	8 17	8 35	p4.5
27	T	3 06	3 22	3.4	9 00	9 17	p4.6
28	F	3 46	4 00	p3.5	9 41	9 57	p4.7
29	S	4 25	4 38	3.5	10 20	10 35	p4.8
30	S	5 03	5 15	3.5	10 59	11 14	p4.8
31	M	5 41	5 52	3.5	11 38	11 52	p4.8

Daylight Time

JUNE

DAY OF MONTH	DAY OF WEEK	CURRENT TURNS: NORTHEAST Flood Starts a.m.	**p.m.**	Kts.	SOUTHWEST Ebb Starts a.m.	**p.m.**	Kts.
1	T	6 19	6 29	3.4	...	12 17	4.7
2	W	6 58	7 08	3.4	12 31	12 56	a4.8
3	T	7 38	7 49	3.3	1 12	1 37	a4.8
4	F	8 21	8 34	3.2	1 55	2 21	a4.7
5	S	9 08	9 25	p3.2	2 41	3 09	a4.6
6	S	9 59	10 21	p3.2	3 33	4 01	a4.6
7	M	10 54	11 22	3.2	4 29	4 58	4.6
8	T	11 52	...	3.3	5 28	5 56	p4.7
9	W	12 24	12 50	p3.5	6 28	6 55	p4.8
10	T	1 24	1 47	p3.6	7 27	7 52	p5.0
11	F	2 23	2 42	p3.8	8 24	8 48	p5.2
12	S	3 19	3 37	p3.9	9 20	9 43	p5.3
13	S	4 14	4 31	p4.0	10 14	10 37	p5.3
14	M	5 08	5 25	p4.0	11 08	11 31	p5.3
15	T	6 03	6 20	3.9	...	12 02	5.0
16	W	6 57	7 16	3.7	12 26	12 56	a5.1
17	T	7 52	8 13	a3.6	1 20	1 50	a5.0
18	F	8 49	9 12	a3.4	2 16	2 46	a4.7
19	S	9 46	10 12	3.2	3 12	3 42	a4.5
20	S	10 44	11 11	3.1	4 09	4 38	a4.3
21	M	11 39	...	3.0	5 06	5 33	4.2
22	T	12 08	12 32	3.0	6 01	6 26	p4.2
23	W	1 00	1 21	p3.1	6 52	7 15	p4.3
24	T	1 49	2 06	3.1	7 40	8 00	p4.4
25	F	2 33	2 49	3.2	8 25	8 44	p4.5
26	S	3 15	3 29	3.3	9 08	9 25	p4.6
27	S	3 56	4 08	p3.4	9 50	10 06	p4.7
28	M	4 35	4 46	p3.5	10 30	10 46	p4.8
29	T	5 14	5 24	p3.5	11 10	11 26	p4.9
30	W	5 53	6 02	3.5	11 49	...	4.7

The Kts. (knots) columns show the **maximum** predicted velocities of the stronger one of the Flood Currents and the stronger one of the Ebb Currents for each day. The letter "a" means the velocity shown should occur **after** the **a.m.** Current Change. The letter "p" means the velocity shown should occur **after** the **p.m.** Current Change (even if next morning). No "a" or "p" means a.m. and p.m. velocities are the same for that day.

Av. Max. Vel.: Fl. 3.4 Kts., Ebb 4.6 Kts.

Max. Flood & Max. Ebb 3 hrs. after Flood Starts & Ebb Starts, ±10 min.

At **City Island** the Current turns 2 hours before Hell Gate. At **Throg's Neck** the Current turns 1 hour before Hell Gate. At **Whitestone Pt.** the Current turns 35 min. before Hell Gate. At **College Pt.** the Current turns 20 min. before Hell Gate.

1999 CURRENT TABLE
HELL GATE, EAST RIVER, N.Y.
(Off Mill Rock – 40-46-42N x 73-56-18W)

JULY
Daylight Time

DAY OF MONTH	DAY OF WEEK	CURRENT TURNS					
		NORTHEAST Flood Starts			SOUTHWEST Ebb Starts		
		a.m.	**p.m.**	Kts.	a.m.	**p.m.**	Kts.
1	T	6 32	**6 41**	p3.5	12 06	**12 29**	a4.9
2	F	7 11	**7 22**	p3.5	12 47	**1 10**	a4.9
3	S	7 53	**8 07**	3.4	1 30	**1 54**	a4.8
4	S	8 38	**8 57**	p3.4	2 16	**2 41**	a4.8
5	M	9 27	**9 53**	3.3	3 07	**3 32**	4.7
6	T	10 21	**10 54**	3.3	4 02	**4 29**	p4.7
7	W	11 20	**11 57**	3.4	5 01	**5 29**	p4.7
8	T	...	**12 21**	3.5	6 03	**6 31**	p4.8
9	F	1 01	**1 23**	p3.6	7 04	**7 32**	p4.9
10	S	2 03	**2 23**	p3.7	8 05	**8 31**	p5.0
11	S	3 03	**3 20**	p3.9	9 03	**9 29**	p5.1
12	M	3 59	**4 16**	p3.9	9 59	**10 24**	p5.1
13	T	4 54	**5 11**	p4.0	10 53	**11 18**	p5.1
14	W	5 46	**6 04**	p3.9	11 46	...	5.0
15	T	6 38	**6 57**	3.8	12 11	**12 37**	a5.0
16	F	7 29	**7 49**	3.6	1 03	**1 29**	a4.9
17	S	8 21	**8 42**	a3.5	1 54	**2 20**	a4.7
18	S	9 12	**9 36**	a3.3	2 46	**3 11**	a4.5
19	M	10 04	**10 30**	3.1	3 38	**4 03**	a4.3
20	T	10 56	**11 24**	3.0	4 30	**4 55**	4.1
21	W	11 48	...	2.9	5 23	**5 47**	4.0
22	T	12 17	**12 38**	p3.0	6 15	**6 38**	p4.1
23	F	1 08	**1 26**	3.0	7 05	**7 27**	p4.2
24	S	1 55	**2 11**	p3.1	7 53	**8 13**	p4.3
25	S	2 40	**2 54**	p3.3	8 38	**8 58**	p4.5
26	M	3 23	**3 35**	p3.4	9 22	**9 41**	p4.6
27	T	4 04	**4 15**	p3.6	10 04	**10 22**	p4.8
28	W	4 44	**4 54**	p3.6	10 44	**11 03**	p4.9
29	T	5 23	**5 33**	p3.7	11 25	**11 45**	p4.9
30	F	6 02	**6 13**	p3.7	...	**12 05**	4.9
31	S	6 41	**6 55**	p3.7	12 27	**12 46**	4.9

AUGUST
Daylight Time

DAY OF MONTH	DAY OF WEEK	CURRENT TURNS					
		NORTHEAST Flood Starts			SOUTHWEST Ebb Starts		
		a.m.	**p.m.**	Kts.	a.m.	**p.m.**	Kts.
1	S	7 22	**7 41**	p3.7	1 10	**1 30**	4.9
2	M	8 07	**8 30**	3.6	1 56	**2 18**	4.8
3	T	8 56	**9 26**	3.5	2 46	**3 10**	p4.8
4	W	9 51	**10 28**	a3.5	3 42	**4 08**	p4.7
5	T	10 52	**11 34**	a3.4	4 42	**5 10**	p4.6
6	F	11 58	...	3.4	5 46	**6 15**	p4.6
7	S	12 41	**1 04**	p3.5	6 50	**7 20**	p4.7
8	S	1 46	**2 07**	p3.7	7 53	**8 21**	p4.8
9	M	2 47	**3 06**	p3.8	8 52	**9 19**	p4.9
10	T	3 43	**4 01**	p3.9	9 47	**10 14**	p5.0
11	W	4 35	**4 54**	p4.0	10 39	**11 06**	p5.0
12	T	5 25	**5 43**	3.9	11 29	**11 55**	4.9
13	F	6 13	**6 32**	p3.9	...	**12 17**	4.8
14	S	6 59	**7 19**	3.7	12 43	**1 04**	a4.8
15	S	7 45	**8 06**	a3.6	1 30	**1 50**	4.6
16	M	8 30	**8 53**	a3.4	2 17	**2 37**	4.4
17	T	9 17	**9 42**	a3.2	3 04	**3 26**	4.2
18	W	10 06	**10 34**	3.0	3 53	**4 16**	4.0
19	T	10 57	**11 27**	2.9	4 45	**5 08**	3.9
20	F	11 50	...	2.9	5 38	**6 01**	p3.9
21	S	12 21	**12 41**	p3.0	6 30	**6 53**	p4.0
22	S	1 13	**1 31**	p3.1	7 21	**7 43**	p4.1
23	M	2 01	**2 17**	p3.3	8 09	**8 30**	p4.3
24	T	2 47	**3 01**	p3.5	8 54	**9 15**	p4.5
25	W	3 30	**3 43**	p3.7	9 37	**9 58**	p4.7
26	T	4 10	**4 24**	p3.8	10 19	**10 41**	p4.9
27	F	4 50	**5 05**	p3.9	11 00	**11 23**	p5.0
28	S	5 30	**5 47**	p4.0	11 41	...	5.0
29	S	6 11	**6 30**	p4.0	12 06	**12 24**	5.0
30	M	6 53	**7 17**	3.9	12 51	**1 09**	5.0
31	T	7 39	**8 08**	a3.8	1 38	**1 58**	4.9

The Kts. (knots) columns show the **maximum** predicted velocities of the stronger one of the Flood Currents and the stronger one of the Ebb Currents for each day. The letter "a" means the velocity shown should occur **after** the **a.m.** Current Change. The letter "p" means the velocity shown should occur **after** the **p.m.** Current Change (even if next morning). No "a" or "p" means a.m. and p.m. velocities are the same for that day.

Av. Max. Vel.: Fl. 3.4 Kts., Ebb 4.6 Kts.

Max. Flood & Max. Ebb 3 hrs. after Flood Starts & Ebb Starts, ±10 min.

See pp. 150-157 for Current Change at other points.

1999 CURRENT TABLE
HELL GATE, EAST RIVER, N.Y.

(Off Mill Rock – 40-46-42N x 73-56-18W)

Daylight Time **Standard Time starts Oct. 31 at 2 a.m.**

SEPTEMBER

DAY OF MONTH	DAY OF WEEK	CURRENT TURNS NORTHEAST Flood Starts a.m.	**p.m.**	Kts.	CURRENT TURNS SOUTHWEST Ebb Starts a.m.	**p.m.**	Kts.
1	W	8 29	**9 04**	a3.7	2 29	**2 52**	p4.8
2	T	9 26	**10 07**	a3.5	3 26	**3 52**	p4.6
3	F	10 31	**11 16**	a3.4	4 28	**4 57**	4.4
4	S	11 41	...	3.4	5 34	**6 05**	p4.4
5	S	12 26	**12 50**	p3.5	6 41	**7 11**	p4.4
6	M	1 32	**1 55**	p3.6	7 44	**8 13**	p4.5
7	T	2 31	**2 53**	p3.8	8 42	**9 10**	p4.7
8	W	3 25	**3 46**	p3.9	9 34	**10 01**	p4.8
9	T	4 15	**4 34**	p4.0	10 23	**10 49**	4.8
10	F	5 00	**5 20**	p4.0	11 09	**11 34**	4.8
11	S	5 44	**6 04**	3.9	11 53	...	4.8
12	S	6 26	**6 46**	3.8	12 18	**12 35**	4.7
13	M	7 07	**7 28**	3.6	1 01	**1 18**	4.6
14	T	7 48	**8 11**	3.4	1 44	**2 01**	4.4
15	W	8 31	**8 56**	3.2	2 28	**2 46**	4.2
16	T	9 16	**9 45**	a3.1	3 15	**3 34**	p4.1
17	F	10 06	**10 38**	2.9	4 05	**4 26**	3.9
18	S	11 00	**11 34**	2.9	4 58	**5 21**	p3.9
19	S	11 56	...	2.9	5 53	**6 16**	p3.9
20	M	12 29	**12 50**	p3.1	6 46	**7 09**	p4.1
21	T	1 22	**1 40**	p3.3	7 36	**7 59**	p4.3
22	W	2 10	**2 27**	p3.5	8 23	**8 46**	p4.5
23	T	2 55	**3 12**	p3.7	9 07	**9 31**	p4.7
24	F	3 37	**3 56**	p3.9	9 50	**10 15**	4.9
25	S	4 19	**4 39**	p4.1	10 32	**10 59**	5.0
26	S	5 01	**5 24**	p4.1	11 16	**11 43**	a5.2
27	M	5 44	**6 10**	4.1	...	**12 01**	5.2
28	T	6 29	**6 59**	4.0	12 30	**12 48**	p5.1
29	W	7 17	**7 52**	a3.9	1 19	**1 39**	p5.0
30	T	8 11	**8 50**	a3.7	2 12	**2 35**	p4.8

OCTOBER

DAY OF MONTH	DAY OF WEEK	CURRENT TURNS NORTHEAST Flood Starts a.m.	**p.m.**	Kts.	CURRENT TURNS SOUTHWEST Ebb Starts a.m.	**p.m.**	Kts.
1	F	9 11	**9 55**	a3.5	3 11	**3 37**	4.5
2	S	10 20	**11 05**	3.3	4 15	**4 44**	p4.3
3	S	11 32	...	3.3	5 23	**5 53**	p4.2
4	M	12 15	**12 41**	p3.4	6 29	**7 00**	p4.3
5	T	1 19	**1 44**	p3.5	7 31	**8 00**	p4.4
6	W	2 16	**2 40**	p3.7	8 26	**8 54**	p4.6
7	T	3 07	**3 30**	p3.8	9 15	**9 42**	4.7
8	F	3 53	**4 15**	p3.9	10 01	**10 26**	a4.8
9	S	4 35	**4 57**	p3.9	10 43	**11 08**	a4.8
10	S	5 15	**5 37**	3.8	11 24	**11 48**	a4.8
11	M	5 54	**6 17**	3.7	...	**12 03**	4.7
12	T	6 33	**6 56**	3.6	12 28	**12 43**	p4.7
13	W	7 11	**7 36**	a3.5	1 09	**1 24**	4.5
14	T	7 52	**8 18**	a3.3	1 51	**2 06**	p4.4
15	F	8 35	**9 05**	a3.1	2 35	**2 52**	p4.2
16	S	9 23	**9 56**	a3.0	3 23	**3 42**	p4.1
17	S	10 16	**10 52**	2.9	4 15	**4 36**	p4.0
18	M	11 13	**11 49**	2.9	5 09	**5 33**	p4.0
19	T	...	**12 11**	3.0	6 03	**6 28**	p4.1
20	W	12 43	**1 05**	p3.3	6 55	**7 21**	p4.4
21	T	1 34	**1 56**	p3.5	7 45	**8 11**	p4.6
22	F	2 21	**2 45**	p3.7	8 32	**8 59**	4.8
23	S	3 07	**3 32**	p3.9	9 18	**9 45**	5.0
24	S	3 52	**4 19**	p4.1	10 04	**10 32**	a5.2
25	M	4 37	**5 07**	4.1	10 50	**11 19**	a5.3
26	T	5 24	**5 56**	4.1	11 38	...	5.3
27	W	6 12	**6 47**	a4.0	12 08	**12 28**	p5.2
28	T	7 04	**7 42**	a3.9	1 00	**1 21**	p5.1
29	F	8 01	**8 42**	a3.7	1 54	**2 18**	p4.8
30	S	9 04	**9 48**	a3.5	2 54	**3 21**	4.5
31	S	*9 13	***9 56**	a3.3	*2 58	***3 27**	4.3

***Standard Time starts**

The Kts. (knots) columns show the **maximum** predicted velocities of the stronger one of the Flood Currents and the stronger one of the Ebb Currents for each day. The letter "a" means the velocity shown should occur **after** the **a.m.** Current Change. The letter "p" means the velocity shown should occur **after** the **p.m.** Current Change (even if next morning). No "a" or "p" means a.m. and p.m. velocities are the same for that day.

Av. Max. Vel.: Fl. 3.4 Kts., Ebb 4.6 Kts.

Max. Flood & Max. Ebb 3 hrs. after Flood Starts & Ebb Starts, ±10 min.

At **City Island** the Current turns 2 hours before Hell Gate. At **Throg's Neck** the Current turns 1 hour before Hell Gate. At **Whitestone Pt.** the Current turns 35 min. before Hell Gate. At **College Pt.** the Current turns 20 min. before Hell Gate.

1999 CURRENT TABLE
HELL GATE, EAST RIVER, N.Y.
(Off Mill Rock – 40-46-42N x 73-56-18W)

Standard Time

NOVEMBER

DAY OF MONTH	DAY OF WEEK	NORTHEAST Flood Starts a.m.	p.m.	Kts.	SOUTHWEST Ebb Starts a.m.	p.m.	Kts.
1	M	10 24	**11 03**	3.2	4 04	**4 35**	4.2
2	T	11 32	**...**	3.3	5 09	**5 39**	p4.3
3	W	12 05	**12 32**	p3.4	6 08	**6 37**	p4.4
4	T	12 59	**1 25**	p3.5	7 01	**7 29**	4.5
5	F	1 48	**2 12**	p3.6	7 49	**8 15**	4.6
6	S	2 31	**2 55**	p3.7	8 32	**8 57**	a4.7
7	S	3 12	**3 35**	3.7	9 13	**9 37**	a4.8
8	M	3 50	**4 14**	3.7	9 52	**10 16**	a4.8
9	T	4 28	**4 52**	3.6	10 30	**10 54**	a4.8
10	W	5 05	**5 30**	3.5	11 08	**11 33**	a4.8
11	T	5 43	**6 09**	3.4	11 47	**...**	4.7
12	F	6 22	**6 51**	a3.3	12 13	**12 28**	p4.6
13	S	7 03	**7 35**	a3.2	12 55	**1 11**	p4.5
14	S	7 49	**8 23**	a3.0	1 40	**1 58**	p4.3
15	M	8 40	**9 15**	2.9	2 28	**2 50**	p4.3
16	T	9 36	**10 11**	2.9	3 21	**3 45**	p4.3
17	W	10 35	**11 06**	p3.1	4 15	**4 42**	p4.3
18	T	11 32	**11 59**	p3.3	5 10	**5 38**	p4.5
19	F	...	**12 27**	3.4	6 03	**6 32**	p4.7
20	S	12 50	**1 20**	p3.6	6 55	**7 25**	4.9
21	S	1 40	**2 11**	3.8	7 46	**8 16**	5.1
22	M	2 29	**3 02**	p4.0	8 36	**9 06**	a5.3
23	T	3 19	**3 53**	4.0	9 26	**9 56**	a5.4
24	W	4 09	**4 45**	a4.1	10 17	**10 48**	a5.4
25	T	5 01	**5 39**	a4.0	11 09	**11 41**	a5.3
26	F	5 56	**6 35**	a3.9	...	**12 04**	5.1
27	S	6 55	**7 34**	a3.6	12 36	**1 01**	p4.9
28	S	7 57	**8 37**	a3.4	1 34	**2 01**	4.6
29	M	9 03	**9 42**	a3.2	2 35	**3 05**	4.4
30	T	10 11	**10 45**	3.1	3 38	**4 08**	4.3

Standard Time

DECEMBER

DAY OF MONTH	DAY OF WEEK	NORTHEAST Flood Starts a.m.	p.m.	Kts.	SOUTHWEST Ebb Starts a.m.	p.m.	Kts.
1	W	11 15	**11 44**	3.1	4 40	**5 10**	4.3
2	T	...	**12 13**	3.2	5 37	**6 06**	4.3
3	F	12 37	**1 05**	p3.3	6 30	**6 57**	4.4
4	S	1 25	**1 51**	3.3	7 17	**7 42**	a4.6
5	S	2 08	**2 34**	3.4	8 00	**8 25**	a4.7
6	M	2 48	**3 13**	3.4	8 41	**9 05**	a4.7
7	T	3 26	**3 52**	3.5	9 20	**9 44**	a4.8
8	W	4 04	**4 30**	3.5	9 59	**10 22**	a4.8
9	T	4 41	**5 08**	a3.5	10 37	**11 01**	a4.9
10	F	5 19	**5 47**	a3.4	11 16	**11 40**	a4.8
11	S	5 57	**6 26**	a3.3	11 55	**...**	4.8
12	S	6 37	**7 08**	a3.3	12 20	**12 37**	p4.7
13	M	7 20	**7 52**	a3.2	1 02	**1 21**	p4.6
14	T	8 08	**8 40**	a3.1	1 46	**2 09**	4.5
15	W	9 01	**9 32**	3.0	2 36	**3 02**	4.5
16	T	9 59	**10 28**	3.1	3 29	**3 59**	4.5
17	F	10 59	**11 25**	p3.3	4 26	**4 58**	4.6
18	S	11 59	**...**	3.3	5 24	**5 56**	a4.8
19	S	12 21	**12 57**	3.5	6 22	**6 54**	4.9
20	M	1 16	**1 53**	3.7	7 18	**7 50**	a5.1
21	T	2 10	**2 48**	3.8	8 13	**8 44**	a5.3
22	W	3 04	**3 41**	a4.0	9 07	**9 37**	a5.4
23	T	3 58	**4 35**	a4.0	10 01	**10 30**	a5.4
24	F	4 52	**5 29**	a4.0	10 54	**11 24**	a5.3
25	S	5 47	**6 24**	a3.8	11 49	**...**	5.2
26	S	6 44	**7 20**	a3.7	12 18	**12 44**	p5.0
27	M	7 43	**8 18**	a3.5	1 13	**1 40**	a4.8
28	T	8 43	**9 17**	a3.2	2 09	**2 38**	a4.6
29	W	9 46	**10 16**	a3.1	3 07	**3 37**	a4.4
30	T	10 47	**11 13**	3.0	4 05	**4 34**	a4.3
31	F	11 44	**...**	3.0	5 01	**5 30**	a4.3

The Kts. (knots) columns show the **maximum** predicted velocities of the stronger one of the Flood Currents and the stronger one of the Ebb Currents for each day. The letter "a" means the velocity shown should occur **after** the **a.m.** Current Change. The letter "p" means the velocity shown should occur **after** the **p.m.** Current Change (even if next morning). No "a" or "p" means a.m. and p.m. velocities are the same for that day.

Av. Max. Vel.: Fl. 3.4 Kts., Ebb 4.6 Kts.

Max. Flood & Max. Ebb 3 hrs. after Flood Starts & Ebb Starts, ±10 min.

See pp. 150-157 for Current Change at other points.

1999 CURRENT TABLE

THE NARROWS, N.Y. HARBOR

(Mid-Channel – 40-36-36N x 74-02-48W)

Standard Time

JANUARY

DAY OF MONTH	DAY OF WEEK	CURRENT TURNS NORTH Flood Starts a.m.	p.m.	Kts.	CURRENT TURNS SOUTH Ebb Starts a.m.	p.m.	Kts.
1	F	2 58	3 52	a2.5	8 54	9 06	a2.5
2	S	3 49	4 41	a2.5	9 45	9 58	a2.6
3	S	4 39	5 29	a2.4	10 34	10 51	a2.6
4	M	5 28	6 17	a2.2	11 23	11 43	a2.5
5	T	6 20	7 07	a2.0	...	12 10	2.3
6	W	7 14	7 58	a1.8	12 34	12 56	p2.2
7	T	8 11	8 49	a1.6	1 24	1 42	p2.0
8	F	9 09	9 40	p1.5	2 16	2 29	p1.8
9	S	10 07	10 30	p1.5	3 09	3 18	p1.7
10	S	11 03	11 18	p1.5	4 06	4 12	p1.7
11	M	11 59	...	1.1	5 03	5 07	p1.7
12	T	12 07	12 53	a1.5	5 57	6 01	1.7
13	W	12 55	1 46	a1.6	6 46	6 51	a1.8
14	T	1 42	2 35	a1.7	7 32	7 38	a1.9
15	F	2 27	3 20	a1.9	8 15	8 23	a2.0
16	S	3 10	4 02	a2.0	8 57	9 07	a2.1
17	S	3 50	4 42	a2.1	9 39	9 51	a2.2
18	M	4 30	5 21	a2.1	10 22	10 36	a2.3
19	T	5 11	6 00	a2.1	11 05	11 22	a2.3
20	W	5 54	6 42	a2.0	11 49	...	2.3
21	T	6 43	7 26	a1.9	12 10	12 34	p2.3
22	F	7 39	8 15	1.8	1 00	1 20	p2.2
23	S	8 41	9 08	p1.9	1 52	2 10	p2.1
24	S	9 45	10 02	p1.9	2 48	3 03	p2.0
25	M	10 48	10 58	p1.9	3 49	4 02	2.0
26	T	11 50	11 56	p2.0	4 53	5 05	2.0
27	W	...	12 51	1.5	5 55	6 06	a2.2
28	T	12 54	1 50	a2.1	6 54	7 04	a2.3
29	F	1 51	2 44	a2.2	7 48	7 59	a2.3
30	S	2 45	3 35	a2.3	8 39	8 51	a2.4
31	S	3 36	4 21	a2.3	9 27	9 41	a2.4

Standard Time

FEBRUARY

DAY OF MONTH	DAY OF WEEK	CURRENT TURNS NORTH Flood Starts a.m.	p.m.	Kts.	CURRENT TURNS SOUTH Ebb Starts a.m.	p.m.	Kts.
1	M	4 25	5 06	a2.2	10 14	10 31	a2.4
2	T	5 12	5 51	a2.1	10 59	11 19	a2.4
3	W	5 59	6 35	a1.9	11 42	...	2.3
4	T	6 48	7 21	a1.7	12 07	12 25	p2.2
5	F	7 40	8 09	1.6	12 53	1 07	p2.0
6	S	8 35	8 58	p1.5	1 40	1 51	p1.8
7	S	9 31	9 48	p1.5	2 29	2 37	p1.7
8	M	10 28	10 38	p1.5	3 21	3 29	p1.6
9	T	11 24	11 28	p1.4	4 17	4 25	1.6
10	W	...	12 20	1.0	5 15	5 23	1.6
11	T	12 19	1 13	a1.5	6 09	6 18	a1.8
12	F	1 09	2 04	a1.6	6 59	7 09	a1.9
13	S	1 58	2 50	a1.8	7 45	7 56	a2.0
14	S	2 44	3 32	a1.9	8 29	8 41	a2.1
15	M	3 28	4 11	a2.1	9 13	9 26	a2.2
16	T	4 10	4 49	a2.2	9 56	10 12	a2.3
17	W	4 53	5 28	a2.2	10 40	10 59	a2.4
18	T	5 37	6 08	a2.1	11 25	11 48	a2.4
19	F	6 26	6 52	2.0	...	12 10	2.4
20	S	7 21	7 43	p2.0	12 38	12 58	2.3
21	S	8 22	8 38	p2.0	1 30	1 47	a2.2
22	M	9 26	9 38	p2.0	2 26	2 42	a2.1
23	T	10 30	10 38	p1.9	3 26	3 42	a2.0
24	W	11 32	11 39	p1.9	4 31	4 46	a2.0
25	T	...	12 33	1.4	5 36	5 51	a2.1
26	F	12 39	1 31	a1.9	6 37	6 51	2.1
27	S	1 37	2 25	a2.0	7 31	7 45	a2.2
28	S	2 32	3 13	a2.1	8 20	8 36	a2.3

The Kts. (knots) columns show the **maximum** predicted velocities of the stronger one of the Flood Currents and the stronger one of the Ebb Currents for each day. The letter "a" means the velocity shown should occur **after** the **a.m.** Current Change. The letter "p" means the velocity shown should occur **after** the **p.m.** Current Change (even if next morning). No "a" or "p" means a.m. and p.m. velocities are the same for that day.

Av. Max. Vel.: Fl. 1.7 Kts., Ebb 2.0 Kts.

Max. Flood 2 hrs. 25 min. after Flood Starts,30 min.

Max. Ebb 3 hrs. 15 min. after Ebb Starts,10 min.

At **The Battery, Desbrosses St., & Chelsea Dock** Current turns 1 1/2 hrs. after the Narrows. At **42nd St.**, the Current turns 1 3/4 hrs. after the Narrows. At **George Washington Br.** the Current turns 2 hrs. after the Narrows.

1999 CURRENT TABLE

THE NARROWS, N.Y. HARBOR

(Mid-Channel – 40-36-36N x 74-02-48W)

Standard Time **Daylight Time starts Apr. 4 at 2 a.m.**

MARCH								APRIL							
DAY OF MONTH	DAY OF WEEK	CURRENT TURNS						DAY OF MONTH	DAY OF WEEK	CURRENT TURNS					
		NORTH Flood Starts			SOUTH Ebb Starts					NORTH Flood Starts			SOUTH Ebb Starts		
		a.m.	**p.m.**	Kts.	a.m.	**p.m.**	Kts.			a.m.	**p.m.**	Kts.	a.m.	**p.m.**	Kts.
1	M	3 22	**3 58**	a2.1	9 06	**9 23**	a2.3	1	T	4 35	**4 50**	p1.9	10 02	**10 27**	2.1
2	T	4 09	**4 40**	a2.1	9 50	**10 09**	a2.3	2	F	5 18	**5 28**	p1.8	10 42	**11 08**	2.1
3	W	4 54	**5 21**	a2.0	10 32	**10 54**	a2.3	3	S	6 00	**6 06**	p1.8	11 22	**11 50**	2.0
4	T	5 38	**6 01**	1.8	11 13	**11 38**	a2.2	4	S	*7 45	***7 46**	p1.7	...	***1 03**	1.9
5	F	6 23	**6 43**	1.7	11 53	**...**	2.1	5	M	8 34	**8 31**	p1.6	1 33	**1 46**	a1.9
6	S	7 11	**7 27**	p1.6	12 21	**12 34**	a2.0	6	T	9 27	**9 22**	p1.5	2 17	**2 31**	a1.8
7	S	8 03	**8 14**	p1.6	1 05	**1 17**	1.8	7	W	10 22	**10 17**	p1.5	3 03	**3 20**	a1.7
8	M	8 58	**9 05**	p1.5	1 51	**2 02**	a1.7	8	T	11 16	**11 13**	p1.4	3 54	**4 14**	a1.6
9	T	9 54	**9 57**	p1.5	2 40	**2 51**	a1.6	9	F	...	**12 08**	1.1	4 50	**5 13**	a1.6
10	W	10 50	**10 50**	p1.4	3 34	**3 47**	a1.6	10	S	12 09	**12 58**	a1.4	5 48	**6 13**	a1.7
11	T	11 44	**11 44**	p1.4	4 31	**4 47**	a1.6	11	S	1 04	**1 47**	a1.5	6 44	**7 09**	a1.9
12	F	...	**12 37**	1.1	5 29	**5 46**	a1.7	12	M	1 58	**2 33**	1.6	7 37	**8 00**	2.0
13	S	12 37	**1 27**	a1.5	6 23	**6 39**	a1.9	13	T	2 50	**3 17**	p1.9	8 25	**8 49**	2.2
14	S	1 28	**2 13**	a1.7	7 12	**7 28**	a2.0	14	W	3 40	**4 00**	p2.2	9 12	**9 37**	p2.4
15	M	2 18	**2 56**	a1.9	7 58	**8 15**	2.1	15	T	4 29	**4 42**	p2.3	9 58	**10 25**	p2.6
16	T	3 04	**3 36**	a2.1	8 43	**9 01**	2.3	16	F	5 16	**5 24**	p2.5	10 46	**11 14**	p2.7
17	W	3 49	**4 15**	a2.2	9 28	**9 48**	2.4	17	S	6 04	**6 09**	p2.5	11 34	**...**	2.5
18	T	4 34	**4 55**	2.2	10 13	**10 36**	2.5	18	S	6 55	**6 58**	p2.4	12 05	**12 25**	a2.7
19	F	5 21	**5 37**	p2.3	10 59	**11 26**	p2.6	19	M	7 50	**7 52**	p2.3	12 58	**1 17**	a2.6
20	S	6 11	**6 23**	p2.2	11 47	**...**	2.4	20	T	8 50	**8 53**	p2.1	1 51	**2 11**	a2.5
21	S	7 06	**7 15**	p2.2	12 17	**12 37**	a2.5	21	W	9 51	**9 58**	p1.9	2 46	**3 08**	a2.3
22	M	8 06	**8 14**	p2.1	1 10	**1 28**	a2.3	22	T	10 52	**11 02**	p1.8	3 44	**4 09**	a2.1
23	T	9 09	**9 17**	p2.0	2 06	**2 24**	a2.2	23	F	11 51	**...**	1.5	4 45	**5 14**	a2.0
24	W	10 12	**10 21**	p1.8	3 05	**3 25**	a2.0	24	S	12 05	**12 47**	a1.7	5 48	**6 20**	a2.0
25	T	11 13	**11 23**	p1.8	4 09	**4 31**	a2.0	25	S	1 05	**1 40**	1.6	6 48	**7 19**	a2.0
26	F	...	**12 12**	1.4	5 14	**5 37**	a2.0	26	M	2 03	**2 31**	1.7	7 41	**8 11**	2.0
27	S	12 24	**1 08**	a1.8	6 15	**6 37**	a2.1	27	T	2 57	**3 17**	p1.9	8 28	**8 57**	2.0
28	S	1 22	**2 00**	a1.8	7 09	**7 31**	a2.1	28	W	3 47	**4 00**	p1.9	9 11	**9 39**	p2.1
29	M	2 17	**2 48**	a1.9	7 57	**8 18**	2.1	29	T	4 32	**4 41**	p1.9	9 52	**10 20**	p2.1
30	T	3 06	**3 31**	1.9	8 41	**9 03**	2.1	30	F	5 15	**5 19**	p1.9	10 31	**11 00**	p2.1
31	W	3 52	**4 11**	1.9	9 22	**9 45**	2.1								

***Daylight Time starts**

The Kts. (knots) columns show the **maximum** predicted velocities of the stronger one of the Flood Currents and the stronger one of the Ebb Currents for each day. The letter "a" means the velocity shown should occur **after** the **a.m.** Current Change. The letter "p" means the velocity shown should occur **after** the **p.m.** Current Change (even if next morning). No "a" or "p" means a.m. and p.m. velocities are the same for that day.

Av. Max. Vel.: Fl. 1.7 Kts., Ebb 2.0 Kts.

Max. Flood 2 hrs. 25 min. after Flood Starts,30 min.

Max. Ebb 3 hrs. 15 min. after Ebb Starts,10 min.

See pp. 150-157 for Current Change at other points.

1999 CURRENT TABLE
THE NARROWS, N.Y. HARBOR
(Mid-Channel – 40-36-36N x 74-02-48W)

MAY
Daylight Time

DAY OF MONTH	DAY OF WEEK	CURRENT TURNS: NORTH Flood Starts a.m.	**p.m.**	Kts.	SOUTH Ebb Starts a.m.	**p.m.**	Kts.
1	S	5 57	**5 56**	p1.9	11 12	**11 41**	p2.1
2	S	6 39	**6 33**	p1.8	11 53	**...**	1.9
3	M	7 23	**7 11**	p1.7	12 22	**12 35**	a2.1
4	T	8 09	**7 53**	p1.7	1 04	**1 19**	a2.0
5	W	8 59	**8 42**	p1.6	1 47	**2 04**	a1.9
6	T	9 51	**9 39**	p1.5	2 32	**2 53**	a1.8
7	F	10 42	**10 37**	p1.5	3 20	**3 45**	a1.7
8	S	11 31	**11 35**	p1.5	4 12	**4 42**	a1.7
9	S	...	**12 18**	1.4	5 08	**5 41**	a1.8
10	M	12 32	**1 05**	p1.6	6 05	**6 39**	1.9
11	T	1 28	**1 52**	p1.8	7 00	**7 33**	2.1
12	W	2 23	**2 39**	p2.0	7 52	**8 24**	p2.3
13	T	3 17	**3 25**	p2.3	8 41	**9 13**	p2.5
14	F	4 08	**4 11**	p2.5	9 30	**10 03**	p2.6
15	S	4 59	**4 58**	p2.6	10 20	**10 54**	p2.7
16	S	5 49	**5 46**	p2.5	11 11	**11 46**	p2.7
17	M	6 40	**6 37**	p2.4	...	**12 04**	2.4
18	T	7 34	**7 33**	p2.3	12 39	**12 58**	a2.7
19	W	8 31	**8 33**	p2.1	1 32	**1 54**	a2.5
20	T	9 30	**9 38**	p1.9	2 26	**2 51**	a2.3
21	F	10 28	**10 41**	p1.7	3 20	**3 50**	a2.1
22	S	11 24	**11 42**	1.6	4 17	**4 53**	a2.0
23	S	...	**12 17**	1.6	5 15	**5 56**	a2.0
24	M	12 41	**1 08**	p1.6	6 13	**6 54**	a2.0
25	T	1 38	**1 57**	p1.7	7 06	**7 46**	a2.0
26	W	2 32	**2 44**	p1.8	7 54	**8 31**	p2.0
27	T	3 23	**3 28**	p1.9	8 38	**9 13**	p2.0
28	F	4 10	**4 10**	p1.9	9 20	**9 54**	p2.0
29	S	4 54	**4 49**	p1.9	10 01	**10 34**	p2.1
30	S	5 36	**5 27**	p1.9	10 43	**11 14**	p2.1
31	M	6 18	**6 04**	p1.9	11 25	**11 56**	p2.1

JUNE
Daylight Time

DAY OF MONTH	DAY OF WEEK	CURRENT TURNS: NORTH Flood Starts a.m.	**p.m.**	Kts.	SOUTH Ebb Starts a.m.	**p.m.**	Kts.
1	T	7 00	**6 42**	p1.8	...	**12 09**	1.8
2	W	7 44	**7 23**	p1.7	12 38	**12 54**	a2.1
3	T	8 30	**8 11**	p1.7	1 21	**1 40**	a2.0
4	F	9 18	**9 06**	p1.6	2 05	**2 27**	a1.9
5	S	10 06	**10 05**	p1.6	2 51	**3 18**	a1.9
6	S	10 53	**11 04**	p1.6	3 39	**4 12**	a1.8
7	M	11 39	**...**	1.6	4 32	**5 11**	a1.9
8	T	12 03	**12 26**	p1.7	5 28	**6 10**	2.0
9	W	1 01	**1 15**	p1.9	6 25	**7 07**	p2.2
10	T	1 59	**2 05**	p2.1	7 20	**8 01**	p2.3
11	F	2 55	**2 56**	p2.3	8 13	**8 53**	p2.5
12	S	3 50	**3 47**	p2.5	9 05	**9 44**	p2.6
13	S	4 42	**4 38**	p2.6	9 57	**10 36**	p2.7
14	M	5 33	**5 28**	p2.5	10 50	**11 28**	p2.7
15	T	6 23	**6 20**	p2.4	11 44	**...**	2.4
16	W	7 15	**7 15**	p2.2	12 20	**12 40**	a2.6
17	T	8 09	**8 13**	p2.0	1 12	**1 35**	a2.5
18	F	9 04	**9 15**	p1.8	2 03	**2 30**	a2.3
19	S	9 59	**10 16**	p1.7	2 53	**3 26**	a2.2
20	S	10 53	**11 16**	a1.6	3 45	**4 24**	a2.0
21	M	11 44	**...**	1.6	4 38	**5 23**	a1.9
22	T	12 13	**12 33**	p1.6	5 33	**6 21**	a1.9
23	W	1 09	**1 22**	p1.6	6 27	**7 14**	1.8
24	T	2 04	**2 10**	p1.7	7 18	**8 01**	p1.9
25	F	2 56	**2 56**	p1.8	8 05	**8 45**	p1.9
26	S	3 45	**3 40**	p1.9	8 50	**9 27**	p2.0
27	S	4 30	**4 21**	p1.9	9 33	**10 08**	p2.0
28	M	5 13	**5 01**	p1.9	10 16	**10 49**	p2.1
29	T	5 55	**5 40**	p1.9	11 00	**11 31**	p2.1
30	W	6 35	**6 19**	p1.9	11 44	**...**	1.8

The Kts. (knots) columns show the **maximum** predicted velocities of the stronger one of the Flood Currents and the stronger one of the Ebb Currents for each day. The letter "a" means the velocity shown should occur **after** the **a.m.** Current Change. The letter "p" means the velocity shown should occur **after** the **p.m.** Current Change (even if next morning). No "a" or "p" means a.m. and p.m. velocities are the same for that day.

Av. Max. Vel.: Fl. 1.7 Kts., Ebb 2.0 Kts.

Max. Flood 2 hrs. 25 min. after Flood Starts,30 min.

Max. Ebb 3 hrs. 15 min. after Ebb Starts,10 min.

At **The Battery, Desbrosses St., & Chelsea Dock** Current turns 1 1/2 hrs. after the Narrows. At **42nd St.**, the Current turns 1 3/4 hrs. after the Narrows. At **George Washington Br.** the Current turns 2 hrs. after the Narrows.

1999 CURRENT TABLE

THE NARROWS, N.Y. HARBOR

(Mid-Channel – 40-36-36N x 74-02-48W)

JULY

Daylight Time

DAY OF MONTH	DAY OF WEEK	CURRENT TURNS: NORTH Flood Starts a.m.	**p.m.**	Kts.	CURRENT TURNS: SOUTH Ebb Starts a.m.	**p.m.**	Kts.
1	T	7 16	**7 00**	p1.8	12 13	**12 29**	a2.1
2	F	7 58	**7 46**	p1.8	12 56	**1 15**	a2.1
3	S	8 42	**8 38**	p1.7	1 39	**2 02**	a2.1
4	S	9 28	**9 37**	p1.7	2 23	**2 52**	a2.0
5	M	10 15	**10 37**	a1.7	3 10	**3 45**	a2.0
6	T	11 03	**11 38**	a1.8	4 01	**4 42**	a1.9
7	W	11 53	**...**	1.9	4 56	**5 43**	2.0
8	T	12 37	**12 45**	p2.0	5 55	**6 43**	2.1
9	F	1 37	**1 39**	p2.1	6 54	**7 40**	p2.3
10	S	2 36	**2 35**	p2.3	7 50	**8 34**	p2.4
11	S	3 32	**3 29**	p2.4	8 45	**9 27**	p2.5
12	M	4 25	**4 22**	p2.5	9 38	**10 18**	p2.6
13	T	5 15	**5 14**	p2.4	10 32	**11 09**	p2.6
14	W	6 04	**6 05**	p2.3	11 26	**...**	2.3
15	T	6 53	**6 57**	p2.1	12 01	**12 20**	a2.6
16	F	7 43	**7 52**	p1.9	12 49	**1 12**	a2.5
17	S	8 34	**8 49**	p1.8	1 36	**2 04**	a2.3
18	S	9 25	**9 47**	a1.7	2 23	**2 56**	a2.1
19	M	10 17	**10 45**	a1.6	3 10	**3 49**	a1.9
20	T	11 08	**11 42**	a1.6	3 59	**4 44**	a1.8
21	W	11 57	**...**	1.5	4 51	**5 41**	a1.7
22	T	12 38	**12 46**	p1.5	5 46	**6 37**	1.7
23	F	1 33	**1 35**	p1.6	6 41	**7 28**	p1.8
24	S	2 26	**2 24**	p1.7	7 33	**8 15**	p1.9
25	S	3 17	**3 11**	p1.8	8 21	**8 58**	p1.9
26	M	4 03	**3 55**	p1.9	9 06	**9 41**	p2.0
27	T	4 46	**4 37**	p2.0	9 50	**10 23**	p2.1
28	W	5 27	**5 18**	p2.0	10 34	**11 05**	p2.2
29	T	6 06	**5 58**	p2.0	11 19	**11 47**	p2.2
30	F	6 44	**6 39**	p1.9	...	**12 04**	2.0
31	S	7 23	**7 24**	p1.9	12 30	**12 50**	a2.2

AUGUST

Daylight Time

DAY OF MONTH	DAY OF WEEK	CURRENT TURNS: NORTH Flood Starts a.m.	**p.m.**	Kts.	CURRENT TURNS: SOUTH Ebb Starts a.m.	**p.m.**	Kts.
1	S	8 04	**8 16**	p1.8	1 13	**1 37**	a2.2
2	M	8 49	**9 13**	a1.8	1 57	**2 27**	a2.1
3	T	9 39	**10 15**	a1.8	2 44	**3 19**	a2.0
4	W	10 31	**11 17**	a1.9	3 34	**4 17**	a2.0
5	T	11 26	**...**	1.9	4 30	**5 19**	2.0
6	F	12 18	**12 23**	p2.0	5 31	**6 21**	p2.1
7	S	1 19	**1 21**	p2.0	6 33	**7 22**	p2.2
8	S	2 18	**2 20**	p2.2	7 33	**8 18**	p2.3
9	M	3 14	**3 16**	p2.3	8 30	**9 11**	p2.4
10	T	4 06	**4 10**	p2.3	9 23	**10 01**	p2.4
11	W	4 55	**5 01**	p2.3	10 16	**10 49**	p2.5
12	T	5 41	**5 50**	p2.2	11 07	**11 37**	p2.5
13	F	6 27	**6 39**	p2.0	11 57	**...**	2.3
14	S	7 12	**7 29**	1.8	12 22	**12 46**	a2.4
15	S	7 59	**8 21**	a1.7	1 07	**1 35**	a2.2
16	M	8 47	**9 16**	a1.7	1 51	**2 22**	a2.1
17	T	9 38	**10 13**	a1.6	2 35	**3 11**	a1.9
18	W	10 29	**11 09**	a1.5	3 21	**4 03**	a1.7
19	T	11 20	**...**	1.5	4 12	**4 58**	1.6
20	F	12 05	**12 11**	p1.4	5 07	**5 55**	1.6
21	S	1 00	**1 02**	p1.5	6 05	**6 50**	p1.7
22	S	1 54	**1 53**	p1.6	7 01	**7 41**	p1.8
23	M	2 44	**2 42**	p1.7	7 53	**8 28**	p1.9
24	T	3 31	**3 29**	p1.8	8 40	**9 12**	p2.0
25	W	4 14	**4 13**	p2.0	9 24	**9 54**	p2.1
26	T	4 54	**4 55**	p2.0	10 08	**10 36**	p2.2
27	F	5 31	**5 37**	p2.1	10 52	**11 19**	p2.3
28	S	6 08	**6 19**	p2.0	11 38	**...**	2.2
29	S	6 46	**7 05**	p2.0	12 02	**12 24**	2.3
30	M	7 27	**7 56**	a2.0	12 46	**1 13**	2.3
31	T	8 13	**8 54**	a2.0	1 32	**2 03**	2.2

The Kts. (knots) columns show the **maximum** predicted velocities of the stronger one of the Flood Currents and the stronger one of the Ebb Currents for each day. The letter "a" means the velocity shown should occur **after** the **a.m.** Current Change. The letter "p" means the velocity shown should occur **after** the **p.m.** Current Change (even if next morning). No "a" or "p" means a.m. and p.m. velocities are the same for that day.

Av. Max. Vel.: Fl. 1.7 Kts., Ebb 2.0 Kts.

Max. Flood 2 hrs. 25 min. after Flood Starts,30 min.

Max. Ebb 3 hrs. 15 min. after Ebb Starts,10 min.

See pp. 150-157 for Current Change at other points.

1999 CURRENT TABLE

THE NARROWS, N.Y. HARBOR

(Mid-Channel – 40-36-36N x 74-02-48W)

Daylight Time **Standard Time starts Oct. 31 at 2 a.m.**

SEPTEMBER

DAY OF MONTH	DAY OF WEEK	CURRENT TURNS NORTH Flood Starts a.m.	**p.m.**	Kts.	SOUTH Ebb Starts a.m.	**p.m.**	Kts.
1	W	9 06	**9 56**	a2.0	2 21	**2 57**	2.1
2	T	10 05	**10 59**	a2.0	3 13	**3 54**	2.0
3	F	11 06	...	1.9	4 10	**4 57**	p2.0
4	S	12 01	**12 07**	p1.9	5 13	**6 02**	p2.1
5	S	1 01	**1 08**	p1.9	6 19	**7 04**	p2.2
6	M	2 00	**2 07**	p2.0	7 21	**8 01**	p2.2
7	T	2 54	**3 04**	p2.1	8 18	**8 53**	p2.3
8	W	3 45	**3 57**	p2.2	9 10	**9 41**	p2.3
9	T	4 32	**4 47**	p2.1	9 59	**10 26**	2.3
10	F	5 16	**5 33**	2.0	10 47	**11 10**	2.3
11	S	5 58	**6 19**	1.9	11 33	**11 53**	2.2
12	S	6 40	**7 05**	a1.9	...	**12 18**	2.2
13	M	7 22	**7 54**	a1.8	12 35	**1 03**	2.1
14	T	8 07	**8 46**	a1.7	1 18	**1 48**	a2.0
15	W	8 56	**9 41**	a1.6	2 01	**2 34**	a1.8
16	T	9 47	**10 37**	a1.5	2 46	**3 23**	1.6
17	F	10 41	**11 33**	a1.4	3 36	**4 15**	p1.6
18	S	11 35	...	1.4	4 31	**5 12**	p1.6
19	S	12 27	**12 28**	p1.4	5 31	**6 10**	p1.7
20	M	1 19	**1 21**	p1.5	6 30	**7 04**	p1.8
21	T	2 08	**2 12**	p1.6	7 24	**7 54**	p2.0
22	W	2 54	**3 01**	p1.8	8 12	**8 39**	p2.1
23	T	3 37	**3 48**	p1.9	8 57	**9 23**	p2.2
24	F	4 17	**4 32**	p2.0	9 42	**10 06**	p2.3
25	S	4 55	**5 16**	2.1	10 26	**10 49**	2.3
26	S	5 32	**6 00**	a2.2	11 12	**11 34**	2.4
27	M	6 12	**6 47**	a2.2	...	**12 01**	2.5
28	T	6 54	**7 39**	a2.2	12 21	**12 50**	p2.5
29	W	7 43	**8 37**	a2.2	1 09	**1 42**	p2.4
30	T	8 40	**9 39**	a2.1	2 00	**2 36**	p2.2

OCTOBER

DAY OF MONTH	DAY OF WEEK	CURRENT TURNS NORTH Flood Starts a.m.	**p.m.**	Kts.	SOUTH Ebb Starts a.m.	**p.m.**	Kts.
1	F	9 43	**10 42**	a2.0	2 55	**3 34**	p2.1
2	S	10 49	**11 43**	a1.9	3 54	**4 36**	p2.0
3	S	11 52	...	1.8	4 59	**5 41**	p2.1
4	M	12 42	**12 54**	p1.8	6 06	**6 44**	p2.1
5	T	1 38	**1 54**	p1.8	7 09	**7 41**	p2.2
6	W	2 31	**2 50**	p1.9	8 05	**8 31**	p2.2
7	T	3 20	**3 42**	p2.0	8 55	**9 17**	2.2
8	F	4 06	**4 31**	a2.0	9 41	**10 00**	2.2
9	S	4 48	**5 16**	a2.0	10 25	**10 42**	2.2
10	S	5 29	**6 00**	a2.0	11 08	**11 23**	a2.2
11	M	6 08	**6 43**	a1.9	11 50	...	2.2
12	T	6 47	**7 29**	a1.8	12 04	**12 33**	p2.1
13	W	7 28	**8 17**	a1.7	12 46	**1 16**	p2.0
14	T	8 13	**9 10**	a1.6	1 29	**2 00**	p1.8
15	F	9 05	**10 05**	a1.5	2 15	**2 46**	p1.7
16	S	10 01	**10 59**	a1.5	3 04	**3 36**	p1.6
17	S	10 58	**11 51**	a1.4	3 58	**4 30**	p1.6
18	M	11 54	...	1.4	4 57	**5 28**	p1.7
19	T	12 41	**12 48**	p1.4	5 56	**6 24**	p1.8
20	W	1 28	**1 41**	p1.5	6 52	**7 16**	p2.0
21	T	2 14	**2 33**	p1.7	7 43	**8 04**	p2.1
22	F	2 57	**3 22**	1.8	8 30	**8 50**	p2.2
23	S	3 38	**4 09**	a2.1	9 16	**9 35**	2.3
24	S	4 19	**4 55**	a2.3	10 02	**10 20**	a2.5
25	M	5 00	**5 41**	a2.4	10 49	**11 07**	a2.6
26	T	5 42	**6 30**	a2.5	11 39	**11 57**	a2.6
27	W	6 28	**7 22**	a2.4	...	**12 30**	2.6
28	T	7 19	**8 20**	a2.3	12 48	**1 22**	p2.5
29	F	8 19	**9 21**	a2.1	1 42	**2 17**	p2.3
30	S	9 24	**10 22**	a2.0	2 38	**3 14**	p2.2
31	S	*9 31	***10 22**	a1.9	*2 39	***3 13**	p2.1

***Standard Time starts**

The Kts. (knots) columns show the **maximum** predicted velocities of the stronger one of the Flood Currents and the stronger one of the Ebb Currents for each day. The letter "a" means the velocity shown should occur **after** the **a.m.** Current Change. The letter "p" means the velocity shown should occur **after** the **p.m.** Current Change (even if next morning). No "a" or "p" means a.m. and p.m. velocities are the same for that day.

Av. Max. Vel.: Fl. 1.7 Kts., Ebb 2.0 Kts.

Max. Flood 2 hrs. 25 min. after Flood Starts,30 min.

Max. Ebb 3 hrs. 15 min. after Ebb Starts,10 min.

At **The Battery, Desbrosses St., & Chelsea Dock** Current turns 1 1/2 hrs. after the Narrows. At **42nd St.**, the Current turns 1 3/4 hrs. after the Narrows. At **George Washington Br.** the Current turns 2 hrs. after the Narrows.

1999 CURRENT TABLE
THE NARROWS, N.Y. HARBOR

(Mid-Channel – 40-36-36N x 74-02-48W)

NOVEMBER (Standard Time)								DECEMBER (Standard Time)							
DAY OF MONTH	DAY OF WEEK	CURRENT TURNS						DAY OF MONTH	DAY OF WEEK	CURRENT TURNS					
		NORTH Flood Starts			SOUTH Ebb Starts					NORTH Flood Starts			SOUTH Ebb Starts		
		a.m.	**p.m.**	Kts.	a.m.	**p.m.**	Kts.			a.m.	**p.m.**	Kts.	a.m.	**p.m.**	Kts.
1	M	10 36	11 18	a1.7	3 43	4 16	p2.0	1	W	11 16	11 43	p1.7	4 27	4 46	p2.0
2	T	11 37	...	1.7	4 50	5 18	p2.1	2	T	...	12 14	1.5	5 30	5 43	p2.0
3	W	12 12	12 37	1.7	5 53	6 15	p2.1	3	F	12 34	1 11	a1.8	6 25	6 34	p2.0
4	T	1 04	1 33	a1.8	6 48	7 05	2.1	4	S	1 22	2 03	a1.9	7 14	7 20	a2.0
5	F	1 52	2 25	a1.9	7 37	7 50	2.1	5	S	2 08	2 52	a1.9	7 58	8 03	a2.0
6	S	2 38	3 13	a2.0	8 21	8 32	a2.1	6	M	2 52	3 37	a2.0	8 39	8 45	a2.1
7	S	3 20	3 57	a2.0	9 02	9 13	a2.1	7	T	3 33	4 20	a2.0	9 19	9 26	a2.1
8	M	4 00	4 40	a2.0	9 43	9 53	a2.1	8	W	4 12	5 02	a1.9	9 59	10 09	a2.1
9	T	4 38	5 22	a1.9	10 24	10 35	a2.1	9	T	4 50	5 43	a1.9	10 39	10 52	a2.1
10	W	5 16	6 06	a1.9	11 05	11 17	a2.1	10	F	5 28	6 26	a1.8	11 21	11 36	a2.1
11	T	5 54	6 52	a1.8	11 47	...	2.0	11	S	6 08	7 11	a1.7	...	12 03	2.0
12	F	6 37	7 41	a1.7	12 01	12 30	p1.9	12	S	6 53	7 58	a1.7	12 21	12 46	p1.9
13	S	7 25	8 33	a1.6	12 47	1 14	p1.8	13	M	7 45	8 45	a1.6	1 08	1 30	p1.9
14	S	8 21	9 24	a1.5	1 35	2 01	p1.7	14	T	8 43	9 32	a1.5	1 57	2 16	p1.8
15	M	9 20	10 13	a1.4	2 27	2 51	p1.7	15	W	9 42	10 18	1.5	2 49	3 07	p1.8
16	T	10 18	11 01	a1.4	3 23	3 46	p1.7	16	T	10 41	11 04	p1.6	3 46	4 01	p1.8
17	W	11 14	11 46	p1.5	4 21	4 42	p1.8	17	F	11 38	11 52	p1.8	4 44	4 58	p2.0
18	T	...	12 09	1.5	5 18	5 37	p2.0	18	S	...	12 35	1.5	5 42	5 54	p2.1
19	F	12 32	1 03	a1.7	6 12	6 28	p2.1	19	S	12 40	1 31	a2.0	6 36	6 48	2.2
20	S	1 17	1 56	a1.9	7 03	7 18	2.2	20	M	1 31	2 26	a2.2	7 28	7 40	a2.4
21	S	2 02	2 46	a2.2	7 51	8 05	a2.4	21	T	2 21	3 17	a2.4	8 19	8 31	a2.5
22	M	2 47	3 36	a2.4	8 39	8 54	a2.5	22	W	3 12	4 07	a2.6	9 10	9 23	a2.6
23	T	3 32	4 24	a2.5	9 28	9 43	a2.7	23	T	4 02	4 57	a2.6	10 01	10 17	a2.7
24	W	4 19	5 14	a2.6	10 19	10 35	a2.7	24	F	4 53	5 47	a2.5	10 53	11 12	a2.7
25	T	5 08	6 06	a2.5	11 11	11 29	a2.7	25	S	5 46	6 39	a2.4	11 45	...	2.6
26	F	6 01	7 01	a2.4	...	12 04	2.6	26	S	6 43	7 34	a2.2	12 07	12 36	p2.5
27	S	7 00	7 59	a2.2	12 25	12 57	p2.4	27	M	7 44	8 30	a2.0	1 02	1 27	p2.3
28	S	8 05	8 58	a2.0	1 22	1 52	p2.2	28	T	8 47	9 25	a1.8	1 58	2 19	p2.1
29	M	9 11	9 56	a1.8	2 21	2 48	p2.1	29	W	9 49	10 18	p1.7	2 56	3 12	p2.0
30	T	10 14	10 50	a1.7	3 23	3 46	p2.0	30	T	10 49	11 10	p1.6	3 57	4 08	p1.9
								31	F	11 47	...	1.3	4 58	5 05	1.8

The Kts. (knots) columns show the **maximum** predicted velocities of the stronger one of the Flood Currents and the stronger one of the Ebb Currents for each day. The letter "a" means the velocity shown should occur **after** the **a.m.** Current Change. The letter "p" means the velocity shown should occur **after** the **p.m.** Current Change (even if next morning). No "a" or "p" means a.m. and p.m. velocities are the same for that day.

Av. Max. Vel.: Fl. 1.7 Kts., Ebb 2.0 Kts.

Max. Flood 2 hrs. 25 min. after Flood Starts,30 min.

Max. Ebb 3 hrs. 15 min. after Ebb Starts,10 min.

See pp. 150-157 for Current Change at other points.

1999 HIGH & LOW WATER THE BATTERY, N.Y. HBR.

40–42N x 74–01W

Standard Time

DAY OF MONTH	DAY OF WEEK	JANUARY HIGH a.m.	Ht.	p.m.	Ht.	LOW a.m.	p.m.
1	F	7 10	5.7	7 44	4.8	1 02	1 49
2	S	7 59	5.7	8 35	4.8	1 54	2 39
3	S	8 49	5.6	9 27	4.7	2 44	3 27
4	M	9 40	5.4	10 21	4.5	3 32	4 14
5	T	10 32	5.1	11 15	4.4	4 19	5 00
6	W	11 24	4.7	...	...	5 06	5 47
7	T	12 07	4.3	12 15	4.4	5 57	6 37
8	F	12 56	4.2	1 03	4.1	6 54	7 31
9	S	1 45	4.1	1 53	3.8	7 56	8 26
10	S	2 34	4.0	2 45	3.6	8 57	9 19
11	M	3 26	4.0	3 43	3.5	9 52	10 07
12	T	4 21	4.1	4 42	3.5	10 43	10 53
13	W	5 12	4.3	5 36	3.6	11 31	11 38
14	T	5 59	4.5	6 22	3.8	...	12 18
15	F	6 40	4.7	7 04	3.9	12 23	1 04
16	S	7 18	4.9	7 42	4.0	1 07	1 48
17	S	7 53	5.0	8 19	4.1	1 51	2 30
18	M	8 29	5.1	8 57	4.2	2 33	3 11
19	T	9 07	5.1	9 38	4.3	3 15	3 51
20	W	9 50	5.0	10 23	4.3	3 56	4 31
21	T	10 38	4.8	11 13	4.4	4 38	5 12
22	F	11 31	4.7	...	...	5 26	5 58
23	S	12 05	4.5	12 25	4.5	6 25	6 54
24	S	12 58	4.5	1 22	4.3	7 37	7 58
25	M	1 54	4.6	2 25	4.1	8 48	9 02
26	T	2 57	4.7	3 34	4.0	9 52	10 01
27	W	4 04	4.8	4 44	4.1	10 51	10 58
28	T	5 10	5.0	5 46	4.3	11 47	11 53
29	F	6 08	5.2	6 41	4.5	...	12 41
30	S	6 59	5.4	7 31	4.7	12 46	1 32
31	S	7 47	5.4	8 19	4.7	1 37	2 20

Standard Time

DAY OF MONTH	DAY OF WEEK	FEBRUARY HIGH a.m.	Ht.	p.m.	Ht.	LOW a.m.	p.m.
1	M	8 33	5.4	9 06	4.7	2 26	3 05
2	T	9 18	5.2	9 54	4.6	3 12	3 48
3	W	10 05	4.9	10 41	4.5	3 56	4 29
4	T	10 52	4.6	11 28	4.4	4 39	5 10
5	F	11 38	4.3	...	...	5 23	5 52
6	S	12 14	4.2	12 24	4.0	6 11	6 37
7	S	12 58	4.1	1 09	3.7	7 07	7 31
8	M	1 44	4.0	1 58	3.5	8 11	8 29
9	T	2 34	3.9	2 55	3.3	9 13	9 25
10	W	3 29	3.9	3 58	3.3	10 08	10 17
11	T	4 29	4.1	5 00	3.5	10 59	11 06
12	F	5 23	4.3	5 52	3.7	11 47	11 54
13	S	6 10	4.6	6 37	4.0	...	12 34
14	S	6 52	4.9	7 16	4.2	12 41	1 19
15	M	7 30	5.1	7 54	4.5	1 27	2 03
16	T	8 09	5.2	8 32	4.7	2 13	2 45
17	W	8 50	5.3	9 14	4.8	2 57	3 26
18	T	9 35	5.2	9 59	4.9	3 41	4 07
19	F	10 24	5.0	10 49	5.0	4 26	4 49
20	S	11 18	4.8	11 43	4.9	5 15	5 34
21	S	...	...	12 14	4.6	6 11	6 28
22	M	12 38	4.9	1 12	4.3	7 19	7 33
23	T	1 36	4.8	2 14	4.1	8 30	8 41
24	W	2 39	4.7	3 23	4.0	9 35	9 44
25	T	3 49	4.7	4 32	4.1	10 35	10 43
26	F	4 57	4.8	5 35	4.3	11 30	11 38
27	S	5 56	5.0	6 28	4.6	...	12 22
28	S	6 47	5.2	7 15	4.8	12 30	1 11

Dates when Ht. of **Low** Water is below Mean Low with Ht. of lowest given for each period and Date of lowest in ( ):

1st - 6th: -1.1' (2nd)
16th - 31st : -0.7' (19th - 20th)
: -1.0' (31st)

1st - 4th: -1.0' (1st)
14th - 21st: -0.9' (17th - 18th)
26th - 28th: -0.6' (28th)

Average Rise and Fall 4.5 ft.

When a high tide exceeds av. ht., the *following* low tide will be lower than av.

1999 HIGH & LOW WATER THE BATTERY, N.Y. HBR.

40–42N x 74–01W

Standard Time **Daylight Time starts Apr. 4 at 2 a.m.**

DAY OF MONTH	DAY OF WEEK	MARCH						DAY OF MONTH	DAY OF WEEK	APRIL					
		HIGH				LOW				HIGH				LOW	
		a.m.	Ht.	p.m.	Ht.	a.m.	p.m.			a.m.	Ht.	p.m.	Ht.	a.m.	p.m.
1	M	7 32	5.2	7 59	4.9	1 20	1 57	1	T	8 31	4.9	8 52	5.2	2 30	2 50
2	T	8 14	5.2	8 41	4.9	2 08	2 40	2	F	9 10	4.8	9 29	5.0	3 10	3 26
3	W	8 55	5.0	9 23	4.9	2 52	3 20	3	S	9 49	4.5	10 06	4.9	3 49	4 01
4	T	9 37	4.8	10 05	4.8	3 33	3 58	4	S	*11 30	4.3	*11 43	4.7	*5 26	*5 32
5	F	10 19	4.6	10 47	4.6	4 13	4 34	5	M	...	...	12 13	4.1	6 03	6 00
6	S	11 02	4.3	11 29	4.4	4 52	5 09	6	T	12 21	4.5	12 56	3.9	6 43	6 24
7	S	11 46	4.0	...	...	5 33	5 43	7	W	1 00	4.4	1 40	3.7	7 32	6 58
8	M	12 10	4.3	12 29	3.8	6 19	6 20	8	T	1 43	4.3	2 28	3.7	8 41	8 32
9	T	12 52	4.1	1 16	3.6	7 20	7 23	9	F	2 34	4.2	3 24	3.7	9 48	9 59
10	W	1 36	4.0	2 06	3.4	8 28	8 37	10	S	3 34	4.2	4 27	3.9	10 45	11 00
11	T	2 29	4.0	3 09	3.4	9 30	9 39	11	S	4 44	4.4	5 30	4.2	11 36	11 54
12	F	3 33	4.1	4 16	3.6	10 23	10 33	12	M	5 50	4.6	6 24	4.6	...	12 24
13	S	4 38	4.3	5 15	3.9	11 13	11 24	13	T	6 46	5.0	7 11	5.1	12 46	1 11
14	S	5 34	4.6	6 03	4.2	...	12 01	14	W	7 35	5.2	7 55	5.6	1 38	1 59
15	M	6 21	4.9	6 45	4.6	12 14	12 47	15	T	8 22	5.4	8 39	5.9	2 29	2 46
16	T	7 04	5.2	7 25	5.0	1 03	1 32	16	F	9 09	5.5	9 24	6.1	3 19	3 32
17	W	7 46	5.4	8 05	5.3	1 51	2 16	17	S	9 59	5.4	10 13	6.1	4 08	4 19
18	T	8 30	5.5	8 48	5.5	2 38	2 59	18	S	10 54	5.3	11 07	5.9	4 57	5 06
19	F	9 17	5.4	9 35	5.6	3 25	3 42	19	M	11 54	5.1	...	...	5 48	5 55
20	S	10 09	5.2	10 27	5.5	4 12	4 26	20	T	12 06	5.7	12 54	4.9	6 43	6 50
21	S	11 06	5.0	11 23	5.4	5 02	5 13	21	W	1 06	5.4	1 54	4.7	7 44	7 53
22	M	...	...	12 05	4.7	5 57	6 08	22	T	2 06	5.2	2 54	4.6	8 49	9 03
23	T	12 21	5.2	1 04	4.5	7 02	7 12	23	F	3 07	4.9	3 55	4.5	9 53	10 09
24	W	1 21	5.0	2 06	4.3	8 11	8 22	24	S	4 11	4.7	4 56	4.6	10 51	11 08
25	T	2 24	4.8	3 11	4.2	9 16	9 28	25	S	5 15	4.6	5 54	4.8	11 43	...
26	F	3 33	4.7	4 18	4.3	10 15	10 27	26	M	6 13	4.7	6 45	5.0	12 01	12 30
27	S	4 40	4.7	5 18	4.5	11 09	11 22	27	T	7 04	4.7	7 30	5.2	12 51	1 15
28	S	5 39	4.8	6 10	4.8	11 59	...	28	W	7 47	4.8	8 09	5.3	1 38	1 58
29	M	6 28	5.0	6 55	5.0	12 13	12 45	29	T	8 28	4.8	8 46	5.4	2 23	2 39
30	T	7 12	5.0	7 36	5.1	1 01	1 29	30	F	9 06	4.7	9 22	5.3	3 06	3 19
31	W	7 52	5.0	8 14	5.2	1 47	2 11								

***Daylight Time starts**

Dates when Ht. of **Low** Water is below Mean Low with Ht. of lowest given for each period and Date of lowest in ( ):

1st - 5th: -0.7' (1st - 2nd)
15th - 22nd: -0.9' (18th - 19th)
29th - 31st: -0.3'

14th - 20th: -0.9' (17th - 18th)

Average Rise and Fall 4.5 ft.

When a high tide exceeds av. ht., the *following* low tide will be lower than av.

1999 HIGH & LOW WATER
THE BATTERY, N.Y. HBR.
40–42N x 74–01W

DAY OF MONTH	DAY OF WEEK	MAY (Daylight Time) HIGH a.m.	Ht.	p.m.	Ht.	LOW a.m.	p.m.	DAY OF MONTH	DAY OF WEEK	JUNE (Daylight Time) HIGH a.m.	Ht.	p.m.	Ht.	LOW a.m.	p.m.
1	S	9 44	4.6	9 56	5.2	3 47	3 56	1	T	10 41	4.2	10 35	5.1	4 44	4 41
2	S	10 23	4.5	10 30	5.1	4 26	4 31	2	W	11 23	4.1	11 11	5.0	5 22	5 14
3	M	11 04	4.3	11 04	4.9	5 04	5 03	3	T	12 07	4.1	11 51	4.8	5 59	5 45
4	T	11 47	4.1	11 39	4.8	5 41	5 32	4	F	...	...	12 50	4.1	6 39	6 22
5	W	...	...	12 30	4.0	6 19	5 58	5	S	12 38	4.8	1 33	4.2	7 25	7 17
6	T	12 18	4.6	1 14	3.9	7 02	6 33	6	S	1 29	4.7	2 19	4.4	8 22	8 44
7	F	1 04	4.5	1 59	3.9	7 58	7 32	7	M	2 23	4.7	3 10	4.6	9 23	9 58
8	S	1 54	4.5	2 48	4.0	9 04	9 19	8	T	3 22	4.6	4 06	4.9	10 20	11 00
9	S	2 52	4.5	3 44	4.2	10 04	10 28	9	W	4 30	4.6	5 08	5.3	11 14	11 56
10	M	3 56	4.5	4 44	4.6	10 57	11 25	10	T	5 38	4.8	6 07	5.7	...	12 06
11	T	5 05	4.7	5 43	5.0	11 47	...	11	F	6 41	4.9	7 02	6.0	12 52	12 58
12	W	6 09	4.9	6 37	5.5	12 20	12 37	12	S	7 38	5.1	7 54	6.3	1 47	1 52
13	T	7 06	5.1	7 27	5.9	1 14	1 26	13	S	8 32	5.2	8 45	6.4	2 41	2 45
14	F	7 58	5.3	8 14	6.2	2 07	2 17	14	M	9 26	5.2	9 38	6.3	3 34	3 37
15	S	8 49	5.4	9 03	6.4	2 59	3 07	15	T	10 23	5.2	10 33	6.1	4 24	4 29
16	S	9 42	5.4	9 54	6.3	3 51	3 57	16	W	11 22	5.1	11 32	5.8	5 14	5 20
17	M	10 39	5.2	10 50	6.1	4 41	4 47	17	T	...	...	12 21	5.0	6 04	6 12
18	T	11 40	5.1	11 50	5.9	5 32	5 37	18	F	12 30	5.5	1 17	5.0	6 57	7 08
19	W	...	...	12 41	5.0	6 25	6 32	19	S	1 25	5.2	2 10	4.9	7 52	8 10
20	T	12 50	5.5	1 39	4.9	7 22	7 33	20	S	2 17	4.8	3 01	4.8	8 49	9 13
21	F	1 48	5.2	2 35	4.8	8 23	8 39	21	M	3 10	4.5	3 53	4.8	9 44	10 13
22	S	2 45	4.9	3 31	4.7	9 24	9 44	22	T	4 04	4.3	4 46	4.8	10 35	11 07
23	S	3 43	4.7	4 28	4.7	10 21	10 43	23	W	5 01	4.2	5 38	4.9	11 22	11 57
24	M	4 42	4.5	5 23	4.8	11 11	11 36	24	T	5 57	4.1	6 27	5.0	...	12 07
25	T	5 40	4.4	6 14	5.0	11 58	...	25	F	6 49	4.2	7 11	5.1	12 44	12 51
26	W	6 33	4.4	7 00	5.1	12 26	12 42	26	S	7 35	4.2	7 51	5.3	1 30	1 34
27	T	7 19	4.5	7 41	5.3	1 12	1 24	27	S	8 17	4.3	8 29	5.3	2 15	2 18
28	F	8 02	4.5	8 19	5.4	1 58	2 07	28	M	8 57	4.3	9 04	5.3	2 59	3 00
29	S	8 41	4.5	8 54	5.4	2 42	2 48	29	T	9 37	4.3	9 39	5.3	3 41	3 41
30	S	9 21	4.4	9 29	5.3	3 24	3 27	30	W	10 16	4.3	10 13	5.2	4 21	4 20
31	M	10 00	4.4	10 02	5.2	4 05	4 05								

Dates when Ht. of **Low** Water is below Mean Low with Ht. of lowest given for each period and Date of lowest in ( ):

13th - 19th: -0.9' (16th - 17th)

11th - 18th: -0.9' (15th)

Average Rise and Fall 4.5 ft.

When a high tide exceeds av. ht., the *following* low tide will be lower than av.

1999 HIGH & LOW WATER
THE BATTERY, N.Y. HBR.
40–42N x 74–01W

DAY OF MONTH	DAY OF WEEK	JULY Daylight Time						DAY OF MONTH	DAY OF WEEK	AUGUST Daylight Time					
		HIGH				LOW				HIGH				LOW	
		a.m.	Ht.	p.m.	Ht.	a.m.	p.m.			a.m.	Ht.	p.m.	Ht.	a.m.	p.m.
1	T	10 57	4.3	10 50	5.1	5 00	4 57	1	S	11 51	4.8	...	...	5 50	6 04
2	F	11 39	4.3	11 32	5.0	5 37	5 34	2	M	12 02	5.1	12 39	5.0	6 30	6 55
3	S	...	...	12 23	4.4	6 15	6 15	3	T	12 54	4.9	1 28	5.1	7 18	8 01
4	S	12 19	4.9	1 07	4.5	6 57	7 07	4	W	1 49	4.7	2 20	5.2	8 16	9 13
5	M	1 10	4.8	1 53	4.7	7 47	8 19	5	T	2 47	4.6	3 17	5.3	9 22	10 20
6	T	2 03	4.7	2 42	4.9	8 46	9 33	6	F	3 52	4.5	4 21	5.4	10 25	11 20
7	W	3 00	4.6	3 38	5.1	9 48	10 38	7	S	5 03	4.5	5 29	5.6	11 24	...
8	T	4 05	4.6	4 39	5.4	10 46	11 37	8	S	6 11	4.7	6 32	5.8	12 17	12 21
9	F	5 16	4.6	5 45	5.7	11 41	...	9	M	7 12	4.9	7 29	5.9	1 12	1 16
10	S	6 23	4.7	6 44	5.9	12 33	12 36	10	T	8 04	5.1	8 19	6.0	2 05	2 10
11	S	7 23	4.9	7 40	6.1	1 29	1 32	11	W	8 54	5.3	9 07	6.0	2 55	3 02
12	M	8 18	5.1	8 32	6.2	2 23	2 26	12	T	9 43	5.3	9 55	5.8	3 43	3 51
13	T	9 11	5.2	9 23	6.2	3 15	3 20	13	F	10 32	5.3	10 43	5.6	4 28	4 38
14	W	10 05	5.2	10 15	6.0	4 05	4 11	14	S	11 22	5.2	11 33	5.3	5 11	5 24
15	T	11 00	5.2	11 09	5.7	4 53	5 00	15	S	...	...	12 12	5.1	5 53	6 10
16	F	11 55	5.1	...	...	5 39	5 49	16	M	12 22	4.9	1 00	5.0	6 36	6 59
17	S	12 04	5.4	12 48	5.0	6 26	6 40	17	T	1 11	4.6	1 46	4.8	7 21	7 54
18	S	12 56	5.1	1 38	4.9	7 15	7 36	18	W	1 59	4.3	2 32	4.7	8 13	8 56
19	M	1 46	4.7	2 26	4.8	8 07	8 36	19	T	2 48	4.1	3 20	4.6	9 10	9 57
20	T	2 34	4.4	3 14	4.7	9 02	9 37	20	F	3 41	3.9	4 13	4.6	10 06	10 52
21	W	3 25	4.2	4 04	4.7	9 55	10 34	21	S	4 42	3.8	5 10	4.6	10 58	11 43
22	T	4 20	4.0	4 57	4.7	10 45	11 25	22	S	5 43	3.9	6 05	4.8	11 47	...
23	F	5 20	3.9	5 50	4.8	11 33	...	23	M	6 37	4.1	6 53	5.0	12 30	12 35
24	S	6 17	4.0	6 40	5.0	12 14	12 18	24	T	7 23	4.3	7 35	5.3	1 16	1 22
25	S	7 07	4.1	7 24	5.1	1 01	1 04	25	W	8 03	4.6	8 13	5.4	2 01	2 08
26	M	7 52	4.3	8 03	5.3	1 47	1 50	26	T	8 39	4.8	8 50	5.5	2 44	2 53
27	T	8 32	4.4	8 40	5.4	2 32	2 34	27	F	9 15	5.0	9 28	5.6	3 25	3 37
28	W	9 10	4.5	9 16	5.4	3 15	3 18	28	S	9 53	5.2	10 08	5.5	4 05	4 21
29	T	9 48	4.6	9 51	5.4	3 55	3 59	29	S	10 34	5.3	10 54	5.4	4 45	5 04
30	F	10 26	4.7	10 30	5.3	4 34	4 40	30	M	11 21	5.4	11 46	5.2	5 24	5 50
31	S	11 07	4.7	11 13	5.2	5 12	5 20	31	T	...	...	12 12	5.4	6 06	6 42

Dates when Ht. of **Low** Water is below Mean Low with Ht. of lowest given for each period and Date of lowest in ( ):

July: 11th - 17th: -0.8' (14th)

August: 9th - 14th: -0.6' (12th)
27th - 31st: -0.3' (28th - 29th)

Average Rise and Fall 4.5 ft.

When a high tide exceeds av. ht., the *following* low tide will be lower than av.

1999 HIGH & LOW WATER THE BATTERY, N.Y. HBR.

40–42N x 74–01W

Daylight Time **Standard Time starts Oct. 31 at 2 a.m.**

SEPTEMBER

Day of Month	Day of Week	High a.m.	Ht.	High p.m.	Ht.	Low a.m.	Low p.m.
1	W	12 42	4.9	1 07	5.4	6 54	7 46
2	T	1 40	4.7	2 03	5.4	7 54	8 56
3	F	2 41	4.6	3 03	5.3	9 03	10 04
4	S	3 46	4.5	4 09	5.3	10 10	11 05
5	S	4 55	4.5	5 18	5.4	11 11	...
6	M	6 01	4.7	6 21	5.5	12 01	12 08
7	T	6 59	5.0	7 16	5.7	12 54	1 02
8	W	7 49	5.3	8 04	5.8	1 44	1 54
9	T	8 36	5.4	8 49	5.7	2 31	2 44
10	F	9 19	5.5	9 32	5.6	3 17	3 31
11	S	10 02	5.5	10 15	5.4	3 59	4 15
12	S	10 46	5.4	11 00	5.1	4 40	4 58
13	M	11 31	5.2	11 47	4.7	5 18	5 40
14	T	...	...	12 17	5.0	5 56	6 23
15	W	12 35	4.4	1 02	4.8	6 34	7 12
16	T	1 23	4.2	1 48	4.6	7 17	8 11
17	F	2 12	4.0	2 35	4.5	8 15	9 16
18	S	3 05	3.8	3 26	4.5	9 23	10 16
19	S	4 03	3.8	4 24	4.5	10 23	11 09
20	M	5 05	3.9	5 24	4.7	11 16	11 57
21	T	6 01	4.2	6 17	4.9	...	12 06
22	W	6 48	4.5	7 02	5.2	12 42	12 54
23	T	7 29	4.9	7 43	5.4	1 26	1 41
24	F	8 06	5.2	8 23	5.6	2 10	2 28
25	S	8 43	5.5	9 03	5.6	2 52	3 15
26	S	9 22	5.7	9 46	5.6	3 35	4 01
27	M	10 05	5.8	10 35	5.4	4 17	4 47
28	T	10 54	5.8	11 31	5.2	4 59	5 36
29	W	11 49	5.7	...	...	5 44	6 29
30	T	12 32	4.9	12 49	5.6	6 35	7 30

OCTOBER

Day of Month	Day of Week	High a.m.	Ht.	High p.m.	Ht.	Low a.m.	Low p.m.
1	F	1 34	4.7	1 50	5.4	7 36	8 39
2	S	2 36	4.6	2 53	5.3	8 47	9 47
3	S	3 40	4.6	3 59	5.2	9 56	10 48
4	M	4 45	4.6	5 05	5.2	10 58	11 42
5	T	5 48	4.8	6 07	5.3	11 54	...
6	W	6 43	5.1	7 00	5.3	12 32	12 46
7	T	7 30	5.3	7 46	5.4	1 20	1 36
8	F	8 13	5.5	8 28	5.4	2 05	2 24
9	S	8 54	5.6	9 09	5.3	2 48	3 09
10	S	9 32	5.5	9 47	5.1	3 29	3 52
11	M	10 11	5.4	10 28	4.8	4 08	4 33
12	T	10 50	5.2	11 12	4.5	4 45	5 13
13	W	11 32	5.0	11 59	4.2	5 19	5 53
14	T	...	...	12 15	4.8	5 52	6 36
15	F	12 49	4.0	1 01	4.6	6 22	7 27
16	S	1 39	3.8	1 47	4.5	6 59	8 32
17	S	2 29	3.8	2 37	4.4	8 27	9 35
18	M	3 23	3.8	3 32	4.4	9 44	10 31
19	T	4 20	3.9	4 33	4.5	10 43	11 20
20	W	5 17	4.2	5 32	4.7	11 36	...
21	T	6 07	4.6	6 24	5.0	12 05	12 25
22	F	6 52	5.1	7 11	5.2	12 50	1 14
23	S	7 33	5.5	7 55	5.4	1 34	2 03
24	S	8 14	5.8	8 40	5.5	2 20	2 53
25	M	8 56	6.1	9 26	5.4	3 05	3 42
26	T	9 42	6.1	10 18	5.3	3 51	4 31
27	W	10 33	6.0	11 17	5.1	4 37	5 21
28	T	11 31	5.9	...	...	5 25	6 14
29	F	12 21	4.8	12 34	5.6	6 18	7 14
30	S	1 24	4.7	1 37	5.4	7 20	8 19
31	S	*1 25	4.6	*1 39	5.1	*7 29	*8 25

***Standard Time starts**

Dates when Ht. of **Low** Water is below Mean Low with Ht. of lowest given for each period and Date of lowest in ( ):

25th - 30th: -0.5' (27th) 23rd - 28th: -0.6' (25th - 27th)

Average Rise and Fall 4.5 ft.

When a high tide exceeds av. ht., the *following* low tide will be lower than av.

1999 HIGH & LOW WATER
THE BATTERY, N.Y. HBR.
40–42N x 74–01W

Standard Time

DAY OF MONTH	DAY OF WEEK	NOVEMBER HIGH a.m.	Ht.	p.m.	Ht.	LOW a.m.	p.m.
1	M	2 26	4.6	2 42	4.9	8 39	9 25
2	T	3 28	4.7	3 45	4.8	9 41	10 19
3	W	4 28	4.8	4 45	4.8	10 37	11 07
4	T	5 21	5.0	5 38	4.9	11 28	11 53
5	F	6 08	5.2	6 24	4.9	...	12 16
6	S	6 50	5.4	7 06	4.9	12 36	1 02
7	S	7 28	5.4	7 45	4.8	1 19	1 46
8	M	8 04	5.4	8 23	4.7	1 59	2 29
9	T	8 41	5.3	9 03	4.5	2 38	3 10
10	W	9 16	5.1	9 43	4.2	3 15	3 49
11	T	9 53	4.9	10 28	4.0	3 49	4 28
12	F	10 32	4.7	11 17	3.8	4 21	5 07
13	S	11 15	4.5	...	...	4 49	5 51
14	S	12 05	3.7	12 01	4.4	5 20	6 45
15	M	12 52	3.7	12 49	4.3	6 11	7 48
16	T	1 40	3.8	1 41	4.3	7 56	8 47
17	W	2 31	3.9	2 39	4.3	9 07	9 39
18	T	3 26	4.2	3 42	4.4	10 04	10 27
19	F	4 22	4.6	4 44	4.6	10 57	11 14
20	S	5 14	5.1	5 40	4.9	11 49	...
21	S	6 02	5.6	6 30	5.1	12 01	12 41
22	M	6 49	5.9	7 19	5.2	12 50	1 32
23	T	7 35	6.1	8 10	5.2	1 39	2 24
24	W	8 24	6.2	9 04	5.1	2 29	3 14
25	T	9 17	6.0	10 03	4.9	3 19	4 05
26	F	10 16	5.8	11 06	4.8	4 09	4 57
27	S	11 19	5.5	...	...	5 02	5 53
28	S	12 08	4.7	12 20	5.2	6 01	6 54
29	M	1 07	4.6	1 19	4.9	7 07	7 57
30	T	2 05	4.5	2 18	4.6	8 15	8 56

Standard Time

DAY OF MONTH	DAY OF WEEK	DECEMBER HIGH a.m.	Ht.	p.m.	Ht.	LOW a.m.	p.m.
1	W	3 02	4.5	3 17	4.4	9 18	9 50
2	T	3 59	4.6	4 16	4.3	10 14	10 38
3	F	4 53	4.8	5 11	4.3	11 05	11 24
4	S	5 42	4.9	6 00	4.3	11 53	...
5	S	6 25	5.0	6 43	4.3	12 07	12 39
6	M	7 04	5.1	7 23	4.3	12 49	1 24
7	T	7 41	5.1	8 02	4.3	1 31	2 06
8	W	8 16	5.1	8 41	4.1	2 11	2 48
9	T	8 52	5.0	9 22	4.0	2 50	3 27
10	F	9 26	4.8	10 02	3.9	3 26	4 06
11	S	10 01	4.7	10 46	3.7	4 00	4 43
12	S	10 39	4.5	11 30	3.7	4 31	5 21
13	M	11 22	4.4	...	...	5 02	6 03
14	T	12 13	3.7	12 09	4.3	5 44	6 56
15	W	12 56	3.8	1 00	4.2	7 01	7 57
16	T	1 44	4.0	1 56	4.2	8 28	8 56
17	F	2 38	4.3	2 59	4.2	9 33	9 50
18	S	3 38	4.6	4 07	4.3	10 31	10 41
19	S	4 39	5.0	5 12	4.5	11 26	11 33
20	M	5 36	5.4	6 10	4.7	...	12 20
21	T	6 29	5.8	7 03	4.8	12 25	1 14
22	W	7 19	6.0	7 56	4.9	1 18	2 07
23	T	8 10	6.0	8 50	4.9	2 11	2 58
24	F	9 04	5.9	9 47	4.8	3 02	3 48
25	S	10 01	5.6	10 47	4.7	3 53	4 38
26	S	11 00	5.3	11 46	4.6	4 45	5 29
27	M	11 58	5.0	...	...	5 39	6 23
28	T	12 41	4.6	12 53	4.6	6 40	7 22
29	W	1 35	4.5	1 47	4.3	7 45	8 20
30	T	2 29	4.4	2 43	4.0	8 48	9 15
31	F	3 24	4.3	3 41	3.8	9 46	10 06

Dates when Ht. of **Low** Water is below Mean Low with Ht. of lowest given for each period and Date of lowest in ( ):

21st - 29th: -0.9' (25th)

18th - 28th: -1.1' (22nd)

Average Rise and Fall 4.5 ft.

When a high tide exceeds av. ht., the *following* low tide will be lower than av.

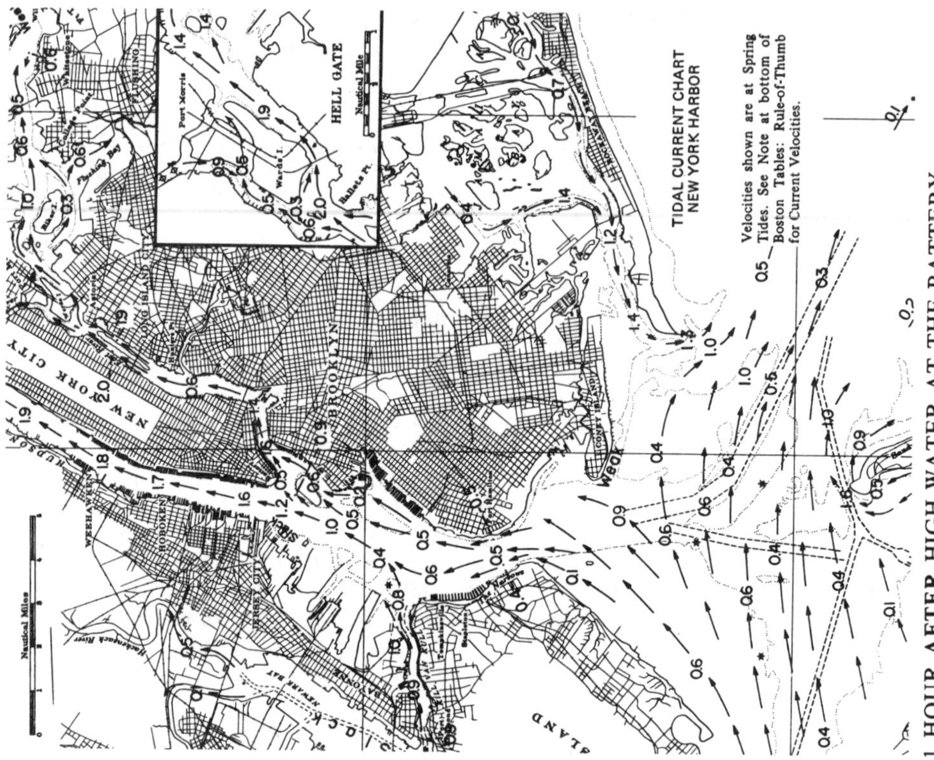

NEW YORK BAY CURRENTS

Old channel north of Ward's Island is non-existent.

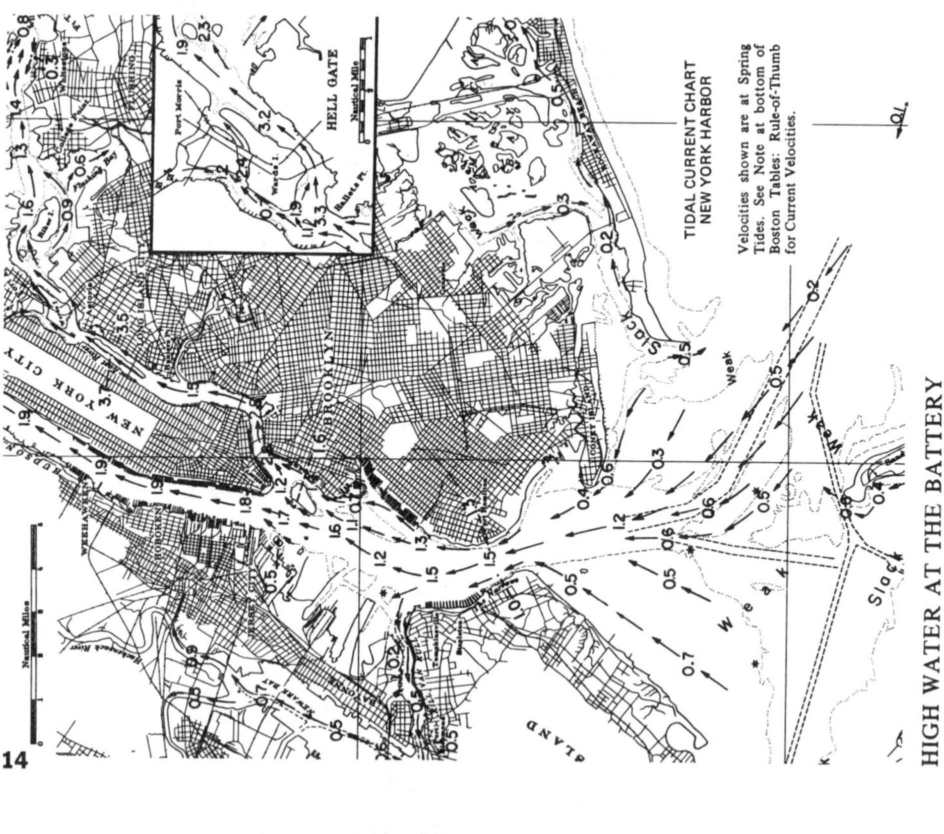

NEW YORK BAY CURRENTS

Old channel north of Ward's Island is non-existent.

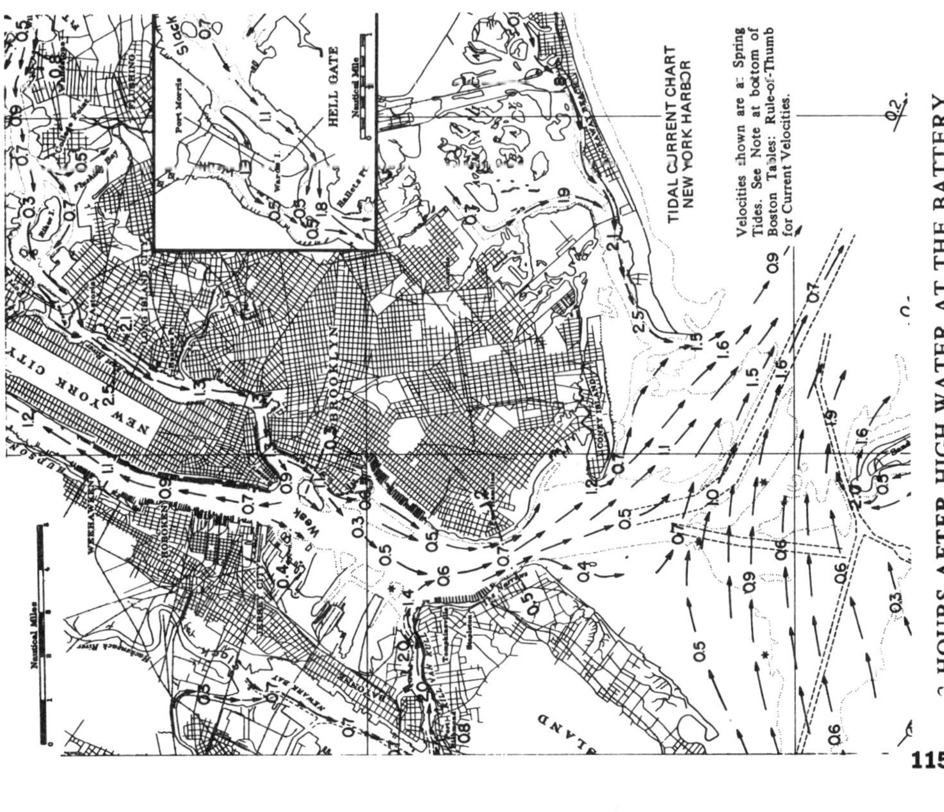

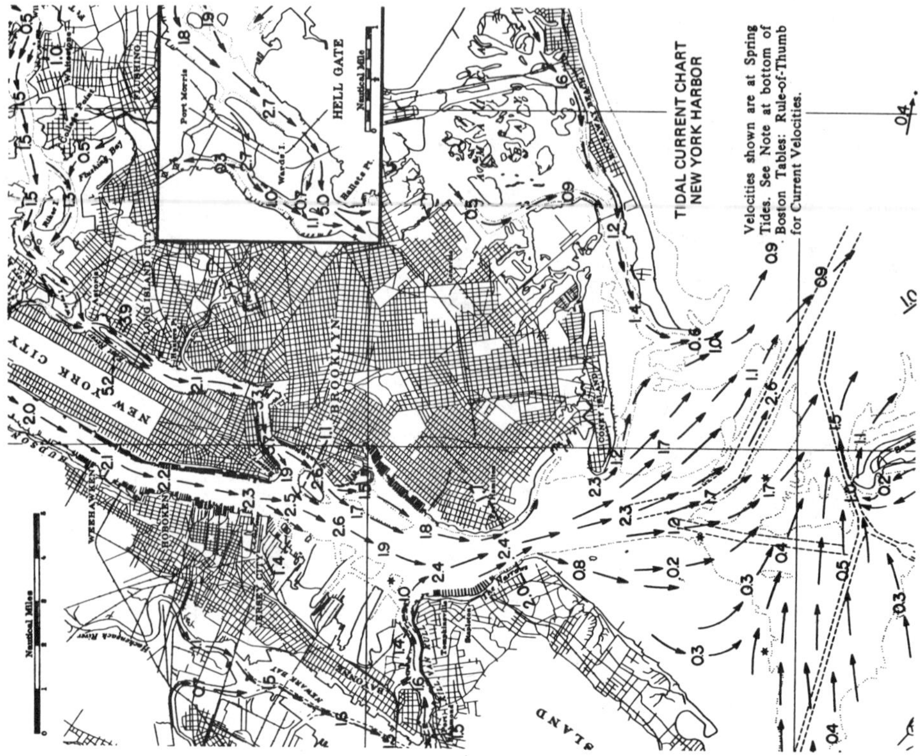

5 HOURS AFTER HIGH WATER AT THE BATTERY

NEW YORK BAY CURRENTS

Old channel north of Ward's Island is non-existent.

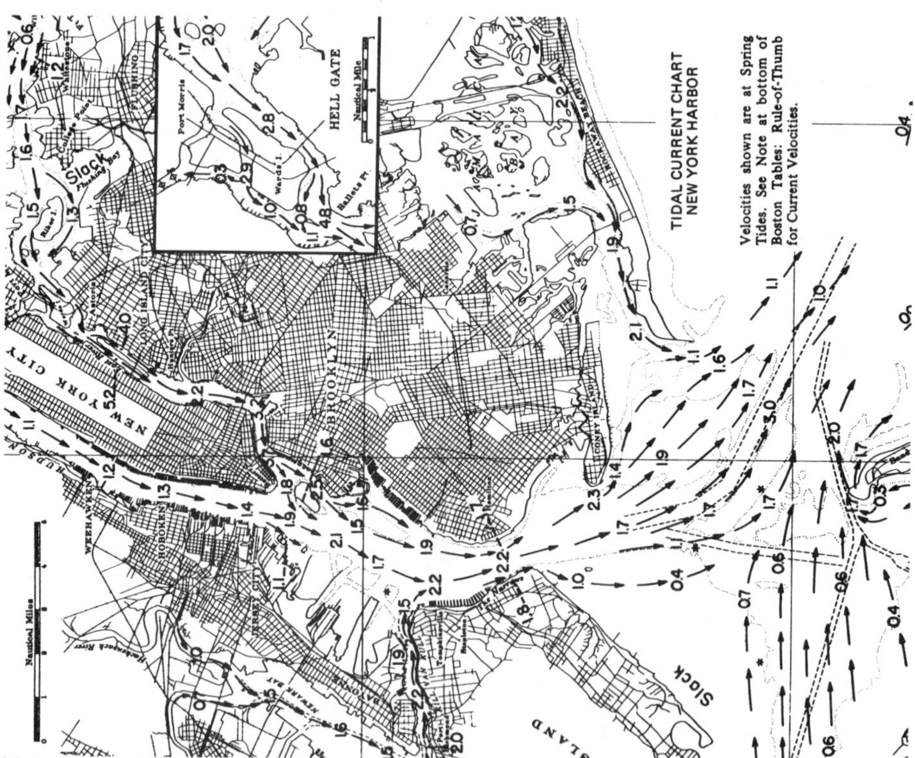

4 HOURS AFTER HIGH WATER AT THE BATTERY

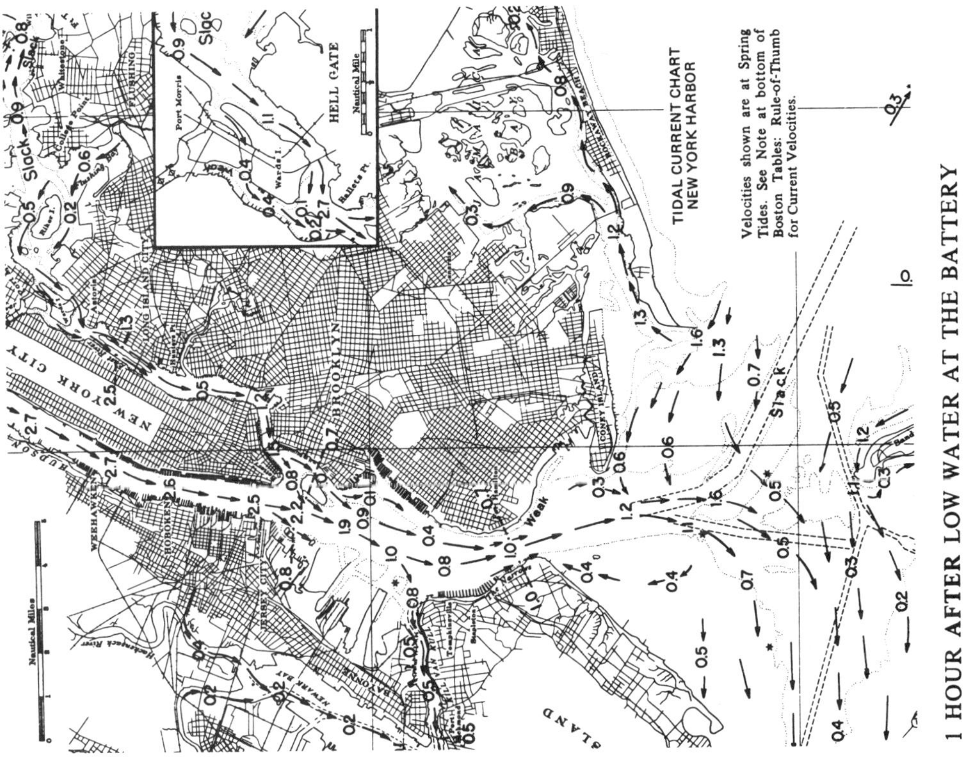

NEW YORK BAY CURRENTS

Old channel north of Ward's Island is non-existent.

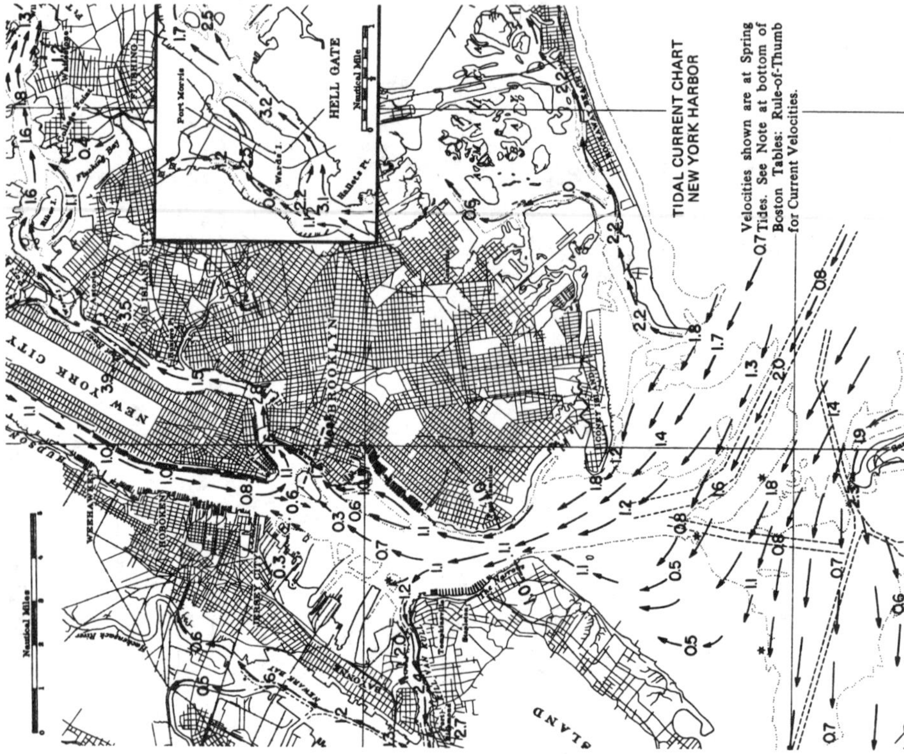

3 HOURS AFTER LOW WATER AT THE BATTERY

NEW YORK BAY CURRENTS

Old channel north of Ward's Island is non-existent.

2 HOURS AFTER LOW WATER AT THE BATTERY

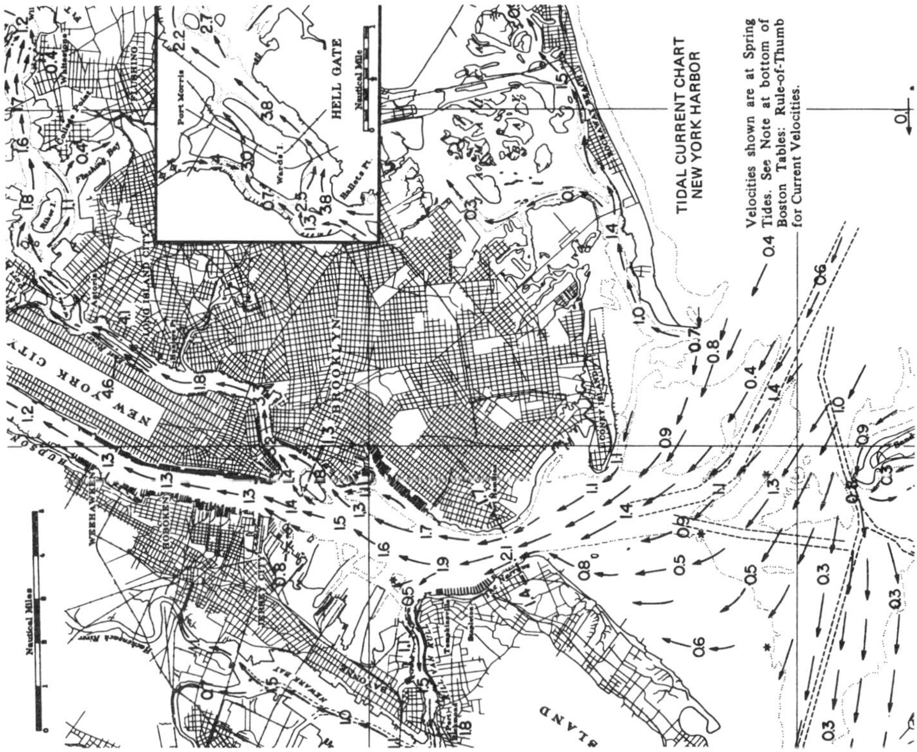

5 HOURS AFTER LOW WATER AT THE BATTERY

NEW YORK BAY CURRENTS

Old channel north of Ward's Island is non-existent.

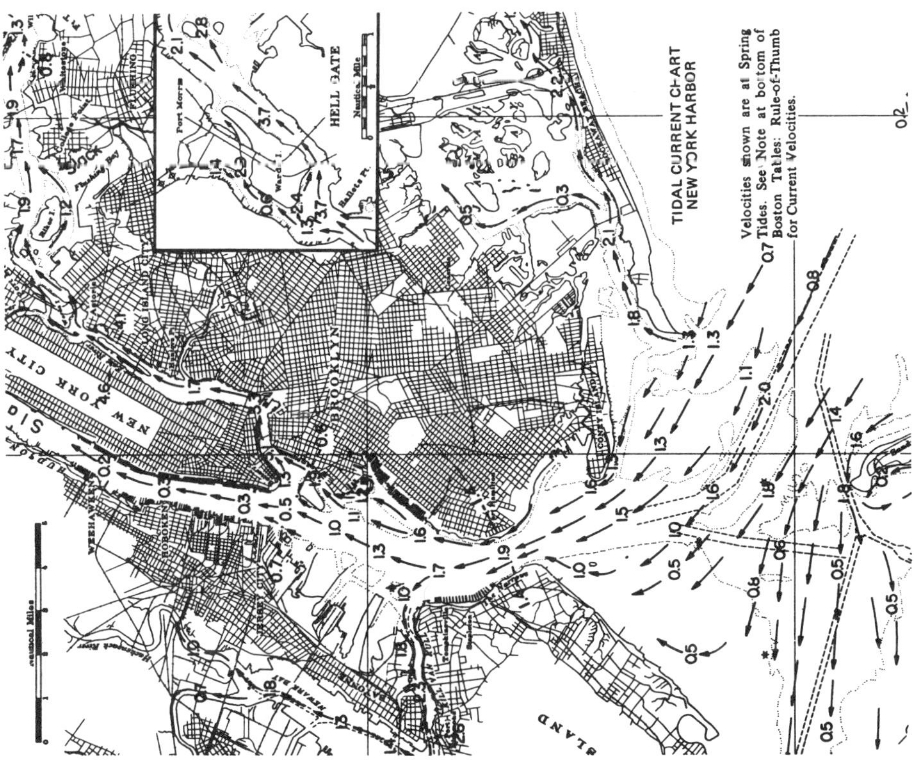

4 HOURS AFTER LOW WATER AT THE BATTERY

1999 HIGH & LOW WATER
SANDY HOOK, N.J.
40–28N x 74–01W

		Standard Time								Standard Time					
DAY OF MONTH	DAY OF WEEK	JANUARY						DAY OF MONTH	DAY OF WEEK	FEBRUARY					
		HIGH				LOW				HIGH				LOW	
		a.m.	Ht.	**p.m.**	Ht.	a.m.	**p.m.**			a.m.	Ht.	**p.m.**	Ht.	a.m.	**p.m.**
1	F	6 46	5.9	**7 17**	5.0	12 33	**1 24**	1	M	8 07	5.6	**8 39**	4.9	2 00	**2 42**
2	S	7 36	6.0	**8 08**	5.0	1 26	**2 15**	2	T	8 52	5.4	**9 25**	4.8	2 47	**3 24**
3	S	8 25	5.8	**8 59**	4.9	2 17	**3 03**	3	W	9 37	5.1	**10 12**	4.7	3 30	**4 04**
4	M	9 14	5.6	**9 51**	4.7	3 06	**3 50**	4	T	10 22	4.8	**10 58**	4.5	4 12	**4 42**
5	T	10 04	5.3	**10 44**	4.5	3 53	**4 34**	5	F	11 07	4.5	**11 44**	4.4	4 54	**5 20**
6	W	10 54	4.9	**11 35**	4.4	4 39	**5 19**	6	S	11 52	4.2	...	...	5 39	**6 01**
7	T	11 44	4.6	...	...	5 27	**6 05**	7	S	12 29	4.2	**12 38**	3.9	6 30	**6 48**
8	F	12 25	4.3	**12 33**	4.2	6 20	**6 56**	8	M	1 15	4.2	**1 27**	3.7	7 30	**7 44**
9	S	1 15	4.2	**1 22**	4.0	7 20	**7 49**	9	T	2 04	4.1	**2 21**	3.5	8 34	**8 44**
10	S	2 03	4.2	**2 12**	3.8	8 22	**8 42**	10	W	2 56	4.2	**3 18**	3.5	9 33	**9 40**
11	M	2 53	4.2	**3 05**	3.7	9 20	**9 32**	11	T	3 51	4.3	**4 18**	3.6	10 26	**10 31**
12	T	3 45	4.3	**4 02**	3.7	10 12	**10 19**	12	F	4 46	4.5	**5 13**	3.8	11 15	**11 21**
13	W	4 36	4.5	**4 56**	3.8	11 01	**11 05**	13	S	5 36	4.8	**6 01**	4.1	...	**12 02**
14	T	5 24	4.7	**5 45**	3.9	11 47	**11 50**	14	S	6 20	5.1	**6 45**	4.4	12 08	**12 48**
15	F	6 07	4.9	**6 30**	4.1	...	**12 33**	15	M	7 02	5.3	**7 26**	4.7	12 56	**1 32**
16	S	6 48	5.1	**7 11**	4.2	12 35	**1 17**	16	T	7 43	5.5	**8 06**	4.9	1 42	**2 15**
17	S	7 26	5.3	**7 50**	4.3	1 19	**2 00**	17	W	8 25	5.5	**8 49**	5.0	2 27	**2 56**
18	M	8 04	5.3	**8 29**	4.4	2 02	**2 41**	18	T	9 09	5.4	**9 34**	5.1	3 12	**3 37**
19	T	8 43	5.3	**9 10**	4.5	2 44	**3 20**	19	F	9 57	5.3	**10 24**	5.1	3 57	**4 18**
20	W	9 25	5.2	**9 55**	4.5	3 26	**4 00**	20	S	10 49	5.0	**11 17**	5.1	4 45	**5 02**
21	T	10 12	5.1	**10 45**	4.6	4 09	**4 40**	21	S	11 45	4.7	...	...	5 39	**5 53**
22	F	11 03	4.9	**11 38**	4.6	4 55	**5 24**	22	M	12 13	5.1	**12 44**	4.5	6 43	**6 54**
23	S	11 59	4.7	...	...	5 50	**6 15**	23	T	1 11	5.0	**1 46**	4.3	7 55	**8 04**
24	S	12 33	4.7	**12 57**	4.5	6 58	**7 17**	24	W	2 13	4.9	**2 51**	4.2	9 06	**9 12**
25	M	1 31	4.8	**1 58**	4.3	8 12	**8 24**	25	T	3 19	4.9	**3 58**	4.3	10 10	**10 15**
26	T	2 32	4.9	**3 04**	4.2	9 22	**9 28**	26	F	4 25	5.0	**5 02**	4.4	11 07	**11 12**
27	W	3 37	5.0	**4 11**	4.3	10 24	**10 28**	27	S	5 25	5.2	**5 58**	4.7	11 59	...
28	T	4 41	5.2	**5 15**	4.4	11 22	**11 24**	28	S	6 18	5.3	**6 47**	4.9	12 05	**12 48**
29	F	5 40	5.4	**6 12**	4.7	...	**12 16**								
30	S	6 33	5.6	**7 04**	4.8	12 19	**1 08**								
31	S	7 21	5.7	**7 52**	4.9	1 11	**1 57**								

Dates when Ht. of **Low** Water is below Mean Low with Ht. of lowest given for each period and Date of lowest in ( ):

1st - 6th: -1.1' (2nd)
16th - 31st: -0.7' (19th - 20th)
: -0.9' (31st)

1st - 4th: -0.9' (1st)
14th - 21st: -0.9' (17th - 18th)
28th: -0.6'

Average Rise and Fall 4.6 ft.

When a high tide exceeds av. ht., the *following* low tide will be lower than av.

1999 HIGH & LOW WATER SANDY HOOK, N.J.

40–28N x 74–01W

Standard Time — **Daylight Time starts Apr. 4 at 2 a.m.**

MARCH

Day of Month	Day of Week	HIGH a.m.	Ht.	HIGH p.m.	Ht.	LOW a.m.	LOW p.m.
1	M	7 04	5.4	7 32	5.1	12 55	1 34
2	T	7 47	5.4	8 15	5.1	1 42	2 16
3	W	8 28	5.3	8 56	5.1	2 26	2 55
4	T	9 09	5.1	9 37	4.9	3 07	3 31
5	F	9 49	4.8	10 18	4.8	3 46	4 05
6	S	10 31	4.5	11 00	4.6	4 24	4 39
7	S	11 14	4.2	11 42	4.5	5 03	5 13
8	M	11 58	3.9	...	...	5 46	5 50
9	T	12 26	4.3	12 46	3.8	6 38	6 40
10	W	1 12	4.2	1 36	3.6	7 43	7 49
11	T	2 04	4.2	2 34	3.6	8 49	8 58
12	F	3 02	4.3	3 36	3.7	9 48	9 57
13	S	4 02	4.5	4 35	4.0	10 40	10 51
14	S	4 59	4.8	5 28	4.4	11 28	11 42
15	M	5 49	5.1	6 15	4.8	...	12 15
16	T	6 35	5.4	6 59	5.2	12 31	1 00
17	W	7 20	5.6	7 42	5.5	1 20	1 45
18	T	8 04	5.7	8 26	5.7	2 08	2 29
19	F	8 51	5.6	9 12	5.8	2 56	3 13
20	S	9 41	5.4	10 03	5.7	3 44	3 57
21	S	10 36	5.2	10 57	5.6	4 33	4 43
22	M	11 34	4.9	11 55	5.4	5 27	5 35
23	T	...	...	12 33	4.6	6 28	6 36
24	W	12 55	5.2	1 35	4.4	7 39	7 47
25	T	1 57	5.0	2 39	4.4	8 49	8 58
26	F	3 01	4.9	3 44	4.4	9 52	10 01
27	S	4 07	4.9	4 45	4.6	10 47	10 57
28	S	5 06	5.0	5 40	4.9	11 36	11 48
29	M	5 58	5.1	6 27	5.1	...	12 22
30	T	6 43	5.2	7 09	5.3	12 36	1 05
31	W	7 24	5.2	7 48	5.4	1 21	1 45

APRIL

Day of Month	Day of Week	HIGH a.m.	Ht.	HIGH p.m.	Ht.	LOW a.m.	LOW p.m.
1	T	8 03	5.1	8 26	5.4	2 03	2 22
2	F	8 41	5.0	9 03	5.3	2 43	2 58
3	S	9 19	4.7	9 40	5.1	3 21	3 31
4	S	*10 59	4.5	*11 18	4.9	*4 58	*5 03
5	M	11 40	4.3	11 56	4.7	5 35	5 35
6	T	...	...	12 23	4.0	6 13	6 08
7	W	12 38	4.6	1 09	3.9	6 58	6 49
8	T	1 24	4.5	1 59	3.8	7 55	7 52
9	F	2 17	4.4	2 55	3.9	9 02	9 13
10	S	3 13	4.5	3 54	4.0	10 06	10 22
11	S	4 16	4.6	4 55	4.3	11 00	11 20
12	M	5 18	4.8	5 52	4.8	11 51	...
13	T	6 15	5.2	6 43	5.3	12 14	12 39
14	W	7 06	5.5	7 31	5.7	1 07	1 27
15	T	7 55	5.7	8 17	6.1	1 58	2 14
16	F	8 43	5.7	9 03	6.3	2 49	3 02
17	S	9 33	5.7	9 52	6.3	3 40	3 49
18	S	10 26	5.5	10 44	6.2	4 30	4 37
19	M	11 22	5.2	11 40	5.9	5 20	5 26
20	T	...	...	12 22	5.0	6 14	6 19
21	W	12 39	5.6	1 22	4.8	7 13	7 20
22	T	1 38	5.3	2 22	4.7	8 19	8 30
23	F	2 38	5.1	3 23	4.6	9 26	9 40
24	S	3 39	4.9	4 23	4.7	10 27	10 43
25	S	4 41	4.8	5 22	4.8	11 20	11 37
26	M	5 39	4.8	6 15	5.0	...	12 07
27	T	6 31	4.8	7 00	5.3	12 27	12 51
28	W	7 16	4.9	7 42	5.4	1 13	1 32
29	T	7 57	4.9	8 20	5.5	1 57	2 11
30	F	8 36	4.9	8 56	5.5	2 39	2 49

***Daylight Time starts**

Dates when Ht. of **Low** Water is below Mean Low with Ht. of lowest given for each period and Date of lowest in ( ):

March:
1st - 5th: -0.6' (1st - 3rd)
15th - 22nd: -0.9' (18th - 19th)
30th - 31st: -0.3'

April:
1st - 2nd: -0.2'
13th - 20th: -0.9' (17th - 18th)

Average Rise and Fall 4.6 ft.

When a high tide exceeds av. ht., the *following* low tide will be lower than av.

1999 HIGH & LOW WATER
SANDY HOOK, N.J.
40–28N x 74–01W

DAY OF MONTH	DAY OF WEEK	MAY HIGH a.m.	Ht.	p.m.	Ht.	LOW a.m.	p.m.	DAY OF MONTH	DAY OF WEEK	JUNE HIGH a.m.	Ht.	p.m.	Ht.	LOW a.m.	p.m.
		Daylight Time								Daylight Time					
1	S	9 14	4.8	**9 32**	5.5	3 19	**3 26**	1	T	10 07	4.4	**10 13**	5.3	4 14	**4 11**
2	S	9 52	4.6	**10 06**	5.3	3 58	**4 01**	2	W	10 47	4.3	**10 50**	5.1	4 51	**4 46**
3	M	10 31	4.5	**10 42**	5.1	4 35	**4 34**	3	T	11 29	4.2	**11 31**	5.0	5 28	**5 22**
4	T	11 11	4.3	**11 18**	5.0	5 12	**5 07**	4	F	...	...	**12 15**	4.2	6 06	**6 01**
5	W	11 54	4.1	**11 59**	4.8	5 49	**5 40**	5	S	12 17	4.9	**1 03**	4.3	6 49	**6 51**
6	T	...	...	**12 40**	4.0	6 29	**6 19**	6	S	1 08	4.9	**1 53**	4.5	7 40	**8 00**
7	F	12 45	4.7	**1 29**	4.1	7 17	**7 14**	7	M	2 03	4.8	**2 46**	4.7	8 40	**9 17**
8	S	1 37	4.7	**2 21**	4.2	8 18	**8 32**	8	T	3 02	4.8	**3 43**	5.0	9 41	**10 24**
9	S	2 34	4.7	**3 18**	4.4	9 22	**9 47**	9	W	4 06	4.8	**4 44**	5.4	10 38	**11 25**
10	M	3 34	4.7	**4 16**	4.7	10 20	**10 51**	10	T	5 10	4.9	**5 43**	5.8	11 32	**...**
11	T	4 38	4.9	**5 15**	5.1	11 13	**11 48**	11	F	6 13	5.1	**6 39**	6.2	12 22	**12 26**
12	W	5 40	5.1	**6 11**	5.6	...	**12 04**	12	S	7 11	5.3	**7 33**	6.4	1 18	**1 20**
13	T	6 38	5.3	**7 03**	6.1	12 43	**12 54**	13	S	8 06	5.4	**8 24**	6.6	2 13	**2 15**
14	F	7 32	5.5	**7 53**	6.4	1 37	**1 45**	14	M	9 00	5.4	**9 16**	6.5	3 07	**3 09**
15	S	8 24	5.6	**8 42**	6.6	2 31	**2 37**	15	T	9 55	5.4	**10 09**	6.3	3 59	**4 01**
16	S	9 16	5.6	**9 33**	6.5	3 24	**3 28**	16	W	10 51	5.2	**11 04**	6.0	4 50	**4 53**
17	M	10 11	5.4	**10 26**	6.3	4 15	**4 18**	17	T	11 49	5.1	**...**	...	5 40	**5 45**
18	T	11 08	5.3	**11 23**	6.0	5 06	**5 10**	18	F	12 01	5.6	**12 45**	5.0	6 30	**6 39**
19	W	...	...	**12 08**	5.1	5 59	**6 03**	19	S	12 55	5.3	**1 39**	4.9	7 23	**7 39**
20	T	12 21	5.7	**1 07**	4.9	6 54	**7 02**	20	S	1 48	5.0	**2 31**	4.9	8 19	**8 42**
21	F	1 19	5.4	**2 04**	4.8	7 55	**8 07**	21	M	2 39	4.7	**3 23**	4.8	9 15	**9 44**
22	S	2 16	5.0	**3 00**	4.8	8 57	**9 15**	22	T	3 31	4.4	**4 14**	4.9	10 06	**10 40**
23	S	3 12	4.8	**3 56**	4.8	9 55	**10 17**	23	W	4 25	4.3	**5 05**	5.0	10 53	**11 30**
24	M	4 08	4.6	**4 51**	4.9	10 46	**11 12**	24	T	5 20	4.2	**5 54**	5.1	11 37	**...**
25	T	5 04	4.5	**5 43**	5.1	11 32	**...**	25	F	6 12	4.3	**6 39**	5.3	12 16	**12 20**
26	W	5 57	4.5	**6 29**	5.3	12 01	**12 15**	26	S	7 00	4.3	**7 21**	5.4	1 02	**1 03**
27	T	6 45	4.6	**7 11**	5.4	12 46	**12 56**	27	S	7 43	4.4	**8 00**	5.5	1 46	**1 46**
28	F	7 29	4.6	**7 50**	5.5	1 30	**1 36**	28	M	8 24	4.5	**8 38**	5.5	2 29	**2 28**
29	S	8 10	4.6	**8 28**	5.6	2 13	**2 16**	29	T	9 04	4.5	**9 14**	5.5	3 11	**3 10**
30	S	8 49	4.6	**9 03**	5.5	2 55	**2 56**	30	W	9 43	4.5	**9 50**	5.4	3 51	**3 50**
31	M	9 28	4.5	**9 38**	5.4	3 35	**3 34**								

Dates when Ht. of **Low** Water is below Mean Low with Ht. of lowest given for each period and Date of lowest in ( ):

12th - 20th: -0.9' (16th - 17th)

10th - 18th: -0.9' (14th - 15th)

Average Rise and Fall 4.6 ft.

When a high tide exceeds av. ht., the *following* low tide will be lower than av.

1999 HIGH & LOW WATER
SANDY HOOK, N.J.
40–28N x 74–01W

Daylight Time

DAY OF MONTH	DAY OF WEEK	JULY					
		HIGH				LOW	
		a.m.	Ht.	p.m.	Ht.	a.m.	p.m.
1	T	10 22	4.4	10 27	5.3	4 29	4 28
2	F	11 04	4.4	11 08	5.2	5 06	5 06
3	S	11 49	4.5	11 55	5.1	5 44	5 47
4	S	...	...	12 37	4.6	6 23	6 36
5	M	12 46	5.0	1 27	4.8	7 09	7 38
6	T	1 40	4.9	2 20	5.0	8 05	8 52
7	W	2 38	4.8	3 16	5.2	9 07	10 02
8	T	3 41	4.7	4 17	5.5	10 09	11 05
9	F	4 48	4.7	5 20	5.8	11 08	...
10	S	5 53	4.9	6 20	6.1	12 04	12 05
11	S	6 54	5.0	7 16	6.3	1 01	1 01
12	M	7 50	5.2	8 09	6.4	1 57	1 57
13	T	8 44	5.3	9 00	6.4	2 50	2 52
14	W	9 37	5.3	9 51	6.2	3 41	3 44
15	T	10 30	5.3	10 42	5.9	4 29	4 34
16	F	11 24	5.2	11 34	5.5	5 15	5 23
17	S	...	...	12 17	5.1	6 01	6 12
18	S	12 25	5.2	1 08	5.0	6 47	7 04
19	M	1 15	4.8	1 56	4.9	7 35	8 02
20	T	2 04	4.5	2 44	4.8	8 27	9 03
21	W	2 53	4.2	3 33	4.8	9 20	10 02
22	T	3 45	4.1	4 24	4.8	10 11	10 55
23	F	4 41	4.0	5 15	4.9	10 59	11 44
24	S	5 37	4.1	6 05	5.1	11 46	...
25	S	6 29	4.2	6 51	5.3	12 31	12 31
26	M	7 16	4.3	7 33	5.4	1 17	1 17
27	T	7 58	4.5	8 12	5.5	2 01	2 02
28	W	8 38	4.6	8 50	5.6	2 44	2 46
29	T	9 17	4.7	9 27	5.6	3 25	3 29
30	F	9 56	4.8	10 06	5.5	4 04	4 10
31	S	10 37	4.8	10 48	5.4	4 42	4 51

Daylight Time

DAY OF MONTH	DAY OF WEEK	AUGUST					
		HIGH				LOW	
		a.m.	Ht.	p.m.	Ht.	a.m.	p.m.
1	S	11 22	4.9	11 35	5.2	5 19	5 34
2	M	...	...	12 11	5.0	5 59	6 23
3	T	12 28	5.0	1 03	5.1	6 43	7 22
4	W	1 23	4.9	1 57	5.3	7 37	8 33
5	T	2 22	4.7	2 55	5.4	8 41	9 45
6	F	3 25	4.6	3 57	5.5	9 48	10 50
7	S	4 32	4.6	5 02	5.7	10 51	11 50
8	S	5 39	4.8	6 04	5.9	11 51	...
9	M	6 42	5.0	7 03	6.1	12 46	12 47
10	T	7 36	5.2	7 54	6.2	1 40	1 43
11	W	8 27	5.4	8 42	6.1	2 31	2 36
12	T	9 16	5.4	9 30	6.0	3 19	3 26
13	F	10 05	5.4	10 16	5.7	4 04	4 13
14	S	10 53	5.3	11 04	5.4	4 46	4 58
15	S	11 42	5.2	11 51	5.0	5 27	5 42
16	M	...	...	12 29	5.0	6 07	6 28
17	T	12 39	4.7	1 16	4.9	6 48	7 19
18	W	1 27	4.4	2 03	4.8	7 34	8 17
19	T	2 15	4.1	2 50	4.7	8 28	9 20
20	F	3 07	4.0	3 41	4.7	9 26	10 18
21	S	4 02	3.9	4 34	4.8	10 22	11 10
22	S	5 01	4.0	5 28	4.9	11 14	11 59
23	M	5 56	4.2	6 18	5.1	...	12 02
24	T	6 45	4.4	7 03	5.4	12 45	12 49
25	W	7 29	4.7	7 44	5.6	1 29	1 36
26	T	8 09	4.9	8 24	5.7	2 13	2 22
27	F	8 48	5.1	9 03	5.7	2 54	3 07
28	S	9 28	5.3	9 44	5.7	3 35	3 50
29	S	10 09	5.4	10 28	5.5	4 14	4 34
30	M	10 55	5.5	11 18	5.3	4 54	5 20
31	T	11 46	5.5	...	...	5 35	6 10

Dates when Ht. of **Low** Water is below Mean Low with Ht. of lowest given for each period and Date of lowest in ( ):

10th - 17th: -0.8' (14th)
30th - 31st: -0.2'

8th - 14th: - 0.6' (11th - 13th)
27th - 31st: -0.4' (29th)

Average Rise and Fall 4.6 ft.

When a high tide exceeds av. ht., the *following* low tide will be lower than av.

1999 HIGH & LOW WATER
SANDY HOOK, N.J.
40–28N x 74–01W

Daylight Time — **Standard Time starts Oct. 31 at 2 a.m.**

DAY OF MONTH	DAY OF WEEK	SEPTEMBER HIGH a.m.	Ht.	p.m.	Ht.	LOW a.m.	p.m.
1	W	12 13	5.1	12 41	5.5	6 21	7 09
2	T	1 12	4.8	1 39	5.4	7 16	8 19
3	F	2 12	4.7	2 39	5.4	8 24	9 31
4	S	3 16	4.6	3 43	5.4	9 35	10 37
5	S	4 22	4.6	4 48	5.5	10 40	11 36
6	M	5 28	4.8	5 51	5.6	11 40	...
7	T	6 27	5.1	6 47	5.8	12 29	12 35
8	W	7 20	5.3	7 37	5.9	1 20	1 28
9	T	8 09	5.5	8 24	5.9	2 08	2 18
10	F	8 53	5.6	9 06	5.7	2 53	3 05
11	S	9 36	5.6	9 49	5.5	3 34	3 49
12	S	10 19	5.4	10 32	5.2	4 14	4 32
13	M	11 03	5.3	11 16	4.9	4 51	5 12
14	T	11 47	5.1	...	...	5 27	5 54
15	W	12 02	4.5	12 32	4.9	6 03	6 38
16	T	12 50	4.3	1 18	4.7	6 42	7 31
17	F	1 39	4.1	2 06	4.6	7 32	8 33
18	S	2 30	3.9	2 56	4.6	8 37	9 37
19	S	3 25	3.9	3 50	4.6	9 43	10 33
20	M	4 23	4.0	4 47	4.8	10 41	11 23
21	T	5 20	4.2	5 40	5.0	11 33	...
22	W	6 11	4.6	6 29	5.3	12 10	12 22
23	T	6 56	4.9	7 14	5.5	12 54	1 09
24	F	7 38	5.3	7 56	5.7	1 38	1 57
25	S	8 19	5.6	8 39	5.8	2 21	2 44
26	S	9 00	5.8	9 22	5.7	3 04	3 31
27	M	9 44	5.9	10 10	5.6	3 46	4 18
28	T	10 31	5.9	11 02	5.3	4 29	5 06
29	W	11 25	5.8	...	...	5 14	5 58
30	T	12 01	5.0	12 23	5.7	6 03	6 57

DAY OF MONTH	DAY OF WEEK	OCTOBER HIGH a.m.	Ht.	p.m.	Ht.	LOW a.m.	p.m.
1	F	1 02	4.8	1 23	5.5	7 01	8 05
2	S	2 04	4.7	2 26	5.4	8 10	9 16
3	S	3 07	4.6	3 29	5.3	9 23	10 22
4	M	4 12	4.7	4 33	5.3	10 30	11 18
5	T	5 14	4.9	5 35	5.3	11 28	...
6	W	6 11	5.1	6 29	5.4	12 09	12 22
7	T	7 01	5.4	7 17	5.5	12 57	1 11
8	F	7 46	5.6	8 01	5.5	1 41	1 58
9	S	8 28	5.7	8 42	5.4	2 23	2 43
10	S	9 07	5.6	9 21	5.2	3 02	3 25
11	M	9 46	5.5	10 01	4.9	3 40	4 06
12	T	10 25	5.3	10 42	4.6	4 16	4 45
13	W	11 05	5.1	11 26	4.4	4 50	5 23
14	T	11 47	4.9	...	...	5 24	6 04
15	F	12 13	4.1	12 32	4.7	6 00	6 49
16	S	1 03	3.9	1 20	4.6	6 42	7 46
17	S	1 54	3.9	2 11	4.5	7 43	8 51
18	M	2 47	3.9	3 05	4.5	8 59	9 51
19	T	3 43	4.0	4 01	4.6	10 05	10 44
20	W	4 39	4.3	4 58	4.8	11 02	11 32
21	T	5 32	4.7	5 52	5.1	11 53	...
22	F	6 22	5.2	6 42	5.4	12 17	12 43
23	S	7 07	5.6	7 29	5.6	1 02	1 33
24	S	7 51	6.0	8 15	5.7	1 48	2 22
25	M	8 35	6.2	9 03	5.6	2 34	3 12
26	T	9 21	6.3	9 53	5.5	3 20	4 02
27	W	10 11	6.2	10 48	5.2	4 08	4 52
28	T	11 06	6.0	11 48	5.0	4 56	5 45
29	F	...	...	11 06	5.7	5 48	6 43
30	S	12 51	4.8	1 08	5.5	6 47	7 48
31	S	12 53	4.7	*1 10	5.2	*6 56	*7 56

***Standard Time starts**

Dates when Ht. of **Low** Water is below Mean Low with Ht. of lowest given for each period and Date of lowest in ( ):

September:
8th - 11th: -0.4' (9th - 10th)
25th - 29th: -0.5' (26th - 28th)

October:
6th - 9th: -0.2'
23rd - 28th: -0.7' (25th - 26th)

Average Rise and Fall 4.6 ft.

When a high tide exceeds av. ht., the *following* low tide will be lower than av.

1999 HIGH & LOW WATER
SANDY HOOK, N.J.
40–28N x 74–01W

Standard Time

DAY OF MONTH	DAY OF WEEK	NOVEMBER HIGH a.m.	Ht.	p.m.	Ht.	LOW a.m.	p.m.
1	M	1 54	4.6	**2 12**	5.0	8 08	**9 00**
2	T	2 55	4.7	**3 13**	4.9	9 15	**9 56**
3	W	3 55	4.9	**4 12**	4.9	10 13	**10 44**
4	T	4 50	5.1	**5 06**	4.9	11 04	**11 29**
5	F	5 38	5.3	**5 54**	5.0	11 52	**...**
6	S	6 22	5.5	**6 37**	5.0	12 11	**12 37**
7	S	7 01	5.6	**7 17**	4.9	12 51	**1 20**
8	M	7 39	5.6	**7 56**	4.8	1 30	**2 01**
9	T	8 16	5.5	**8 35**	4.6	2 08	**2 41**
10	W	8 52	5.3	**9 14**	4.4	2 45	**3 20**
11	T	9 29	5.1	**9 55**	4.2	3 20	**3 58**
12	F	10 08	4.9	**10 40**	4.0	3 54	**4 36**
13	S	10 50	4.7	**11 28**	3.8	4 29	**5 17**
14	S	11 37	4.5	**...**	...	5 07	**6 04**
15	M	12 18	3.8	**12 27**	4.5	5 58	**7 01**
16	T	1 08	3.9	**1 20**	4.4	7 10	**8 03**
17	W	2 01	4.1	**2 16**	4.5	8 26	**9 00**
18	T	2 56	4.4	**3 15**	4.6	9 29	**9 52**
19	F	3 52	4.8	**4 14**	4.8	10 25	**10 40**
20	S	4 46	5.2	**5 11**	5.0	11 18	**11 28**
21	S	5 37	5.7	**6 04**	5.2	...	**12 10**
22	M	6 26	6.1	**6 55**	5.4	12 18	**1 03**
23	T	7 14	6.3	**7 45**	5.4	1 08	**1 55**
24	W	8 03	6.4	**8 38**	5.3	1 58	**2 47**
25	T	8 55	6.2	**9 34**	5.1	2 49	**3 38**
26	F	9 51	6.0	**10 34**	4.9	3 41	**4 30**
27	S	10 50	5.7	**11 36**	4.8	4 34	**5 25**
28	S	11 51	5.3	**...**	...	5 31	**6 25**
29	M	12 36	4.7	**12 50**	5.0	6 36	**7 28**
30	T	1 34	4.6	**1 47**	4.7	7 45	**8 30**

Standard Time

DAY OF MONTH	DAY OF WEEK	DECEMBER HIGH a.m.	Ht.	p.m.	Ht.	LOW a.m.	p.m.
1	W	2 31	4.6	**2 45**	4.5	8 52	**9 25**
2	T	3 28	4.7	**3 42**	4.4	9 50	**10 14**
3	F	4 22	4.9	**4 37**	4.4	10 42	**10 58**
4	S	5 11	5.0	**5 27**	4.4	11 28	**11 40**
5	S	5 55	5.2	**6 12**	4.4	...	**12 13**
6	M	6 35	5.3	**6 53**	4.4	12 20	**12 56**
7	T	7 13	5.3	**7 33**	4.4	1 01	**1 38**
8	W	7 50	5.3	**8 11**	4.3	1 40	**2 19**
9	T	8 27	5.2	**8 51**	4.2	2 19	**2 58**
10	F	9 01	5.0	**9 29**	4.0	2 56	**3 35**
11	S	9 38	4.9	**10 10**	3.9	3 32	**4 12**
12	S	10 17	4.7	**10 54**	3.8	4 07	**4 49**
13	M	11 00	4.5	**11 41**	3.9	4 44	**5 29**
14	T	11 48	4.4	**...**	...	5 27	**6 16**
15	W	12 30	4.0	**12 40**	4.4	6 27	**7 12**
16	T	1 21	4.2	**1 36**	4.3	7 44	**8 14**
17	F	2 16	4.4	**2 36**	4.4	8 56	**9 12**
18	S	3 14	4.8	**3 40**	4.5	9 58	**10 06**
19	S	4 14	5.2	**4 43**	4.6	10 55	**11 00**
20	M	5 11	5.6	**5 42**	4.8	11 50	**11 53**
21	T	6 05	6.0	**6 37**	5.0	...	**12 45**
22	W	6 57	6.2	**7 30**	5.1	12 47	**1 39**
23	T	7 48	6.2	**8 24**	5.1	1 41	**2 31**
24	F	8 40	6.1	**9 19**	5.0	2 34	**3 22**
25	S	9 35	5.9	**10 16**	4.9	3 26	**4 12**
26	S	10 31	5.5	**11 14**	4.8	4 18	**5 03**
27	M	11 28	5.1	**...**	...	5 12	**5 56**
28	T	12 11	4.7	**12 23**	4.8	6 10	**6 52**
29	W	1 06	4.6	**1 17**	4.4	7 14	**7 51**
30	T	1 59	4.5	**2 11**	4.2	8 20	**8 47**
31	F	2 53	4.5	**3 07**	4.0	9 20	**9 38**

Dates when Ht. of **Low** Water is below Mean Low with Ht. of lowest given for each period and Date of lowest in ( ):

November: 19th - 28th: -1.0' (23rd - 24th)

December: 7th - 9th: -0.2'
18th - 28th: -1.0' (22nd - 24th)

Average Rise and Fall 4.6 ft.

When a high tide exceeds av. ht., the *following* low tide will be lower than av.

1999 CURRENT TABLE
DELAWARE BAY ENTRANCE

38 - 48 - 00N x 75 - 01 - 00W

JANUARY

Standard Time

DAY OF MONTH	DAY OF WEEK	CURRENT TURNS: NORTHWEST Flood Starts			SOUTHEAST Ebb Starts		
		a.m.	**p.m.**	Kts.	a.m.	**p.m.**	Kts.
1	F	2 35	**3 41**	a2.0	9 01	**9 27**	a1.9
2	S	3 29	**4 34**	a2.0	9 54	**10 20**	a1.9
3	S	4 20	**5 25**	a1.9	10 44	**11 10**	a1.9
4	M	5 10	**6 14**	a1.8	11 32	**11 58**	a1.7
5	T	5 58	**7 00**	a1.7	...	**12 19**	1.6
6	W	6 46	**7 44**	a1.5	12 45	**1 04**	p1.4
7	T	7 33	**8 26**	a1.3	1 31	**1 48**	p1.3
8	F	8 24	**9 07**	a1.2	2 19	**2 33**	p1.1
9	S	9 19	**9 47**	p1.1	3 09	**3 22**	1.0
10	S	10 19	**10 28**	p1.1	4 02	**4 12**	1.0
11	M	11 18	**11 10**	p1.2	4 53	**5 04**	1.0
12	T	12 14	**11 54**	p1.3	5 43	**5 56**	a1.1
13	W	1 08	**...**	1.0	6 33	**6 48**	a1.2
14	T	12 40	**1 58**	a1.3	7 22	**7 40**	a1.3
15	F	1 28	**2 43**	a1.4	8 09	**8 29**	a1.4
16	S	2 15	**3 25**	a1.5	8 52	**9 14**	a1.5
17	S	3 00	**4 06**	a1.6	9 35	**9 58**	a1.6
18	M	3 47	**4 46**	a1.6	10 17	**10 40**	a1.6
19	T	4 32	**5 26**	a1.7	10 58	**11 23**	a1.6
20	W	5 20	**6 08**	a1.7	11 40	**...**	1.6
21	T	6 10	**6 51**	a1.6	12 08	**12 25**	p1.5
22	F	7 05	**7 38**	1.5	12 57	**1 12**	p1.5
23	S	8 03	**8 29**	p1.6	1 49	**2 04**	p1.4
24	S	9 08	**9 24**	p1.6	2 47	**3 03**	p1.4
25	M	10 18	**10 23**	p1.6	3 49	**4 07**	a1.4
26	T	11 27	**11 24**	p1.7	4 52	**5 11**	a1.4
27	W	...	**12 33**	1.4	5 55	**6 16**	a1.5
28	T	12 25	**1 36**	a1.7	6 57	**7 20**	a1.7
29	F	1 25	**2 34**	a1.8	7 56	**8 19**	a1.7
30	S	2 22	**3 28**	a1.8	8 50	**9 13**	a1.8
31	S	3 15	**4 17**	a1.9	9 40	**10 02**	a1.8

FEBRUARY

Standard Time

DAY OF MONTH	DAY OF WEEK	CURRENT TURNS: NORTHWEST Flood Starts			SOUTHEAST Ebb Starts		
		a.m.	**p.m.**	Kts.	a.m.	**p.m.**	Kts.
1	M	4 04	**5 03**	a1.8	10 27	**10 48**	a1.8
2	T	4 50	**5 47**	a1.8	11 10	**11 31**	a1.7
3	W	5 35	**6 26**	a1.6	11 51	**...**	1.5
4	T	6 18	**7 01**	a1.5	12 13	**12 31**	p1.4
5	F	7 00	**7 34**	a1.3	12 55	**1 10**	1.2
6	S	7 44	**8 07**	p1.2	1 37	**1 50**	1.1
7	S	8 33	**8 44**	p1.1	2 22	**2 34**	1.0
8	M	9 29	**9 26**	p1.1	3 11	**3 23**	a1.0
9	T	10 28	**10 13**	p1.1	4 03	**4 17**	a1.0
10	W	11 27	**11 05**	p1.2	4 57	**5 12**	a1.0
11	T	...	**12 23**	0.9	5 50	**6 08**	a1.1
12	F	12 01	**1 18**	a1.3	6 43	**7 03**	a1.2
13	S	12 51	**2 07**	a1.4	7 35	**7 56**	a1.4
14	S	1 46	**2 52**	a1.5	8 23	**8 44**	a1.5
15	M	2 38	**3 35**	a1.6	9 09	**9 30**	a1.6
16	T	3 28	**4 17**	a1.7	9 53	**10 15**	a1.7
17	W	4 18	**4 59**	a1.8	10 37	**11 01**	a1.7
18	T	5 09	**5 41**	a1.8	11 22	**11 48**	a1.7
19	F	6 00	**6 27**	1.7	...	**12 08**	1.6
20	S	6 57	**7 14**	p1.7	12 38	**12 57**	1.5
21	S	7 55	**8 06**	p1.7	1 31	**1 49**	a1.5
22	M	8 59	**9 03**	p1.6	2 29	**2 48**	a1.4
23	T	10 08	**10 06**	p1.6	3 32	**3 53**	a1.4
24	W	11 18	**11 10**	p1.6	4 38	**5 00**	a1.4
25	T	...	**12 23**	1.3	5 44	**6 06**	a1.5
26	F	12 14	**1 24**	a1.6	6 48	**7 09**	a1.6
27	S	1 15	**2 20**	a1.7	7 46	**8 07**	a1.6
28	S	2 11	**3 09**	a1.7	8 38	**8 57**	a1.7

The Kts. (knots) columns show the **maximum** predicted velocities of the stronger one of the Flood Currents and the stronger one of the Ebb Currents for each day. The letter "a" means the velocity shown should occur **after** the **a.m.** Current Change. The letter "p" means the velocity shown should occur **after** the **p.m.** Current Change (even if next morning). No "a" or "p" means a.m. and p.m. velocities are the same for that day.

Av. Max. Vel.: Fl. 1.8 Kts., Ebb 1.9 Kts.

Max. Flood & Max. Ebb 3 hrs. 5 min. after Flood Starts & Ebb Starts, ±15 min.

See pp. 150-157 for Current Change at other points.

1999 CURRENT TABLE
DELAWARE BAY ENTRANCE

38 - 48 - 00N x 75 - 01 - 00W

Standard Time **Daylight Time starts Apr. 4 at 2 a.m.**

MARCH

DAY OF MONTH	DAY OF WEEK	NORTHWEST Flood Starts a.m.	NORTHWEST Flood Starts **p.m.**	NORTHWEST Flood Starts Kts.	SOUTHEAST Ebb Starts a.m.	SOUTHEAST Ebb Starts **p.m.**	SOUTHEAST Ebb Starts Kts.
1	**M**	3 02	**3 55**	a1.7	9 24	**9 41**	a1.7
2	**T**	3 48	**4 36**	a1.7	10 05	**10 23**	a1.7
3	**W**	4 31	**5 12**	a1.7	10 44	**11 02**	a1.6
4	**T**	5 12	**5 46**	a1.6	11 21	**11 41**	a1.5
5	**F**	5 51	**6 15**	a1.5	11 58	...	1.4
6	**S**	6 30	**6 44**	1.3	12 19	**12 34**	a1.3
7	**S**	7 10	**7 16**	p1.2	12 57	**1 12**	a1.2
8	**M**	7 55	**7 52**	p1.2	1 38	**1 53**	a1.1
9	**T**	8 44	**8 36**	p1.1	2 23	**2 40**	a1.0
10	**W**	9 41	**9 26**	p1.1	3 14	**3 34**	a1.0
11	**T**	10 40	**10 22**	p1.2	4 09	**4 31**	a1.0
12	**F**	11 39	**11 21**	p1.2	5 06	**5 29**	a1.1
13	**S**	...	**12 34**	1.0	6 03	**6 26**	a1.2
14	**S**	12 20	**1 26**	a1.4	6 58	**7 21**	a1.3
15	**M**	1 19	**2 15**	a1.5	7 51	**8 13**	a1.5
16	**T**	2 16	**3 00**	a1.7	8 41	**9 02**	a1.6
17	**W**	3 10	**3 45**	a1.8	9 28	**9 50**	1.7
18	**T**	4 03	**4 29**	1.8	10 15	**10 38**	1.8
19	**F**	4 57	**5 15**	1.9	11 02	**11 27**	1.8
20	**S**	5 50	**6 02**	p1.9	11 51	...	1.7
21	**S**	6 47	**6 52**	p1.8	12 19	**12 42**	a1.7
22	**M**	7 46	**7 46**	p1.7	1 12	**1 36**	a1.6
23	**T**	8 49	**8 46**	p1.6	2 10	**2 36**	a1.5
24	**W**	9 58	**9 51**	p1.5	3 15	**3 43**	a1.4
25	**T**	11 06	**10 59**	p1.4	4 24	**4 52**	a1.4
26	**F**	...	**12 09**	1.2	5 32	**5 57**	a1.4
27	**S**	12 04	**1 07**	a1.5	6 35	**6 58**	a1.5
28	**S**	1 04	**1 59**	a1.5	7 31	**7 51**	a1.5
29	**M**	1 58	**2 45**	a1.6	8 19	**8 37**	a1.5
30	**T**	2 47	**3 25**	a1.6	9 01	**9 18**	a1.5
31	**W**	3 30	**4 01**	a1.6	9 39	**9 56**	1.5

APRIL

DAY OF MONTH	DAY OF WEEK	NORTHWEST Flood Starts a.m.	NORTHWEST Flood Starts **p.m.**	NORTHWEST Flood Starts Kts.	SOUTHEAST Ebb Starts a.m.	SOUTHEAST Ebb Starts **p.m.**	SOUTHEAST Ebb Starts Kts.
1	**T**	4 11	**4 33**	a1.6	10 15	**10 32**	a1.5
2	**F**	4 49	**5 02**	1.5	10 51	**11 09**	1.4
3	**S**	5 27	**5 31**	p1.5	11 27	**11 45**	1.3
4	**S**	*7 04	***7 01**	p1.4	...	***1 03**	1.2
5	**M**	7 42	**7 34**	p1.3	1 21	**1 41**	a1.3
6	**T**	8 23	**8 11**	p1.2	1 59	**2 20**	a1.2
7	**W**	9 09	**8 56**	p1.2	2 41	**3 05**	a1.1
8	**T**	10 01	**9 48**	p1.2	3 30	**3 56**	a1.1
9	**F**	10 59	**10 48**	p1.2	4 25	**4 54**	a1.1
10	**S**	11 56	**11 50**	p1.3	5 24	**5 54**	a1.1
11	**S**	...	**12 51**	1.1	6 23	**6 51**	a1.2
12	**M**	12 54	**1 45**	a1.4	7 21	**7 48**	1.3
13	**T**	1 57	**2 36**	1.5	8 17	**8 43**	1.5
14	**W**	2 57	**3 25**	1.7	9 11	**9 35**	p1.7
15	**T**	3 54	**4 12**	p1.9	10 02	**10 26**	p1.8
16	**F**	4 49	**5 00**	p2.0	10 52	**11 16**	p1.9
17	**S**	5 44	**5 49**	p2.1	11 42	...	1.7
18	**S**	6 39	**6 39**	p2.0	12 07	**12 34**	a1.9
19	**M**	7 36	**7 32**	p1.9	1 00	**1 28**	a1.8
20	**T**	8 34	**8 28**	p1.7	1 55	**2 24**	a1.7
21	**W**	9 37	**9 29**	p1.5	2 53	**3 24**	a1.6
22	**T**	10 42	**10 36**	p1.4	3 58	**4 31**	a1.4
23	**F**	11 48	**11 45**	p1.3	5 06	**5 39**	a1.4
24	**S**	...	**12 47**	1.2	6 12	**6 42**	a1.3
25	**S**	12 49	**1 41**	1.3	7 12	**7 39**	a1.4
26	**M**	1 48	**2 29**	1.4	8 05	**8 28**	a1.4
27	**T**	2 41	**3 11**	1.4	8 50	**9 11**	1.4
28	**W**	3 28	**3 48**	p1.5	9 30	**9 50**	1.4
29	**T**	4 11	**4 21**	p1.5	10 08	**10 27**	p1.5
30	**F**	4 50	**4 52**	p1.5	10 45	**11 03**	p1.5

***Daylight Time starts**

The Kts. (knots) columns show the **maximum** predicted velocities of the stronger one of the Flood Currents and the stronger one of the Ebb Currents for each day. The letter "a" means the velocity shown should occur **after** the **a.m.** Current Change. The letter "p" means the velocity shown should occur **after** the **p.m.** Current Change (even if next morning). No "a" or "p" means a.m. and p.m. velocities are the same for that day.

Av. Max. Vel.: Fl. 1.8 Kts., Ebb 1.9 Kts.

Max. Flood & Max. Ebb 3 hrs. 5 min. after Flood Starts & Ebb Starts, ±15 min.

See pp. 150-157 for Current Change at other points.

1999 CURRENT TABLE
DELAWARE BAY ENTRANCE
38 - 48 - 00N x 75 - 01 - 00W

MAY
Daylight Time

DAY OF MONTH	DAY OF WEEK	NORTHWEST Flood Starts a.m.	NORTHWEST Flood Starts **p.m.**	NORTHWEST Flood Starts Kts.	SOUTHEAST Ebb Starts a.m.	SOUTHEAST Ebb Starts **p.m.**	SOUTHEAST Ebb Starts Kts.
1	S	5 28	**5 21**	p1.5	11 22	**11 39**	p1.5
2	S	6 05	**5 52**	p1.5	11 59	...	1.2
3	M	6 41	**6 26**	p1.4	12 14	**12 37**	a1.4
4	T	7 19	**7 00**	p1.4	12 50	**1 15**	a1.4
5	W	7 58	**7 40**	p1.3	1 28	**1 54**	a1.3
6	T	8 41	**8 27**	p1.2	2 09	**2 37**	a1.2
7	F	9 29	**9 19**	p1.2	2 54	**3 26**	a1.2
8	S	10 22	**10 20**	p1.2	3 47	**4 22**	a1.1
9	S	11 18	**11 26**	p1.3	4 46	**5 22**	a1.2
10	M	...	**12 12**	1.2	5 46	**6 21**	1.2
11	T	12 31	**1 07**	p1.5	6 45	**7 19**	1.4
12	W	1 37	**1 59**	p1.7	7 44	**8 16**	p1.6
13	T	2 39	**2 52**	p1.9	8 41	**9 11**	p1.8
14	F	3 39	**3 44**	p2.0	9 37	**10 04**	p1.9
15	S	4 36	**4 35**	p2.1	10 31	**10 56**	p2.0
16	S	5 31	**5 27**	p2.1	11 24	**11 49**	p2.0
17	M	6 28	**6 19**	p2.0	...	**12 18**	1.6
18	T	7 23	**7 13**	p1.9	12 42	**1 13**	a1.9
19	W	8 20	**8 10**	p1.7	1 38	**2 09**	a1.7
20	T	9 19	**9 10**	p1.5	2 35	**3 07**	a1.6
21	F	10 20	**10 15**	1.3	3 35	**4 11**	a1.4
22	S	11 20	**11 23**	p1.3	4 39	**5 16**	a1.3
23	S	...	**12 16**	1.2	5 40	**6 15**	a1.3
24	M	12 26	**1 05**	1.2	6 35	**7 08**	1.2
25	T	1 24	**1 49**	p1.3	7 24	**7 55**	1.2
26	W	2 17	**2 28**	p1.4	8 10	**8 38**	p1.3
27	T	3 05	**3 05**	p1.5	8 53	**9 19**	p1.4
28	F	3 48	**3 39**	p1.5	9 34	**9 57**	p1.4
29	S	4 29	**4 12**	p1.5	10 15	**10 35**	p1.5
30	S	5 07	**4 46**	p1.5	10 55	**11 12**	p1.5
31	M	5 43	**5 20**	p1.5	11 34	**11 49**	p1.5

JUNE
Daylight Time

DAY OF MONTH	DAY OF WEEK	NORTHWEST Flood Starts a.m.	NORTHWEST Flood Starts **p.m.**	NORTHWEST Flood Starts Kts.	SOUTHEAST Ebb Starts a.m.	SOUTHEAST Ebb Starts **p.m.**	SOUTHEAST Ebb Starts Kts.
1	T	6 20	**5 57**	p1.5	...	**12 13**	1.1
2	W	6 57	**6 36**	p1.4	12 26	**12 51**	a1.4
3	T	7 36	**7 18**	p1.4	1 03	**1 31**	a1.4
4	F	8 17	**8 05**	p1.3	1 43	**2 13**	a1.3
5	S	9 01	**8 58**	p1.3	2 27	**3 01**	a1.3
6	S	9 50	**9 59**	p1.3	3 16	**3 56**	a1.2
7	M	10 43	**11 06**	1.3	4 13	**4 55**	a1.3
8	T	11 38	...	1.4	5 13	**5 55**	1.3
9	W	12 13	**12 32**	p1.6	6 14	**6 54**	p1.5
10	T	1 19	**1 29**	p1.8	7 15	**7 52**	p1.7
11	F	2 24	**2 24**	p1.9	8 16	**8 50**	p1.8
12	S	3 26	**3 20**	p2.1	9 16	**9 46**	p1.9
13	S	4 23	**4 15**	p2.1	10 12	**10 40**	p2.0
14	M	5 19	**5 09**	p2.1	11 08	**11 33**	p2.0
15	T	6 14	**6 02**	p2.0	...	**12 02**	1.6
16	W	7 09	**6 57**	p1.9	12 27	**12 57**	a1.9
17	T	8 02	**7 51**	p1.7	1 20	**1 50**	a1.8
18	F	8 55	**8 47**	p1.5	2 13	**2 44**	a1.6
19	S	9 49	**9 46**	1.3	3 06	**3 40**	a1.4
20	S	10 42	**10 50**	1.2	4 00	**4 38**	a1.3
21	M	11 32	**11 52**	a1.2	4 55	**5 35**	a1.2
22	T	...	**12 17**	1.2	5 47	**6 26**	1.1
23	W	12 50	**12 57**	p1.3	6 36	**7 13**	1.1
24	T	1 45	**1 36**	p1.3	7 24	**8 00**	p1.2
25	F	2 36	**2 16**	p1.4	8 13	**8 44**	p1.3
26	S	3 21	**2 55**	p1.4	9 00	**9 27**	p1.4
27	S	4 03	**3 35**	p1.5	9 45	**10 08**	p1.5
28	M	4 42	**4 15**	p1.5	10 29	**10 47**	p1.5
29	T	5 20	**4 54**	p1.5	11 10	**11 26**	p1.5
30	W	5 57	**5 34**	p1.5	11 50	...	1.2

The Kts. (knots) columns show the **maximum** predicted velocities of the stronger one of the Flood Currents and the stronger one of the Ebb Currents for each day. The letter "a" means the velocity shown should occur **after** the **a.m.** Current Change. The letter "p" means the velocity shown should occur **after** the **p.m.** Current Change (even if next morning). No "a" or "p" means a.m. and p.m. velocities are the same for that day.
Av. Max. Vel.: Fl. 1.8 Kts., Ebb 1.9 Kts.
Max. Flood & Max. Ebb 3 hrs. 5 min. after Flood Starts & Ebb Starts, ±15 min.
See pp. 150-157 for Current Change at other points.

1999 CURRENT TABLE

DELAWARE BAY ENTRANCE

38 - 48 - 00N x 75 - 01 - 00W

Daylight Time

JULY

DAY OF MONTH	DAY OF WEEK	CURRENT TURNS					
		NORTHWEST Flood Starts			SOUTHEAST Ebb Starts		
		a.m.	**p.m.**	Kts.	a.m.	**p.m.**	Kts.
1	T	6 34	**6 17**	p1.5	12 04	**12 29**	a1.5
2	F	7 12	**7 00**	p1.5	12 42	**1 10**	a1.5
3	S	7 51	**7 49**	p1.4	1 22	**1 52**	a1.4
4	S	8 34	**8 43**	p1.4	2 05	**2 39**	a1.4
5	M	9 21	**9 43**	a1.4	2 52	**3 32**	a1.3
6	T	10 12	**10 50**	a1.5	3 46	**4 31**	a1.3
7	W	11 08	**11 59**	a1.6	4 46	**5 32**	1.3
8	T	...	**12 06**	1.7	5 49	**6 33**	p1.5
9	F	1 06	**1 04**	p1.8	6 53	**7 33**	p1.6
10	S	2 11	**2 03**	p1.9	7 57	**8 33**	p1.8
11	S	3 13	**3 02**	p2.0	8 59	**9 31**	p1.9
12	M	4 11	**3 59**	p2.0	9 58	**10 27**	p1.9
13	T	5 06	**4 54**	p2.0	10 52	**11 19**	p1.9
14	W	5 59	**5 47**	p1.9	11 46	**...**	1.6
15	T	6 49	**6 39**	p1.8	12 10	**12 37**	a1.9
16	F	7 38	**7 29**	p1.7	1 00	**1 26**	a1.7
17	S	8 24	**8 19**	p1.5	1 47	**2 14**	a1.6
18	S	9 08	**9 11**	1.3	2 32	**3 02**	a1.4
19	M	9 51	**10 08**	a1.2	3 18	**3 52**	a1.2
20	T	10 34	**11 08**	a1.2	4 06	**4 45**	a1.1
21	W	11 15	**...**	1.2	4 56	**5 37**	1.0
22	T	12 08	**12 01**	p1.2	5 47	**6 28**	p1.0
23	F	1 04	**12 41**	p1.2	6 39	**7 18**	p1.1
24	S	1 58	**1 27**	p1.3	7 32	**8 07**	p1.2
25	S	2 48	**2 14**	p1.4	8 24	**8 55**	p1.3
26	M	3 33	**3 01**	p1.4	9 14	**9 40**	p1.4
27	T	4 13	**3 47**	p1.5	10 00	**10 22**	p1.5
28	W	4 52	**4 31**	p1.6	10 43	**11 02**	p1.6
29	T	5 29	**5 16**	p1.6	11 25	**11 42**	p1.6
30	F	6 07	**6 00**	p1.6	...	**12 06**	1.3
31	S	6 45	**6 48**	p1.6	12 21	**12 47**	a1.6

Daylight Time

AUGUST

DAY OF MONTH	DAY OF WEEK	CURRENT TURNS					
		NORTHWEST Flood Starts			SOUTHEAST Ebb Starts		
		a.m.	**p.m.**	Kts.	a.m.	**p.m.**	Kts.
1	S	7 25	**7 38**	1.5	1 02	**1 31**	a1.5
2	M	8 08	**8 31**	a1.5	1 45	**2 18**	a1.5
3	T	8 54	**9 31**	a1.5	2 32	**3 11**	a1.4
4	W	9 47	**10 38**	a1.6	3 26	**4 10**	1.3
5	T	10 44	**11 47**	a1.6	4 27	**5 12**	p1.4
6	F	11 45	**...**	1.6	5 32	**6 16**	p1.5
7	S	12 55	**12 48**	p1.7	6 38	**7 20**	p1.6
8	S	2 00	**1 50**	p1.8	7 43	**8 22**	p1.7
9	M	3 02	**2 51**	p1.9	8 47	**9 21**	p1.8
10	T	3 58	**3 48**	p1.9	9 45	**10 15**	p1.9
11	W	4 50	**4 41**	p1.9	10 38	**11 05**	p1.9
12	T	5 39	**5 31**	p1.9	11 27	**11 51**	p1.8
13	F	6 25	**6 19**	p1.8	...	**12 13**	1.5
14	S	7 07	**7 05**	p1.6	12 35	**12 57**	a1.7
15	S	7 46	**7 50**	1.4	1 16	**1 39**	a1.5
16	M	8 21	**8 36**	a1.3	1 56	**2 22**	a1.3
17	T	8 55	**9 24**	a1.2	2 37	**3 06**	a1.2
18	W	9 31	**10 18**	a1.2	3 20	**3 54**	1.0
19	T	10 12	**11 17**	a1.1	4 08	**4 46**	p1.0
20	F	10 59	**...**	1.1	5 01	**5 40**	p1.0
21	S	12 16	**12 01**	p1.1	5 57	**6 34**	p1.1
22	S	1 12	**12 43**	p1.2	6 52	**7 28**	p1.2
23	M	2 05	**1 37**	p1.3	7 48	**8 20**	p1.3
24	T	2 53	**2 30**	p1.4	8 41	**9 09**	p1.4
25	W	3 37	**3 21**	p1.5	9 29	**9 54**	p1.5
26	T	4 18	**4 10**	p1.6	10 14	**10 37**	p1.6
27	F	4 57	**4 58**	p1.7	10 57	**11 18**	p1.6
28	S	5 36	**5 46**	p1.7	11 40	**...**	1.5
29	S	6 16	**6 36**	p1.7	12 01	**12 23**	a1.6
30	M	6 58	**7 28**	a1.7	12 42	**1 09**	a1.6
31	T	7 42	**8 21**	a1.7	1 28	**1 58**	1.5

The Kts. (knots) columns show the **maximum** predicted velocities of the stronger one of the Flood Currents and the stronger one of the Ebb Currents for each day. The letter "a" means the velocity shown should occur **after** the **a.m.** Current Change. The letter "p" means the velocity shown should occur **after** the **p.m.** Current Change (even if next morning). No "a" or "p" means a.m. and p.m. velocities are the same for that day.

Av. Max. Vel.: Fl. 1.8 Kts., Ebb 1.9 Kts.

Max. Flood & Max. Ebb 3 hrs. 5 min. after Flood Starts & Ebb Starts, ±15 min.

See pp. 150-157 for Current Change at other points.

1999 CURRENT TABLE
DELAWARE BAY ENTRANCE
38 - 48 - 00N x 75 - 01 - 00W

Daylight Time — **Standard Time starts Oct. 31 at 2 a.m.**

SEPTEMBER

DAY OF MONTH	DAY OF WEEK	NORTHWEST Flood Starts a.m.	p.m.	Kts.	SOUTHEAST Ebb Starts a.m.	p.m.	Kts.
1	W	8 31	9 21	a1.7	2 17	2 51	1.4
2	T	9 25	10 28	a1.6	3 12	3 51	p1.4
3	F	10 26	11 38	a1.6	4 14	4 57	p1.4
4	S	11 31	...	1.6	5 22	6 04	p1.4
5	S	12 45	12 38	p1.6	6 29	7 10	p1.5
6	M	1 49	1 42	p1.7	7 35	8 13	p1.6
7	T	2 48	2 43	p1.7	8 37	9 10	p1.7
8	W	3 41	3 38	p1.8	9 32	10 01	p1.7
9	T	4 29	4 28	p1.8	10 20	10 46	p1.7
10	F	5 13	5 15	p1.8	11 04	11 27	p1.7
11	S	5 53	5 59	p1.7	11 46	...	1.5
12	S	6 30	6 41	a1.6	12 07	12 26	a1.6
13	M	7 02	7 21	a1.5	12 44	1 04	a1.4
14	T	7 33	8 02	a1.4	1 22	1 43	a1.3
15	W	8 04	8 45	a1.3	2 00	2 23	1.1
16	T	8 39	9 33	a1.2	2 41	3 07	p1.1
17	F	9 20	10 27	a1.1	3 26	3 57	p1.0
18	S	10 09	11 26	a1.1	4 19	4 52	p1.0
19	S	11 05	...	1.1	5 16	5 49	p1.0
20	M	12 23	12 03	p1.2	6 14	6 46	p1.1
21	T	1 17	1 02	p1.3	7 10	7 41	p1.2
22	W	2 07	2 00	p1.4	8 05	8 33	p1.4
23	T	2 54	2 57	p1.5	8 56	9 22	p1.5
24	F	3 38	3 50	p1.6	9 43	10 08	p1.6
25	S	4 20	4 41	p1.7	10 28	10 52	p1.7
26	S	5 02	5 31	1.8	11 13	11 37	1.7
27	M	5 46	6 23	a1.9	...	12 01	1.7
28	T	6 31	7 18	a1.9	12 23	12 48	p1.7
29	W	7 20	8 12	a1.9	1 12	1 39	p1.6
30	T	8 11	9 12	a1.8	2 04	2 34	p1.5

OCTOBER

DAY OF MONTH	DAY OF WEEK	NORTHWEST Flood Starts a.m.	p.m.	Kts.	SOUTHEAST Ebb Starts a.m.	p.m.	Kts.
1	F	9 09	10 19	a1.6	3 01	3 36	p1.5
2	S	10 12	11 27	a1.5	4 05	4 44	p1.4
3	S	11 22	...	1.5	5 15	5 53	p1.4
4	M	12 33	12 30	p1.5	6 23	6 59	p1.5
5	T	1 33	1 34	p1.5	7 27	8 00	p1.5
6	W	2 29	2 33	p1.6	8 25	8 54	p1.6
7	T	3 19	3 26	p1.6	9 15	9 40	p1.6
8	F	4 02	4 14	1.6	9 59	10 21	p1.6
9	S	4 42	4 58	1.6	10 39	11 00	1.5
10	S	5 17	5 39	a1.6	11 18	11 37	a1.5
11	M	5 49	6 18	a1.6	11 54	...	1.5
12	T	6 19	6 56	a1.5	12 13	12 30	p1.4
13	W	6 49	7 33	a1.4	12 51	1 08	p1.3
14	T	7 21	8 13	a1.3	1 29	1 46	p1.2
15	F	7 58	8 57	a1.2	2 09	2 27	p1.1
16	S	8 40	9 45	a1.1	2 52	3 14	p1.1
17	S	9 29	10 40	a1.1	3 42	4 07	p1.0
18	M	10 27	11 36	a1.1	4 38	5 05	p1.1
19	T	11 29	...	1.1	5 37	6 03	p1.1
20	W	12 29	12 31	p1.2	6 33	6 59	p1.2
21	T	1 20	1 33	p1.4	7 28	7 54	p1.4
22	F	2 10	2 32	p1.5	8 21	8 47	p1.5
23	S	2 58	3 30	1.6	9 12	9 37	1.6
24	S	3 45	4 24	a1.8	10 00	10 26	a1.8
25	M	4 31	5 18	a1.9	10 49	11 15	a1.9
26	T	5 19	6 11	a2.0	11 38	...	1.9
27	W	6 08	7 07	a2.0	12 05	12 29	p1.9
28	T	7 00	8 02	a1.9	12 58	1 23	p1.8
29	F	7 56	9 02	a1.8	1 52	2 20	p1.6
30	S	8 55	10 06	a1.6	2 51	3 21	p1.5
31	S	*9 00	*10 11	a1.5	*2 56	*3 29	p1.4

***Standard Time starts**

The Kts. (knots) columns show the **maximum** predicted velocities of the stronger one of the Flood Currents and the stronger one of the Ebb Currents for each day. The letter "a" means the velocity shown should occur **after** the **a.m.** Current Change. The letter "p" means the velocity shown should occur **after** the **p.m.** Current Change (even if next morning). No "a" or "p" means a.m. and p.m. velocities are the same for that day.

Av. Max. Vel.: Fl. 1.8 Kts., Ebb 1.9 Kts.

Max. Flood & Max. Ebb 3 hrs. 5 min. after Flood Starts & Ebb Starts, ±15 min.

See pp. 150-157 for Current Change at other points.

1999 CURRENT TABLE
DELAWARE BAY ENTRANCE

38 - 48 - 00N x 75 - 01 - 00W

		NOVEMBER Standard Time								DECEMBER Standard Time					
		CURRENT TURNS								CURRENT TURNS					
		NORTHWEST Flood Starts			SOUTHEAST Ebb Starts					NORTHWEST Flood Starts			SOUTHEAST Ebb Starts		
DAY OF MONTH	DAY OF WEEK	a.m.	p.m.	Kts.	a.m.	p.m.	Kts.	DAY OF MONTH	DAY OF WEEK	a.m.	p.m.	Kts.	a.m.	p.m.	Kts.
1	M	10 11	11 14	a1.4	4 05	4 37	p1.4	1	W	11 00	11 39	1.3	4 50	5 09	p1.3
2	T	11 19	...	1.4	5 12	5 40	p1.4	2	T	...	12 01	1.2	5 47	6 02	p1.3
3	W	12 11	12 21	p1.4	6 12	6 38	p1.4	3	F	12 26	12 58	a1.4	6 37	6 51	a1.3
4	T	1 02	1 19	1.4	7 06	7 28	1.4	4	S	1 10	1 49	a1.4	7 23	7 37	a1.3
5	F	1 48	2 10	a1.5	7 53	8 12	1.4	5	S	1 49	2 36	a1.5	8 04	8 19	a1.4
6	S	2 29	2 57	a1.5	8 34	8 52	a1.5	6	M	2 25	3 18	a1.5	8 43	9 00	a1.4
7	S	3 06	3 39	a1.6	9 12	9 30	a1.5	7	T	3 00	3 57	a1.5	9 21	9 40	a1.5
8	M	3 38	4 18	a1.6	9 49	10 08	a1.5	8	W	3 34	4 34	a1.5	9 59	10 20	a1.5
9	T	4 10	4 56	a1.6	10 25	10 46	a1.5	9	T	4 09	5 10	a1.5	10 36	11 00	a1.5
10	W	4 40	5 32	a1.5	11 01	11 24	a1.4	10	F	4 44	5 46	a1.5	11 13	11 39	a1.4
11	T	5 12	6 09	a1.4	11 38	...	1.4	11	S	5 21	6 22	a1.4	11 51	...	1.4
12	F	5 48	6 47	a1.4	12 02	12 16	p1.3	12	S	6 01	7 00	a1.3	12 19	12 29	p1.3
13	S	6 27	7 27	a1.3	12 42	12 55	p1.2	13	M	6 47	7 40	a1.3	12 59	1 10	p1.3
14	S	7 09	8 11	a1.2	1 24	1 39	p1.2	14	T	7 36	8 25	a1.2	1 42	1 55	p1.2
15	M	7 59	9 00	a1.2	2 10	2 28	p1.1	15	W	8 31	9 14	1.2	2 32	2 46	p1.2
16	T	8 56	9 52	a1.2	3 03	3 23	p1.1	16	T	9 35	10 06	p1.3	3 28	3 42	p1.2
17	W	9 59	10 45	1.2	4 01	4 20	p1.2	17	F	10 41	10 59	p1.5	4 25	4 41	p1.3
18	T	11 04	11 37	1.3	4 58	5 18	p1.3	18	S	11 47	11 53	p1.7	5 22	5 41	1.3
19	F	...	12 08	1.4	5 53	6 15	p1.4	19	S	...	12 51	1.5	6 20	6 42	a1.5
20	S	12 29	1 11	a1.6	6 48	7 12	1.5	20	M	12 50	1 54	a1.9	7 18	7 43	a1.7
21	S	1 20	2 11	a1.8	7 43	8 08	a1.7	21	T	1 47	2 54	a2.0	8 15	8 41	a1.9
22	M	2 12	3 09	a1.9	8 36	9 02	a1.8	22	W	2 42	3 50	a2.1	9 10	9 38	a2.0
23	T	3 04	4 04	a2.1	9 28	9 55	a2.0	23	T	3 39	4 46	a2.1	10 05	10 33	a2.0
24	W	3 56	4 59	a2.1	10 20	10 49	a2.0	24	F	4 33	5 40	a2.1	10 59	11 29	a2.0
25	T	4 49	5 54	a2.1	11 13	11 44	a1.9	25	S	5 29	6 34	a2.0	11 54	...	1.9
26	F	5 43	6 50	a2.0	...	12 09	1.8	26	S	6 26	7 27	a1.8	12 24	12 48	p1.7
27	S	6 40	7 48	a1.8	12 40	1 05	p1.7	27	M	7 22	8 21	a1.6	1 19	1 41	p1.6
28	S	7 40	8 47	a1.6	1 38	2 04	p1.6	28	T	8 22	9 15	a1.4	2 15	2 36	p1.4
29	M	8 45	9 48	a1.5	2 40	3 07	p1.4	29	W	9 26	10 08	1.3	3 14	3 32	p1.3
30	T	9 53	10 46	1.3	3 46	4 10	p1.4	30	T	10 31	10 56	p1.3	4 14	4 27	p1.2
								31	F	11 33	11 41	p1.3	5 09	5 18	1.1

The Kts. (knots) columns show the **maximum** predicted velocities of the stronger one of the Flood Currents and the stronger one of the Ebb Currents for each day. The letter "a" means the velocity shown should occur **after** the **a.m.** Current Change. The letter "p" means the velocity shown should occur **after** the **p.m.** Current Change (even if next morning). No "a" or "p" means a.m. and p.m. velocities are the same for that day.

Av. Max. Vel.: Fl. 1.8 Kts., Ebb 1.9 Kts.

Max. Flood & Max. Ebb 3 hrs. 5 min. after Flood Starts & Ebb Starts, ±15 min.

See pp. 150-157 for Current Change at other points.

See Chesapeake & Delaware Canal Current Tables, Pages 134-139.

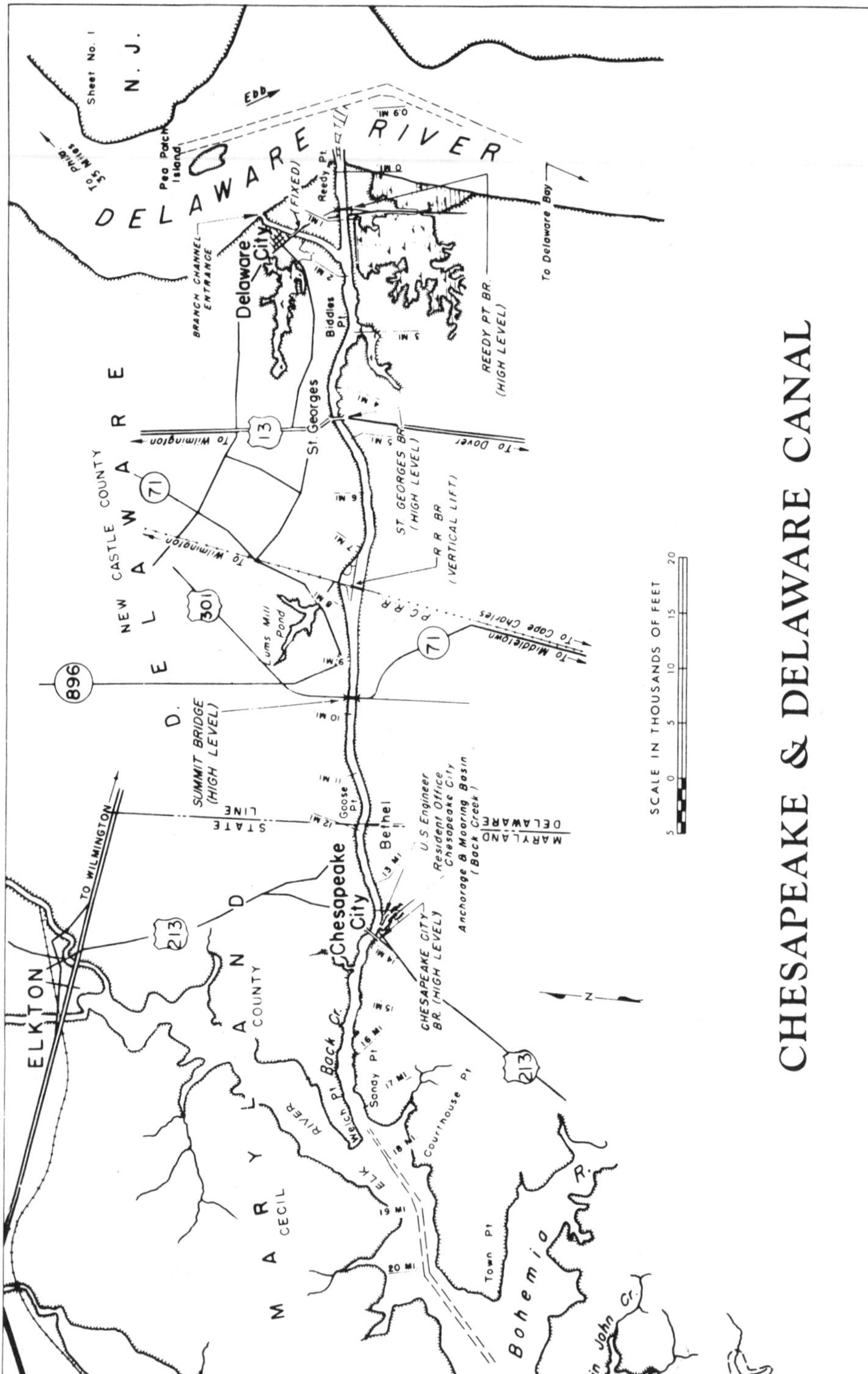

CHESAPEAKE & DELAWARE CANAL REGULATIONS

(Traffic Dispatcher is located at Chesapeake City)

Philadelphia District Engineer issues notices periodically showing available channel depths and navigation conditions.

Projected Channel dimensions are 35 ft. deep and 450 ft. wide. (The branch to Delaware City is 8 ft. deep and 50 ft. wide.) The distance from the Delaware River Ship Channel to the Elk River is 19.1 miles.

1. Traffic controls, located at Reedy Point and Old Town Point Wharf flash green when Canal is open, flash red when it is closed.
2. Vessel identification and monitoring are performed by TV cameras at Reedy Point and Old Town Point Wharf.
3. Vessels, Tugs and Tows are required to have radiotelephones as applied to the following:
   a. Power vessels of 300 gross tons and upward.
   b. Vessels of 100 gross tons and upward carrying 1 or more passengers for hire.
   c. Every towing vessel of 26 feet or over.
4. Vessels listed in 3. will not enter the Canal until radio communication is made with the dispatcher and clearance is received. Ships' captains will tell the dispatcher the estimated time of passing Reedy Point or Town Point. Communication is to be established on Channel 13 (156.65 MHz). Dispatcher also monitors Channel 16 (156.8 MHz) to respond to emergencies.
5. A westbound vessel must be able to pass Reedy Is. or Pea Patch Is. within 60 min. of receiving clearance; an eastbound vessel must be able to pass Arnold Point within 60 min. If passage is not made within these 60 min., a new clearance must be solicited. Vessels must also report to the dispatcher the time of passing the outer end of the jetties at Reedy Point and Old Town Point Wharf.
6. Maximum overall length of vessels and tows is 886 feet. Maximum overall length of tows and tugs is 760 feet. Vessels exceeding this are required to have operable bow thrusters.
7. Maximum combined extreme breadth of vessels meeting and overtaking each other is 190 feet.
8. Vessels of all types are required to travel at a safe speed to avoid damage by suction or wash to wharves, landings, other boats, etc. Operators of yachts, motorboats, etc. are cautioned that large, deep-draft ocean-going and commercial vessels ply the Canal and to moor or anchor well away from the main ship channel.
9. Vessels proceeding *with* the current shall have the right-of-way but all small pleasure craft shall relinquish the right-of-way to deeper draft vessels which have a limited maneuvering ability.
10. Vessels under sail will not be permitted in the Canal.
11. Vessels difficult to handle must use the Canal during daylight hours and must have tug assistance. They should clear Reedy Point Bridge (going east) or Chesapeake City Bridge (going west) before dark.

Anchorage and wharfage facilities for small vessels only are at Chesapeake City and permission to use them for more than 24 hours must be obtained in advance from the dispatcher.

The 4 Highway Bridges are high level and fixed and the 1 vertical lift RR bridge has a clearance when closed of 45 ft. at MHW.

Normal tide range is 5.4 ft. at Delaware R. end of the Canal and 2.6 ft. at Chesapeake City. Local mean low water at Courthouse Pt. is 2.5 ft. and decreases gradually eastward to 0.6 ft. at Delaware R. (See pp. 24 and 25 for times of High Water in this area.)
Note: A violent northeast storm may raise tide 4 to 5 ft. above normal in the Canal; a westerly storm may cause low tide to fall slightly below normal at Chesapeake City and as much as 4.0 ft. below normal at Reedy Point.

1999 CURRENT TABLE
CHESAPEAKE & DELAWARE CANAL

At Chesapeake City

Standard Time

JANUARY

DAY OF MONTH	DAY OF WEEK	CURRENT TURNS					
		EAST Flood Starts			WEST Ebb Starts		
		a.m.	**p.m.**	Kts.	a.m.	**p.m.**	Kts.
1	F	2 42	**3 53**	p2.4	8 19	**10 27**	a2.5
2	S	3 42	**4 45**	p2.4	9 13	**11 18**	a2.5
3	S	4 39	**5 34**	p2.4	10 05	**11 59**	a2.4
4	M	5 34	**6 21**	p2.3	10 57	**...**	2.3
5	T	6 28	**7 06**	p2.2	12 48	**-A-**	1.6
6	W	7 22	**7 48**	p2.2	1 28	**12 38**	p2.1
7	T	8 16	**8 28**	p2.1	2 04	**1 31**	p1.9
8	F	9 10	**9 07**	p2.0	2 39	**2 27**	p1.8
9	S	10 04	**9 46**	p2.0	3 13	**3 27**	a1.8
10	S	10 56	**10 28**	1.9	3 50	**4 31**	a1.9
11	M	11 47	**11 14**	1.9	4 30	**5 35**	a2.0
12	T	...	**12 37**	2.0	5 12	**6 37**	a2.1
13	W	12 01	**1 25**	p2.1	5 55	**7 34**	a2.1
14	T	12 48	**2 10**	p2.1	6 40	**8 27**	a2.2
15	F	1 33	**2 54**	p2.2	7 23	**9 15**	a2.2
16	S	2 14	**3 34**	p2.2	8 06	**9 58**	a2.3
17	S	2 54	**4 12**	p2.2	8 49	**10 38**	a2.3
18	M	3 34	**4 48**	p2.2	9 32	**11 14**	a2.3
19	T	4 16	**5 22**	p2.2	10 17	**11 49**	a2.3
20	W	5 02	**5 57**	p2.2	11 05	**...**	2.2
21	T	5 54	**6 35**	p2.1	12 24	**12 01**	p2.1
22	F	6 52	**7 15**	p2.1	1 01	**12 54**	p2.0
23	S	7 57	**8 00**	p2.1	1 42	**1 58**	1.9
24	S	9 10	**8 50**	2.0	2 27	**3 09**	a2.0
25	M	10 27	**9 49**	2.0	3 16	**4 29**	a2.1
26	T	11 41	**10 58**	a2.1	4 10	**5 53**	a2.2
27	W	...	**12 48**	2.2	5 08	**7 11**	a2.3
28	T	12 17	**1 50**	p2.2	6 09	**8 18**	a2.3
29	F	1 31	**2 47**	p2.3	7 10	**9 16**	a2.4
30	S	2 37	**3 39**	p2.3	8 09	**10 06**	a2.4
31	S	3 35	**4 29**	p2.3	9 05	**10 51**	a2.4

Standard Time

FEBRUARY

DAY OF MONTH	DAY OF WEEK	CURRENT TURNS					
		EAST Flood Starts			WEST Ebb Starts		
		a.m.	**p.m.**	Kts.	a.m.	**p.m.**	Kts.
1	M	4 29	**5 15**	p2.3	9 58	**11 32**	a2.3
2	T	5 19	**5 57**	p2.2	10 48	**11.59**	a2.2
3	W	6 06	**6 37**	p2.2	11 36	**...**	p2.1
4	T	6 52	**7 13**	p2.1	12 41	**12 23**	p2.0
5	F	7 36	**7 45**	2.0	1 11	**1 11**	p1.9
6	S	8 20	**8 15**	p2.0	1 41	**2 00**	a1.9
7	S	9 07	**8 47**	1.9	2 14	**2 53**	a1.9
8	M	9 57	**9 24**	1.9	2 51	**3 50**	a1.9
9	T	10 51	**10 09**	a1.9	3 34	**4 50**	a2.0
10	W	11 46	**11 00**	a2.0	4 20	**5 52**	a2.0
11	T	...	**12 39**	2.0	5 09	**6 51**	a2.1
12	F	12 01	**1 31**	p2.1	6 00	**7 46**	a2.1
13	S	12 50	**2 19**	p2.1	6 51	**8 35**	a2.2
14	S	1 42	**3 03**	p2.2	7 41	**9 19**	a2.2
15	M	2 32	**3 45**	p2.2	8 31	**9 58**	a2.3
16	T	3 20	**4 24**	p2.2	9 21	**10 34**	a2.3
17	W	4 09	**5 01**	p2.2	10 11	**11 10**	a2.2
18	T	4 58	**5 38**	p2.2	11 04	**11 46**	a2.2
19	F	5 51	**6 16**	2.1	11 59	**...**	2.1
20	S	6 48	**6 56**	2.1	12 25	**12 58**	a2.1
21	S	7 50	**7 41**	a2.1	1 07	**2 02**	a2.1
22	M	8 59	**8 34**	a2.1	1 54	**3 13**	a2.2
23	T	10 12	**9 38**	a2.1	2 46	**4 30**	a2.2
24	W	11 25	**11 00**	a2.1	3 45	**5 48**	a2.2
25	T	...	**12 32**	2.1	4 50	**6 58**	a2.2
26	F	12 22	**1 33**	p2.2	5 57	**7 57**	a2.2
27	S	1 32	**2 29**	p2.2	7 04	**8 49**	a2.2
28	S	2 32	**3 20**	p2.2	8 06	**9 33**	a2.2

A also at 11:47 a.m. 2.2

The Kts. (knots) columns show the **maximum** predicted velocities of the stronger one of the Flood Currents and the stronger one of the Ebb Currents for each day.
The letter "a" means the velocity shown should occur **after** the **a.m.** Current Change. The letter "p" means the velocity shown should occur **after** the **p.m.** Current Change (even if next morning). No "a" or "p" means a.m. and p.m. velocities are the same for that day.

Av. Max. Vel.: Fl. 2.0 Kts., Ebb 1.9 Kts.

Max. Flood 3 hrs. 10 min. after Flood Starts ±45 min.

Max. Ebb 2 hrs. 45 min. after Ebb Starts ±45 min.

See pp. 150-157 for Current Change at other points.

1999 CURRENT TABLE
CHESAPEAKE & DELAWARE CANAL
At Chesapeake City

Standard Time **Daylight Time starts April 4 at 2 a.m.**

MARCH

DAY OF MONTH	DAY OF WEEK	CURRENT TURNS EAST Flood Starts a.m.	**p.m.**	Kts.	CURRENT TURNS WEST Ebb Starts a.m.	**p.m.**	Kts.
1	M	3 25	**4 07**	p2.2	9 03	**10 13**	a2.2
2	T	4 14	**4 50**	p2.2	9 54	**10 48**	a2.1
3	W	4 59	**5 30**	p2.2	10 42	**11 19**	a2.1
4	T	5 40	**6 05**	2.1	11 27	**11 48**	a2.0
5	F	6 17	**6 37**	a2.1	...	**12 09**	1.9
6	S	6 53	**7 03**	a2.1	12 15	**12 52**	a2.0
7	S	7 29	**7 28**	a2.0	12 45	**1 36**	a2.0
8	M	8 08	**7 57**	a2.0	1 19	**2 22**	a2.0
9	T	8 54	**8 34**	a2.0	1 59	**3 13**	a2.0
10	W	9 46	**9 21**	a2.0	2 43	**4 08**	a2.0
11	T	10 44	**10 15**	a2.0	3 33	**5 05**	a2.0
12	F	11 44	**11 16**	a2.0	4 27	**6 03**	a2.0
13	S	...	**12 42**	2.1	5 24	**6 58**	a2.1
14	S	12 19	**1 36**	p2.1	6 22	**7 47**	a2.1
15	M	1 20	**2 25**	p2.2	7 20	**8 31**	a2.2
16	T	2 17	**3 12**	p2.2	8 18	**9 11**	a2.2
17	W	3 10	**3 55**	2.2	9 15	**9 50**	a2.2
18	T	4 01	**4 37**	2.2	10 11	**10 28**	p2.2
19	F	4 53	**5 17**	a2.3	11 08	**11 07**	p2.2
20	S	5 45	**5 59**	a2.3	11 59	**11 49**	p2.3
21	S	6 41	**6 43**	a2.3	...	**1 06**	1.8
22	M	7 41	**7 33**	a2.2	12 35	**2 09**	a2.3
23	T	8 46	**8 34**	a2.2	1 26	**3 17**	a2.2
24	W	9 55	**9 53**	a2.1	2 23	**4 26**	a2.2
25	T	11 05	**11 15**	a2.1	3 27	**5 32**	a2.1
26	F	...	**12 10**	2.1	4 38	**6 32**	a2.1
27	S	12 26	**1 09**	p2.1	5 51	**7 24**	a2.0
28	S	1 27	**2 03**	p2.1	7 01	**8 10**	a2.0
29	M	2 21	**2 53**	p2.1	8 03	**8 50**	a2.0
30	T	3 10	**3 38**	2.1	8 59	**9 26**	2.0
31	W	3 54	**4 20**	a2.2	9 49	**9 58**	2.0

APRIL

DAY OF MONTH	DAY OF WEEK	CURRENT TURNS EAST Flood Starts a.m.	**p.m.**	Kts.	CURRENT TURNS WEST Ebb Starts a.m.	**p.m.**	Kts.
1	T	4 34	**4 58**	a2.2	10 34	**10 27**	p2.1
2	F	5 10	**5 32**	a2.2	11 17	**10 54**	p2.1
3	S	5 43	**6 00**	a2.2	11 56	**11 23**	p2.1
4	S	*7 14	***7 23**	a2.2	...	***1 35**	1.7
5	M	7 45	**7 46**	a2.1	12 55	**2 14**	a2.1
6	T	8 20	**8 17**	a2.1	1 31	**2 56**	a2.1
7	W	9 02	**8 56**	a2.0	2 12	**3 41**	a2.1
8	T	9 50	**9 45**	a2.0	2 59	**4 31**	a2.0
9	F	10 44	**10 44**	a2.0	3 51	**5 23**	a2.0
10	S	11 44	**11 50**	a2.0	4 49	**6 17**	a2.0
11	S	...	**12 45**	2.1	5 52	**7 08**	a2.0
12	M	1 00	**1 45**	p2.1	6 58	**7 56**	a2.0
13	T	2 04	**2 41**	p2.1	8 05	**8 41**	2.0
14	W	3 03	**3 33**	a2.2	9 10	**9 24**	p2.2
15	T	3 58	**4 22**	a2.3	10 13	**10 06**	p2.3
16	F	4 51	**5 09**	a2.4	11 14	**10 48**	p2.4
17	S	5 44	**5 56**	a2.4	-A-	**12 14**	1.9
18	S	6 37	**6 44**	a2.4	...	**1 13**	1.8
19	M	7 32	**7 36**	a2.4	12 19	**2 12**	a2.4
20	T	8 30	**8 37**	a2.3	1 09	**3 12**	a2.3
21	W	9 31	**9 19**	a2.2	2 03	**4 12**	a2.2
22	T	10 35	**11 06**	a2.2	3 04	**5 10**	a2.1
23	F	11 38	**...**	2.1	4 13	**6 05**	a2.0
24	S	12 16	**12 39**	p2.1	5 28	**6 56**	a1.9
25	S	1 18	**1 36**	p2.1	6 44	**7 41**	a1.9
26	M	2 13	**2 29**	p2.1	7 54	**8 22**	p1.9
27	T	3 02	**3 17**	2.1	8 56	**9 00**	p2.0
28	W	3 47	**4 02**	a2.1	9 50	**9 34**	p2.1
29	T	4 29	**4 44**	a2.2	10 39	**10 05**	p2.1
30	F	5 06	**5 22**	a2.3	11 23	**10 36**	p2.2

***Daylight Time starts**

A also at 11:32 p.m. 2.4

The Kts. (knots) columns show the **maximum** predicted velocities of the stronger one of the Flood Currents and the stronger one of the Ebb Currents for each day.
The letter "a" means the velocity shown should occur **after** the **a.m.** Current Change. The letter "p" means the velocity shown should occur **after** the **p.m.** Current Change (even if next morning). No "a" or "p" means a.m. and p.m. velocities are the same for that day.
Av. Max. Vel.: Fl. 2.0 Kts., Ebb 1.9 Kts.
Max. Flood 3 hrs. 10 min. after Flood Starts ±45 min.
Max. Ebb 2 hrs. 45 min. after Ebb Starts ±45 min.
See pp. 150-157 for Current Change at other points.

1999 CURRENT TABLE
CHESAPEAKE & DELAWARE CANAL

At Chesapeake City

Daylight Time

MAY

DAY OF MONTH	DAY OF WEEK	CURRENT TURNS					
		EAST Flood Starts			WEST Ebb Starts		
		a.m.	**p.m.**	Kts.	a.m.	**p.m.**	Kts.
1	S	5 41	**5 56**	a2.3	11 59	**11 06**	p2.2
2	S	6 12	**6 22**	a2.3	-A-	**12 43**	1.6
3	M	6 41	**6 44**	a2.2	...	**1 19**	1.6
4	T	7 11	**7 09**	a2.2	12 12	**1 56**	a2.1
5	W	7 44	**7 43**	a2.2	12 50	**2 34**	a2.1
6	T	8 22	**8 26**	a2.1	1 33	**3 15**	a2.1
7	F	9 07	**9 20**	a2.1	2 21	**3 59**	a2.1
8	S	9 57	**10 23**	a2.1	3 16	**4 45**	a2.0
9	S	10 52	**11 34**	a2.1	4 18	**5 33**	a2.0
10	M	11 51	**...**	2.1	5 26	**6 21**	1.9
11	T	12 45	**12 52**	p2.1	6 39	**7 08**	p2.1
12	W	1 50	**1 52**	2.1	7 54	**7 55**	p2.2
13	T	2 50	**2 51**	a2.3	9 06	**8 41**	p2.4
14	F	3 46	**3 47**	a2.4	10 13	**9 28**	p2.5
15	S	4 40	**4 42**	a2.5	11 17	**10 15**	p2.5
16	S	5 34	**5 37**	a2.5	-B-	**12 17**	1.7
17	M	6 27	**6 34**	a2.5	...	**1 14**	1.7
18	T	7 22	**7 35**	a2.4	12 01	**2 10**	a2.4
19	W	8 16	**8 42**	a2.4	12 48	**3 03**	a2.3
20	T	9 12	**9 52**	a2.3	1 46	**3 54**	a2.2
21	F	10 09	**10 59**	a2.2	2 49	**4 43**	a2.1
22	S	11 05	**...**	2.1	3 59	**5 29**	a1.9
23	S	12 01	**12 01**	p2.1	5 15	**6 12**	1.8
24	M	12 58	**12 54**	p2.0	6 30	**6 52**	p1.9
25	T	1 49	**1 45**	2.0	7 38	**7 30**	p2.0
26	W	2 36	**2 33**	a2.1	8 39	**8 07**	p2.1
27	T	3 19	**3 19**	a2.2	9 33	**8 42**	p2.2
28	F	4 00	**4 02**	a2.2	10 22	**9 17**	p2.2
29	S	4 38	**4 41**	a2.3	11 07	**9 51**	p2.2
30	S	5 13	**5 15**	a2.3	11 48	**10 26**	p2.2
31	M	5 45	**5 43**	a2.3	-C-	**12 26**	1.5

Daylight Time

JUNE

DAY OF MONTH	DAY OF WEEK	CURRENT TURNS					
		EAST Flood Starts			WEST Ebb Starts		
		a.m.	**p.m.**	Kts.	a.m.	**p.m.**	Kts.
1	T	6 15	**6 08**	a2.3	-D-	**1 02**	1.5
2	W	6 44	**6 38**	a2.2	...	**1 36**	1.5
3	T	7 16	**7 17**	a2.2	12 18	**2 11**	a2.2
4	F	7 52	**8 05**	a2.2	1 02	**2 47**	a2.1
5	S	8 32	**9 02**	a2.1	1 53	**3 26**	a2.1
6	S	9 18	**10 09**	a2.1	2 50	**4 08**	a2.0
7	M	10 09	**11 20**	a2.1	3 56	**4 53**	1.9
8	T	11 04	**...**	2.1	5 08	**5 39**	p2.1
9	W	12 31	**12 03**	p2.1	6 27	**6 27**	p2.2
10	T	1 36	**1 06**	a2.2	7 47	**7 16**	p2.4
11	F	2 37	**2 11**	a2.3	9 04	**8 07**	p2.5
12	S	3 34	**3 16**	a2.4	10 13	**8 58**	p2.5
13	S	4 30	**4 20**	a2.5	11 15	**9 51**	p2.5
14	M	5 24	**5 23**	a2.5	-E-	**12 12**	1.6
15	T	6 17	**6 27**	a2.5	-F-	**1 05**	1.6
16	W	7 09	**7 30**	a2.4	...	**1 55**	1.6
17	T	7 59	**8 34**	a2.4	12 34	**2 41**	a2.3
18	F	8 49	**9 37**	a2.3	1 33	**3 25**	a2.2
19	S	9 37	**10 38**	a2.2	2 35	**4 05**	a2.0
20	S	10 25	**11 35**	a2.1	3 42	**4 44**	1.8
21	M	11 13	**...**	2.1	4 53	**5 22**	p1.9
22	T	12 28	**12 02**	p2.0	6 05	**5 59**	p2.0
23	W	1 17	**12 52**	2.0	7 12	**6 38**	p2.1
24	T	2 03	**1 41**	a2.0	8 13	**7 16**	p2.2
25	F	2 47	**2 29**	a2.1	9 08	**7 56**	p2.2
26	S	3 29	**3 14**	a2.2	9 58	**8 35**	p2.3
27	S	4 09	**3 56**	a2.2	10 43	**9 14**	p2.3
28	M	4 46	**4 33**	a2.3	11 25	**9 53**	p2.3
29	T	5 20	**5 05**	a2.3	11 59	**10 32**	p2.3
30	W	5 52	**5 38**	a2.3	-G-	**12 38**	1.5

A also at 11:38 p.m. 2.2 **B** also at 11:04 p.m. 2.5 **C** also at 11:01 p.m. 2.2
D also at 11:38 p.m. 2.2 **E** also at 10:44 p.m. 2.5 **F** also at 11:38 p.m. 2.4
G also at 11:12 p.m. 2.2

The Kts. (knots) columns show the **maximum** predicted velocities of the stronger one of the Flood Currents and the stronger one of the Ebb Currents for each day.
The letter "a" means the velocity shown should occur **after** the **a.m.** Current Change. The letter "p" means the velocity shown should occur **after** the **p.m.** Current Change (even if next morning). No "a" or "p" means a.m. and p.m. velocities are the same for that day.
Av. Max. Vel.: Fl. 2.0 Kts., Ebb 1.9 Kts.
Max. Flood 3 hrs. 10 min. after Flood Starts ±45 min.
Max. Ebb 2 hrs. 45 min. after Ebb Starts ±45 min.
See pp. 150-157 for Current Change at other points.

1999 CURRENT TABLE

CHESAPEAKE & DELAWARE CANAL

At Chesapeake City

JULY

Daylight Time

DAY OF MONTH	DAY OF WEEK	CURRENT TURNS EAST Flood Starts a.m.	**p.m.**	Kts.	CURRENT TURNS WEST Ebb Starts a.m.	**p.m.**	Kts.
1	T	6 21	**6 15**	a2.2	...	**1 10**	1.5
2	F	6 52	**6 59**	a2.2	12 01	**1 42**	a2.2
3	S	7 25	**7 50**	a2.2	12 42	**2 15**	a2.1
4	S	8 03	**8 49**	a2.1	1 35	**2 51**	a2.0
5	M	8 46	**9 55**	a2.1	2 35	**3 31**	p2.0
6	T	9 33	**11 06**	a2.1	3 43	**4 14**	p2.1
7	W	10 25	**...**	2.0	4 58	**5 02**	p2.2
8	T	12 16	**-A-**	2.1	6 21	**5 52**	p2.3
9	F	1 22	**12 30**	a2.2	7 45	**6 46**	p2.4
10	S	2 25	**1 43**	a2.3	9 01	**7 41**	p2.5
11	S	3 24	**2 59**	a2.4	10 07	**8 38**	p2.5
12	M	4 19	**4 10**	a2.4	11 05	**9 36**	p2.5
13	T	5 13	**5 16**	a2.5	11 57	**10 32**	p2.4
14	W	6 03	**6 18**	a2.4	-B-	**12 44**	1.7
15	T	6 52	**7 17**	a2.4	...	**1 28**	1.7
16	F	7 37	**8 15**	a2.3	12 25	**2 07**	a2.2
17	S	8 21	**9 11**	a2.3	1 21	**2 44**	a2.1
18	S	9 01	**10 05**	a2.2	2 19	**3 19**	1.9
19	M	9 41	**10 57**	a2.1	3 20	**3 52**	p1.9
20	T	10 20	**11 47**	a2.0	4 24	**4 28**	p2.0
21	W	11 02	**...**	2.0	5 30	**5 06**	p2.1
22	T	12 36	**11 48**	a2.0	6 35	**5 47**	p2.1
23	F	1 24	**12 39**	a2.0	7 37	**6 30**	p2.2
24	S	2 10	**1 32**	a2.1	8 33	**7 15**	p2.2
25	S	2 55	**2 24**	a2.2	9 25	**8 00**	p2.3
26	M	3 38	**3 11**	a2.2	10 11	**8 44**	p2.3
27	T	4 17	**3 55**	a2.2	10 53	**9 28**	p2.3
28	W	4 54	**4 36**	a2.3	11 30	**10 11**	p2.3
29	T	5 27	**5 17**	a2.3	11 59	**10 56**	p2.2
30	F	5 58	**6 00**	a2.2	-C-	**12 34**	1.7
31	S	6 28	**6 47**	a2.2	...	**1 04**	1.8

AUGUST

Daylight Time

DAY OF MONTH	DAY OF WEEK	CURRENT TURNS EAST Flood Starts a.m.	**p.m.**	Kts.	CURRENT TURNS WEST Ebb Starts a.m.	**p.m.**	Kts.
1	S	7 01	**7 39**	a2.2	12 33	**1 37**	a2.1
2	M	7 37	**8 37**	a2.1	1 28	**2 14**	2.0
3	T	8 18	**9 42**	a2.1	2 30	**2 54**	p2.1
4	W	9 03	**10 51**	2.0	3 38	**3 39**	p2.2
5	T	9 55	**11 59**	p2.1	4 56	**4 30**	p2.3
6	F	10 57	**...**	1.9	6 20	**5 25**	p2.3
7	S	1 10	**12 12**	a2.2	7 42	**6 24**	p2.4
8	S	2 13	**1 38**	a2.3	8 52	**7 26**	p2.4
9	M	3 12	**2 59**	a2.3	9 52	**8 29**	p2.4
10	T	4 07	**4 08**	a2.4	10 44	**9 30**	p2.4
11	W	4 58	**5 08**	a2.4	11 30	**10 28**	p2.3
12	T	5 45	**6 04**	a2.4	-D-	**12 11**	1.8
13	F	6 30	**6 57**	a2.3	...	**12 49**	1.8
14	S	7 11	**7 47**	a2.3	12 18	**1 23**	a2.1
15	S	7 48	**8 35**	a2.2	1 10	**1 55**	a2.0
16	M	8 22	**9 22**	a2.1	2 02	**2 25**	p2.0
17	T	8 53	**10 09**	a2.0	2 56	**2 57**	p2.0
18	W	9 24	**10 58**	1.9	3 52	**3 33**	p2.0
19	T	10 01	**11 48**	p2.0	4 51	**4 14**	p2.1
20	F	10 46	**...**	1.8	5 52	**4 59**	p2.1
21	S	12 38	**-E-**	2.0	6 53	**5 47**	p2.1
22	S	1 28	**12 40**	a2.1	7 50	**6 37**	p2.2
23	M	2 16	**1 40**	a2.1	8 42	**7 28**	p2.2
24	T	3 01	**2 36**	a2.2	9 28	**8 18**	p2.2
25	W	3 43	**3 27**	a2.2	10 09	**9 08**	p2.2
26	T	4 22	**4 15**	a2.2	10 45	**9 57**	p2.2
27	F	4 57	**5 01**	a2.2	11 18	**10 47**	p2.2
28	S	5 31	**5 47**	a2.2	11 50	**11 38**	p2.1
29	S	6 03	**6 35**	a2.2	...	**12 22**	2.0
30	M	6 37	**7 28**	a2.1	12 32	**12 58**	p2.1
31	T	7 14	**8 25**	2.1	1 29	**1 37**	p2.1

A also at 11:24 a.m. 2.0 **B** also at 11:29 p.m. 2.4 **C** also at 11:43 p.m. 2.2
D also at 11:24 p.m. 2.2 **E** also at 11:40 a.m. 1.8

The Kts. (knots) columns show the **maximum** predicted velocities of the stronger one of the Flood Currents and the stronger one of the Ebb Currents for each day.
The letter "a" means the velocity shown should occur **after** the **a.m.** Current Change. The letter "p" means the velocity shown should occur **after** the **p.m.** Current Change (even if next morning). No "a" or "p" means a.m. and p.m. velocities are the same for that day.
Av. Max. Vel.: Fl. 2.0 Kts., Ebb 1.9 Kts.
Max. Flood 3 hrs. 10 min. after Flood Starts ±45 min.
Max. Ebb 2 hrs. 45 min. after Ebb Starts ±45 min.
See pp. 150-157 for Current Change at other points.

1999 CURRENT TABLE
CHESAPEAKE & DELAWARE CANAL

At Chesapeake City

Daylight Time — **Standard Time starts Oct. 31 at 2 a.m.**

SEPTEMBER

DAY OF MONTH	DAY OF WEEK	CURRENT TURNS: EAST Flood Starts a.m.	**p.m.**	Kts.	WEST Ebb Starts a.m.	**p.m.**	Kts.
1	W	7 55	**9 28**	p2.1	2 31	**2 20**	p2.2
2	T	8 41	**10 37**	p2.1	3 41	**3 09**	p2.3
3	F	9 38	**11 48**	p2.1	4 58	**4 04**	p2.3
4	S	10 50	...	1.9	6 18	**5 06**	p2.3
5	S	12 56	**12 24**	a2.2	7 32	**6 12**	p2.3
6	M	1 59	**1 53**	a2.2	8 33	**7 20**	p2.3
7	T	2 56	**3 03**	a2.3	9 26	**8 27**	p2.3
8	W	3 48	**4 02**	a2.3	10 12	**9 30**	p2.2
9	T	4 37	**4 56**	a2.3	10 53	**10 27**	p2.2
10	F	5 21	**5 45**	a2.3	11 29	**11 21**	p2.1
11	S	6 03	**6 31**	a2.2	...	**12 03**	2.0
12	S	6 40	**7 14**	a2.2	12 10	**12 33**	2.0
13	M	7 13	**7 55**	2.1	12 58	**1 01**	p2.0
14	T	7 41	**8 35**	2.0	1 44	**1 31**	p2.1
15	W	8 06	**9 16**	p2.0	2 31	**2 04**	p2.1
16	T	8 34	**10 01**	p2.0	3 20	**2 42**	p2.1
17	F	9 11	**10 50**	p2.0	4 12	**3 25**	p2.1
18	S	9 58	**11 43**	p2.0	5 07	**4 13**	p2.1
19	S	10 55	...	1.7	6 05	**5 06**	p2.1
20	M	12 37	**12 01**	a2.0	7 00	**6 01**	p2.1
21	T	1 29	**1 08**	a2.1	7 51	**6 58**	p2.1
22	W	2 17	**2 10**	a2.1	8 37	**7 55**	p2.1
23	T	3 02	**3 06**	a2.2	9 17	**8 52**	p2.2
24	F	3 44	**3 57**	a2.2	9 54	**9 47**	p2.1
25	S	4 23	**4 45**	a2.2	10 29	**10 43**	p2.1
26	S	5 00	**5 34**	2.2	11 05	**11 38**	a2.2
27	M	5 37	**6 24**	p2.2	11 41	...	2.2
28	T	6 14	**7 16**	p2.2	12 35	**12 21**	p2.3
29	W	6 54	**8 14**	p2.2	1 34	**1 04**	p2.3
30	T	7 38	**9 16**	p2.2	2 37	**1 52**	p2.3

OCTOBER

DAY OF MONTH	DAY OF WEEK	CURRENT TURNS: EAST Flood Starts a.m.	**p.m.**	Kts.	WEST Ebb Starts a.m.	**p.m.**	Kts.
1	F	8 31	**10 24**	p2.1	3 46	**2 45**	p2.3
2	S	9 41	**11 32**	p2.1	4 58	**3 46**	p2.2
3	S	11 17	...	1.8	6 08	**4 54**	p2.2
4	M	12 38	**12 47**	a2.1	7 10	**6 07**	p2.1
5	T	1 38	**1 58**	a2.2	8 04	**7 21**	p2.1
6	W	2 33	**2 57**	a2.2	8 50	**8 29**	p2.1
7	T	3 23	**3 50**	a2.2	9 32	**9 30**	p2.0
8	F	4 10	**4 38**	a2.2	10 09	**10 25**	2.0
9	S	4 53	**5 23**	2.2	10 43	**11 15**	a2.1
10	S	5 32	**6 03**	p2.2	11 14	**11 59**	a2.1
11	M	6 07	**6 41**	p2.2	11 42	...	2.1
12	T	6 37	**7 17**	p2.1	12 44	**12 11**	p2.1
13	W	7 00	**7 51**	p2.1	1 26	**12 43**	p2.1
14	T	7 22	**8 27**	p2.1	2 07	**1 17**	p2.1
15	F	7 52	**9 06**	p2.0	2 50	**1 57**	p2.1
16	S	8 31	**9 52**	p2.0	3 36	**2 41**	p2.1
17	S	9 20	**10 42**	p2.0	4 26	**3 31**	p2.0
18	M	10 20	**11 36**	p2.0	5 18	**4 27**	p2.0
19	T	11 29	...	1.7	6 09	**5 27**	p2.0
20	W	12 31	**12 42**	a2.1	6 57	**6 31**	p2.0
21	T	1 24	**1 47**	a2.1	7 42	**7 35**	p2.0
22	F	2 13	**2 45**	a2.2	8 24	**8 39**	2.0
23	S	3 00	**3 39**	2.2	9 04	**9 41**	a2.1
24	S	3 45	**4 30**	p2.3	9 44	**10 41**	a2.3
25	M	4 28	**5 21**	p2.3	10 24	**11 40**	a2.4
26	T	5 10	**6 13**	p2.4	11 06	...	2.4
27	W	5 53	**7 07**	p2.3	12 40	**12 01**	p2.4
28	T	6 40	**8 04**	p2.3	1 39	**12 38**	p2.4
29	F	7 33	**9 04**	p2.2	2 41	**1 30**	p2.3
30	S	8 40	**10 07**	p2.2	3 43	**2 28**	p2.2
31	S	*9 08	***10 11**	p2.1	*3 46	***2 33**	p2.1

***Standard Time starts**

The Kts. (knots) columns show the **maximum** predicted velocities of the stronger one of the Flood Currents and the stronger one of the Ebb Currents for each day.
The letter "a" means the velocity shown should occur **after** the **a.m.** Current Change. The letter "p" means the velocity shown should occur **after** the **p.m.** Current Change (even if next morning). No "a" or "p" means a.m. and p.m. velocities are the same for that day.
Av. Max. Vel.: Fl. 2.0 Kts., Ebb 1.9 Kts.
Max. Flood 3 hrs. 10 min. after Flood Starts ±45 min.
Max. Ebb 2 hrs. 45 min. after Ebb Starts ±45 min.
See pp. 150-157 for Current Change at other points.

1999 CURRENT TABLE
CHESAPEAKE & DELAWARE CANAL

At Chesapeake City

NOVEMBER

Standard Time

DAY OF MONTH	DAY OF WEEK	CURRENT TURNS EAST Flood Starts a.m.	**p.m.**	Kts.	CURRENT TURNS WEST Ebb Starts a.m.	**p.m.**	Kts.
1	M	10 35	**11 13**	p2.1	4 44	**3 46**	p2.0
2	T	11 47	...	1.8	5 38	**5 04**	p1.9
3	W	12 10	**12 48**	a2.1	6 25	**6 19**	p1.9
4	T	1 03	**1 42**	a2.1	7 08	**7 26**	1.9
5	F	1 52	**2 31**	2.1	7 47	**8 24**	a2.0
6	S	2 37	**3 16**	p2.2	8 23	**9 17**	a2.1
7	S	3 19	**3 57**	p2.2	8 56	**10 05**	a2.2
8	M	3 58	**4 35**	p2.2	9 27	**10 48**	a2.2
9	T	4 32	**5 11**	p2.2	9 58	**11 29**	a2.2
10	W	4 59	**5 44**	p2.2	10 29	**11 59**	a2.2
11	T	5 21	**6 16**	p2.2	11 02	...	2.2
12	F	5 45	**6 48**	p2.1	12 45	**-A-**	1.5
13	S	6 17	**7 23**	p2.1	1 23	**12 19**	p2.1
14	S	6 58	**8 03**	p2.0	2 04	**1 04**	p2.1
15	M	7 50	**8 48**	p2.0	2 47	**1 55**	p2.0
16	T	8 52	**9 38**	p2.0	3 32	**2 53**	p2.0
17	W	10 03	**10 32**	p2.1	4 19	**3 57**	p1.9
18	T	11 16	**11 27**	p2.1	5 05	**5 07**	p1.9
19	F	...	**12 24**	2.0	5 50	**6 18**	a2.0
20	S	12 22	**1 25**	p2.2	6 35	**7 29**	a2.2
21	S	1 15	**2 21**	p2.3	7 20	**8 36**	a2.3
22	M	2 07	**3 16**	p2.4	8 05	**9 40**	a2.4
23	T	2 57	**4 09**	p2.4	8 51	**10 41**	a2.5
24	W	3 48	**5 03**	p2.4	9 39	**11 40**	a2.5
25	T	4 39	**5 58**	p2.4	10 29	...	2.5
26	F	5 35	**6 53**	p2.3	12 37	**-B-**	1.6
27	S	6 40	**7 50**	p2.3	1 33	**12 17**	p2.3
28	S	7 56	**8 47**	p2.2	2 27	**1 17**	p2.2
29	M	9 14	**9 44**	p2.1	3 19	**2 24**	p2.0
30	T	10 26	**10 40**	p2.1	4 09	**3 38**	p1.9

DECEMBER

Standard Time

DAY OF MONTH	DAY OF WEEK	CURRENT TURNS EAST Flood Starts a.m.	**p.m.**	Kts.	CURRENT TURNS WEST Ebb Starts a.m.	**p.m.**	Kts.
1	W	11 30	**11 34**	p2.1	4 56	**4 55**	p1.8
2	T	...	**12 27**	1.9	5 39	**6 08**	a1.9
3	F	12 25	**1 19**	a2.1	6 20	**7 13**	a2.0
4	S	1 14	**2 06**	2.1	6 59	**8 10**	a2.1
5	S	2 00	**2 49**	p2.2	7 36	**9 01**	a2.2
6	M	2 43	**3 30**	p2.2	8 12	**9 47**	a2.2
7	T	3 22	**4 09**	p2.2	8 47	**10 30**	a2.2
8	W	3 56	**4 45**	p2.2	9 21	**11 10**	a2.2
9	T	4 24	**5 18**	p2.2	9 56	**11 47**	a2.2
10	F	4 48	**5 48**	p2.2	10 31	...	2.2
11	S	5 15	**6 18**	p2.1	12 21	**-C-**	1.5
12	S	5 50	**6 49**	p2.1	12 55	**12 01**	p2.1
13	M	6 34	**7 24**	p2.1	1 30	**12 36**	p2.1
14	T	7 27	**8 05**	p2.1	2 08	**1 28**	p2.0
15	W	8 29	**8 51**	p2.0	2 48	**2 28**	p2.0
16	T	9 38	**9 41**	p2.0	3 31	**3 34**	p1.9
17	F	10 51	**10 36**	p2.0	4 16	**4 47**	a2.0
18	S	11 59	**11 34**	2.1	5 04	**6 04**	a2.1
19	S	...	**1 05**	2.2	5 53	**7 20**	a2.3
20	M	12 34	**2 06**	p2.3	6 44	**8 31**	a2.4
21	T	1 34	**3 04**	p2.4	7 35	**9 36**	a2.5
22	W	2 34	**4 00**	p2.4	8 28	**10 35**	a2.5
23	T	3 34	**4 54**	p2.4	9 21	**11 30**	a2.5
24	F	4 35	**5 47**	p2.4	10 16	...	2.5
25	S	5 38	**6 39**	p2.4	12 22	**-D-**	1.6
26	S	6 43	**7 31**	p2.3	1 11	**12 09**	p2.3
27	M	7 51	**8 21**	p2.2	1 57	**1 09**	p2.1
28	T	8 58	**9 11**	p2.1	2 42	**2 14**	p1.9
29	W	10 01	**10 01**	p2.1	3 24	**3 23**	1.8
30	T	11 01	**10 51**	p2.0	4 06	**4 34**	a1.9
31	F	11 56	**11 41**	p2.0	4 48	**5 44**	a1.9

A also at 11:39 a.m. 2.2 **B** also at 11:21 a.m. 2.4 **C** also at 11:09 a.m. 2.2
D also at 11:11 a.m. 2.4

The Kts. (knots) columns show the **maximum** predicted velocities of the stronger one of the Flood Currents and the stronger one of the Ebb Currents for each day.
The letter "a" means the velocity shown should occur **after** the **a.m.** Current Change. The letter "p" means the velocity shown should occur **after** the **p.m.** Current Change (even if next morning). No "a" or "p" means a.m. and p.m. velocities are the same for that day.
Av. Max. Vel.: Fl. 2.0 Kts., Ebb 1.9 Kts.
Max. Flood 3 hrs. 10 min. after Flood Starts ±45 min.
Max. Ebb 2 hrs. 45 min. after Ebb Starts ±45 min.
See pp. 150-157 for Current Change at other points.

1999 HIGH WATER
BALTIMORE, MD.
At Ft. McHenry 39–16N x 76–35W

DAY OF MONTH	DAY OF WEEK	JANUARY (Standard Time) a.m.	Ht.	p.m.	Ht.	DAY OF WEEK	FEBRUARY (Standard Time) a.m.	Ht.	p.m.	Ht.	DAY OF WEEK	MARCH (Standard Time) a.m.	Ht.	p.m.	Ht.	DAY OF MONTH
1	F	5 32	0.7	**6 39**	1.4	M	7 02	0.7	**7 56**	1.2	M	6 02	0.9	**6 53**	1.2	1
2	S	6 25	0.7	**7 28**	1.4	T	7 50	0.8	**8 37**	1.1	T	6 47	1.0	**7 33**	1.1	2
3	S	7 17	0.7	**8 15**	1.4	W	8 37	0.8	**9 16**	1.0	W	7 30	1.0	**8 10**	1.1	3
4	M	8 08	0.7	**9 01**	1.3	T	9 25	0.8	**9 54**	0.9	T	8 12	1.0	**8 44**	1.0	4
5	T	9 01	0.7	**9 46**	1.2	F	10 14	0.8	**10 32**	0.8	F	8 54	1.1	**9 18**	0.9	5
6	W	9 54	0.7	**10 29**	1.1	S	11 07	0.8	**11 11**	0.8	S	9 38	1.1	**9 53**	0.9	6
7	T	10 50	0.7	**11 13**	0.9	S	12 03	0.8	**11 54**	0.7	S	10 24	1.0	**10 30**	0.8	7
8	F	11 50	0.7	**11 58**	0.8	M	...	...	**1 03**	0.8	M	11 15	1.0	**11 13**	0.8	8
9	S	...	...	**12 52**	0.8	T	12 43	0.6	**2 04**	0.9	T	...	...	**12 11**	1.0	9
10	S	12 44	0.7	**1 52**	0.8	W	1 36	0.6	**3 01**	0.9	W	12 03	0.8	**1 10**	1.0	10
11	M	1 32	0.7	**2 51**	0.9	T	2 32	0.6	**3 54**	1.0	T	1 01	0.7	**2 11**	1.1	11
12	T	2 21	0.6	**3 44**	0.9	F	3 28	0.6	**4 42**	1.1	F	2 02	0.7	**3 08**	1.1	12
13	W	3 11	0.6	**4 33**	1.0	S	4 20	0.6	**5 26**	1.1	S	3 02	0.8	**4 00**	1.1	13
14	T	4 00	0.6	**5 17**	1.1	S	5 08	0.7	**6 06**	1.2	S	3 57	0.9	**4 47**	1.2	14
15	F	4 46	0.5	**5 57**	1.1	M	5 54	0.7	**6 46**	1.2	M	4 47	1.0	**5 31**	1.2	15
16	S	5 31	0.5	**6 36**	1.2	T	6 39	0.8	**7 24**	1.2	T	5 34	1.1	**6 14**	1.3	16
17	S	6 14	0.6	**7 13**	1.2	W	7 24	0.9	**8 04**	1.2	W	6 20	1.2	**6 55**	1.2	17
18	M	6 57	0.6	**7 51**	1.2	T	8 12	0.9	**8 45**	1.1	T	7 06	1.3	**7 37**	1.2	18
19	T	7 42	0.6	**8 29**	1.2	F	9 02	1.0	**9 27**	1.0	F	7 54	1.4	**8 20**	1.2	19
20	W	8 29	0.7	**9 08**	1.1	S	9 56	1.1	**10 13**	1.0	S	8 44	1.4	**9 06**	1.1	20
21	T	9 20	0.7	**9 50**	1.1	S	10 55	1.1	**11 04**	0.9	S	9 37	1.4	**9 54**	1.0	21
22	F	10 15	0.8	**10 35**	1.0	M	11 59	1.1	...	...	M	10 35	1.4	**10 48**	1.0	22
23	S	11 16	0.8	**11 24**	0.9	T	12 01	0.8	**1 08**	1.2	T	11 37	1.4	**11 49**	0.9	23
24	S	...	...	**12 22**	0.9	W	1 03	0.7	**2 18**	1.2	W	...	...	**12 45**	1.4	24
25	M	12 18	0.8	**1 30**	1.0	T	2 10	0.7	**3 25**	1.2	T	12 57	0.9	**1 55**	1.4	25
26	T	1 17	0.7	**2 38**	1.1	F	3 16	0.7	**4 26**	1.3	F	2 07	0.9	**3 03**	1.3	26
27	W	2 20	0.6	**3 42**	1.2	S	4 17	0.8	**5 21**	1.3	S	3 13	1.0	**4 04**	1.3	27
28	T	3 23	0.6	**4 41**	1.3	S	5 12	0.8	**6 09**	1.2	S	4 11	1.1	**4 58**	1.3	28
29	F	4 23	0.6	**5 36**	1.3						M	5 03	1.2	**5 46**	1.2	29
30	S	5 20	0.6	**6 26**	1.3						T	5 48	1.2	**6 27**	1.2	30
31	S	6 13	0.7	**7 13**	1.3						W	6 30	1.3	**7 04**	1.1	31

Dates when Ht. of **Low** Water is below Mean Low with Ht. of lowest given for each period and Date of lowest in ( ):

1st - 6th: -0.5' (1st - 3rd)
10th - 31st: -0.4 (17th - 18th)
: -0.5' (29th - 31st)

1st - 4th: -0.4' (1st - 2nd)
10th - 28th: -0.3' (14th - 18th)
: -0.3' (26th - 28th)

1st - 3rd: -0.3' (1st)

Average Rise and Fall 1.2 ft.

When a high tide exceeds av. ht., the *following* low tide will be lower than av.

1999 HIGH WATER
BALTIMORE, MD.
At Ft. McHenry 39–16N x 76–35W

DAY OF MONTH	DAY OF WEEK	APRIL (Daylight Time starts Apr. 4 at 2 a.m.)				DAY OF WEEK	MAY (Daylight Time)				DAY OF WEEK	JUNE (Daylight Time)				DAY OF MONTH
		a.m.	Ht.	**p.m.**	Ht.		a.m.	Ht.	**p.m.**	Ht.		a.m.	Ht.	**p.m.**	Ht.	
1	T	7 09	1.4	**7 38**	1.1	S	8 24	1.7	**8 37**	1.1	T	9 12	1.8	**9 24**	1.1	1
2	F	7 47	1.4	**8 11**	1.0	S	9 00	1.7	**9 12**	1.1	W	9 48	1.8	**10 08**	1.1	2
3	S	8 26	1.4	**8 43**	1.0	M	9 36	1.6	**9 48**	1.1	T	10 26	1.8	**10 56**	1.2	3
4	S	*10 05	1.4	***10 18**	1.0	T	10 14	1.6	**10 30**	1.1	F	11 07	1.7	**11 50**	1.2	4
5	M	10 46	1.3	**10 57**	0.9	W	10 55	1.6	**11 17**	1.1	S	11 52	1.6	**...**	...	5
6	T	11 31	1.3	**11 41**	0.9	T	11 40	1.5	**...**	...	S	12 49	1.3	**12 42**	1.6	6
7	W	...	...	**12 21**	1.3	F	12 11	1.1	**12 30**	1.5	M	1 50	1.4	**1 37**	1.5	7
8	T	12 34	0.9	**1 17**	1.3	S	1 12	1.1	**1 25**	1.4	T	2 51	1.5	**2 35**	1.4	8
9	F	1 35	0.9	**2 17**	1.3	S	2 16	1.2	**2 24**	1.4	W	3 50	1.7	**3 35**	1.4	9
10	S	2 38	1.0	**3 16**	1.3	M	3 16	1.3	**3 22**	1.4	T	4 44	1.9	**4 34**	1.3	10
11	S	3 39	1.1	**4 12**	1.3	T	4 12	1.5	**4 18**	1.4	F	5 37	2.0	**5 31**	1.3	11
12	M	4 35	1.2	**5 04**	1.3	W	5 04	1.6	**5 12**	1.3	S	6 28	2.1	**6 27**	1.3	12
13	T	5 25	1.3	**5 52**	1.3	T	5 54	1.8	**6 03**	1.3	S	7 18	2.2	**7 22**	1.3	13
14	W	6 14	1.5	**6 39**	1.3	F	6 43	1.9	**6 54**	1.3	M	8 07	2.2	**8 17**	1.3	14
15	T	7 01	1.6	**7 24**	1.3	S	7 32	2.0	**7 44**	1.3	T	8 57	2.2	**9 13**	1.3	15
16	F	7 48	1.7	**8 10**	1.3	S	8 21	2.1	**8 35**	1.2	W	9 46	2.1	**10 10**	1.3	16
17	S	8 37	1.8	**8 57**	1.2	M	9 11	2.1	**9 28**	1.2	T	10 36	2.0	**11 09**	1.4	17
18	S	9 27	1.8	**9 46**	1.2	T	10 03	2.0	**10 24**	1.2	F	11 26	1.9	**...**	...	18
19	M	10 19	1.8	**10 39**	1.1	W	10 56	1.9	**11 25**	1.2	S	12 11	1.4	**12 17**	1.7	19
20	T	11 15	1.7	**11 37**	1.1	T	11 52	1.8	**...**	...	S	1 14	1.5	**1 09**	1.6	20
21	W	...	...	**12 15**	1.7	F	12 30	1.2	**12 50**	1.7	M	2 17	1.6	**2 03**	1.4	21
22	T	12 41	1.1	**1 20**	1.6	S	1 38	1.3	**1 51**	1.6	T	3 16	1.7	**2 57**	1.3	22
23	F	1 51	1.1	**2 27**	1.5	S	2 44	1.4	**2 52**	1.4	W	4 11	1.7	**3 50**	1.3	23
24	S	3 01	1.2	**3 32**	1.4	M	3 45	1.5	**3 50**	1.3	T	5 00	1.8	**4 41**	1.2	24
25	S	4 04	1.3	**4 33**	1.3	T	4 39	1.6	**4 43**	1.3	F	5 45	1.9	**5 29**	1.2	25
26	M	4 59	1.4	**5 26**	1.3	W	5 27	1.7	**5 30**	1.2	S	6 26	1.9	**6 14**	1.2	26
27	T	5 48	1.5	**6 13**	1.2	T	6 10	1.8	**6 13**	1.2	S	7 04	2.0	**6 57**	1.2	27
28	W	6 31	1.6	**6 53**	1.2	F	6 49	1.8	**6 52**	1.1	M	7 40	2.0	**7 39**	1.2	28
29	T	7 11	1.6	**7 30**	1.1	S	7 27	1.9	**7 29**	1.1	T	8 15	2.0	**8 21**	1.2	29
30	F	7 48	1.7	**8 04**	1.1	S	8 02	1.9	**8 06**	1.1	W	8 49	2.0	**9 03**	1.2	30
31						M	8 37	1.9	**8 44**	1.1						31

***Daylight Time starts**

Average Rise and Fall 1.2 ft.

When a high tide exceeds av. ht., the *following* low tide will be lower than av.

1999 HIGH WATER
BALTIMORE, MD.
At Ft. McHenry 39–16N x 76–35W

DAY OF MONTH	DAY OF WEEK	JULY (Daylight Time) a.m.	Ht.	p.m.	Ht.	DAY OF WEEK	AUGUST (Daylight Time) a.m.	Ht.	p.m.	Ht.	DAY OF WEEK	SEPTEMBER (Daylight Time) a.m.	Ht.	p.m.	Ht.	DAY OF MONTH
1	T	9 24	1.9	9 47	1.3	S	10 14	1.8	11 03	1.6	W	11 17	1.6	...	...	1
2	F	10 01	1.9	10 35	1.3	M	10 54	1.8	11 58	1.7	T	12 34	1.9	12 10	1.5	2
3	S	10 39	1.8	11 27	1.4	T	11 39	1.6	...	...	F	1 38	2.0	1 12	1.4	3
4	S	11 21	1.7	...	...	W	12 58	1.8	12 30	1.5	S	2 44	2.0	2 24	1.3	4
5	M	12 24	1.5	12 07	1.6	T	2 01	1.9	1 28	1.4	S	3 49	2.1	3 39	1.3	5
6	T	1 24	1.6	12 59	1.5	F	3 05	2.0	2 34	1.3	M	4 50	2.1	4 49	1.4	6
7	W	2 25	1.7	1 56	1.4	S	4 07	2.1	3 45	1.3	T	5 45	2.1	5 52	1.4	7
8	T	3 26	1.9	2 58	1.4	S	5 06	2.2	4 54	1.3	W	6 35	2.0	6 47	1.5	8
9	F	4 26	2.0	4 04	1.3	M	6 02	2.2	5 59	1.4	T	7 21	2.0	7 37	1.6	9
10	S	5 21	2.1	5 07	1.3	T	6 52	2.2	6 56	1.4	F	8 01	1.9	8 23	1.7	10
11	S	6 14	2.2	6 08	1.3	W	7 39	2.2	7 50	1.5	S	8 39	1.8	9 08	1.7	11
12	M	7 05	2.3	7 07	1.3	T	8 24	2.1	8 41	1.6	S	9 15	1.7	9 52	1.8	12
13	T	7 54	2.3	8 03	1.4	F	9 06	2.0	9 31	1.6	M	9 48	1.6	10 36	1.8	13
14	W	8 41	2.2	8 58	1.4	S	9 45	1.9	10 20	1.7	T	10 22	1.5	11 23	1.8	14
15	T	9 27	2.1	9 52	1.5	S	10 23	1.8	11 10	1.7	W	10 58	1.4	...	...	15
16	F	10 12	2.0	10 47	1.5	M	11 01	1.6	11 59	1.7	T	12 12	1.8	-A-	...	16
17	S	10 56	1.8	11 43	1.6	T	11 39	1.5	...	...	F	1 05	1.7	12 29	1.3	17
18	S	11 40	1.7	...	...	W	12 56	1.8	12 21	1.4	S	2 01	1.7	1 29	1.2	18
19	M	12 41	1.6	12 24	1.6	T	1 53	1.8	1 09	1.3	S	2 58	1.7	2 39	1.2	19
20	T	1 40	1.7	1 10	1.4	F	2 51	1.8	2 06	1.3	M	3 51	1.8	3 47	1.3	20
21	W	2 38	1.7	2 01	1.3	S	3 46	1.8	3 11	1.3	T	4 40	1.8	4 47	1.3	21
22	T	3 35	1.8	2 55	1.3	S	4 37	1.9	4 15	1.3	W	5 25	1.8	5 39	1.4	22
23	F	4 27	1.9	3 52	1.2	M	5 23	1.9	5 13	1.3	T	6 07	1.8	6 26	1.5	23
24	S	5 15	1.9	4 48	1.2	T	6 05	2.0	6 05	1.3	F	6 46	1.8	7 11	1.6	24
25	S	5 58	2.0	5 41	1.2	W	6 44	2.0	6 51	1.4	S	7 25	1.8	7 56	1.7	25
26	M	6 38	2.0	6 30	1.2	T	7 21	2.0	7 35	1.5	S	8 04	1.7	8 41	1.8	26
27	T	7 15	2.0	7 15	1.3	F	7 57	2.0	8 19	1.6	M	8 44	1.7	9 29	1.9	27
28	W	7 50	2.0	7 59	1.3	S	8 33	1.9	9 03	1.7	T	9 26	1.6	10 19	1.9	28
29	T	8 25	2.0	8 42	1.4	S	9 10	1.9	9 50	1.7	W	10 11	1.5	11 14	2.0	29
30	F	9 00	2.0	9 26	1.5	M	9 49	1.8	10 40	1.8	T	11 01	1.4	...	...	30
31	S	9 36	1.9	10 13	1.5	T	10 31	1.7	11 34	1.9						31

A also at 11:39 a.m. 1.3

Average Rise and Fall 1.2 ft.

When a high tide exceeds av. ht., the *following* low tide will be lower than av.

1999 HIGH WATER
BALTIMORE, MD.
At Ft. McHenry 39–16N x 76–35W

		Standard Time starts Oct. 31 at 2 a.m.					Standard Time					Standard Time				
DAY OF MONTH	DAY OF WEEK	OCTOBER				DAY OF WEEK	NOVEMBER				DAY OF WEEK	DECEMBER				DAY OF MONTH
		a.m.	Ht.	**p.m.**	Ht.		a.m.	Ht.	**p.m.**	Ht.		a.m.	Ht.	**p.m.**	Ht.	
1	F	12 13	2.0	**12 01**	1.3	M	12 57	1.7	**1 19**	1.1	W	1 26	1.2	**2 12**	1.0	1
2	S	1 16	1.9	**1 07**	1.2	T	2 00	1.6	**2 31**	1.2	T	2 23	1.1	**3 15**	1.1	2
3	S	2 22	1.9	**2 23**	1.2	W	3 00	1.5	**3 35**	1.3	F	3 15	1.0	**4 10**	1.2	3
4	M	3 27	1.9	**3 38**	1.3	T	3 53	1.4	**4 30**	1.4	S	4 03	0.9	**4 59**	1.3	4
5	T	4 28	1.8	**4 45**	1.4	F	4 41	1.3	**5 19**	1.5	S	4 46	0.8	**5 43**	1.3	5
6	W	5 23	1.8	**5 44**	1.5	S	5 23	1.2	**6 03**	1.5	M	5 25	0.8	**6 24**	1.3	6
7	T	6 12	1.7	**6 35**	1.6	S	6 01	1.2	**6 44**	1.6	T	6 02	0.8	**7 02**	1.3	7
8	F	6 55	1.7	**7 21**	1.6	M	6 35	1.1	**7 23**	1.6	W	6 38	0.7	**7 38**	1.3	8
9	S	7 35	1.6	**8 05**	1.7	T	7 09	1.1	**8 01**	1.6	T	7 15	0.7	**8 15**	1.3	9
10	S	8 09	1.5	**8 45**	1.7	W	7 41	1.0	**8 38**	1.5	F	7 53	0.7	**8 50**	1.3	10
11	M	8 41	1.4	**9 25**	1.7	T	8 16	1.0	**9 16**	1.5	S	8 34	0.7	**9 27**	1.3	11
12	T	9 13	1.3	**10 05**	1.7	F	8 56	0.9	**9 56**	1.5	S	9 20	0.7	**10 05**	1.2	12
13	W	9 47	1.3	**10 47**	1.7	S	9 41	0.9	**10 39**	1.4	M	10 12	0.7	**10 47**	1.2	13
14	T	10 23	1.2	**11 31**	1.7	S	10 35	0.9	**11 25**	1.4	T	11 10	0.7	**11 31**	1.1	14
15	F	11 06	1.2	...	...	M	11 36	0.9	...	...	W	...	...	**12 12**	0.8	15
16	S	12 19	1.6	**12 01**	1.1	T	12 15	1.3	**12 43**	0.9	T	12 20	1.0	**1 17**	0.9	16
17	S	1 11	1.6	**1 01**	1.1	W	1 06	1.3	**1 49**	1.0	F	1 11	1.0	**2 19**	1.0	17
18	M	2 05	1.6	**2 11**	1.1	T	1 59	1.3	**2 50**	1.1	S	2 05	0.9	**3 18**	1.1	18
19	T	2 59	1.6	**3 19**	1.2	F	2 50	1.2	**3 45**	1.2	S	3 00	0.9	**4 14**	1.3	19
20	W	3 51	1.6	**4 19**	1.2	S	3 41	1.2	**4 36**	1.4	M	3 55	0.8	**5 07**	1.4	20
21	T	4 39	1.6	**5 12**	1.4	S	4 30	1.1	**5 26**	1.5	T	4 50	0.8	**5 59**	1.5	21
22	F	5 24	1.5	**6 01**	1.5	M	5 18	1.1	**6 15**	1.6	W	5 44	0.8	**6 49**	1.6	22
23	S	6 08	1.5	**6 48**	1.6	T	6 06	1.1	**7 04**	1.7	T	6 38	0.7	**7 40**	1.6	23
24	S	6 51	1.5	**7 34**	1.7	W	6 56	1.0	**7 54**	1.8	F	7 32	0.7	**8 30**	1.5	24
25	M	7 34	1.4	**8 21**	1.8	T	7 47	1.0	**8 45**	1.7	S	8 28	0.7	**9 20**	1.4	25
26	T	8 18	1.4	**9 10**	1.9	F	8 41	0.9	**9 38**	1.7	S	9 26	0.7	**10 11**	1.3	26
27	W	9 05	1.3	**10 01**	1.9	S	9 40	0.9	**10 33**	1.6	M	10 27	0.8	**11 02**	1.2	27
28	T	9 55	1.2	**10 55**	1.9	S	10 44	0.9	**11 30**	1.5	T	11 31	0.8	**11 54**	1.0	28
29	F	10 50	1.1	**11 53**	1.8	M	11 53	0.9	...	...	W	...	...	**12 37**	0.8	29
30	S	11 53	1.1	...	...	T	12 28	1.3	**1 04**	0.9	T	12 47	0.9	**1 43**	0.9	30
31	S	12 54	1.8	***12 04**	1.1						F	1 40	0.8	**2 46**	0.9	31

***Standard Time starts**

Dates when Ht. of **Low** Water is below Mean Low with Ht. of lowest given for each period and Date of lowest in ( ):

October	November	December
	22nd - 26th: -0.3' (24th)	18th - 27th: -0.5' (22nd - 24th) 31st: -0.2'

Average Rise and Fall 1.2 ft.

When a high tide exceeds av. ht., the *following* low tide will be lower than av.

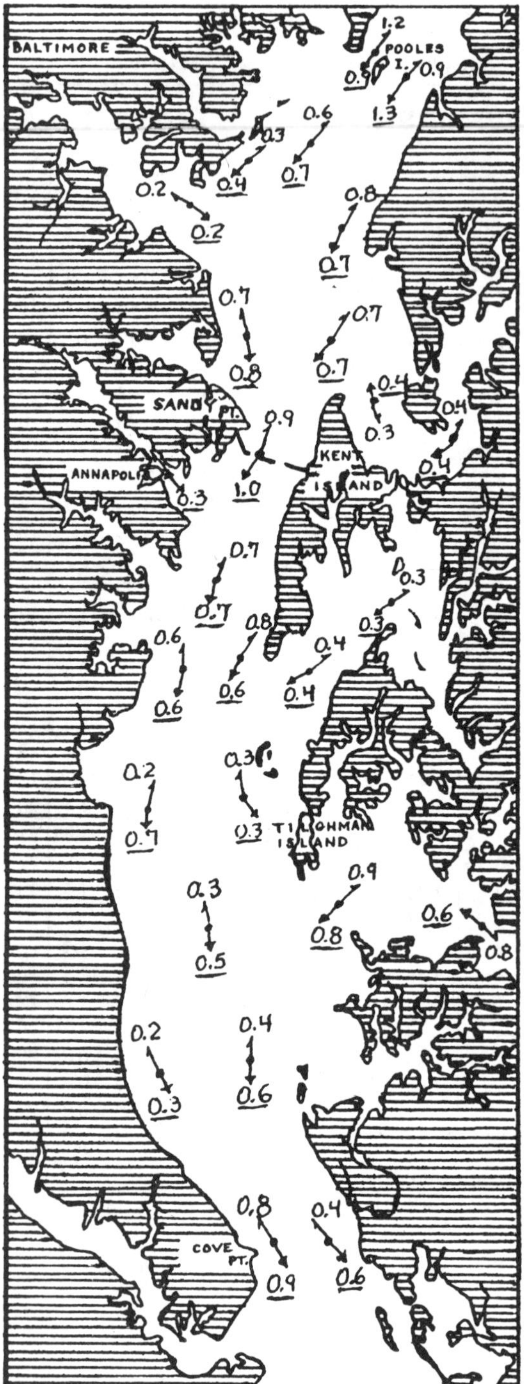

UPPER CHESAPEAKE BAY CURRENTS

On this Current Diagram, the arrows denote maximum velocities. Refer to the four areas listed below for the specific times. Double-headed arrows are for Ebb and the velocities are underlined. Single-headed arrows are for Flood and velocities are not underlined.

All times below are in hours and relate to the time of High Water at Baltimore.

West of Pooles I.:
- Flood begins 3½ before
- Flood max. 1½ before (1.2 kts.)
- Ebb begins 2½ after
- Ebb max. 4½ after (0.9 kts.)

Sandy Point:
- Flood begins 3½ before
- Flood max. 1½ before (0.9 kts.)
- Ebb begins 1½ after
- Ebb max. 4½ after (1.0 kts.)

off Tilghman I.:
- Flood begins 5½ before
- Flood max. 3½ before (0.3 kts.)
- Ebb begins ½ after
- Ebb max. 3½ after (0.7 kts.)

off Cove Pt.:
- Flood begins 6½ before
- Flood max. 4½ before (0.9 kts.)
- Ebb begins ½ before
- Ebb max. 1½ after (0.8 kts.)

Note: From the beginning of the Flood Current at Cove Pt. until the Ebb Current begins off Baltimore, a northbound vessel will have over 8 hrs. of Fair Tide. A vessel bound southward from Sandy Pt. can expect only 4 hrs. of Fair Tide.

RELATIONSHIP OF HIGH WATER AND EBB CURRENT

Many people wonder why the times of High Water and the start of Ebb Current at the mouths of bays and inlets are not simultaneous. After doing a rough survey of tide and current data, involving principally the New Jersey Coast and south shore of Long Island, we have made diagrams of what we believe happens hour by hour. Picture the rising Tide, borne by the Flood Current, as a long wave. The wave enters the inlet and the crest reaches its maximum height in or at the inlet. But, the body of water inside the inlet - in the bay - has yet to be filled and the Flood Current continues to pour water through the inlet for a good period *after* the crest has already passed the inlet. The Ebb Current will not start until the level of the water in the ocean is lower than the water in the bay. This does not necessarily apply to the mouths of small bays with wide entrances. The narrowness of the inlet and the size of the bay are the controlling factors. Picture a siphon effect.

1. 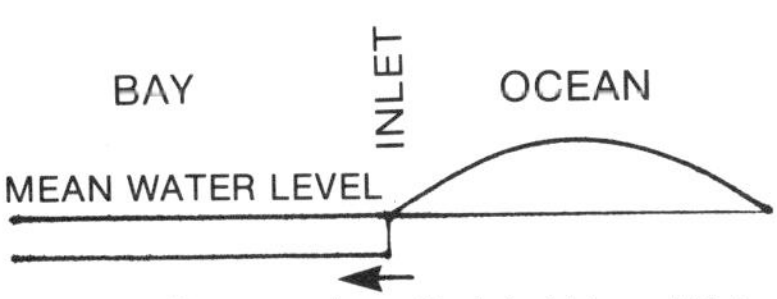

Tide on Ocean edge of Inlet at Mean Water;
Tide rising in Bay; Current Flooding.

2. 

Tide rising in Bay; Current Flooding;
Crest approaching Inlet.

3. 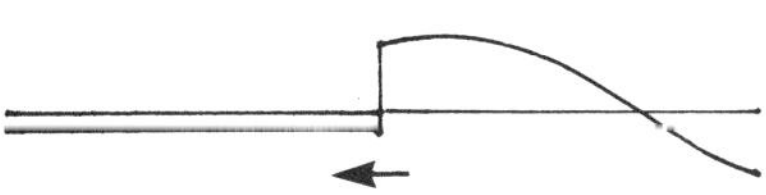

Tide rising in Bay; Current Flooding;
Crest approaching Inlet.

4. 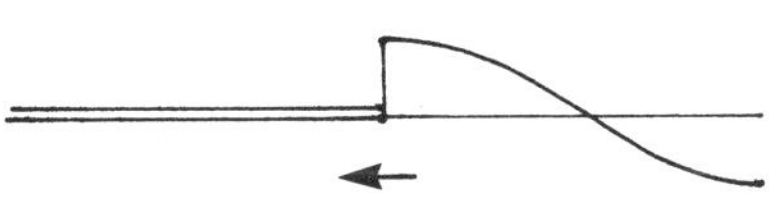

Tide rising in Bay; Current Flooding;
Crest at Inlet (HW at Inlet).

5. 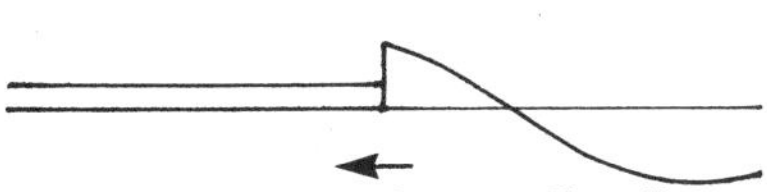

Tide rising in Bay; Current Flooding;
Crest has passed Inlet.

6. 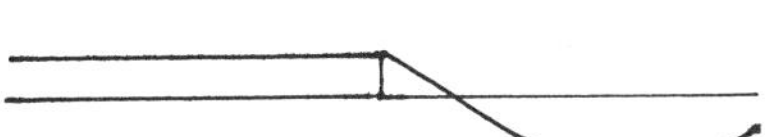

High Water in Bay; Ebb Current about to start.

7. 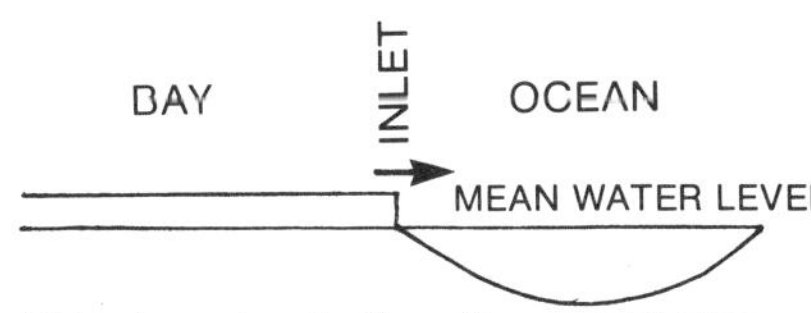

Tide dropping in Bay; Current Ebbing;
Mean Water at Ocean edge of Inlet.

8. 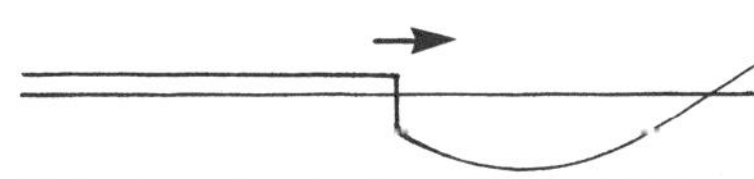

Tide dropping in Bay; Current Ebbing.

9. 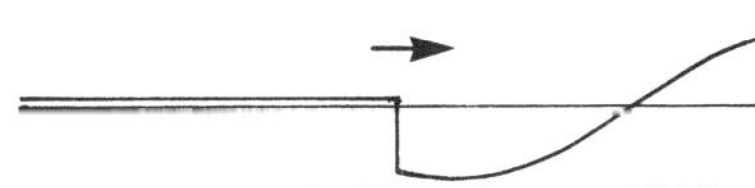

Tide dropping in Bay; Current Ebbing.

10. 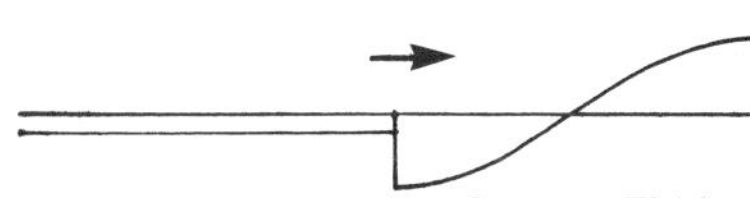

Tide dropping in Bay; Current Ebbing;
Low Water at Ocean edge of Inlet.

11. 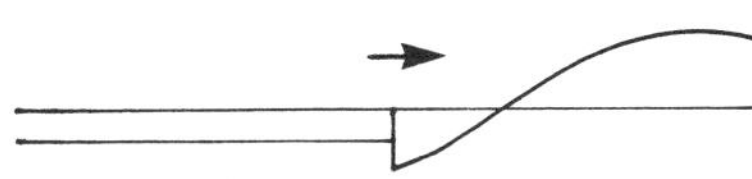

Tide dropping in Bay; Current Ebbing;
Tide rising at Ocean edge of Inlet.

12. 

Low Water in Bay; Flood Current about to start.

1999 HIGH WATER
MIAMI HARBOR ENTR., FL
25° - 46′ N x 80° - 08′ W

DAY OF MONTH	DAY OF WEEK	JANUARY Standard Time a.m.	Ht.	p.m.	Ht.	DAY OF WEEK	FEBRUARY Standard Time a.m.	Ht.	p.m.	Ht.	DAY OF WEEK	MARCH Standard Time a.m.	Ht.	p.m.	Ht.	DAY OF MONTH
1	F	7 23	3.1	7 31	2.9	M	8 42	2.8	8 55	2.6	M	7 39	2.7	7 54	2.6	1
2	S	8 13	3.2	8 22	2.9	T	9 25	2.7	9 40	2.5	T	8 20	2.7	8 36	2.6	2
3	S	9 01	3.1	9 12	2.8	W	10 06	2.6	10 25	2.4	W	8 58	2.6	9 17	2.6	3
4	M	9 48	3.0	10 01	2.7	T	10 47	2.5	11 09	2.3	T	9 35	2.6	9 57	2.5	4
5	T	10 35	2.9	10 52	2.6	F	11 27	2.3	11 55	2.2	F	10 11	2.5	10 36	2.4	5
6	W	11 21	2.7	11 43	2.4	S	...	...	12 09	2.2	S	10 47	2.3	11 17	2.3	6
7	T	...	...	12 08	2.5	S	12 44	2.1	12 54	2.1	S	11 25	2.2	...	...	7
8	F	12 36	2.3	12 57	2.3	M	1 38	2.0	1 44	1.9	M	12 01	2.2	12 06	2.1	8
9	S	1 33	2.2	1 48	2.2	T	2 36	2.0	2 39	1.9	T	12 50	2.1	12 54	2.0	9
10	S	2 31	2.1	2 39	2.1	W	3 35	2.0	3 35	1.9	W	1 44	2.0	1 48	1.9	10
11	M	3 28	2.1	3 32	2.1	T	4 31	2.1	4 31	2.0	T	2 45	2.0	2 50	1.9	11
12	T	4 22	2.2	4 22	2.1	F	5 22	2.2	5 23	2.1	F	3 46	2.1	3 53	2.0	12
13	W	5 12	2.3	5 10	2.2	S	6 08	2.3	6 11	2.2	S	4 41	2.2	4 51	2.2	13
14	T	5 57	2.4	5 55	2.2	S	6 51	2.5	6 57	2.4	S	5 32	2.4	5 44	2.4	14
15	F	6 39	2.5	6 38	2.3	M	7 32	2.6	7 41	2.5	M	6 18	2.5	6 33	2.6	15
16	S	7 20	2.6	7 20	2.4	T	8 13	2.7	8 26	2.6	T	7 02	2.7	7 20	2.8	16
17	S	7 59	2.6	8 01	2.5	W	8 54	2.7	9 11	2.7	W	7 45	2.8	8 07	2.9	17
18	M	8 38	2.7	8 44	2.5	T	9 36	2.7	9 59	2.7	T	8 28	2.9	8 54	3.0	18
19	T	9 18	2.7	9 28	2.5	F	10 20	2.7	10 49	2.7	F	9 13	2.9	9 42	3.0	19
20	W	9 59	2.7	10 14	2.5	S	11 08	2.6	11 43	2.6	S	9 59	2.9	10 33	3.0	20
21	T	10 42	2.6	11 04	2.5	S	11 59	2.5	...	...	S	10 48	2.8	11 27	2.8	21
22	F	11 29	2.5	11 59	2.5	M	12 42	2.5	12 57	2.4	M	11 41	2.6	...	...	22
23	S	...	...	12 21	2.5	T	1 47	2.4	2 01	2.3	T	12 26	2.7	12 41	2.5	23
24	S	1 00	2.4	1 18	2.4	W	2 57	2.4	3 11	2.2	W	1 31	2.6	1 47	2.4	24
25	M	2 06	2.4	2 20	2.3	T	4 06	2.4	4 19	2.3	T	2 40	2.5	2 59	2.3	25
26	T	3 14	2.4	3 26	2.3	F	5 09	2.5	5 22	2.4	F	3 50	2.4	4 09	2.3	26
27	W	4 21	2.5	4 30	2.4	S	6 05	2.5	6 18	2.5	S	4 52	2.5	5 12	2.4	27
28	T	5 22	2.6	5 31	2.5	S	6 54	2.6	7 08	2.6	S	5 46	2.6	6 06	2.5	28
29	F	6 18	2.7	6 27	2.6						M	6 33	2.6	6 53	2.6	29
30	S	7 10	2.8	7 19	2.6						T	7 14	2.7	7 35	2.7	30
31	S	7 57	2.8	8 09	2.6						W	7 52	2.7	8 14	2.7	31

Dates when Ht. of **Low** Water is below Mean Low with Ht. of lowest given for each period and Date of lowest in ( ):

January	February	March
1st - 5th: -0.6' (1st - 3rd)	1st - 4th: -0.6' (1st)	1st: -0.5'
16th - 31st: -0.4' (18th - 20th)	12th - 28th: -0.6' (17th)	14th - 23rd: -0.6' (18th - 20th)
31st: -0.7'	: -0.4' (26th - 28th)	

Average Rise and Fall 2.5 ft.

When a high tide exceeds av. ht., the *following* low tide will be lower than av.

1999 HIGH WATER
MIAMI HARBOR ENTR., FL

25° - 46′ N x 80° - 08′ W

Daylight Time starts Apr. 4 at 2 a.m. — Daylight Time — Daylight Time

DAY OF MONTH	DAY OF WEEK	APRIL a.m.	Ht.	p.m.	Ht.	DAY OF WEEK	MAY a.m.	Ht.	p.m.	Ht.	DAY OF WEEK	JUNE a.m.	Ht.	p.m.	Ht.	DAY OF MONTH
1	T	8 27	2.7	8 52	2.7	S	9 29	2.6	10 01	2.7	T	10 11	2.4	10 50	2.6	1
2	F	9 02	2.6	9 28	2.7	S	10 03	2.5	10 37	2.7	W	10 50	2.3	11 29	2.5	2
3	S	9 36	2.5	10 05	2.6	M	10 38	2.4	11 15	2.6	T	11 32	2.3	...	...	3
4	S	*11 10	2.4	*11 43	2.5	T	11 15	2.4	11 54	2.5	F	12 11	2.5	12 19	2.2	4
5	M	11 47	2.3	...	...	W	11 56	2.3	...	...	S	12 56	2.4	1 12	2.2	5
6	T	12 24	2.4	12 27	2.2	T	12 37	2.4	12 42	2.2	S	1 46	2.4	2 12	2.2	6
7	W	1 09	2.3	1 13	2.1	F	1 26	2.4	1 36	2.2	M	2 42	2.4	3 16	2.3	7
8	T	2 01	2.2	2 07	2.1	S	2 20	2.3	2 39	2.2	T	3 40	2.4	4 20	2.5	8
9	F	3 00	2.2	3 12	2.1	S	3 19	2.3	3 46	2.3	W	4 39	2.5	5 23	2.7	9
10	S	4 00	2.2	4 17	2.2	M	4 17	2.4	4 49	2.5	T	5 35	2.7	6 20	2.9	10
11	S	4 59	2.3	5 19	2.3	T	5 13	2.6	5 48	2.7	F	6 31	2.8	7 15	3.0	11
12	M	5 52	2.5	6 15	2.6	W	6 07	2.7	6 43	2.9	S	7 25	2.9	8 08	3.2	12
13	T	6 42	2.7	7 08	2.8	T	6 58	2.9	7 35	3.1	S	8 18	3.0	9 01	3.2	13
14	W	7 29	2.9	7 57	3.0	F	7 48	3.0	8 26	3.3	M	9 10	3.0	9 52	3.2	14
15	T	8 16	3.0	8 46	3.2	S	8 38	3.1	9 17	3.3	T	10 04	2.9	10 44	3.1	15
16	F	9 02	3.1	9 35	3.3	S	9 28	3.1	10 08	3.3	W	10 58	2.8	11 37	3.0	16
17	S	9 49	3.1	10 25	3.3	M	10 20	3.0	11 01	3.2	T	11 53	2.7	...	...	17
18	S	10 38	3.0	11 17	3.2	T	11 13	2.9	11 55	3.1	F	12 30	2.8	12 51	2.5	18
19	M	11 30	2.9	...	...	W	...	...	12 10	2.8	S	1 24	2.7	1 51	2.4	19
20	T	12 11	3.0	12 25	2.8	T	12 52	2.9	1 11	2.6	S	2 20	2.5	2 53	2.3	20
21	W	1 10	2.9	1 26	2.6	F	1 52	2.7	2 16	2.5	M	3 15	2.4	3 54	2.3	21
22	T	2 13	2.7	2 34	2.4	S	2 54	2.6	3 24	2.4	T	4 09	2.3	4 52	2.3	22
23	F	3 20	2.6	3 45	2.4	S	3 55	2.5	4 29	2.4	W	5 00	2.3	5 44	2.3	23
24	S	4 27	2.5	4 51	2.4	M	4 52	2.4	5 28	2.4	T	5 47	2.2	6 31	2.4	24
25	S	5 27	2.5	5 55	2.5	T	5 43	2.4	6 19	2.5	F	6 31	2.3	7 14	2.5	25
26	M	6 19	2.6	6 46	2.6	W	6 28	2.4	7 04	2.5	S	7 12	2.3	7 54	2.5	26
27	T	7 04	2.6	7 32	2.7	T	7 08	2.4	7 44	2.6	S	7 52	2.3	8 33	2.6	27
28	W	7 44	2.6	8 12	2.7	F	7 46	2.5	8 22	2.7	M	8 31	2.3	9 11	2.6	28
29	T	8 20	2.6	8 49	2.8	S	8 22	2.5	8 59	2.7	T	9 10	2.4	9 49	2.6	29
30	F	8 55	2.6	9 25	2.8	S	8 58	2.5	9 36	2.7	W	9 49	2.4	10 26	2.6	30
31						M	9 34	2.4	10 13	2.7						31

***Daylight Time starts**

Dates when Ht. of **Low** Water is below Mean Low with Ht. of lowest given for each period and Date of lowest in ( ):

April	May	June
1st - 2nd : -0.2' 14th - 20th: -0.7' (16th - 17th)	12th - 19th: -0.7' (15th - 17th)	10th - 17th: -0.8' (13th - 14th) 28th - 30th: -0.2'

Average Rise and Fall 2.5 ft.

When a high tide exceeds av. ht., the *following* low tide will be lower than av.

1999 HIGH WATER
MIAMI HARBOR ENTR., FL
25° - 46′ N x 80° - 08′ W

DAY OF MONTH	DAY OF WEEK	JULY Daylight Time a.m.	Ht.	p.m.	Ht.	DAY OF WEEK	AUGUST Daylight Time a.m.	Ht.	p.m.	Ht.	DAY OF WEEK	SEPTEMBER Daylight Time a.m.	Ht.	p.m.	Ht.	DAY OF MONTH
1	T	10 30	2.4	**11 05**	2.6	S	11 41	2.6	...	...	W	12 28	3.0	**1 09**	3.0	1
2	F	11 13	2.3	**11 46**	2.5	M	12 04	2.7	**12 32**	2.6	T	1 23	2.9	**2 11**	2.9	2
3	S	11 59	2.3	...	...	T	12 51	2.6	**1 28**	2.6	F	2 25	2.8	**3 19**	2.9	3
4	S	12 29	2.5	**12 51**	2.3	W	1 44	2.6	**2 29**	2.6	S	3 33	2.8	**4 28**	2.9	4
5	M	1 17	2.4	**1 49**	2.3	T	2 43	2.6	**3 35**	2.6	S	4 43	2.8	**5 34**	3.0	5
6	T	2 10	2.4	**2 51**	2.4	F	3 47	2.6	**4 42**	2.7	M	5 49	2.9	**6 33**	3.1	6
7	W	3 08	2.4	**3 56**	2.5	S	4 53	2.6	**5 46**	2.8	T	6 48	3.0	**7 25**	3.2	7
8	T	4 09	2.5	**5 00**	2.6	S	5 56	2.7	**6 45**	2.9	W	7 41	3.1	**8 12**	3.2	8
9	F	5 11	2.6	**6 02**	2.8	M	6 57	2.8	**7 41**	3.0	T	8 31	3.2	**8 57**	3.3	9
10	S	6 09	2.7	**6 59**	2.9	T	7 51	2.9	**8 30**	3.1	F	9 15	3.3	**9 37**	3.2	10
11	S	7 07	2.8	**7 53**	3.0	W	8 43	3.0	**9 18**	3.1	S	9 59	3.2	**10 16**	3.2	11
12	M	8 02	2.9	**8 46**	3.1	T	9 33	3.0	**10 03**	3.1	S	10 40	3.2	**10 54**	3.1	12
13	T	8 56	2.9	**9 36**	3.1	F	10 21	3.0	**10 47**	3.0	M	11 22	3.1	**11 33**	2.9	13
14	W	9 49	2.9	**10 26**	3.0	S	11 08	2.9	**11 30**	2.9	T	...	...	**12 04**	2.9	14
15	T	10 41	2.8	**11 14**	2.9	S	11 54	2.8	...	...	W	12 12	2.8	**12 48**	2.8	15
16	F	11 32	2.7	...	...	M	12 12	2.7	**12 42**	2.6	T	12 55	2.7	**1 37**	2.7	16
17	S	12 02	2.8	**12 25**	2.6	T	12 56	2.6	**1 31**	2.5	F	1 43	2.6	**2 32**	2.6	17
18	S	12 51	2.6	**1 19**	2.4	W	1 41	2.5	**2 24**	2.4	S	2 38	2.5	**3 32**	2.6	18
19	M	1 39	2.5	**2 14**	2.3	T	2 31	2.3	**3 21**	2.4	S	3 39	2.5	**4 32**	2.7	19
20	T	2 30	2.3	**3 12**	2.2	F	3 25	2.3	**4 19**	2.4	M	4 41	2.5	**5 26**	2.8	20
21	W	3 21	2.2	**4 09**	2.2	S	4 22	2.3	**5 15**	2.4	T	5 37	2.7	**6 15**	2.9	21
22	T	4 13	2.2	**5 04**	2.2	S	5 17	2.3	**6 06**	2.5	W	6 28	2.9	**6 59**	3.0	22
23	F	5 04	2.2	**5 55**	2.3	M	6 09	2.4	**6 52**	2.7	T	7 15	3.1	**7 41**	3.2	23
24	S	5 53	2.2	**6 42**	2.4	T	6 56	2.6	**7 35**	2.8	F	8 00	3.3	**8 22**	3.3	24
25	S	6 40	2.3	**7 25**	2.5	W	7 41	2.7	**8 15**	2.9	S	8 44	3.4	**9 03**	3.4	25
26	M	7 24	2.3	**8 06**	2.6	T	8 24	2.9	**8 54**	3.0	S	9 29	3.6	**9 46**	3.5	26
27	T	8 06	2.4	**8 45**	2.7	F	9 07	3.0	**9 33**	3.1	M	10 15	3.6	**10 30**	3.4	27
28	W	8 47	2.5	**9 23**	2.7	S	9 50	3.1	**10 13**	3.1	T	11 04	3.6	**11 17**	3.4	28
29	T	9 28	2.6	**10 01**	2.7	S	10 35	3.1	**10 54**	3.1	W	11 56	3.5	...	...	29
30	F	10 10	2.6	**10 40**	2.7	M	11 22	3.1	**11 39**	3.0	T	12 09	3.3	**12 52**	3.3	30
31	S	10 54	2.6	**11 20**	2.7	T	...	...	**12 13**	3.1						31

Dates when Ht. of **Low** Water is below Mean Low with Ht. of lowest given for each period and Date of lowest in ( ):

8th - 16th: -0.7' (12th - 13th)
28th - 29th: -0.2'

7th - 13th: -0.4' (9th - 11th)

Average Rise and Fall 2.5 ft.

When a high tide exceeds av. ht., the *following* low tide will be lower than av.

1999 HIGH WATER
MIAMI HARBOR ENTR., FL
25° - 46′ N x 80° - 08′ W

DAY OF MONTH	DAY OF WEEK	OCTOBER a.m.	Ht.	OCTOBER p.m.	Ht.	DAY OF WEEK	NOVEMBER a.m.	Ht.	NOVEMBER p.m.	Ht.	DAY OF WEEK	DECEMBER a.m.	Ht.	DECEMBER p.m.	Ht.	DAY OF MONTH
		Standard Time starts Oct. 31 at 2 a.m.					Standard Time					Standard Time				
1	F	1 06	3.1	**1 55**	3.2	M	2 12	3.0	**2 54**	3.1	W	3 01	2.8	**3 26**	2.8	1
2	S	2 12	3.0	**3 04**	3.1	T	3 23	3.0	**3 57**	3.1	T	4 04	2.8	**4 21**	2.8	2
3	S	3 23	2.9	**4 14**	3.1	W	4 28	3.1	**4 53**	3.1	F	5 00	2.8	**5 09**	2.8	3
4	M	4 36	3.0	**5 19**	3.1	T	5 23	3.2	**5 41**	3.2	S	5 48	2.9	**5 52**	2.8	4
5	T	5 41	3.1	**6 16**	3.2	F	6 11	3.2	**6 23**	3.2	S	6 31	2.9	**6 32**	2.8	5
6	W	6 39	3.2	**7 06**	3.3	S	6 54	3.3	**7 02**	3.2	M	7 10	3.0	**7 09**	2.8	6
7	T	7 29	3.3	**7 50**	3.3	S	7 33	3.3	**7 38**	3.2	T	7 47	3.0	**7 45**	2.8	7
8	F	8 14	3.4	**8 30**	3.3	M	8 10	3.3	**8 13**	3.1	W	8 23	3.0	**8 21**	2.7	8
9	S	8 56	3.4	**9 09**	3.3	T	8 47	3.3	**8 49**	3.0	T	9 00	3.0	**8 59**	2.7	9
10	S	9 34	3.4	**9 44**	3.3	W	9 23	3.2	**9 23**	3.0	F	9 36	2.9	**9 36**	2.6	10
11	M	10 12	3.4	**10 19**	3.2	T	10 00	3.1	**10 00**	2.8	S	10 13	2.8	**10 16**	2.5	11
12	T	10 50	3.3	**10 55**	3.1	F	10 39	3.0	**10 41**	2.7	S	10 53	2.7	**11 00**	2.5	12
13	W	11 29	3.2	**11 33**	2.9	S	11 22	2.9	**11 26**	2.6	M	11 35	2.6	**11 49**	2.4	13
14	T	...	...	**12 11**	3.0	S	...	...	**12 08**	2.8	T	...	...	**12 22**	2.6	14
15	F	12 13	2.8	**12 56**	2.9	M	12 19	2.6	**1 01**	2.8	W	12 46	2.4	**1 14**	2.5	15
16	S	1 00	2.7	**1 47**	2.8	T	1 20	2.6	**1 57**	2.7	T	1 48	2.4	**2 10**	2.6	16
17	S	1 55	2.6	**2 45**	2.8	W	2 24	2.6	**2 54**	2.8	F	2 52	2.6	**3 08**	2.6	17
18	M	2 58	2.6	**3 45**	2.8	T	3 27	2.8	**3 50**	2.9	S	3 54	2.7	**4 06**	2.7	18
19	T	4 03	2.7	**4 42**	2.9	F	4 25	3.0	**4 42**	3.0	S	4 53	2.9	**5 02**	2.9	19
20	W	5 03	2.9	**5 33**	3.0	S	5 19	3.2	**5 32**	3.2	M	5 48	3.1	**5 57**	3.0	20
21	T	5 58	3.1	**6 21**	3.2	S	6 10	3.4	**6 21**	3.3	T	6 42	3.3	**6 50**	3.1	21
22	F	6 47	3.3	**7 06**	3.3	M	7 00	3.6	**7 10**	3.4	W	7 34	3.4	**7 43**	3.2	22
23	S	7 35	3.5	**7 51**	3.5	T	7 50	3.7	**8 00**	3.5	T	8 25	3.4	**8 36**	3.1	23
24	S	8 21	3.7	**8 35**	3.6	W	8 40	3.7	**8 50**	3.5	F	9 16	3.4	**9 29**	3.1	24
25	M	9 09	3.8	**9 21**	3.6	T	9 31	3.7	**9 43**	3.4	S	10 08	3.2	**10 24**	3.0	25
26	T	9 57	3.8	**10 08**	3.6	F	10 24	3.5	**10 39**	3.2	S	11 01	3.1	**11 21**	2.8	26
27	W	10 47	3.8	**10 58**	3.5	S	11 20	3.4	**11 39**	3.0	M	11 55	2.9	...	...	27
28	T	11 40	3.7	**11 53**	3.4	S	...	...	**12 20**	3.2	T	12 21	2.6	**12 51**	2.7	28
29	F	...	...	**12 37**	3.5	M	12 44	2.9	**1 22**	3.0	W	1 24	2.5	**1 48**	2.5	29
30	S	12 53	3.2	**1 39**	3.3	T	1 53	2.8	**2 25**	2.9	T	2 28	2.4	**2 45**	2.4	30
31	S	1 00	3.1	***1 46**	3.2						F	3 31	2.4	**3 41**	2.3	31

***Standard Time starts**

Dates when Ht. of **Low** Water is below Mean Low with Ht. of lowest given for each period and Date of lowest in ( ):

October	November	December
	23rd - 25th: - 0.3'	19th - 26th: -0.7' (23rd)

Average Rise and Fall 2.5 ft.

When a high tide exceeds av. ht., the *following* low tide will be lower than av.

TIMES OF CURRENT CHANGE AT VARIOUS POINTS

(See Note at bottom of Boston Tables: Rule-of-Thumb for Current Velocities.)

CURRENTS IN THE GULF OF MAINE - In the Gulf of Maine, on the western side, the Flood Current splits at Cape Ann, Mass., and floods north and east along the shore towards the Bay of Fundy. At the same time, on the eastern side of the Gulf, at the southern tip of Nova Scotia, the Flood Current runs to the west and then north and eastwards along the shore into the Bay of Fundy. The Ebb Current is just the reverse. In addition to these large principal currents, along the Maine Coast, at least at the mouths of principal bays, there is a shoreward set during the Flood and an offshore set during the Ebb, although this set is of considerably less velocity.

West of Mount Desert, the average along-shore current is rarely more than a knot but the farther east one goes, the greater are the average velocities to be expected, up to 2 knots or more. When heading west, therefore, start off at the time shown for High Water in your area (see p. 19) and have a fair Ebb current for 6 hours. Headed east, start at the time for Low Water in your area (about 6 1/2 hours after High Water) and carry the beneficial Flood current. East of Schoodic Point, the average currents are up to 2 knots and taking advantage of them will save considerable time and fuel.

Off shore, in the Gulf of Maine, unlike the along-shore currents that come to dead slack and *reverse*, there are so-called *rotary* currents. These currents constantly change direction in a clockwise flow completing the circle in about 12 1/2 hours. The maximum currents are when it is flooding in the northeasterly direction or ebbing in a southwesterly direction; minimum currents occur halfway between. There is no slack water.

Entering the Bay of Fundy through Grand Manan Channel, one finds that the average velocities are from 1-2 1/2 knots, although in the narrower channels off the Bay, velocities are higher (Friar Roads at Eastport has average velocities of 3 knots of more). The Current in the Bay Floods to the Northeast and Ebbs to the Southwest.

In using this table it should be borne in mind that actual Times of Slack or Maximum occasionally differ from the predicted times by as much as half an hour and in rare instances as much as an hour. Referring the Time of Current Change at the subordinate stations listed below, to the predicted Current Change at the reference station gives the *approximate* time only. To make sure, therefore, of getting the full advantage of a favorable current or slack water, the navigator should reach the entrance or strait at least half an hour before the predicted time. (This is basically the same precautionary note found in the U.S. Tidal Currents Table Book.)

Figures shown below are average maximum velocities in knots. We have omitted places having an average maximum velocity of less than 1 knot. To find the Time of Current Change (Start of Flood and Start of Ebb) at a selected point, refer to the table heading that particular section (in bold type) and add or subtract the time listed.

	TIME DIFFERENCES Flood Starts; Ebb Starts H.M.	MAXIMUM FLOOD Dir.(true) in degrees	MAXIMUM FLOOD Av. Max. in knots	MAXIMUM EBB Dir.(true) in degrees	MAXIMUM EBB Av. Max. in knots
MAINE COAST – **based on Boston Tables, pp. 12-17** (Fl. starts at Low Water; Ebb starts at High Water)					
Isle Au Haut, 0.8 mi. E of Richs Pt.	-0 05	335	1.4	140	1.5
Damariscotta R., off Cavis Pt.	+1 00	350	0.8	215	1.1
Sheepscot R., off Barter Is.	+1 00	005	1.7	200	1.8
Lowe Pt., NE of, Sasanoa R.	+1 20	325	3.0	150	1.8
Lower Hell Gate, Knubble Bay*	+1 25	290	3.0	155	3.5

* Velocities up to 9.0 kts. have been observed in the vicinity of the Boilers.

TIMES OF CURRENT CHANGE AT VARIOUS POINTS

(See Note at bottom of Boston Tables: Rule-of-Thumb for Current Velocities.)

	TIME DIFFERENCES	MAXIMUM FLOOD		MAXIMUM EBB	
	Flood Starts; Ebb Starts H.M.	Dir.(true) in degrees	Av. Max. in knots	Dir.(true) in degrees	Av. Max. in knots
KENNEBEC RIVER – **based on Boston Tables, pp. 12-17** (Fl. starts at Low Water; Ebb starts at High Water)					
Hunniwell Pt., NE of	+2 05	330	2.4	150	2.9
Bald Head, 0.3 mi. SW of	+2 15	320	1.6	155	2.3
Bluff Head, W of	+2 35	015	2.3	185	3.4
Fiddler Ledge, N of	+2 40	265	1.9	115	2.6
Doubling Pt., S of	+2 30	300	2.6	125	3.0
Lincoln Ledge, E. of	+2 35	000	1.9	175	2.8
Bath, 0.2 mi. S of bridge	+2 40	005	1.0	175	1.5
CASCO BAY – **based on Boston Tables, pp. 12-17** (Fl. starts at Low Water; Ebb starts at High Water)					
Broad Sound, W. of Eagle Is.	+0 45	010	0.9	170	1.3
Hussey Sound, 0.5 mi. S of Overset Is.	+0 35	340	1.1	130	1.2
Portland Hbr. Ent., SW of Cushing Is.	+0 35	320	1.0	155	1.1
Portland Bridge, Center of draw	+1 15	225	0.9	050	1.0
PORTSMOUTH HARBOR – **based on Boston Tables, pp. 12-17** (Fl. starts at Low Water; Ebb starts at High Water)					
Kitts Rocks, 0.2 mi. W of	+2 05	325	0.8	175	1.6
Portsmouth Hbr. Ent., off Wood Is.	+2 05	355	1.2	195	1.8
Fort Point	+2 10	350	1.5	130	2.0
Seavey Is., N of	+2 20	260	1.4	080	1.8
Clarks Is., S of	+2 20	260	2.1	080	3.1
Seavey Is., S of	+2 20	260	3.0	090	3.8
Henderson Pt., W of	+2 35	340	2.6	170	2.3
Off Gangway Rock	+2 35	280	2.1	110	3.0
Badgers Is., SW of	+2 35	330	3.3	125	3.7
MASSACHUSETTS COAST – **based on Boston Tables, pp. 12-17** (Fl. starts at Low Water; Ebb starts at High Water)					
Merrimac River Entr.	+1 25	285	2.2	105	1.4
Newburyport	+1 55	290	1.5	100	1.4
Plum Is. Sound Entr.	+1 00	315	1.6	185	1.5
Gloucester Hbr. Blynman Canal Entr.	+0 05	310	3.0	130	3.3
Hypocrite Channel	+0 30	255	0.9	070	1.0
Nantasket Roads Entr.	+0 05	260	1.4	085	1.5
BOSTON HARBOR – **based on Boston Tables, pp. 12-17** (Fl. starts at Low Water; Ebb starts at High Water)					
Nantasket Rds.	+0 05	250	1.3	040	1.9
Hull Gut	+0 05	155	1.2	350	1.8
Weymouth Back R.	-0 10	165	1.2	345	1.2
W. of Bumkin Is.	0 00	130	0.6	325	0.8
Worlds End, N of	+0 35	085	1.0	275	1.2
CAPE COD BAY – **based on Boston Tables, pp. 12-17** (Fl. starts at Low Water; Ebb starts at High Water)					
Race Point, 7 mi. N of	+0 15	290	1.5	–	1.5
Race Point, 1 mi. NW of	+0 10	225	1.0	060	0.9
Barnstable Harbor	+0 35	190	1.2	005	1.4
Manomet Point	+0 20	155	1.1	010	0.9
Gurnet Point, 1 mi. E of	+0 10	250	1.4	–	1.0
Farnham Rock, 1 mi. E of	-0 05	180	1.1	010	0.9

Important: See Footnote page 157

TIMES OF CURRENT CHANGE AT VARIOUS POINTS

(See Note at bottom of Boston Tables: Rule-of-Thumb for Current Velocities.)

	TIME DIFFERENCES Flood Starts; Ebb Starts H.M.	MAXIMUM FLOOD Dir.(true) in degrees	MAXIMUM FLOOD Av. Max. in knots	MAXIMUM EBB Dir.(true) in degrees	MAXIMUM EBB Av. Max. in knots
NANTUCKET SOUND – **based on Pollock Rip Channel Tables, pp. 44-49**					
Pollock Rip Channel, E end	-0 20	55	2.0	210	1.8
POLLOCK RIP CHANNEL at Butler Hole - See table, pp. 44-49					
Monomoy Point, 0.2 mi. W of	+0 10	170	1.7	345	2.0
Halfmoon Shoal, 3.5 mi. E of	+1 10	090	1.1	295	1.0
Great Point, 0.5 mi. W of	F+0 25, E+1 15	030	1.1	195	1.2
Tuckernuck Shoal, off E end	+1 15	115	0.9	285	0.9
Nantucket Hbr. entr. chan.	F+3 20, E+2 45	170	1.2	350	1.5
Muskeget Is. chan., 1 mi. NE of	F+1 30, E+1 00	110	1.1	295	1.5
Muskeget Rock, 1.3 mi. SW of	+1 05	025	1.3	190	1.0
Muskeget Channel	+1 35	020	3.8	200	3.3
Betw. Long Shoal-Norton Shoal	+1 30	100	1.4	260	1.1
Cape Poge Lt., 1.7 mi. SSE of	+0 55	025	1.6	215	1.3
Cross Rip Channel	+1 50	090	1.3	270	0.9
Cape Poge, 3.2 mi. NE of	+2 40	095	1.6	300	1.2
Betw. Broken Gr.-Horseshoe Sh.	F+1 45, E+1 15	105	1.1	275	0.9
Point Gammon, 1.2 mi. S of	+1 10	105	1.1	260	1.0
Lewis Bay Entr. Channel	+2 45	005	0.9	185	1.3
Betw. Wreck Shoal-Eldridge Shoal	+1 45	060	1.7	245	1.4
Hedge Fence Lighted Gong Buoy 22	+2 45	110	1.4	270	1.2
Betw. East Chop-Squash Meadow	+1 55	130	1.4	330	1.8
East Chop, 1 mi. N of	+2 30	115	2.2	295	2.2
West Chop, 0.8 mi. N of	F+2 50, E+2 20	095	3.1	280	3.0
Betw. Hedge Fence-L'hommedieu Shoal	+2 15	105	2.1	275	2.2
Waquoit Bay Entr.	+3 30	350	1.5	205	1.4
L'hommedieu Shoal, N of W end	+2 20	080	2.3	270	2.3
Nobska Point, 1.8 mi. E of	+2 05	065	2.3	240	1.7
VINEYARD SOUND – **based on Pollock Rip Channel, pp. 44-49**					
West Chop, 0.2 mi. W of	F+1 20, E+1 50	060	2.7	240	1.4
Nobska Point, 1 mi. SE of	+2 30	070	2.6	260	2.4
Norton Point, 0.5 mi. N of	+2 00	050	3.4	240	2.4
Tarpaulin Cove, 1.5 mi. E of	F+2 50, E+2 10	055	1.9	230	2.3
Robinson's Hole, 1.2 mi. SE of	+2 20	060	1.9	240	2.1
Gay Head, 1.5 mi. NW of	+1 35	010	2.0	250	2.0
Gay Head, 3 mi. N of	+2 05	075	1.1	255	1.2
VINEYARD SOUND-BUZZARDS BAY – **based on Cape Cod Canal, pp. 30-35**					
WOODS HOLE	**See table, pp. 36-41**				
Robinson's Hole, S end	+1 15	160	0.8	340	1.0
Robinson's Hole, Middle	+1 15	145	2.8	315	2.9
Robinson's Hole, N end	Fl+1 55, E+0 50	160	1.0	340	1.2
Quick's Hole, S end	Fl+2 20, E+1 15	140	1.9	300	2.0
Quick's Hole, Middle	Fl+2 20, E+1 25	165	2.5	340	2.2
Quick's Hole, N end	Fl+2 40, E+1 45	165	2.0	000	2.6
Canapitsit Channel	Fl+2 05, E+1 00	155	2.6	310	1.7
BUZZARDS BAY – **based on Pollock Rip Channel, pp. 44-49**					
Westport River Entr.	F+0 10, E-0 25	290	2.2	110	2.5
Gooseberry Nk., 2 mi. SSE of Buzz Bay Entr.- rotary current, no slack water. Av. max. 0.6 kts, approx. dir. 52° true at 3 1/2 hrs. after Fl. starts at Poll. Rip. Ave. max. 0.5 kts, approx. dir. 232° true 2 1/2 hrs. after Ebb starts at Poll. Rip.					
Betw. Ribbon Reef-Sow &Pigs Rf.	F-0 20, E-2 45	060	0.8	235	1.2
Penikese Is., 0.8 mi. NW of	F-1 40, E-0 55	050	1.2	255	1.1
Betw. Gull Is.-Nashawena Is.	-2 10	090	0.9	245	1.1
Abiels Ledge, 0.4 mi. S of	F+0 25, E+0 05	035	0.8	215	1.0
Dumpling Rocks, 0.2 mi. SE of	-1 40	065	0.8	190	1.1
CAPE COD CANAL	**See table, pp. 30-35**				

Important: See Footnote page 157

TIMES OF CURRENT CHANGE AT VARIOUS POINTS

(See Note at bottom of Boston Tables: Rule-of-Thumb for Current Velocities.)

	TIME DIFFERENCES Flood Starts; Ebb Starts H.M.	MAXIMUM FLOOD Dir.(true) in degrees	MAXIMUM FLOOD Av. Max. in knots	MAXIMUM EBB Dir.(true) in degrees	MAXIMUM EBB Av. Max. in knots
NARRAGANSETT BAY – **based on Pollock Rip Channel, pp. 44-49**					
Tiverton, Stone Bridge	F-3 00, E-2 25	010	2.7	190	2.7
Tiverton, RR Bridge	F-3 25, E-2 50	000	2.3	180	2.4
Castle Hill, W of	F-0 05, E-1 05	000	0.7	210	1.2
Bull Point, E of	-1 10	000	1.2	205	1.5
Rose Is., NE of	F-2 00, E-1 15	340	0.8	165	1.0
Rose Is., W of	F-0 40, E-1 20	000	0.7	170	1.0
Dyer Is., W of	-1 00	025	0.8	215	1.0
Mount Hope Bridge	-1 10	045	1.1	230	1.4
Kickamuit R., Mt. Hope Bay	F-2 05, E-1 20	000	1.4	190	1.7
Beavertail Point, 0.8 mi NW of	F-0 10, E-1 30	005	0.5	190	1.0
Betw. Dutch Is.-Beaver Head	-1 55	030	1.0	235	1.0
Dutch Is., W of	-1 25	015	1.3	205	1.2
Warren R., Warren	-0 20	000	1.0	170	0.9
India Pt. RR Bridge, Seekonk R.	-1 40	020	1.0	180	1.4
BLOCK ISLAND SOUND – **based on The Race, pp. 70-75**					
Pt. Judith Pond Entr.	-3 20	350	1.8	185	1.5
Sandy Pt., Block Is. 1.5 mi N of	F-0 20, E-1 00	315	1.9	065	2.1
Lewis Pt., 1.0 mi. SW of	F-1 35, E-0 35	300	1.9	135	1.8
Lewis Pt., 1.5 mi. W of	F-1 30, E-0 45	320	1.4	170	1.7
Southwest Ledge	-0 20	320	1.5	140	2.1
Watch Hill Pt., 2.2 mi. E of	F-0 35, E+0 35	260	1.2	085	0.7
Montauk Pt., 1.2 mi. E of	F-1 30, E-0 50	345	2.8	160	2.8
Montauk Pt., 1 mi. NE of	F-2 00, E-1 10	355	2.4	145	1.9
Betw. Shagwong Reef-Cerberus Shoal	-0 35	240	1.9	055	1.8
Betw. Cerberus Shoal-Fishers Is.	F-0 55, E+0 10	265	1.3	095	1.3
Gardiners Is., 3 mi. NE of	-0 35	305	0.9	140	1.0
GARDINERS BAY etc. – **based on The Race, pp. 70-75**					
Goff Point, 0.4 mi. NW of	-1 45	225	1.2	010	1.6
Acabonack Hbr. Ent., 0.6 mi. ESE of	-1 30	345	1.4	140	1.2
Gardiners Pt. Ruins, 1.1 mi. N of	-0 20	270	1.2	065	1.8
Betw. Gardiners Point-Plum Is.	-0 35	265	1.4	080	1.6
Jennings Pt., 0.2 mi. NNW of	+0 25	290	1.6	055	1.5
Cedar Pt., 0.2 mi. W of	F-0 20, E+0 20	195	1.8	005	1.6
North Haven Peninsula, N of	+0 15	230	2.4	035	2.1
Paradise Pt., 0.4 mi. E of	+0 25	145	1.5	345	1.5
Little Peconic Bay entr.	+0 35	240	1.6	015	1.5
Robins Is., 0.5 mi. S of	+0 35	245	1.7	065	0.6
FISHERS ISLAND SOUND – **based on The Race, pp. 70-75**					
Napatree Point, 0.7 mi. SW of	-0 55	285	1.7	115	2.2
Little Narragansett Bay entr.	-2 00	090	1.3	270	1.3
Ram Island Reef, S of	-0 45	255	1.3	090	1.6
LONG ISLAND SOUND – **based on The Race, pp. 70-75**					
THE RACE (near Valiant Rock) – See table, pp. 70-75		295	2.9	100	3.5
Race Point, 0.4 mi. SW of	F-0 25, E-0 45	290	2.6	135	3.5
Little Gull Is., 0.5 mi. NE of	-0 20	000	3.3	105	3.1
Little Gull Is., 1.1 mi. ENE of	-0 05	300	4.0	130	4.7
Great Gull Is., 0.7 mi. WSW of	-0 40	300	2.6	135	3.2
Plum Gut	-1 10	325	3.5	125	4.3
Goshen Pt., 1.9 mi. SSE of	-1 05	285	1.2	060	1.6
Little Gull Is., 0.8 mi. NNW of	F+0 15, E-2 30	260	1.9	045	2.9
Bartlett Reef, 0.2 mi. S of	F-2 00, E-1 00	255	1.4	090	1.3
Twotree Is. Channel	-0 55	265	1.2	100	1.6

Important: See Footnote page 157

TIMES OF CURRENT CHANGE AT VARIOUS POINTS

(See Note at bottom of Boston Tables: Rule-of-Thumb for Current Velocities.)

	TIME DIFFERENCES Flood Starts; Ebb Starts H.M.	MAXIMUM FLOOD Dir.(true) in degrees	MAXIMUM FLOOD Av. Max. in knots	MAXIMUM EBB Dir.(true) in degrees	MAXIMUM EBB Av. Max. in knots
LONG ISLAND SOUND – **based on The Race, pp. 70-75**					
Black Point, 0.8 mi. S of	-0 35	265	1.2	075	1.4
Betw. Black Pt.-Plum Is.	+0 25	235	2.1	075	2.4
Plum Is., 0.8 mi. NNW of	F+0 05,E-1 15	245	1.7	065	2.4
Hatchett Pt., 1.1 mi. WSW of	F-2 35, E-0 50	240	1.3	045	1.2
Conn. River RR Drawbridge	+0 40	000	1.0	200	1.0
Saybrook Bkwtr., 1.5 mi. SE of	F-1 30, E-0 55	260	1.9	070	2.0
Mulford Pt., 3.1 mi. NW of	-0 05	270	1.9	065	2.3
Cornfield Point, 3 mi. S of	F-0 55, E-0 05	255	2.0	095	1.7
Cornfield Point, 1 mi. S of	-1 00	295	1.4	110	1.6
Six Mile Reef, 2 mi. E of	-0 20	235	1.6	040	2.1
Kelsey Point, 1 mi. S of	-1 15	250	2.0	120	1.5
Sachem Head, 1 mi. SSE of	-0 35	255	1.1	065	1.0
New Haven Harbor entr.	F-1 10, E-0 35	320	1.4	150	0.9
Housatonic R., Milford Pt., 0.2 mi. W of	+0 05	330	1.2	135	1.2
Point No Point, 2 mi. S of	-0 20	250	1.3	075	1.2
Port Jefferson Harbor entr.	+0 20	150	2.6	325	1.9
Crane Neck Point, 0.5 mi. NW of	F-0 45, E-1 40	255	1.3	015	1.5
Eatons Neck Pt., 1.3 mi. N of	+0 15	285	1.4	075	1.4
Lloyd Point, 1.3 mi. NNW of	+1 20	255	1.0	055	1.0
EAST RIVER – **based on Hell Gate, pp. 96-101**					
Cryders Pt., 0.4 mi. NNW of	-0 30	110	1.3	285	1.1
College Pt. Rf., NW of	F-0 35, E-0 00	074	1.5	261	1.4
Rikers Is. Chann. off La Guardia Field	+0 05	090	1.1	260	1.3
Hunts Point, SW of	0 00	110	1.7	280	1.3
S. Brother Is. NW of	- 0 10	054	1.5	252	1.2
Off Winthrop Ave., Astoria	0 00	040	3.4	220	2.5
Mill Rock, NE of	-0 25	105	2.3	290	0.6
Mill Rock, W of	-0 15	000	1.2	180	1.0
HELL GATE (off Mill Rock) – See table, pp. 96-101					
Roosevelt Is., W of, off 75th St.	-0 05	035	3.8	215	4.7
Roosevelt Is., E of, off 36th Ave.	-0 10	030	3.5	210	3.4
Roosevelt Is., W of, off 67th St.	+0 10	010	3.6	230	4.0
Off 19th St. (Pier 67)	-0 10	355	1.8	180	1.9
Williamsburg Br., 0.3 mi. N of	-0 05	020	2.7	220	2.9
Brooklyn Bridge, 0.1 mi. SW of	-0 10	045	2.9	220	3.5
Buttermilk Channel	-0 10	050	1.8	220	2.4
LONG ISLAND, South Coast – **based on The Narrows, pp. 102-107**					
Shinnecock Inlet	F-0 05, E-0 30	350	2.5	170	2.3
Fire Is. Inl., 0.5 mi. off Oak Bch	F-0 05, E+0 30	080	2.4	245	2.4
Jones Inlet	F-1 15, E-0 50	035	3.1	215	2.6
East Rockaway Inlet	F-1 45, E-1 05	040	2.2	225	2.3
JAMAICA BAY – **based on The Narrows, pp. 102-107**					
Rockaway Inlet	-1 45	085	1.8	245	2.7
Barren Is., E of	-2 00	005	1.2	190	1.7
Beach Channel (bridge)	F-1 50, E-0 55	060	1.9	225	2.0
Grass Hassock Channel	-1 10	050	1.0	230	1.0
NEW YORK HARBOR ENTRANCE – **based on The Narrows, pp. 102-107**					
Ambrose Channel	F-1 00, E-0 25	303	1.6	123	1.7
Norton Pt., WSW of	F-0 15, E+0 25	341	1.0	166	1.2
THE NARROWS (midchannel) – See table, pp. 102-107					

Important: See Footnote page 157

TIMES OF CURRENT CHANGE AT VARIOUS POINTS

(See Note at bottom of Boston Tables: Rule-of-Thumb for Current Velocities.)

	TIME DIFFERENCES Flood Starts; Ebb Starts H.M.	MAXIMUM FLOOD Dir.(true) in degrees	Av. Max. in knots	MAXIMUM EBB Dir.(true) in degrees	Av. Max. in knots
NEW YORK HARBOR, Upper Bay – **based on The Narrows, pp. 102-107**					
Bay Ridge, W of	F-0 11, E+0 40	354	1.4	185	1.5
Red Hook Channel	F-1 05, E-0 10	355	1.0	170	0.7
Robbins Reef Light, E of	+0 10	015	1.3	205	1.6
Red Hook, 1 mi. W of	+0 45	025	1.3	205	2.3
Statue of Liberty, E of	+0 55	030	1.4	205	1.9
HUDSON RIVER, Midchannel – **based on The Narrows, pp. 102-107**					
The Battery, NW of	+1 30	015	1.5	195	2.3
Desbrosses St.	+1 35	010	1.5	–	2.3
Chelsea Docks	+1 30	020	1.7	185	2.0
Ninety-sixth St.	+1 50	030	1.7	–	2.3
George Washington Bridge	+1 45	020	1.6	200	2.2
Spuyten Duyvil	+2 00	020	1.6	–	2.1
Riverdale	+2 10	015	1.4	200	2.0
Dobbs Ferry	+2 30	010	1.3	–	1.7
Tarrytown	+2 40	000	1.1	–	1.5
West Point, off Duck Is.	+3 40	010	1.0	185	1.1
NEW YORK HARBOR, Lower Bay – **based on The Narrows, pp. 102-107**					
Sandy Hook Chan.	F-1 35, E-1 05	285	1.6	095	1.9
Sandy Hook Chan., 0.4 mi. W of N. tip	-1 55	235	2.0	050	1.6
Coney Is. Lt., 1.5 mi. SSE of	F-1 25, E-1 00	310	1.1	125	1.3
Rockaway Inlet Jetty, 1 mi. SW of	F-2 15, E-1 30	285	1.2	140	1.4
Coney Is. Channel, W end	F-1 20, E-0 25	295	1.1	100	1.2
SANDY HOOK BAY – **based on The Narrows, pp. 102-107**					
Highlands Bridge, Shrewsbury R.	+0 25	170	2.6	–	2.5
Sea Bright Bridge, Shrewsbury R.	+0 55	185	1.4	–	1.7
RARITAN RIVER – **based on The Narrows, pp. 102-107**					
Washington Canal, N entr.	-1 20	240	1.5	060	1.5
South River Entr.	F-1 55, E-0 25	180	1.1	000	1.0
ARTHUR KILL & KILL VAN KULL – **based on The Narrows, pp. 102-107**					
Perth Amboy	F-1 25, E-0 05	010	0.9	190	1.0
Tottenville	-0 50	025	1.4	240	1.5
Tufts Pt.-Smoking Pt.	-0 35	110	1.2	265	1.2
Carteret	-0 20	015	1.1	175	1.3
Tremley Point	F-0 40, E+0 05	025	1.1	205	1.5
Elizabethport	+0 20	090	1.4	260	1.1
Bayonne Bridge	F-1 15, E-0 50	260	1.8	080	1.7
Bergen Pt., East Reach	-1 35	275	1.1	095	1.2
New Brighton	F-1 45, E-1 25	260	1.3	070	1.9
NEW JERSEY COAST – **based on Del. Bay Entr., pp. 126-131**					
Manasquan Inlet	F-0 45, E-1 05	300	1.7	120	1.8
Manasquan R. Hy. Br. Main Ch.	F-0 40, E-1 10	230	2.2	050	2.1
***Pt. Pleasant Canal, north bridge**	F+1 45, E+0 55	170	1.8	350	2.0
Barnegat Inlet	F+1 00, E+0 25	270	2.2	090	2.5
Manahawkin Bridge	+2 30	030	1.1	210	0.9
McCrie Shoal	-0 35	280	1.3	100	1.4
Cape May Harbor entr.	F-1 45, E-1 05	335	1.8	150	2.2
Cape May Canal, E end	-1 45	310	1.9	130	1.9
Cape May Canal, W end	-1 45	265	0.9	090	0.9

* Waters are extremely turbulent.

Important: See Footnote page 157

TIMES OF CURRENT CHANGE AT VARIOUS POINTS

(See Note at bottom of Boston Tables: Rule-of-Thumb for Current Velocities.)

	TIME DIFFERENCES Flood Starts; Ebb Starts H.M.	MAXIMUM FLOOD Dir.(true) in degrees	MAXIMUM FLOOD Av. Max. in knots	MAXIMUM EBB Dir.(true) in degrees	MAXIMUM EBB Av. Max. in knots
DELAWARE BAY & RIVER – **based on Del. Bay Entr., pp. 126-131**					
Cape May Channel	-1 10	305	1.5	150	2.3
DELAWARE BAY ENTR. – See Table, pp. 126-131					
Cape Henlopen, 0.3 mi. N of	-0 30	300	2.0	125	2.3
Cape Henlopen, 2 mi. NE of	+0 15	315	2.0	145	2.3
Cape Henlopen, 5 mi. N of	F+0 20, E+0 50	345	2.0	175	1.9
Mispillion River Mouth	F+2 35, E+1 55	025	1.5	190	1.0
Fourteen Ft. Bk. Lt., 1.2 mi. E of	F+0 10, E+0 35	340	1.3	175	1.5
Bay Shore Channel, NW of Fishing Creek	-0 30	005	1.1	185	1.1
Maurice River entr.	+1 00	010	1.1	190	1.0
Kelly Island, 1.5 mi. E of	+0 50	350	0.9	165	1.2
Elbow of Cross Ledge, 0.8 mi. S of	F+1 30, E+0 55	330	1.3	165	2.2
False Egg Is. Pt., 2 mi. off	+0 25	340	1.1	160	1.3
Ben Davis Pt. chan., abreast of	+2 00	320	1.6	145	2.3
Cohansey R., 0.5 mi. above entr.	+1 35	075	1.2	255	1.4
Arnold Point	+2 15	355	2.0	155	2.1
Smyrna River entr.	+1 55	250	1.2	070	1.5
Stony Point, W of	F+3 25, E+2 30	325	1.5	150	1.9
Appoquinimink R. entr.	+2 25	230	1.0	050	1.2
Reedy Is., off end of pier	+3 00	025	2.4	195	2.6
Alloway Creek entr., 0.2 mi. above	+2 20	130	2.1	325	2.1
Reedy Point, 0.4 mi. E of	F+3 20, E+2 55	335	1.8	165	2.3
Salem River entr.	+3 40	060	1.5	245	1.6
Bulkhead Shoal Chan., off Del. City	+3 10	310	2.1	140	2.1
Pea Patch Is., chan., E of	+3 30	320	2.3	150	2.3
New Castle, chan., opp.	F+4 05, E+3 35	050	1.9	230	2.4
CHESAPEAKE BAY – **based on The Race, pp. 70-75** (over 90% correlation within 15 min. throughout year)					
Cape Henry Light, 1 mi. N of	+0 45	280	1.1	090	2.0
Cape Henry Light, 1.8 mi. N of	+0 40	290	1.2	100	1.5
Chesapeake Bay entr.	+0 45	305	1.0	125	1.5
Cape Henry Light, 4.6 mi. N of	F-0 15, E+0 35	295	1.3	105	1.3
Cape Charles Light, 9.5 mi. WSW of	+0 55	320	1.5	125	1.4
Cape Henry Light, 8.3 mi. NW of	+0 20	330	1.0	135	1.1
Thimble Shoal Channel	+0 05	310	1.4	095	1.3
Tail of the Horseshoe	+0 20	300	0.9	110	1.0
Chesapeake Channel	+0 30	335	1.8	145	1.5
Fisherman Is., 1.8 mi. S of	0 00	320	1.6	120	1.4
Old Plantation Flats Lt., W of	+1 55	005	1.2	175	1.3
York Spit Channel	+1 40	010	0.8	195	1.1
Wolf Trap Lt., 0.5 mi. W of	+1 50	015	1.0	190	1.2
Stingray Point, 5.5 mi. E of	F+2 35, E+3 35	345	1.0	180	0.9
Stingray Point, 12.5 mi. E of	+2 25	030	1.0	175	0.8
Smith Point Lt., 6.0 mi. N of	+4 35	350	0.4	135	1.0
Cove Point - See Chesapeake Bay Current Diagram, p. 144					
Poole Island - See Chesapeake Bay Current Diagram, p. 144					
Worton Point - See Chesapeake Bay Current Diagram, p. 144					
CHESAPEAKE & DELAWARE CANAL - See table, pp. 134-139					
C & D CANAL POINTS – **based on C&D Canal, pp. 134-139**					
Back Creek Entrance	+0 45	055	1.0	240	1.2
Reedy Point Bridge	-0 45	085	2.6	265	2.1

Important: See Footnote page 157

TIMES OF CURRENT CHANGE AT VARIOUS POINTS

(See Note at bottom of Boston Tables: Rule-of-Thumb for Current Velocities.)

	TIME DIFFERENCES	MAXIMUM FLOOD		MAXIMUM EBB	
	Flood Starts; Ebb Starts H.M.	Dir.(true) in degrees	Av. Max. in knots	Dir.(true) in degrees	Av. Max. in knots
HAMPTON ROADS – **based on The Race, pp. 70-75** (over 90% correlation within 15 min. throughout year)					
Fort Wool, 0.9 mi. NE of	-0 50	265	1.0	080	1.8
Fort Wool, 0.2 mi. NW of	-1 00	240	1.3	050	2.0
Old Point Comfort, 0.2 mi. S of	-0 35	240	1.7	075	1.4
Willoughby Spit, 0.8 mi. NW of	-1 35	260	0.7	040	1.0
Sewells Point, chan., W of	-0 50	195	0.9	000	1.2
Newport News, chan., middle	F-0 35, E 0 00	245	1.1	075	1.2
NC, SC, GA & FL, outer coast – **based on Hell Gate, pp. 96-101** (over 90% correlation within 15 min. throughout year)					
Hatteras Inlet	+1 05	305	2.1	150	2.0
Ocracoke Inlet Chan. Entr.	+0 50	000	1.7	145	2.4
Beaufort Inlet Approach	F+0 20, E-0 30	000	0.3	160	1.4
Cape Fear R. Bald Head	-0 45	035	2.2	190	2.9
Winyah Bay Entrance	0 00	320	1.9	140	2.0
North Santee R. Entrance	-0 55	010	1.5	165	1.8
South Santee R. Entrance	F-1 25, E-0 40	045	1.5	240	1.6
Charleston Hbr. Entr., betw. jetties	-1 25	320	1.8	120	1.8
Charleston Hbr., off Ft. Sumter	-1 10	340	1.4	130	1.5
Charleston Hbr. S. ch. 0.8 mi. ENE of Ft. Johnson	-1 10	275	0.8	115	2.6
Charleston Hbr., Drum Is. (bridge)	-1 00	020	1.2	185	2.0
North Edisto River Entr.	-0 20	330	2.9	140	3.7
South Edisto River Entr.	-1 20	350	1.8	145	2.2
Ashepoo R. off Jefford Cr. Entr.	-0 25	015	1.5	195	1.6
Port Royal Sd., SE Chan. Entr.	F-2 10, E-1 15	310	1.3	150	1.6
Hilton Head	-1 00	325	1.8	145	1.8
Beaufort River Entrance	-0 50	005	1.4	175	1.9
Vernon R. 1.2 mi. S of Possum Pt.	-1 00	325	1.1	165	1.7
Betw. Egg Is. & Raccoon Key	-0 40	290	1.6	125	1.9
St. Catherines Sound Entr.	-1 00	280	1.9	085	2.1
Sapelo Sound Entr.	-1 25	295	2.1	100	2.5
Doboy Sound Entr.	-1 10	305	2.0	110	2.0
Altamaha Sd., 1 mi. SE of Onemile Cut	F-0 05, E-1 45	270	1.0	090	1.9
St. Simons Sound Entr.	-0 50	300	2.1	110	1.9
St. Andrews Sound Entr.	-0 50	270	2.1	105	2.2
Cumberland Sd., St. Marys Entr.	-0 35	275	2.3	090	2.6
Cumberland Sd., 0.2 mi. NW of Beach Cr. Entr.	-0 45	340	1.5	165	2.2
Nassau Sd., 1 mi. N of Sawpit Cr. Entr.	-0 10	310	1.7	135	1.7
FLORIDA EAST COAST – **based on The Narrows, pp. 102-107** (over 90% correlation within 15 min. throughout year)					
St. Johns R. Entr. betw. jetties	Fl+1 10, E+0 45	275	1.9	100	2.3
Mayport	Fl+1 05, E+0 40	210	2.2	025	3.1
St. Johns Bluff	Fl+1 15, E+0 50	245	1.6	060	2.2
FLORIDA EAST COAST – **based on Hell Gate, pp. 96-101** (over 90% correlation within 15 min. throughout year)					
Fort Pierce Inlet	+1 00	250	2.6	070	3.1
Lake Worth Inlet, betw. jetties	0 00	275	2.4	095	3.6
Miami Hbr., Bakers Haulover Cut	0 00	270	2.9	090	2.5
Miami Hbr. Entr., betw. jetties	+0 10	290	1.9	125	2.1

Note: Velocities shown are from U.S. Gov't. figures. It is obvious, however, to local mariners and other observers, that coastal inlets may have far greater velocities than indicated here. Strong winds and opposing tides can cause even more dangerous conditions, and great caution should be used.

DURATION OF SLACK WATER

The tables below give, for various maximum currents, the approx. period of time during which weak currents not exceeding 0.5 knot will be encountered. This duration includes the last of the Flood or Ebb and the beginning of the following Ebb or Flood; i.e.: half of the duration will be before and half after the time of Slack Water.

Duration of Weak Current for Cape Cod Canal, Hell Gate, Ches. & Del. Canal:

If the Max Curr. will be:	Total Period of Time with Vel. not more than 0.5 knot
1.0 kts	89 min.
1.5 kts.	52 min.
2.0 kts.	36 min.
3.0 kts.	22 min.
4.0 kts.	17 min.
5.0 kts.	13 min.

Duration of Weak Current for all other places:

If the Max. Curr. will be:	Total Period of Time with Vel. not more than 0.5 knot:
1.0 kts.	120 min.
1.5 kts.	78 min.
2.0 kts.	58 min.
3.0 kts.	38 min.
4.0 kts.	29 min.
5.0 kts.	23 min.

TIDAL HEIGHTS AND DEPTHS

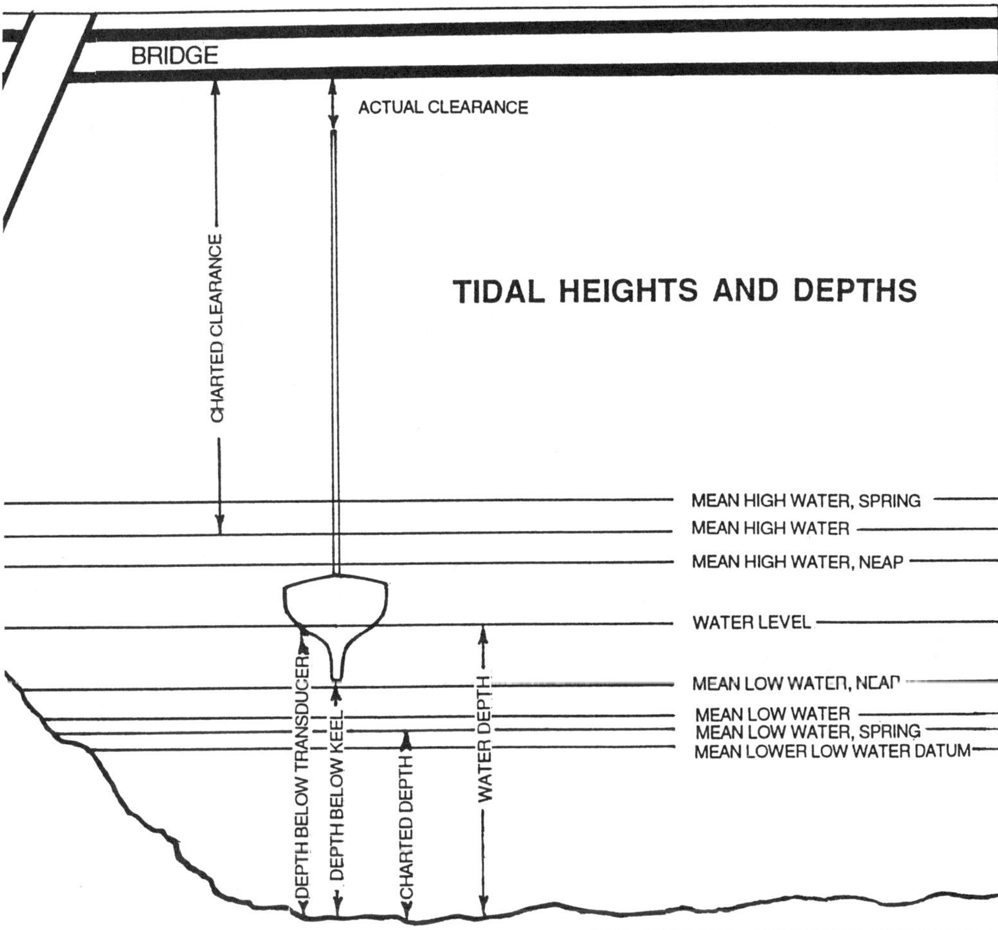

SPRING TIDES - tides of increased range, occurring twice a month, around the times of the new and the full moons.

NEAP TIDES - tides of decreased range, occurring twice a month, around the times of the waning and the waxing half moons.

DIURNAL INEQUALITY - the difference in height of the two daily low waters or of the two daily high waters. It is a result of the moon's (and to a lesser extent, the sun's) changing declination.

MEAN HIGH WATER, SPRING - the mean of high water heights of spring tides

MEAN HIGH WATER - the mean of all high water heights; the charted clearance of bridges, etc. is measured from this height

MEAN HIGH WATER, NEAP - the mean of high water heights of neap tides

MEAN LOW WATER, NEAP - the mean of low water heights of neap tides

MEAN LOW WATER - the mean of all low water heights

MEAN LOW WATER, SPRING - the mean of low water heights of spring tides

MEAN LOWER LOW WATER DATUM - the mean of the lower low water heights; charted depths originate from this reference height or datum

CHARACTERISTICS OF LIGHT SIGNALS
(see footnote on next page for abbreviations used.)

Fixed **F**
Light continuous and steady.

Single Occulting **Oc**
Eclipse (darkness period) of shorter duration than light period.

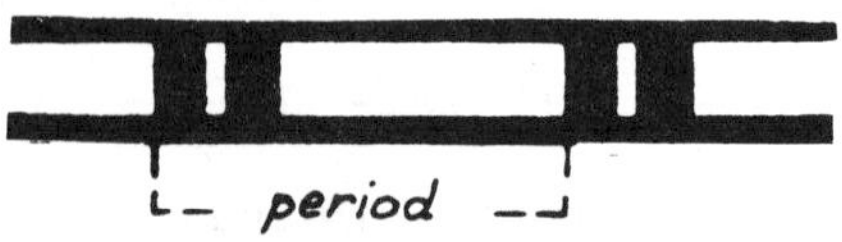

Group-Occulting **Oc(2)**
Group of eclipses specified in numbers, regularly repeated.

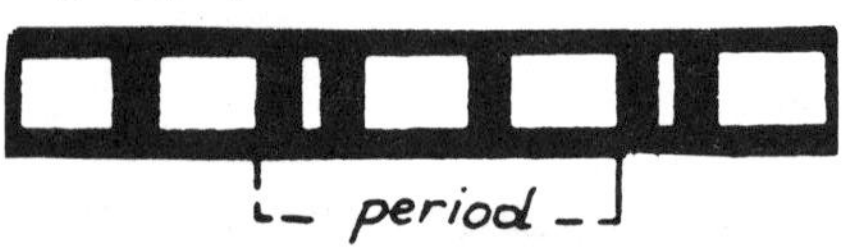

Composite group-occulting **Oc (2+1)**
Similar to group-occulting except for groups having different number of eclipses.

Isophase **Iso**
Equal periods of light and darkness.

Flashing **Fl**
Period of light is shorter than period of darkness.

Group-flashing **Fl(2)**
Flashing light with specified number regularly repeated.

Composite group-flashing **Fl(3+1)**
Successive groups have different number of flashes.

Quick **Q**
Flash is regularly repeated.

Interrupted Quick **IQ**
Sequence of flashes is interrupted by longer periods of darkness.

Morse Code **Mo(Y)**
Group of flashes represented by dots and dashes of letter.

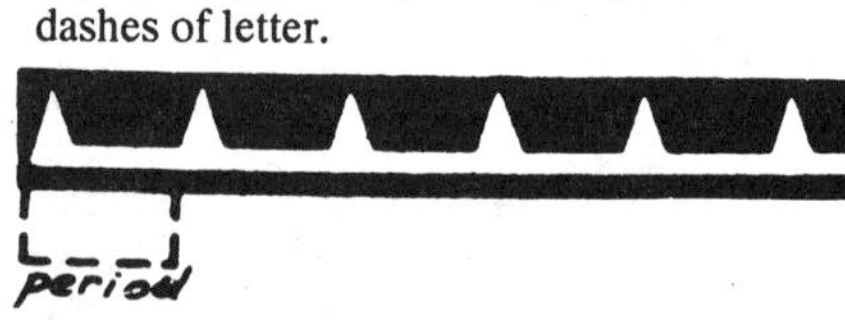

Fixed and Flashing **FFl**
Fixed light is of lower intensity than flashing light.

Alternating **AlWR**
Light shows alternating different colors.

NOTICES TO MARINERS: To keep informed of the many important changes currently planned by the Coast Guard, contact C.G. District Aids to Nav. Dept., and request the local mailing of weekly Notices to Mariners.
When reporting deficiencies in navigational aids, contact nearest C.G. unit and give official name of the aid.

PRINCIPAL LIGHTS AND FOG SIGNALS

NOVA SCOTIA, EAST COAST

NORTH CANSO, W. side of N. entr. to Strait of Canso – Fl. W. Obscured S. of 120°. Ht. 36.7 m. (120'), Rge. 13 mi.
CRANBERRY IS., off Cape Canso, S. part of is. – Fl. W. ev. 15 sec. 2 Horns 2 bl. ev. 60 sec. Horns point 066° and 141°. Ht. 16.9 m. (56'), Rge. 23 mi. Racon (B).
WHITE HEAD ISLAND, SW side of is. – Fl. W. ev. 5 sec. Horn 1 bl. ev. 60 sec. Horn points 190°. Ht. 18.2 m. (60'), Rge. 12 mi.
COUNTRY IS., S. side of is. – Fl. W. ev. 20 sec. Horn 1 bl. ev. 60 sec. Horn points 180°. Ht. 13.7 m. (45'), Rge. 14 mi.
LISCOMB IS., near Cranberry Pt. – Fl. W. ev. 10 sec. Horn 1 bl. ev. 30 sec. Horn points 144°. Ht. 21.9 m. (72'), Rge. 14 mi.
BEAVER IS., E. end of is. – Gp. Fl. W. 3 fl. ev. 20 sec. Horn 1 bl. ev. 60 sec. Horn points 144°. Ht. 19.9 m. (66'), Rge. 14 mi.
SHIP HARBOUR, on Wolf Pt. – Fl. G. ev. 12 sec. Ht. 22.9 m. (76'), Rge. 6 mi.
OWLS HEAD, at end of head – Fl. W. ev. 4 sec. Ht. 25.6 m. (84'), Rge. 8 mi.
EGG IS., center of is. – Fl. W. ev. 15 sec. Ht. 25.6 m. (84'), Rge. 9 mi.
JEDDORE ROCK, summit of rock – Fl. W. ev. 12 sec. Ht. 29.3 m. (97'), Rge. 8 mi.
SAMBRO HARBOR LIGHT & WHISTLE BUOY "NS," S. of SW breaker. Halifax Hbr. appr. – Fl. W. Mo (A) RaRef.
CHEBUCTO HEAD, on summit, Halifax Hbr. appr. – Fl. W. ev. 20 sec. Horn 2 bl. ev. 60 sec. Horn points 113°. Ht. 49.4 m. (163'), Rge. 14 mi. Racon (Z).
HALIFAX ALPHA LT. & WHISTLE BUOY "HA," Halifax appr. – RW vertical stripes, marked "HA," Mo (A), Racon (B).
SAMBRO IS., Halifax Hbr. appr. – Fl. W. ev. 5 sec. 2 Horns 3 bl. ev. 60 sec. Horns point 068° and 178°. Ht. 47 m. (155'), Rge. 21 mi. Radiobeacon.
KETCH HARBOUR BUOY "HE 19," Halifax Hbr. appr. – Fl. G. Whistle. RaRef.
BETTY IS., on Brig. Pt. – Fl. W. ev. 15 sec. Horn 1 bl. ev. 60 sec. Horn points S. Ht. 19 m. (63'), Rge. 14 mi.
E. IRONBOUND IS., ctr. of is. – Iso. W. ev. 6 sec. 2 Horns 1 bl. ev. 30 sec. Ht. 44.5 m. (147'), Rge. 12 mi.
PEARL IS., off St. Margaret & Mahone Bays – Fl. W. ev. 10 sec. Ht. 19.0 m. (63'), Rge. 8 mi.
EAST POINT ISLAND, Mahone Bay – F.G. Ht. 9 m. (30'), Rge. 7 mi.
CROSS IS., E. pt. of is. – Fl. W. ev. 10 sec. Horn 2 bl. ev. 60 sec. Ht. 24.9 m. (82'), Rge. 14 mi.
W. IRONBOUND IS., Entr. to La Have R. – Fl. W. ev. 12 sec. Ht. 24.3 m. (80'), Rge. 8 mi.
MOSHER IS., W. side Entr. to La Have R. – F.W. Horn 1 bl. ev. 20 sec. Ht. 23.3 m. (77'), Rge. 13 mi.
CHERRY COVE, betw. Little Hbr. & Back Cove – F.G. Horn 1 bl. ev. 30 sec. Ht. 9.4 m. (31'), Rge. 13 mi.
MEDWAY HEAD, W. side entr. to Pt. Medway – Fl. W. ev. 12 sec. Ht. 24.2 m. (80'), Rge. 13 mi.
COFFIN IS., S. pt. of is. – Fl. W. ev. 4 sec. Ht. 18.6 m. (60'), Rge. 7 mi.
WESTERN HEAD, W. side entr. to Liverpool Bay – Fl. W. ev. 15 sec. Horn 1 bl. ev. 60 sec. Horn points 104°. Ht. 16.9 m. (55'), Rge. 14 mi.
LITTLE HOPE IS., N. end of is. – Fl. W. ev. 10 sec. Ht. 30.2 m. (100'), Rge. 8 mi.
LOCKEPORT, on Gull Rock, entr. to hbr. – Occ. W. ev. 10 sec. Horn 1 bl. ev. 30 sec. Horn points 130.5°. Ht. 17.1 m. (56'), Rge. 14 mi.
CAPE ROSEWAY, near SE pt. of McNutt – Fl. W. ev. 10 sec. Ht. 33.1 m. (109'), Rge. 14 mi.

Abbreviations used in this text: Lt., Light; LNB, Large Navigation Buoy; Pt., Point; Is., Island; Sh., Shoal; Ch., Channel; Entr., Entrance; Qk., Quick; Fl., Flashing; F., Fixed; Occ., Occulting; Alt., Alternating; W., White; R., Red; G., Green; Y., Yellow; RaRef., Radar Reflector; Mo (A) Morse Code "A," Mo (U) Morse Code "U"; Iso. W., Isophase White; Red sector(s) of Lights warn of dangerous angle of approach. Bearings and ranges are from the observer to the aid.

See p. 235-237 for Radiobeacon Stations and Racon characteristics.

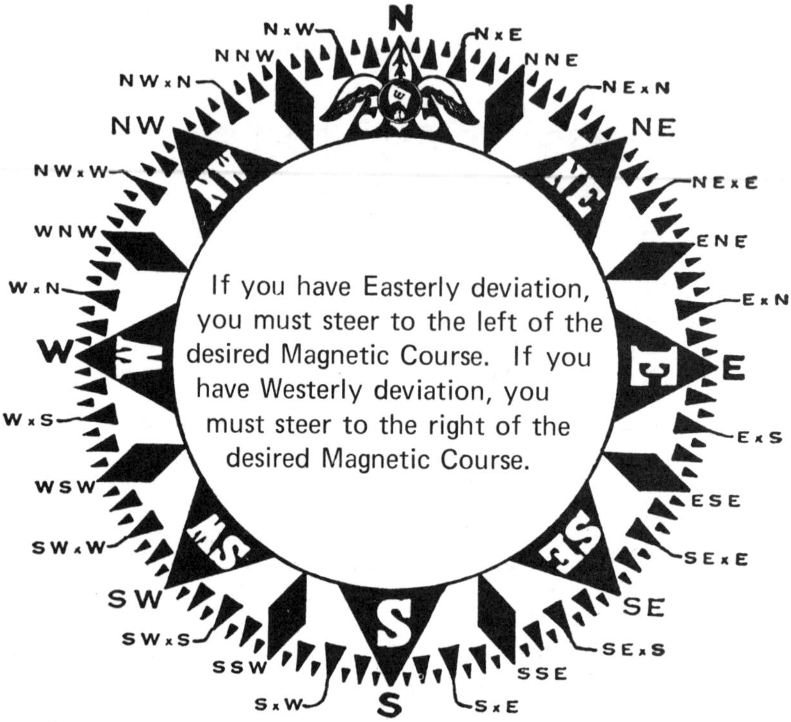

Table for Turning Compass Points into Degrees, and the Contrary

MERCHANT MARINE PRACTICE

NORTH	0	EAST	90	SOUTH	180	WEST	270
N. ¼ E.	2¾	E. ¼ S.	93	S. ¼ W.	182¾	W. ¼ N.	272¾
N. ½ E.	5½	E. ½ S.	95½	S. ½ W.	185½	W. ½ N.	275½
N. ¾ E.	8¼	E. ¾ S.	98½	S. ¾ W.	188½	W. ¾ N.	278½
N. by E.	11¼	E. by S.	101¼	S. by W.	191¼	W. by N.	281¼
N. by E. ¼ E. .	14	E. by S. ¼ S. ..	104	S. by W. ¼ W.	194	W. by N. ¼ N.	284
N. by E. ½ E. .	17	E. by S. ½ S. ..	107	S. by W. ½ W.	197	W. by N. ½ N.	287
N. by E. ¾ E. .	20	E. by S. ¾ S. ..	109¾	S. by W. ¾ W.	199¾	W. by N. ¾ N.	290
N. N. E.	22½	E. S. E.	112½	S. S. W.	202½	W. N. W.	292½
N. E. by N. ¾ N.	25	S. E. by E. ¾ E.	115¼	S. W. by S. ¾ S.	205¼	N. W. by W. ¾ W.	295¼
N. E. by N. ½ N.	28	S. E. by E. ½ E.	118	S. W. by S. ½ S.	208	N. W. by W. ½ W.	298
N. E. by N. ¼ N.	31	S. E. by E. ¼ E.	121	S. W. by S. ¼ S.	211	N. W. by W. ¼ W.	301
N. E. by N.	33¾	S. E. by E.	123¾	S. W. by S.	213¾	N. W. by W. ..	303¾
N. E. ¾ N.	36½	S. E. ¾ E.	126½	S. W. ¾ S.	216½	N. W. ¾ W. ...	306½
N. E. ½ N.	39½	S. E. ½ E.	129½	S. W. ½ S.	219½	N. W. ½ W. ...	309½
N. E. ¼ N.	42	S. E. ¼ E.	132¼	S. W. ¼ S.	222¼	N. W. ¼ W. ...	312¼
N. E.	45	S. E.	135	S. W.	225	N. W.	315
N. E. ¼ E.	47¾	S. E. ¼ S.	137¾	S. W. ¼ W. ...	227¾	N. W. ¼ N. ...	317¾
N. E. ½ E.	50½	S. E. ½ S.	140½	S. W. ½ W. ...	230½	N. W. ½ N. ...	320½
N. E. ¾ E.	53½	S. E. ¾ S.	143½	S. W. ¾ W. ...	233½	N. W. ¾ N. ...	323½
N. E. by E.	56¼	S. E. by S.	146¼	S. W. by W. ...	236¼	N. W. by N. ...	326¼
N. E. by E. ¼ E.	59	S. E. by S. ¼ S.	149	S. W. by W. ¼ W.	239	N. W. by N. ¼ N.	329
N. E. by E. ½ E.	62	S. E. by S. ½ S.	152	S. W. by W. ½ W.	242	N. W. by N. ½ N.	332
N. E. by E. ¾ E.	64¾	S. E. by S. ¾ S.	154¾	S. W. by W. ¾ W.	244¾	N. W. by N. ¾ N.	334¾
E. N. E.	67½	S. S. E.	157½	W. S. W.	247½	N. N. W.	337½
E. by N. ¾ N...	70¼	S. by E. ¾ E...	160¼	W. by S. ¾ S. .	250¼	N. by W. ¾ W.	340¼
E. by N. ½ N. .	73	S. by E. ½ E...	163	W. by S. ½ S. .	253	N. by W. ½ W.	343
E. by N. ¼ N. .	75	S. by E. ¼ E...	166	W. by S. ¼ S. .	256	N. by W. ¼ W.	346
E. by N.	78¾	S. by E.	168¾	W. by S.	258¾	N. by W.	348¾
E. ¾ N.	81½	S. ¾ E.	171½	W. ¾ S.	261½	N. ¾ W.	351½
E. ½ N.	84½	S. ½ E.	174½	W. ½ S.	264½	N. ½ W.	854½
E. ¼ N.	87	S. ¼ E.	177¼	W. ¼ S.	267¼	N. ¼ W.	357¼
East	90	South	180	West	270	North	360

CAPE NEGRO IS., on SE end of is. – Gp. Fl. W. 2 fl. ev. 15 sec. Horn 1 bl. ev. 60 sec. Ht. 27.7 m. (91'), Rge. 16 mi.
THE SALVAGES, SE end of is. – Fl. W. ev. 12 sec. Horn 3 bl. ev. 60 sec. Horn points 116°. Ht. 16.8 m. (55'), Rge. 15 mi.
BACCARO POINT, E. side entr. to Barrington Bay – Gp. Fl. W. 3 fl. ev. 10 sec. Horn 1 bl. ev. 20 sec. Ht. 16 m. (53'), Rge. 16 mi. Horn points 200°.
CAPE SABLE, on cape – Fl. W. ev. 5 sec. Horn 1 bl. ev. 60 sec. Horn points 150°. Ht. 29.5 m. (97'), Rge. 22 mi. Racon (C).
WEST HEAD, Cape Sable Is. – F.R. Horn 2 bl. ev. 60 sec. Horn points 254°. Ht. 15.6 m. (51'), Rge. 6 mi.
OUTER ISLAND, on S. pt. of Outer is. – Fl. W. ev. 10 sec. Horn 1 bl. ev. 20 sec. Horn points 200°. Ht. 14 m. (46'), Rge. 14 mi.
SEAL IS., S. pt. of is. – Fl. W. ev. 10 sec. Emerg. Lt. Horn 3 bl. ev. 60 sec. Horn points 183°. Ht. 31.1 m. (103'), Radiobeacon.

NOVA SCOTIA, WEST COAST

PEASE IS., S. pt. of one of Tusket Is. – Fl. W. ev. 6 sec. Horn 2 bl. ev. 60 sec. Horn points 130°. Ht. 16 m. (53'), Rge. 9 mi.
CAPE FORCHU, E. cape S. pt. Yarmouth Sd. – Gp. Fl. W. 2 fl. ev. 10 sec. Ht. 37.5 m. 124', Rge. 22 mi.
LURCHER SHOAL WEST LIGHT & WHISTLE BUOY "NK," SW of S. end of shoal – Vert. stripes marked "NK," Fl. W. Mo (A), Racon (K).
CAPE ST. MARY, E. side of St. Mary Bay – Fl. W. ev. 5 sec. Horn 1 bl. ev. 60 sec. Horn points 251.5°. Ht. 31.8 m (105'). Rge. 13 mi.
BRIER IS., on W. side of is. R. & W. Tower – Gp. Fl. W. 3 fl. ev. 18 sec. 2 Horns 2 bl. ev. 60 sec. Horns point 270° and 315°. Ht. 18.3 m. (60'), Rge. 17 mi.
BOARS HEAD, W. side of N. entr. to Petit Passage – Fl. W. ev. 12 sec. Horn 3 bl. ev. 60 sec. Horn points 315°. Ht. 18.3 m. (60'), Rge. 12 mi.
PRIM PT., Digby Gut, W. pt. of entr. to Annapolis Basin – Iso. W. ev. 6 sec. Horn 1 bl. ev. 30 sec. Horn points 318°. Ht. 24.8 m. (82'), Rge. 13 mi.
ILE HAUTE, on highest pt. – Fl. W. ev. 4 sec. Ht. 108.5 m. (358')

NEW BRUNSWICK COAST

GRINDSTONE IS., W. end of is. – Fl. W. ev. 4 sec. Ht. 18.6 m. (61'), Rge. 7 mi.
CAPE ENRAGE, at pitch of cape – Fl. G. ev. 5 sec. Horn 3 bl. ev. 60 sec. Horn points 220°. Ht. 38 m. (125'), Rge. 13 mi.
QUACO, tower on head – Fl. W. ev. 10 sec. Horn 1 bl. ev. 30 sec. Horn points 130°. Ht. 28.7 m. (95'), Rge. 14 mi.
CAPE SPENCER, R. & W. tower – Fl. W. ev. 10 sec. Horn 3 bl. ev. 60 sec. Horn points 165°. Ht. 61.6 m. (203'), Rge. 14 mi.
PARTRIDGE IS., Saint John Harbour – Fl. W. ev. 7.5 sec. Horn 2 bl. ev. 60 sec. Horn points 165°. Ht. 35.3 m. (116'), Rge. 17 mi. Radiobeacon.
MUSQUASH, E. side entr. to Musquash Hbr. – Fl. W. ev. 3 sec. Horn 1 bl. ev. 60 sec. Horn points 180°. Ht. 36.6 m. (121'), Rge. 14 mi.
PT. LEPREAU, on point – Fl. W. ev. 5 sec. Horn 3 bl. ev. 60 sec. Horn points 190°. Ht. 25.6 m. (84'), Rge. 14 mi.
SW WOLF IS., on SE pt. – Fl. W. ev. 10 sec. Ht. 38.7 m. (128'), Rge. 15 mi.
PEA PT., E. side entr. to Letang Hbr. – F.W. visible 251° thru N & E to 161°. Horn 2 bl. ev. 60 sec. Horn points 180°. Ht. 17.3 m. (57'), Rge. 12 mi.
HEAD HARBOUR, outer rock of E. Quoddy Head – F.R. Horn 1 bl. ev. 60 sec. Horn points 116°. Ht. 17.6 m. (58'), Rge. 13 mi.
SWALLOWTAIL, NE pt. of Grand Manan – Occ. W. ev. 6 sec. Visible 153° thru S & W to 349°. Horn 1 bl. ev. 20 sec. Horn points 100°. Ht. 37 m. (122'), Rge. 12 mi.
GREAT DUCK IS., S. end of is. – Fl. W. ev. 10 sec. Horn 1 bl. ev. 60 sec. Horn points 120°. Ht.

For abbreviations see footnote p. 161.

15.3 m. (50'), Rge. 15 mi.

FUNDY ENTR. S. LIGHT & WHISTLE BUOY "M" – Racon (B). 44-13-36 N x 66- 50-12 W. Mo (A), RW vertical stripes, marked Fundy Entr. South "M."

GANNET ROCK, S. of Grand Manan – Fl. W. ev. 6 sec. Horn 3 bl. ev. 60 sec. Horn points 162°. Ht. 28.2 m. (93'), Rge. 19 mi. Racon (G).

SOUTHWEST HEAD, S. end of Grand Manan – Fl. W. ev. 10 sec. Horn 1 bl. ev. 60 sec. Horn points 240°. Ht. 47.5 m. (157'), Rge. 14 mi.

MACHIAS SEAL IS., on is. summit – Fl. W. ev. 3 sec. Horn 2 bl. ev. 60 sec. Horn points 270°. Ht. 25 m. (83'), Rge. 17 mi.

MAINE

WEST QUODDY HEAD, Entr. Quoddy Roads – Gp. Fl. W. 2 fl. ev. 15 sec. Horn 2 bl. ev. 30 sec. Ht. 83', Rge. 18 mi.

LITTLE RIVER, at mouth of river. NR on skeleton tower – Fl. W. ev. 6 sec. Horn 1 bl. ev. 10 sec. Ht. 56', Rge. 13 mi.

LIBBY ISLAND, Entr. Machias Bay – Gp. Fl. W. 2 fl. ev. 20 sec. Obscured from 208°-220°. Horn 1 bl. ev. 15 sec. Ht. 91', Rge. 25 mi. Emerg. Lt. of reduced intensity.

MOOSE PEAK, E. end Mistake Is. – Fl. W. ev. 30 sec. Horn 2 bl. ev. 30 sec. Ht. 72', Rge. 25 mi.

PETIT MANAN, E. pt. of is. – Fl. W. ev. 10 sec. Horn 1 bl. ev. 30 sec. Ht. 123', Rge. 26 mi.

MOUNT DESERT, 20 mi. S. of island – Fl. W. ev. 15 sec. Horn 2 bl. ev. 30 sec. Ht. 75', Rge. 18 mi.

GREAT DUCK ISLAND, S. end of island – Fl. R. ev. 5 sec. Horn 1 bl. ev. 15 sec. Ht. 67', Rge. 19 mi.

EGG ROCK, Frenchman Bay – Fl. R. ev. 5 sec. Horn 2 bl. ev. 30 sec. Ht. 64', Rge. 14 mi.

BAKER ISLAND, SW Entr. Somes Sound – Fl. W. ev. 10 sec. Ht. 105', Rge. 10 mi.

BASS HARBOR HEAD, SW pt. Mt. Desert Is. – Occ. R. ev. 4 sec. Ht. 56', Rge. 13 mi.

BLUE HILL BAY #3, on Green Is. – Fl. G. ev. 4 sec. Ht. 25', Rge. 5 mi. SG on tower.

BURNT COAT HARBOR, on white tower connected to dwelling – Occ. W. ev. 4 sec. Ht. 75', Rge. 9 mi.

HALIBUT ROCKS, Jericho Bay – Fl. W. ev. 6 sec. Horn 1 bl. ev. 10 sec. Ht. 25', Rge. 6 mi. NR on tower.

CROTCH ISLAND #21, Deer Is. Thorofare – Fl. G. ev. 4 sec. Ht. 20', Rge. 5 mi.

SADDLEBACK LEDGE, Isle au Haut Bay – Fl. W. ev. 6 sec. Ht. 54', Rge. 9 mi., Horn 1 bl. ev. 10 sec.

ISLE AU HAUT, Isle au Haut Bay – Fl. R. ev. 4 sec. W. Sector: 034°-060°. Ht. 48', Rge. R. Fl. 6 mi., W. sector 8 mi.

DEER ISLAND THOROFARE, W. end of thorofare – Fl. W. ev. 6 sec. Horn 1 bl. ev. 15 sec. Ht. 52', Rge. 8 mi.

GOOSE ROCKS, E. Entr. Fox Is. Thorofare – Fl. R. ev. 6 sec. W. Sector: 301°-304°. Horn 1 bl. ev. 10 sec. Ht. 51', Rge. R. Fl. 11 mi., W. Sector 12 mi. No light in daytime when horn operating.

EAGLE ISLAND, E. Penobscot Bay – Fl. W. ev. 4 sec. Ht. 106', Rge. 9 mi.

GREEN LEDGE #4, E. Penobscot Bay – Fl. R. ev. 6 sec. Ht. 31', Rge. 5 mi.

HERON NECK, E. Entr. Hurricane Sound – F.R. W. Sector: 030°-063°. Horn 1 bl. ev. 30 sec. Ht. 92', Rge. F.R. 10 mi., W. Sector 13 mi.

MATINICUS ROCK, Penobscot Bay Appr. – Fl. W. ev. 10 sec. Ht. 90', Rge. 20 mi.

TWO-BUSH ISLAND, Two-Bush Chann. – Fl. W. ev. 5 sec. R. Sector: 061°-247°. Light on continuously. Horn 1 bl. ev. 15 sec. Ht. 65', Rge. R. Sector 18 mi., W. Fl. 22 mi. Emerg. Lt.

WHITEHEAD, W. side of S. entr. Muscle Ridge Chann. – Occ. G. ev. 4 sec. Horn 2 bl. ev. 30 sec. Ht. 75', Rge. 10 mi.

MONROE ISLAND, W. Penobscot Bay – Fl. W. ev. 4 sec. Ht. 29', Rge. 6 mi.

OWL'S HEAD, S. side Rockland Entr. – F.W. Horn 2 bl. ev. 20 sec. Ht. 100', Rge. 16 mi. Obscured from 324°-354° by Monroe Island.

ROCKLAND BREAKWATER, S. end of breakwater – Fl. W. ev. 5 sec. Horn 1 bl. ev. 15 sec. Ht. 39', Rge. 17 mi.

For abbreviations see footnote p. 161.

The Ship's Bell Code

Telling time by ship's bell has a romantic background that goes back hundreds of years. It is based on the workday routine of the ship's crew. A ship at sea requires a constant watch throughout the whole twenty-four hours of the day. To divide the duty, the day is broken up into six watches of four hours each and the crew into three divisions, or watches.

Each division of the crew stands two four hour watches a day. In order to rotate the duty, so that a division does not have the same watch day in and day out, the 4 to 8 watch in the afternoon is divided into two watches known as the dog watches.

The Mid-Watch - Midnight to 4 A.M.
The Morning Watch - 4 A.M. to 8 A.M.
The Forenoon Watch - 8 A.M. to 12 M.
The Afternoon Watch - 12 M. to 4 P.M.
The 1st Dog Watch - 4 P.M. to 6 P.M.
The 2nd Dog Watch - 6 P.M. to 8 P.M.
The First Watch - 8 P.M. to Midnight

To apprise the crew of the time, the ship's bell was struck by the watch officer at half hour intervals, the first half hour being one bell, the first hour two bells, hour and a half three bells, and so on up to eight bells, denoting time to relieve the watch. By this method of timekeeping eight bells marks 4, 8, or 12 o'clock.

8 Bells	4:00	8:00	12:00
1 Bell	4:30	8:30	12:30
2 Bells	5:00	9:00	1:00
3 Bells	5:30	9:30	1:30
4 Bells	6:00	10:00	2:00
5 Bells	6:30	10:30	2:30
6 Bells	7:00	11:00	3:00
7 Bells	7:30	11:30	3:30

Courtesy of Chelsea Clock Co., Chelsea, MA

LOWELL ROCK #2, Rockport Entr. – Fl. R. ev. 6 sec. Ht. 25', Rge. 5 mi.
BROWNS HEAD, W. Entr. Fox Is. Thorofare – F.W. 2 R. Sectors: 001°-050°; 061°-091°. Horn 1 bl. ev. 10 sec. Ht. 39', Rge. R. sector 11 mi., F.W. 14 mi.
CURTIS ISLAND, S. side Camden Entr. – Occ.G. ev. 4 sec. Ht. 52', Rge. 6 mi.
NORTHEAST POINT #2, Camden Entr. – Fl. R. ev. 4 sec. Ht. 20', Rge. 5 mi.
STEELS LEDGE MONUMENT #4, E. side Belfast Entr. – Fl. R. ev. 4 sec. Ht. 24', Rge. 4 mi.
DICE HEAD, N. side Entr. to Castine – Fl. W. ev. 6 sec. Ht. 27', Rge. 11 mi.
FORT POINT, W. side Entr. to Penobscot R. – F.W. Horn 1 bl. ev. 10 sec. Ht. 88', Rge. 15 mi.
MARSHALL POINT, E. side of Pt. Clyde Hbr. S. Entr. – F.W. Horn 1 bl. ev. 10 sec. Ht. 30', Rge. 13 mi.
MONHEGAN ISLAND, Penobscot Bay – Fl. W. ev. 15 sec. Ht. 178', Rge. 20 mi.
MANANA ISLAND FOG SIGNAL STATION, Penobscot Bay – Horn 2 bl. ev. 60 sec.
FRANKLIN ISLAND, Muscongus Bay – Fl. W. ev. 6 sec. Ht. 57', Rge. 8 mi.
PEMAQUID POINT, W. side Muscongus Bay Entr. – Fl. W. ev. 6 sec. Ht. 79', Rge. 14 mi.
RAM ISLAND, Fisherman Is. Passage S. side – Iso. R. ev. 6 sec. 2 W. Sectors: 258°-261°; 30°-46°. Covers fairway. Horn 1 bl. ev. 30 sec. Ht. 36', Rge. R. Fl. 9 mi., W. Sector 11 mi.
BURNT ISLAND, Boothbay Hbr. W. side Entr. – Fl. R. ev. 6 sec., 2 W. Sectors: 307°-316°; 355°-008°. Covers fairways. Horn 1 bl. ev. 10 sec. Ht. 61', Rge. R. Fl. 12 mi., W. sectors 15 mi.
THE CUCKOLDS, Boothbay – Gp. Fl. W. 2 fl. ev. 6 sec. Horn 1 bl. ev. 15 sec. Ht. 59', Rge. 12 mi. Emerg. Lt.
SEGUIN, 2 mi. S. of Kennebec R. mouth – F.W. Horn 2 bl. ev. 20 sec. Ht. 180', Rge. 18 mi.
HENDRICKS HEAD, Sheepscot R. mouth E. side – F.W. R. Sector: 180°-000°. Ht. 43', Rge. R. sector 7 mi., F.W. 9 mi.
POND ISLAND, Kennebec R. mouth W. side – Iso. W. ev. 6 sec. Horn 2 bl. ev. 30 sec. Ht. 52', Rge. 9 mi.
FULLER ROCK, off Cape Small – Fl. W. ev. 4 sec. Ht. 39', Rge. 6 mi.
WHALEBOAT, Broad Sd., Casco Bay – Fl. W. ev. 6 sec. Rge. 6 mi. Ht. 47'.
COW ISLAND, Portland to Merepoint – Fl. W. ev. 6 sec. Ht. 23'. Rge. 8 mi. RaRef.
COUSINS ISLAND #5, Portland to Merepoint – Fl. W. ev. 2.5 sec. Ht. 24', Rge. 5 mi.
HALFWAY ROCK, midway betw. Cape Small Pt. and Cape Eliz. – Fl. R. ev. 5 sec. Horn 2 bl. ev. 30 sec. Ht. 77', Rge. 19 mi.
PORTLAND LIGHTED HORN BUOY "P", Portland Hbr. Appr. – Mo (A) W. Horn 1 bl. ev. 30 sec. Ht. 42', Rge. 6 mi. Racon "M." passing lt. Fl. W. ev. 2.5 sec.
RAM ISLAND LEDGE, N. side of Portland Hbr. Entr. – Gp. Fl. W. 2 fl. ev. 6 sec. Horn 1 bl. ev. 10 sec. Ht. 77', Rge. 12 mi. Emerg. Lt. of reduced intensity.
CAPE ELIZABETH, S. of Portland Hbr. Entr. – Gp. Fl. W. 4 fl. ev. 15 sec. Horn 2 bl. ev. 60 sec. Ht. 129', Rge. 27 mi.
PORTLAND HEAD, SW side Portland Hbr. Entr. – Fl. W. ev. 4 sec. Horn 1 bl. ev. 15 sec. Ht. 101', Rge. 24 mi.
SPRING POINT LEDGE, Portland main Chann. W. side – Fl. W. ev. 6 sec. 2 R. Sectors. 2 W. Sectors: 331°-337° and 074°-288°. Horn 1 bl. ev. 10 sec. Ht. 54', Rge. R. sectors 9 mi., W. Fl. 12 mi.
WOOD ISLAND, S. Entr. Wood Is. Hbr. N. side – Alt. W. and G. ev. 10 sec. Horn 2 bl. ev. 30 sec. Ht. 71', Rge. W. 16 mi., G. 14 mi.
GOAT ISLAND, Cape Porpoise Hbr. Entr. – Fl. W. ev. 6 sec. Horn 1 bl. ev. 15 sec. Ht. 38', Rge. 12 mi.
CAPE NEDDICK, On N. side of Nubble – Iso. R. ev. 6 sec. Horn 1 bl. ev. 10 sec. Ht. 88', Rge. 13 mi.
BOON ISLAND, 6.5 mi. off coast – Fl. W. ev. 5 sec. Horn 1 bl. ev. 10 sec. Ht. 137', Rge. 19 mi.

NEW HAMPSHIRE

WHALEBACK, Portsmouth Entr. NE side – Gp. Fl. W. 2 fl. ev. 10 sec. Horn 2 bl. ev. 30 sec. Ht. 59', Rge. 24 mi.
PORTSMOUTH HARBOR (NEWCASTLE), on Fort Point – F. G. Horn 1 bl. ev. 10 sec. Ht. 52', Rge. 12 mi. Radiobeacon.

For abbreviations see footnote p. 161.

ISLES OF SHOALS, 5.5 mi. off coast – Fl. W. ev. 15 sec. Horn 1 bl. ev. 30 sec. Ht. 82', Rge. 20 mi.

MASSACHUSETTS

NEWBURYPORT HARBOR, N. end of Plum Is. – Gp. Occ. G. 2 fl. ev. 15 sec. Obscured from 165°-192° and from 313°-344°. Ht. 50', Rge. 10 mi.

IPSWICH, Ipswich Entr. S. side – Occ. W. ev. 4 sec. Ht. 30', Rge. 9 mi.

ROCKPORT BREAKWATER #6, W. side Entr. Rockport inner hbr. – Fl. R. ev. 4 sec. Ht. 32', Rge. 5 mi.

ANNISQUAM HARBOR, E. side Entr. – Fl. W. ev. 7.5 sec. R. Sector: 180°-217°. Horn 2 bl. ev. 60 sec. Ht. 45', Rge. R. sector 11 mi., W. Fl. 14 mi.

STRAITSMOUTH, Rockport Entr. S. side – Fl. G. ev. 6 sec. Horn 1 bl. ev. 15 sec. Ht. 46', Rge. 6 mi.

CAPE ANN, E. side Thacher Is. – Fl. R. ev. 5 sec. Horn 2 bl. ev. 60 sec. Ht. 166', Rge. 19 mi.

EASTERN POINT, Gloucester Entr. E. side – Fl. W. ev. 5 sec. Ht. 57', Rge. 24 mi.

GLOUCESTER BREAKWATER, W. end – Occ. R. ev. 4 sec. Horn 1 bl. ev. 10 sec. Ht. 45', Rge. 10 mi.

BAKERS ISLAND, Salem Channel – Alt. Fl. W. and R. ev. 20 sec. Horn 1 bl. ev. 30 sec. Ht. 111', Rge. R. Fl. 13 mi., W. Fl. 16 mi.

HOSPITAL POINT RANGE, Beverly Cove W. side – F.W. Ht. 70'.

MARBLEHEAD, N. point Marblehead Neck – F.G. Ht. 130', Rge. 7 mi.

THE GRAVES, Boston Hbr. S. Chann. Entr. – Gp. Fl. W. 2 fl. ev. 12 sec. Horn 2 bl. ev. 20 sec. Ht. 98', Rge. 24 mi.

DEER ISLAND, President Roads, Boston Hbr. – Alt. W. and R. ev. 10 sec. R. Sector: 198°-222°. Obscured 112°-186°. Horn 1 bl. ev. 10 sec. Ht. 53', Rge. R. Fl. 10 mi., W. Fl. 14 mi.

DEERISLAND DANGER LT., F.R., R. Sector 198°-122°T.

LONG ISLAND HEAD, President Roads, Boston Hbr. – Fl. W. ev. 2.5 sec. Ht. 120', Rge. 6 mi.

BOSTON RW LIGHTED HORN BUOY "B," Boston Hbr. Entr. – Mo (A). Horn 2 bl. ev. 60 sec. Ht. 42', Rge. 7 mi., Racon (B).

BOSTON RW LTD. WH. BY. "NC" – MO (A) W, Rge. 6 mi., Racon (N).

MINOTS LEDGE, Boston Hbr. Entr. S. side – Gp. Fl. W. ev. 45 sec. Fl. 1-4-3. Horn 1 bl. ev. 10 sec. Ht. 85', Rge. 10 mi.

BOSTON, SE side Little Brewster Is. – Fl. W. ev. 10 sec. Horn 1 bl. ev. 30 sec. Ht. 102', Rge. 27 mi.

PLYMOUTH (GURNET), N. side Entr. to hbr. – Gp. Fl. W. 3 fl. ev. 30 sec. R. Sector: 323°-352°. Horn 1 bl. ev. 15 sec. Ht. 102', Rge. R. sector 14 mi., W. Fl. 17 mi.

RACE POINT, NW Point of Cape Cod – Obscured 220°-292° True. Fl. W. ev. 10 sec. Horn 2 bl. ev. 60 sec. Ht. 41', Rge. 16 mi. Emerg. Lt. of reduced intensity.

WOOD END, Entr. to Provincetown – Fl. R. ev. 10 sec. Horn 1 bl. ev. 30 sec. Ht. 45', Rge. 13 mi.

LONG POINT, Provincetown Entr. SW side – Occ. G. ev. 4 sec. Horn 1 bl. ev. 15 sec. Ht. 36', Rge. 8 mi.

CAPE COD CANAL BREAKWATER #6, E. Entr. – Fl. R. ev. 5 sec. Horn 1 bl. ev. 15 sec. Ht. 43', Rge. 17 mi.

HIGHLAND LIGHT, NE side of Cape Cod – Fl. W. ev. 5 sec. Horn 1 bl. ev. 15 sec. Ht. 183', Rge. 18 mi.

NAUSET BEACH, E. side of Cape Cod – Alt. Fl. W. and R. 2 fl. ev. 10 sec. (Tides divide and run in opposite directions abreast of light). Ht. 114', Rge. W. 23 mi., R. 20 mi.

CHATHAM, W. side of hbr. – Gp. Fl. W. 2 fl. ev. 10 sec. Ht. 80', Rge. 24 mi. Radiobeacon.

CHATHAM INLET BAR GUIDE LT., Fl. Y. ev. 2.5 sec. Ht. 62', Rge. 11 mi.

POLLOCK RIP ENTRANCE LIGHTED WHISTLE BUOY "PR" – R. & W. vert. striped buoy, Mo (A) W. Rge. 6 mi. RaRef.

POLLOCK RIP CHANNEL LIGHTED BUOY #8 – FL. R. ev. 6 sec. Rge. 4 mi. RaRef.

For abbreviations see footnote p. 161.

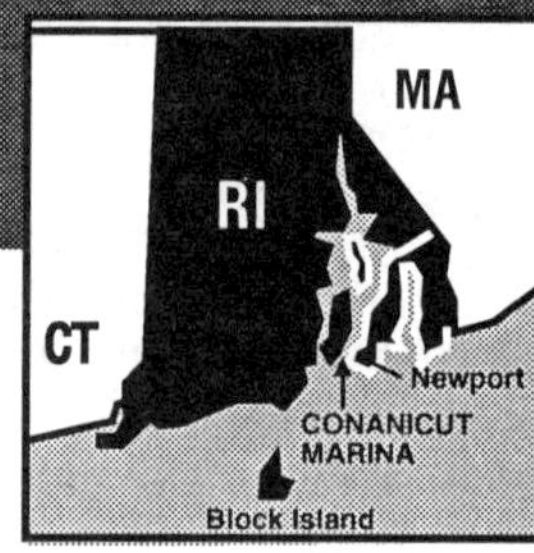

CONANICUT MARINA

- At the entrance of Narragansett Bay
- Quaint village location, ships' store
- Launch service, fuel, showers
- Reservable moorings & dockage
- *New service!* 200 ton marine railway
- Group rendezvous welcome!

CONANICUT MARINE SERVICES INC.

Monitoring marine channels 09 & 71.

One Ferry Wharf • Jamestown, RI • 401- 423-1556 ◆ FAX 423-7152

NANTUCKET SOUND

GREAT ROUND SHOAL CH. ENTR. LTD. WHISTLE BUOY "GRS" – R & W vert. stripes, Fl. W. Mo (A). Rge. 6 mi. Racon (G).
GREAT POINT, NANTUCKET, N. end of is. White tower, Ht. 70'. – Fl. W. ev. 5 sec. R. sector 084°-106°. Rge. W. 14 mi., R. 12 mi.
SANKATY HEAD, E. end of is. – Fl. W. ev. 7.5 sec. Ht. 158', Rge. 24 mi.
NANTUCKET EAST BREAKWATER #3, Outer Entr. to hbr. – Fl. G. ev. 4 sec. Horn 1 bl. ev. 10 sec. Ht. 30', Rge. 9 mi.
BRANT POINT, Hbr. Entr. W. side – Occ. R. ev. 4 sec. Horn 1 bl. ev. 10 sec. Ht. 26', Rge. 10 mi.
CAPE POGE, NE point of Chappaquiddick Is. – Fl. W. ev. 6 sec. Ht. 65', Rge. 9 mi.
EDGARTOWN HARBOR, Inner end of hbr. W. side – Fl. R. ev. 6 sec. Ht. 45', Rge. 5 mi.
EAST CHOP, E. side Vineyard Haven Hbr. Entr. – Iso. G. ev. 12 sec. Ht. 79', Rge. 9 mi.
WEST CHOP, W. side Vineyard Haven Hbr. Entr. – Occ. W. ev. 4 sec. R. Sector: 281°-331°. Horn 1 bl. ev. 30 sec. Ht. 84', Rge. R. Sector 11 mi., W. 15 mi.
NOBSKA POINT, Woods Hole E. Entr. – Fl. W. ev. 6 sec. R. Sector: 263°-289°. Horn 2 bl. ev. 30 sec. Ht. 87', Rge. R. Sector 12 mi., W. Fl. 16 mi. Radiobeacon.
TARPAULIN COVE, SE side Naushon Is. – Fl. W. ev. 6 sec. Ht. 78', Rge. 9 mi.
CUTTYHUNK, SW point of is. – Qk. Fl. W. Ht. 63', Rge. 8 mi.
GAY HEAD, W. point of Martha's Vineyard – Alt. W. and R. ev. 15 sec. Ht. 170', Rge. W. Fl. 24 mi., R. Fl. 20 mi. Obscured 342-359° by Nomans Land.
NANTUCKET SHOALS LTD. HORN BUOY "N," R. & W. – Mo (A) W Rge. 6 mi. Ht. 40'. Horn 1 bl. ev. 30 sec. Racon (N).

BUZZARDS BAY

NARRAGANSETT - BUZZARDS BAY APPR. LTD. WHISTLE BUOY "A" – RW vert. stripe with top mark. Mo (A) W. Racon (N).
BUZZARDS BAY ENTRANCE LIGHT, W. Entr. – Fl. W. ev. 2.5 sec. Horn 2 bl. ev. 30 sec. Ht. 67', Rge. 17 mi. Racon (B).
DUMPLING ROCKS #7, off Round Hill Pt. – Fl. G. ev. 6 sec. Ht. 52', Rge. 8 mi.
NEW BEDFORD WEST BARRIER LIGHT – Qk. Fl. G. Horn 1 bl. ev. 10 sec. Ht. 48', Rge. 8 mi.
NEW BEDFORD EAST BARRIER LIGHT – Qk. Fl. R. Ht. 48', Rge. 5 mi.
CLEVELAND LEDGE, Cape Cod Canal Appr. E. side of S. Entr. Fl. W. ev. 10 sec. Horn 1 bl. ev. 15 sec. Ht. 74', Rge. 17 mi. Racon (C).
WESTPORT HARBOR ENTR. #7, W. side – Fl. G. ev. 6 sec. Ht. 35', Rge. 9 mi.

RHODE ISLAND

SAKNNNET LT., Fl. W. ev. 6 sec. R. sector 195°- 350°. Ht. 58', Rge. W. Fl. 7 mi., R. sector 5 mi.
SAKONNET BREAKWATER #2, Entr. to hbr. – Fl. R. ev. 4 sec. Ht. 34', Rge. 8 mi.
NARRAGANSETT BAY ENTR. LTD. HORN BUOY "NB" - R. & W. stripes, F.W. and Fl. W. ev. 4 sec. Horn 1 bl. ev. 15 sec. Racon (B). Passing lt. Mo (A).
BEAVERTAIL, Narrag. Bay E. passage – Fl. W. ev. 6 sec. Horn 1 bl. ev. 30 sec. Ht. 64', Rge. 15 mi.
CASTLE HILL, Narrag. Bay E. passage – Iso. R. ev. 6 sec. Horn 1 bl. ev. 10 sec. Ht. 40', Rge. 12 mi.
FORT ADAMS #2, Narrag. Bay E. passage – Fl. R. ev. 6 sec. Horn 1 bl. ev. 15 sec. Ht. 32', Rge. 7 mi.
GOAT ISLAND SHOAL #5, Newport Hbr. – Fl. G. ev. 6 sec. Horn 2 bl. ev. 20 sec. Ht. 35', Rge. 6 mi.
NEWPORT HARBOR, N. end of breakwater – F.G. Ht. 33', Rge. 11 mi.
ROSE IS. LT., Fl W. ev. 6 sec. Ht. 48', Rge. 8 mi. (private aid).
PRUDENCE ISLAND (SANDY POINT), Narrag. Bay E. passage – Fl. G. ev. 6 sec. Ht. 28', Rge. 6 mi.
HOG ISLAND SHOAL, N. side Entr. to Mt. Hope Bay – Iso. W. ev. 6 sec. Horn 2 bl. ev. 30 sec. Ht. 54', Rge. 12 mi.

For abbreviations see footnote p. 161.

MUSSELBED SHOALS #6A, Mt. Hope Bay chann. – Fl. R. ev. 6 sec. Ht. 26', Rge. 6 mi.
CASTLE ISLAND #2, N. of Hog Is. – Fl. R. ev. 6 sec. Ht. 25', Rge. 5 mi.
BRISTOL HARBOR #4 – F.R. Ht. 25', Rge. 11 mi.
CONIMICUT, Providence R. Appr. – Fl. W. ev. 2.5 sec. R. Sector: 322°-349°. Horn 2 bl. ev. 30 sec. Ht. 58', Rge. R. sector 13 mi., W. Fl. 15 mi.
BULLOCK POINT "BP," Prov. R. – Occ. W. ev. 4 sec. Ht. 29', Rge. 6 mi.
POMHAM ROCK LIGHT, Prov. R. – F.R. Ht. 54', Rge. 6 mi.
PROVIDENCE RIVER CHANNEL LIGHT #42, off rock – Iso. R. ev. 6 sec. Ht. 31', Rge. 4 mi.
WICKFORD HARBOR #1, Narrag. Bay W. passage – Fl. G. ev. 6 sec. Ht. 20', Rge. 6 mi.
WARWICK, Greenwich Bay Appr. – Occ. G. ev. 4 sec. Horn 1 bl. ev. 15 sec. Ht. 66', Rge. 12 mi.
POINT JUDITH, Block Is. Sd. Entr. – Gp. Occ. W. 3 eclipses ev. 15 sec. Horn 1 bl. ev. 15 sec. Ht. 65', Rge. 16 mi.
PT. JUDITH HARBOR OF REFUGE, W. Entr. Lt. #3 – Fl. G. ev. 6 sec. Horn 1 bl. ev. 30 sec. Ht. 35', Rge. 6 mi.
BLOCK ISLAND NORTH, N. end of is. – Fl. W. ev. 5 sec. Ht. 58', Rge. 13 mi.
BLOCK ISLAND SOUTHEAST, SE end of is. – Fl. G. ev. 5 sec. Ht. 261', Rge. 20 mi. Horn 1 bl. ev. 30 sec.

FISHERS ISLAND SOUND

WATCH HILL LIGHT, Fishers Is. Sd. E. Entr. – Alt. W. and R. ev. 5 sec. Horn 1 bl. ev. 30 sec. Ht. 61', Rge. R. Fl. 14 mi., W. Fl. 16 mi.
LATIMER REEF, Fishers Is. Sd. main chann. – Fl. W. ev. 6 sec. Bell 2 strokes ev. 15 sec. Ht. 55', Rge. 9 mi.
N. DUMPLING, Fishers Is. Sd. main chann. – F.W. Horn 1 bl. ev. 30 sec. R. Sector: 257°-023°. Ht. 94', Rge. R. sector 7 mi., F.W. 9 mi.
STONINGTON OUTER BREAKWATER #4 – Fl. R. ev. 4 sec. Horn 1 bl. ev. 10 sec. Ht. 46', Rge. 5 mi.

LONG ISLAND SOUND, NORTH SIDE

RACE ROCK, SW end of Fishers Is. – Fl. R. ev. 10 sec. Horn 2 bl. ev. 30 sec. Ht. 67', Rge. 19 mi.
BARTLETT REEF, S. end of reef – Fl. W. ev. 5 sec. Horn 2 bl. ev. 60 sec. Ht. 35', Rge. 12 mi.
NEW LONDON LEDGE, N. London Entr. E. side – Gp. Fl. W. Alt. R. Fl. ev. 30 sec. (3 W. Fl., 1 R. Fl.) Horn 2 bl. ev. 20 sec. Ht. 58', Rge. W. Fl. 17 mi., R. Fl. 14 mi.
NEW LONDON HARBOR, W. side Entr. – Iso. W. ev. 6 sec. R. Sector: 000°-041°. Ht. 89', Rge. R. sector 14 mi. W. Fl. 17 mi.
SAYBROOK BREAKWATER, W. jetty – Fl. G. ev. 6 sec. Horn 1 bl. ev. 30 sec. Ht. 58'. Rge. 11 mi.
LYNDE POINT, Conn. R. mouth W. side – F.W. Ht. 71', Rge. 14 mi.
FALKNER ISLAND, off Guilford Hbr. – Fl. W. ev. 10 sec. Ht. 94', Rge. 13 mi.
BRANFORD REEF, SE Entr. New Haven – Fl. W. ev. 6 sec. Ht. 30', Rge. 7 mi.
SOUTHWEST LEDGE, E. side Entr. New Haven – Fl. R. ev. 5 sec. Horn 1 bl. ev. 15 sec. Ht. 57', Rge. 13 mi.
STRATFORD POINT, W. side Entr. Housatonic R. – Gp. Fl. W. 2 fl. ev. 20 sec. If disabled, emerg. lt. 1 fl. ev. 6 sec. Ht. 52', Rge. 16 mi. Radiobeacon.
STRATFORD SHOAL, Middle Ground – Fl. W. ev. 5 sec. Horn 1 bl. ev. 15 sec. Ht. 60', Rge. 13 mi. Racon (Y).
TONGUE POINT, at Bridgeport Brkwtr. – Fl. G. ev. 4 sec. Ht. 31', Rge. 5 mi.
PENFIELD REEF, S. side Entr. to Black Rock – Fl. R. ev. 6 sec. Horn 1 bl. ev. 15 sec. Ht. 51', Rge. 15 mi.
PECK LEDGE #7, E. Appr. to Norwalk – Fl. W. ev. 4 sec. Ht. 61', Rge. 7 mi.
GREENS LEDGE, W. end of ledge – Alt. Fl. W. and R. ev. 24 sec. Horn 2 bl. ev. 20 sec. Ht. 62', Rge. R. Fl. 15 mi., W. Fl. 18 mi.
STAMFORD HARBOR, Ledge Obstruction Light – Fl. W. ev. 4 sec. Ht. 80'.
STAMFORD HBR. WEST BRKWTR. LT. #3 – Iso. G. ev. 6 sec. Horn 1 bl. ev. 15 sec. Ht. 47', Rge. 5 mi.

For abbreviations see footnote p. 161.

GREAT CAPTAIN ISLAND, SE pt. of is. – Alt. Fl. W. and R. ev. 12 sec. Horn 1 bl. ev. 15 sec. If disabled, emerg. lt. Fl. W. ev. 6 sec. Ht. 62', Rge: R. Fl. 12 mi., W. Fl. 15 mi.

LARCHMONT HARBOR #2, S. end of breakwater – Fl. R. ev. 4 sec. Ht. 19', Rge. 4 mi.

LONG ISLAND SOUND, SOUTH SIDE

LITTLE GULL ISLAND, E. Entr. L.I. Sd. – Fl. W. 2 fl. ev. 15 sec. Horn 1 bl. ev. 15 sec. Ht. 91', Rge. 18 mi.

PLUM GUT LT. – on 14' brick shed 55' above MHW. Fl. W. ev. 2.5 sec. Rge. 5 mi.

PLUM ISLAND, W. end of is. – Siren 1 bl. ev. 15 sec. (Maintained by U.S. Agr. Dept.)

ORIENT POINT, outer end of Oyster Pond Reef – Fl. W. ev. 5 sec. Horn 2 bl. ev. 30 sec. Ht. 64', Rge. 17 mi.

HORTON POINT, NW point of Horton Neck – Fl. G. ev. 10 sec. Ht. 103', Rge. 14 mi.

MATTITUCK BREAKWATER "MI" – Fl. W. ev. 4 sec. Ht. 25', Rge. 6 mi.

OLD FIELD POINT – Alt. Fl. R. and G. 2 fl. ev. 24 sec. Ht. 74', Rge. 12 mi.

EATONS NECK, E. side Entr. Huntington Bay – F. W. Horn 1 bl. ev. 30 sec. Ht. 144', Rge. 18 mi.

COLD SPRING HBR. LT., on pt. of shoal – F.W., R. Sector 039°-125°, Ht. 37', Rge. W. Sector 9 mi., R. Sector 7 mi.

GLEN COVE BREAKWATER #5, E. side Entr. to hbr. – Fl. G. ev. 4 sec. Ht. 24', Rge. 4 mi.

NEW YORK HARBOR & APPROACHES

EXECUTION ROCKS – Fl. W. ev. 10 sec. Horn 1 bl. ev. 15 sec. Ht. 62', Rge. 16 mi. Racon (X).

HART ISLAND #46, off S. end of is. – Fl. R. ev. 4 sec. Ht. 23', Rge. 6 mi.

STEPPING STONES, outer end of reef – Occ. G. ev. 4 sec. Ht. 46', Rge. 8 mi.

THROGS NECK, Fort Schuyler – F. R. Ht. 60', Rge. 11 mi.

WHITESTONE POINT #1, East R. main chann. – F.G. Bell 1 stroke ev. 15 sec. Ht. 48', Rge. 10 mi.

HELL GATE #15, East R. Hallets Pt. – Fl. G. ev. 2.5 sec. Ht. 33', Rge. 4 mi.

MILL ROCK SOUTH #16, East R., main chann. – Fl. R. ev. 4 sec. Ht. 37', Rge. 4 mi.

GOVERNORS ISLAND EXTENSION, SW pt. of is. – F. R. Horn 1 bl. ev. 15 sec. Ht. 47', Rge. 9 mi.

GOVERNORS ISLAND – NW pt of is. – 2 F.R. arranged vertically. Horn 2 bl. ev. 20 sec. Ht. 75', Rge. 7 mi.

VERRAZANO-NARROWS BRIDGE FOG SIGNAL – 2 Horns on Bridge. 1 bl. ev. 15 sec.

CONEY ISLAND, N.Y. Hbr. main chann. – Fl. R. ev. 5 sec. Ht. 75', Rge. 16 mi.

ROMER SHOAL, N.Y. Hbr. S. Appr. – Gp. Fl. W. 2 fl. ev. 15 sec. If disabled, emerg. lt. Fl. W. ev. 6 sec. Horn 1 bl. ev. 15 sec. Ht. 54', Rge. 16 mi.

WEST BANK, Ambrose Chann. outer sect. – Iso. W. ev. 6 sec. R. Sector: 001°-184°. Horn 2 bl. ev. 20 sec. Ht. 69', Rge. 7 mi.

OLD ORCHARD SHOAL, N.Y. Hbr. – Fl. W. ev. 6 sec. R. Sector: 087°-203°. Ht. 51', Rge. W. Fl. 7 mi. R. sector 5 mi.

SANDY HOOK, White tower on N. part of Sandy Hook – F. W. Ht. 88', Rge. 19 mi. Radiobeacon.

SANDY HOOK CH. RANGE FRONT LT. – Qk. W., G., and R. sectors. Ht. 30' Racon (C).

SCOTLAND LIGHTED HORN BUOY "S," Sandy Hook Chann. Approach – R. & W. vert. stripes. Fl. W. Mo (A). Horn 1 bl. ev. 30 sec. Rge. 7 mi. Racon (M).

AMBROSE LIGHT STATION – Fl. W. ev. 5 sec. Horn 1 bl. ev. 15 sec. Ht. 136', Rge. 24 mi. Racon (N). Emerg. Lt. of reduced intensity.

LONG ISLAND, OUTSIDE

MONTAUK POINT, E. end of L.I. – Fl. W. ev. 5 sec. Horn 1 bl. ev. 15 sec. If disabled, emerg. Lt. Fl. W. ev. 5 sec. Ht. 168', Rge. 24 mi. Radiobeacon.

SHINNECOCK, W. side of Inlet – Gp. Fl. R. 2 fl. ev. 15 sec. Horn 1 bl. ev. 15 sec. Ht. 75', Rge. 10 mi.

For abbreviations see footnote p. 161.

The Hull Lifesaving Museum - celebrating its 20th birthday

For centuries, the only entrance to Boston Harbor lay along the narrow passageway between the Brewster Islands, marked by Boston Light, and the town of Hull–a seven mile long peninsula pointing north and west toward the City. A sailing ship caught during one of the region's fierce northeasters was in imminent danger of wrecking on Hull's rock-strewn shore, and many ships foundered there.

Because Hull was the scene of so many shipwrecks, it was among the first to have an effective lifesaving service, with surfboats designed for this purpose and crews on constant alert to man them. The Point Allerton U.S. Life Saving Station, headquarters of the Hull Lifesaving Museum, is one of the few lifesaving stations open as a museum in a major metropolitan area. It stands directly across the water from Boston Light, and has been restored to look much as it did when built in 1889.

Near the end of the 19th century, the first Keeper of the Point Allerton Station was Captain Joshua James. The most decorated of U.S. lifesavers, he personally saved more than 600 people from drowning at sea.

Captain James personified a lifesaving tradition that the Hull Lifesaving Museum is dedicated to preserve. It took courage to launch a lifeboat into a raging surf and skill to do so successfully, but most of all it took motivation–caring for human lives. It was skill, courage and caring that made Hull's early townspeople outstanding life savers. These characteristics form the philosophic core of the Hull Lifesaving Museum's purposes and programs.

The Museum is an actual U.S. Lifesaving Station restored as a living museum. It is fully furnished and equipped as if its 19th century crew had just run for their lifeboats.

Among the many important artifacts in the Museum's collection are the 1888 surfboat *Nantasket*, designed by Joshua James' brother Samuel, which is on long-term loan from the Mariners Museum in Virginia. One also finds a 1930s beach cart carrying rescue apparatus, a completely rigged breeches buoy and a three-dimensional scale model of Boston Harbor.

HULL LIFESAVING MUSEUM INC.

P.O. Box 221 • 1117 Nantasket Ave. Hull, MA 02045

Tel: (617)925-LIFE

JONES INLET, end of breakwater – Fl. R. ev. 2.5 sec. Ht. 33', Rge. 5 mi.
FIRE ISLAND, 2 mi. E. of inlet – BW tower. Fl. W. ev. 7.5 sec. Ht. 167', Rge. 24 mi.
ROCKAWAY POINT BREAKWATER LT. #4, end of breakwater – Fl. R. ev. 4 sec. Ht. 34', Rge. 5 mi.

NEW JERSEY

HIGHLANDS – Occ. W. ev. 10 sec. (6 sec. lt.) Ht. 246', Rge. 14 mi.
BARNEGAT INLET, S. breakwater #7 – Qk. G. Ht. 37', Rge. 5 mi., Horn 1 bl. ev. 30 sec.
BARNEGAT LIGHTED HORN BUOY "B" – Fl. Y. ev. 6 sec. Horn 1 bl. ev. 30 sec. Rge. 7 mi. Racon (B).
ABSECON INLET BREAKWATER Lt. #7 – Fl. G. ev. 4 sec. Ht. 29', Rge. 6 mi.
HEREFORD INLET, S. side – Fl. W. ev. 10 sec. Ht. 57', Rge. 24 mi.
FIVE FATHOM BANK LIGHTED BUOY "F," Delaware Bay Entr. – Fl. Y. ev. 2.5 sec. Ht. 36', Rge. 6 mi. Emerg. Lt. of reduced intensity.
CAPE MAY – Fl. W. ev. 15 sec. Ht. 165', Rge. 24 mi.

NEW JERSEY, DELAWARE AND MARYLAND

DELAWARE LIGHTED BUOY "D," Del. Bay – Fl. Y. ev. 6 sec. If disabled, emerg. Lt. Fl. W. ev. 4 sec. Ht. 36', Rge. 7 mi. Racon (K).
BROWN SHOAL LT., on steel tower – Fl. W. ev. 2.5 sec. Ht. 23', Rge. 7 mi. Racon (B).
BRANDYWINE SHOAL, Del. Bay main chann. 9 mi. from S. end of shoal – Fl. W. ev. 10 sec. R. Sector: 151°-338°, 2 sec. Fl., 2 sec. eclipse, 6 sec. Fl., 2 sec. eclipse. Horn 1 bl. ev. 15 sec. Ht. 60', Rge. R. Sector 17 mi., W. Fl. 19 mi.
HARBOR OF REFUGE, Del. Bay main chann., S. end of breakwater – Fl. W. ev. 5 sec. 2 R. Sectors: 325°-351°; 127°-175°. Horn 2 bl. ev. 30 sec. Ht. 72', Rge. R. sector 16 mi., W. Fl. 19 mi.
FOURTEEN FOOT BANK, Del. Bay main chann. W. side – Fl. W. ev. 9 sec. R. Sector: 332.5°-151°. Horn 1 bl. ev. 30 sec. Ht. 59', Rge. R. sector 10 mi., W. Fl. 13 mi.
MIAH MAULL SHOAL, Del Bay main chann. – Occ. W. ev. 4 sec. R. Sector: 137.5°-333°. Horn 1 bl. ev. 10 sec. (Mar. 15-Dec. 15). Ht. 59', Rge. R. sector 12 mi., W. Fl. 15 mi. Racon (M).
ELBOW OF CROSS LEDGE, Del. Bay main chann. – Iso. W. ev. 6 sec. R. Sector: 155°-325°. Horn 2 bl. ev. 20 sec. (Mar. 15-Dec. 15). Ht. 61', Rge. R. sector 11 mi., W. Fl. 14 mi.
SHIP JOHN SHOAL, Del. Bay main chann. – Fl. W. ev. 5 sec. R. Sector: 138°-321°. Horn 1 bl. ev. 15 sec. Ht. 50', Rge. R. sector 12 mi., W. Fl. 16 mi. Racon (O)
FENWICK ISLAND – Occ. W. ev. 13 sec. Ht. 83', Rge. 8 mi.
OCEAN CITY INLET, on end of jetty – Iso. W. ev. 6 sec. Horn 1 bl. ev. 15 sec. Ht. 28', Rge. 9 mi.

VIRGINIA

ASSATEAGUE, S. side of is. – Gp. Fl. W. 2 fl. ev. 5 sec. Ht. 154', Rge. 22 mi.
GREAT MACHIPONGO ENTR. #5, S. side – Fl. G. ev. 4 sec. Ht. 19', Rge. 4 mi.
CAPE CHARLES, N. side of Entr. to Ches. Bay – Fl. W. ev. 5 sec. Ht. 180°, Rge. 18 mi.
CHESAPEAKE LIGHT TOWER, off Entr. to Ches. Bay – Gp. Fl. W. 2 fl. ev. 15 sec. Horn 1 bl. ev. 30 sec. If disabled, emerg. Lt. Qk. Fl. W. Ht. 117', Rge. 18 mi. Racon (N).
CHESAPEAKE BAY ENTR. JCT. LIGHTED GONG BUOY "CBJ" – Gp. Fl. G. 3 fl. ev. 6 sec. Rge. 5 mi.
CAPE HENRY, S. side of Entr. to Ches. Bay – Fl. W. Mo (U), ev. 20 sec. R. Sector: 154°-233°. Ht. 164', Rge. R. sector 15 mi., W. Fl. 17 mi. Emerg. Lt. Radiobeacon.

CHESAPEAKE BAY

THIMBLE SHOAL, Hampton Roads – Fl. W. ev. 10 sec. Horn 1 bl. ev. 30 sec. Ht. 55', Rge. 18 mi.
OLD POINT COMFORT, N. side Entr. to Hampton Roads – Gp. Fl. R. 2 fl. ev. 12 sec. W. Sector: 265°-038°. If disabled, emerg. Lt. F. W. Ht. 54', Rge. R. Fl. 14 mi., W. sector 16 mi.
YORK SPIT, N. side Entr. to York R. – On pile, Fl. W. ev. 6 sec. Ht. 30', Rge. 8 mi.

For abbreviations see footnote p. 161.

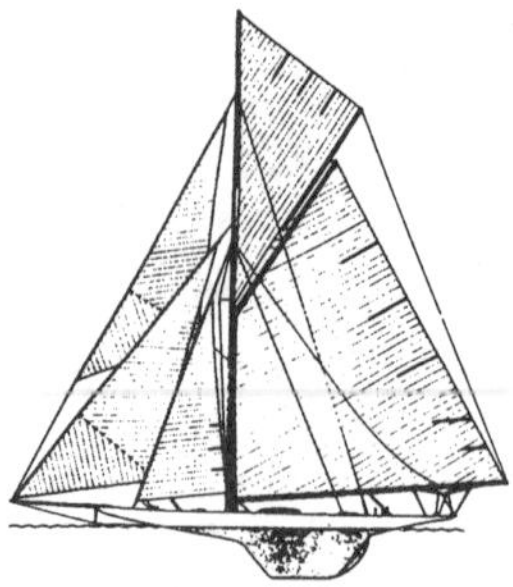

RELIANCE

The Largest Cup Defender

Captain Nat's legacy lives on at the Herreshoff Marine Museum America's Cup Hall of Fame

Herreshoff Marine Museum – America's Cup Hall of Fame
One Burnside Street P.O. Box 450
Bristol, Rhode Island 02809-0450

Only superlatives adequately describe RELIANCE, the largest single-masted vessel ever built. Her topmast sprit towered 189 feet 6 inches above water; it took sixty-six men to race her; there was enough square footage in her sails for eight modern 12-Meter boats; her mainsheet measured 1,000 feet in length; and she carried over 100 tons of lead ballast in her keel. - *from Herreshoff of Bristol by Maynard Bray and Carlton Pinheiro, WoodenBoat Publications*

In the 1903 America's Cup races RELIANCE defeated SHAMROCK III, designed by William Fife, Jr. for Sir Thomas Lipton, Royal Ulster Yacht Club 3 races to 0. RELIANCE was scrapped in 1913 at Robins Shipyard, South Brooklyn.

This museum is dedicated to the preservation, exhibition and interpretation of the accomplishments of the Herreshoff Mfg. Co. and the influence of America's Cup competition for the purposes of education, research, and the inspiration of excellence in the international maritime community.

Inspired by the donation of the 60' power launch, Thania, the Herreshoff Marine Museum was founded by Sidney and Rebecca Herreshoff in 1971.

Over a 27 year period, a preeminent display of wooden Herreshoff craft has been assembled. These yachts, dating from 1859 to 1945, are being restored for sailing and exhibition.

The America's Cup Hall of Fame is a tribute to the 146 year tradition of the world's most distinguished international competition. Exhibits include the Hall of Honor, the fully rigged 1992 IACC yacht DEFIANT, defender and challenger half models, artifacts and historic photographs.

★NANTUCKET SOUNDER★

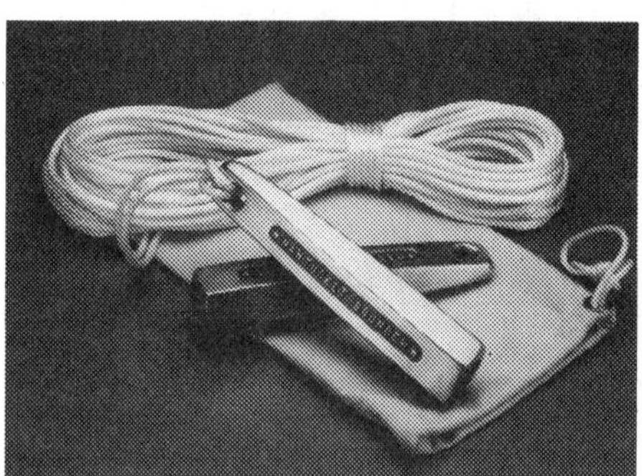

The NANTUCKET SOUNDER measures water depth to 60' without batteries. The 3-lb. solid bronze weight is safer and more durable than lead, and its hexagonal cross-section means no rolling on deck. A cavity in the base can be used for taking bottom samples.

The complete package includes:

- ☆ 3-lb bronze weight
- ☆ 60' braided line with a knot every fathom
- ☆ custom Sunbrella™ storage bag
- ☆ brochure with instructions, history, and lore

If you are looking for an unusual gift for a sailor, the NANTUCKET SOUNDER is a perfect choice. It's even engravable. Reasonably priced at $55.00 (plus S/H).

Publishers of **ELDRIDGE**

ROBERT E. WHITE INSTRUMENTS, INC.

1-617-742-3045 Eastern MA
1-800-992-3045 U.S. and Canada
A/E, M/C and VISA accepted

www.robertwhite.com
34 Commercial Wharf
Boston, MA 02110

OLD PLANTATION FLATS, York Spit Chann. – On pile, Fl. W. ev. 4 sec. Ht. 39', Rge. 7 mi.
WOLF TRAP, York Spit Chann. – Fl. W. ev. 15 sec. Ht. 52', Rge. 14 mi.
STINGRAY POINT, Ches. Chann. – Fl. W. ev. 4 sec. Ht. 34', Rge. 8 mi.
WINDMILL POINT, Ches. Chann. – On pile. Fl. W. ev. 6 sec. 2 R. Sectors: 293°-082° and 091°-113°. Ht. 31', Rge. R. sector 6 mi., W. Fl. 8 mi.
TANGIER SOUND, Ches. Chann. – Fl. W. ev. 6 sec. R. Sector: 115°-193°. Ht. 45', Rge. R. sector 7 mi., W. Fl. 10 mi.
SMITH POINT, Ches. Chann. – Fl. W. ev. 10 sec. R. Sector: 003°-156°. Horn 1 bl. ev. 15 sec. Ht. 52', Rge. R. sector 18 mi., W. Fl. 22 mi.
POINT LOOKOUT, Ches. Chann. – Gp. Fl. W., 2 fl. ev. 5 sec. Horn 1 bl. ev. 30 sec. (Sept. 15-June 1). Ht. 39', Rge. 8 mi.
HOLLAND ISLAND BAR, Ches. Chann. – Fl. W. ev. 2.5 sec. Horn 1 bl. ev. 30 sec. (Sept. 15-June 1). Ht. 37', Rge. 7 mi.
POINT NO POINT, Ches. Chann. – Fl. W. ev. 6 sec. Horn 1 bl. ev. 30 sec. (Sept. 15-June 1). Ht. 52', Rge. 9 mi.
HOOPER ISLAND, Ches. Chann. – Fl. W. ev. 6 sec. Horn 1 bl. ev. 30 sec. Ht. 63', Rge. 10 mi.
DRUM POINT #4, Ches. Chann. – Fl. R. ev. 2.5 sec. Ht. 22', Rge. 5 mi.
COVE POINT, Ches. Chann. – Fl. W. ev. 10 sec. (Obscured from 040°-110°.) Horn 1 bl. ev. 15 sec. Ht. 45', Rge. 19 mi.
SHARPS ISLAND, Ches. Chann. – Fl. W. ev. 6 sec. R. Sector: 159°-262° (Obscured from 067°-263°). Ht. 54', Rge. R. sector 7 mi., W. Fl. 9 mi.
BLOODY POINT BAR, Ches. Chann. – Fl. W. ev. 6 sec. 2 R. Sectors 003°-022° and 183°-202°. Horn 1 bl. ev. 30 sec. (Sept. 15-June 1.) Ht. 54', Rge. W. Fl. 9 mi., R. Sectors 7 mi.
THOMAS POINT SHOAL, Ches. Chann. – Fl. W. ev. 5 sec. 2 R. Sectors: 011°-051°; 096°-202°. Horn 1 bl. ev. 15 sec. Ht. 43', Rge. R. sectors 11 mi., W. Fl. 13 mi.
SANDY POINT SHOAL, Ches. Chann. – Fl. W. ev. 6 sec. Horn 1 bl. ev. 30 sec. (Sept. 15-June 1). Ht. 51', Rge. 9 mi.
CHESAPEAKE BAY BRIDGE Fog Signal, on main channel span – Horn 1 bl. ev. 15 sec. Racon (K) on south span.

NORTH CAROLINA

CURRITUCK BEACH, on beach – Fl. W. ev. 20 sec. Ht. 158', Rge. 18 mi.
BODIE ISLAND, on island – Gp. Fl. W. 2 fl. ev. 30 sec. Ht. 156', Rge. 18 mi.
OREGON INLET – Fl. W. ev. 10 sec. Ht. 54', Rge. 15 mi.
CAPE HATTERAS, on beach – Fl. W. ev. 7.5 sec. Ht. 191', Rge. 24 mi.
DIAMOND SHOAL, on tower – Fl. W. ev. 10 sec. Horn 1 bl. ev. 30 sec. Ht. 125', Rge. 18 mi.
HATTERAS INLET, tower on shore – Iso. W. ev. 6 sec. Ht. 88', Rge. 10 mi.
OCRACOKE ISLAND, on W. part of island – F.W. Ht. 75', Rge. 14 mi.
CAPE LOOKOUT, on N. pt. of cape – Fl. W. ev. 15 sec. Ht. 156', Rge. 25 mi.
OAK ISLAND, on SE pt. of island – Gp. Fl. W. 4 fl. ev. 10 sec. Ht. 169', Rge. 24 mi.
FRYING PAN SHOALS, on tower – Fl. W. ev. 6 sec. Horn 1 bl. ev. 30 sec. Ht. 118', Rge. 19 mi.

SOUTH CAROLINA

GEORGETOWN, E. side Entr. to Winyah Bay – Gp. Fl. W. 2 fl. ev. 15 sec. Ht. 85', Rge. 15 mi.
CHARLESTON, S. side of Sullivans Is. – Gp. Fl. W. 2 fl. ev. 30 sec. Ht. 163', Rge. 26 mi. Radiobeacon.

GEORGIA

TYBEE, NE end of is. – F. W. Ht. 144', Rge. 19 mi.
ST. SIMONS ISLAND, N. side Entr. to St. Simons Sd. – F.W. and Fl. W. ev. 60 sec. Ht. 104', Rge.: F. W. 18 mi., Fl. W. 23 mi.
CUMBERLAND ISLAND, Entr. to St. Mary's Sd. – Fl. W. ev. 6 sec. Rge. 8 mi. Ht. 35'.
AMELIA ISLAND, 2 mi. from N. end of is. – Fl. W. ev. 10 sec. R. Sector: 344°-360°. Ht. 107', Rge. R. sector 19 mi., W. Fl. 23 mi.

For abbreviations see footnote p. 161.

FLORIDA

ST. JOHNS, on shore – Gp. Fl. W. 4 fl. ev. 20 sec. Obscured 179°-354°. Ht. 83', Rge. 22 mi.

ST. AUGUSTINE, N. end of Anastasia Is. – F. W. and Fl. W. ev. 30 sec. Ht. 161', Rge. F. W. 19 mi., Fl. W. 24 mi.

PONCE DE LEON INLET APPR., S. side on inlet – Fl. W. ev. 10 sec. Ht. 159', Rge. 17 mi.

CAPE CANAVERAL, on Cape – Gp. Fl. W. 2 fl. ev. 20 sec. Ht. 137', Rge. 24 mi. Radiobeacon.

SEBASTIAN INLET N. JETTY – Fl. W. ev. 4 sec. R. Sector: 104°-154°. Ht. 27'.

JUPITER INLET, N. side of inlet – Gp. Fl. W. 2 fl. ev. 30 sec. Obscured 231°-234°. Ht. 146', Rge. 25 mi.

HILLSBORO INLET, N. side of inlet – Fl. W. ev. 20 sec. Obscured 015°-186°. Ht. 136', Rge. 28 mi.

PORT EVERGLADES APPR., on roof of bldg. – Fl. W. ev. 5 sec. Obscured 030°-180°. Ht. 349', Rge. 17 mi.

MIAMI LTD. BUOY "M" – E. end of Miami Beach, R. and W. vert. stripes. Mo (A) W, Rge. 7 mi. Racon (M).

MIAMI – Gp. Fl. W. 2 fl. ev. 30 sec. Obscured 065°-195°. Ht. 240', Rge. 13 mi. Radiobeacon.

FOWEY ROCKS, Hawk Chann. – Fl. W. ev. 10 sec. 2 R. Sectors. W. Sectors: 188°-359°, 022°-180°. Ht. 110', Rge. R. sector 10 mi. W. Fl. 15 mi.

CARYSFORT REEF, outer line of reefs – Gp. Fl. W. 3 fl. ev. 60 sec. 3 R. Sectors. W. Sectors: 211°-018°; 049°-087°; 145°-184°, Ht. 100', Rge. R. sectors 13 mi., W. Fl. 15 mi.

ALLIGATOR REEF, outer line of reefs – Gp. Fl. W. 4 fl. ev. 60 sec. 2 R. sectors: 223°-249°; 047°-068°, Ht. 136', Rge. R. sectors 12 mi., W. Fl. 18 mi.

SOMBRERO KEY, outer line of reefs – Gp. Fl. W. 5 fl. ev. 60 sec. 3 W. Sectors: 222°-238°; 264°-066°; 094°-163°. Ht. 142', Rge. R. sectors 12 mi., W. Fl. 15 mi.

AMERICAN SHOAL, outer line of reefs – Gp. Fl. W. 3 fl. ev. 15 sec. 2 W. Sectors: 270°-067°; 125°-242°. Obscured 90°-125°. Ht. 109', Rge.: R. sectors 8 mi., W Fl. 10 mi.

SAND KEY, Gp. Fl. W. 2 fl. ev. 15 sec. Ht. 40', Rge. 13 mi.

DRY TORTUGAS, on Loggerhead Key – Fl. W. ev. 20 sec. Ht. 151', Rge. 20 mi.

BERMUDA – APPROACH LIGHTS FROM SEAWARD

NORTH ROCK BEACON – Gp. Fl. W. 4 fl. ev. 20 sec. RaRef. Ht. 60', Rge. 12 mi. 32-28-30 N. - 64-46-00 W.

NORTH EAST BREAKER BEACON, Red tower on red tripod base reading "Northeast." Fl. W. ev. 2.5 sec. Ht. 45', Rge. 12 mi. RaRef., Racon (N). 32-28-42 N - 64-40-54 W.

KITCHEN SHOAL – Gp. Fl. W. 3 Fl. ev. 15 sec. Rge. 12 mi. Ht. 45'. R. & W. horizont. stripes, Red "Kitchen" on White background; 32-26-00 N - 64-37-36 W. Has RaRef.

MILLS BREAKER BUOY – 3 Qk. Fl. W. ev. 5 sec. Rge. 5 mi. Black "Mills" on yellow background. 32-23-54 N - 64-36-48 W.

SPIT BUOY – 3 Qk. Fl. W. ev. 10 sec. Rge. 5 mi. Black "Spit" on yellow. 32-22-36 N - 64-38-24 W.

ST. DAVID'S ISLAND LIGHT – F. R. and G. Sectors below Gp. Fl. W. 2 fl. ev. 20 sec. Ht. 212', Rge. 15 mi. 32-21-48 N - 64-39-00 W. Your bearing from seaward of G. Sectors is 221°-276° true; remaining Sector is R. and partially obscured by land 044°-135° true.

KINDLEY FIELD AERO BEACON – Alt. Gp. Fl. 2 W. Fl. and 1 G. Fl. ev. 10 sec. Ht. 140', Rge. 15 mi. 32-21-57 N - 64-40-33 W.

GIBB'S HILL LIGHTHOUSE – Fl. W. ev. 10 sec. Ht. 354', Rge. 26 mi. 32-15-06 N - 64-50-00 W.

CHUB HEADS – 9 Qk. Fl. W. ev. 15 sec. 60' structure (22' black tripod with 35' yellow & black tower). Rge. 12 mi. RaRef. 32-17-24 N - 64-58-42 W. 7.64 mi. from Gibbs Hill Racon (C).

EASTERN BLUE CUT BEACON – Black letters on W. background. Fl. W. Mo (U) ev. 10 sec. Ht. 60', Rge. 12 mi. 32-24-00 N - 64-52-36 W.

ST. GEORGE'S HBR. ENTR. – Qk. Fl. G. is 3' above an inner G. Lt. on N. side of Town Cut Chann. visible 260°-266°; Rge. 10 mi. 32-22-07 N - 64-39-09 W.

For abbreviations see footnote p. 161.

Foregoing information checked to date, September, 1998. See page 4 for free Supplement in May 1999.

Please Read This!

We have streamlined the following section (Courses and Distances from Norfolk to Moose Peak, ME) making this piloting guide, we hope, less confusing to follow.

We have included alongside each aid its Latitude and Longitude in degrees, minutes and seconds to make your piloting fit with Loran and GPS or DGPS usage.

The Local Coast Guard Districts (our most reliable source for up-to-date navigation aids information) have told us firmly that providing coordinates in seconds instead of fractions of minutes is risky, on the grounds that all locations of buoys are approximate and to consider locations exact is unrealistic and could lead to trouble, so caution is advised.

The Latitude and Longitude given in this section are from the Local Coast Guard District, the U.S. Light List and Local Notices to Mariners. Where the Light List omits these coordinates, we have plotted these from the latest charts. You will find that some have tenths of minutes and not seconds.

It is important to remember that any "new" government publication may be months old before it gets into the hands of the mariner and changes may have occurred meantime.

The prudent pilot or navigator should never rely on Loran or GPS and DGPS alone but always have a chart and plotting tools handy and lay out his courses before starting out.

The coordinates of a lighthouse or a beacon, fixed objects, can be calculated very precisely but buoys are simply moored items, swinging this way and that according to the currents at the time so their positions are not exact (also occasionally they are relocated by the Coast Guard).

* * * * * * * * * *

Abbreviations used in this text:

Lt: Light, L.H.: Lighthouse, By.: Buoy, Sh.: Shoal, Ch.: Channel, Rge.: Range, abm.: abeam, ahd.: ahead, ast.: astern, Qk.: Quick, Fl.: Flashing, F.: Fixed, W.: White, R.: Red, G.: Green, O.: Orange, Y.: Yellow, RW: Red & White (mid-channel Fairway Buoys), RG: Red & Green, GR: Green & Red, Wh.: Whistle, Hn.: Horn, Mo (A): Flash White Morse Code "A" (• –).

See p. 215 for colored diagram.

Courses given herein are from mark to mark unless otherwise indicated.

COURSES AND DISTANCES

CHESAPEAKE LT. STATION TO BALTIMORE VIA DREDGED CHANNEL
Charts 12221, 12225, 12230, 12263, 12278

COURSES AS GIVEN ARE FOR VARIATION FOR CURRENT YEAR	Lat.	Long.	Courses °M.	Courses °T.	Naut. Miles	Reverse of Courses °M.	Reverse of Courses °T.
*Chesapeake Lt. Sta., Racon (N)	36-54.3	75-42.8					
To R. #2 Fl. R. 2.5 sec.	36.49.7	75-46.7	225	215	5.7	45	35
To R. #10 Qk. R.	36-52.4	75-52.1	311	301	5.1	131	121
To R #12 Fl. R. 2.5 sec.	36-53.3	75-53.1	334	324	1.1	154	144
To R. #16 Qk. R.	36-55.1	75-55.2	328	318	2.7	148	138
To GR "CBJ" 3 Fl. G. 6 sec. Gong	36-56.1	75-57.5	312	302	2.0	132	122
To R. #4 Fl. R. 4 sec.	36-57.8	75-59.4	326	316	2.2	146	136
To R. #6 Qk. R.	36-59.3	76-00.3	344	334	1.8	164	154
To R. #12 Qk. R. Bell (thru Bay Bridge)	37-01.1	76-02.7	324	314	2.6	144	134
To R. #22 Qk. R. Bell (York Spit Ch.)	37-08.4	76-08.6	337	327	8.8	157	147
To G. #29 Qk. G.	37-11.2	76-09.5	356	346	2.9	176	166
To G. #31 Fl. G. 4 sec.	37-12.5	76-08.9	28	18	1.4	208	198
To Wolf Trap L.H. abm. .3 mi.	37-23.4	76-11.4	2	352	11.1	182	172
To G. #49 Fl. G. 4 sec. Gong (Rapp. Spit Ch.)	37-34.9	76-03.7	37	28	12.8	217	208
To G. #57 Fl. G. 4 sec.	37-39.3	76.08.3	328	319	5.7	148	139
To R. #62 Fl. R. 4 sec. Bell Racon (Y)	37-46.5	76-10.3	357	348	7.3	177	168
To Y "SP" Fl. Y. 2.5 sec. Bell (1.5 mi. abm. Smith Pt. Lt.)	37-52.8	76-09.2	18	9	6.4	198	189
To Cedar Pt. RW "HI" Mo (A)	38-18.7	76-18.7	354	344	27.0	174	164
To R. #78 Fl. R. 4 sec. Bell Racon (Q)	38-33.3	76-25.7	349	339	15.7	169	159
To G. #83 Fl. G. 4 sec. Gong	38-45.1	76-26.7	6	356	11.8	186	176
To R. #90 Fl. R. 2.5 sec. (Bridge Appr.)	38-58.3	76-23.3	21	11	13.4	201	191
Thru Bay Bridge							
To Sandy Pt. L.H. abm. .5 mi.	39-01.0	76-22.3	29	19	2.6	209	199
▶ **Into Baltimore:**							
Sandy Pt. L.H. abm.	39-01.0	76-22.3					
To R. #10 Fl. R. 2.5 sec. (Craighill Ch.)	39-04.5	76-23.6	359	348	3.6	179	168
To G. #15 Qk. G.	39-07.0	76-23.8	8	357	2.6	188	177
To G. #21 Qk. G.	39-08.5	76-24.5	353	343	1.5	173	163
To G. #25 Qk. G.	39-10.2	76-25.8	339	329	2.1	159	149
To R. #14 Qk. R.	39-12.0	76-30.1	307	297	3.7	127	117
Thence to Baltimore			331	321	4.0	151	141

SANDY POINT LIGHT TO CHES. & DEL. CANAL
Charts 12273, 12277

	Lat.	Long.	Courses °M.	Courses °T.	Naut. Miles	Reverse of Courses °M.	Reverse of Courses °T.
▶ **To Chesapeake & Delaware Canal**							
.5 mi. off Sandy Pt. L.H.	39-01.0	76-22.4					
To G. #13 Fl. G.	39.09.8	76-18.3	31	21	9.4	211	201
To G. #19 Fl. G.	39-11.5	76-16.5	50	40	1.3	230	220
To G. #21 Qk. G.	39-11.6	76-16.0	83	73	.5	263	253
To R. #30 Fl. R. 4 sec.	39-13.0	76-14.8	41	31	1.5	221	211
To R. #34 Qk. R.	39-15.8	76-14.1	20	10	2.8	200	190
To R. #40 Qk. R.	39-18.8	76-13.1	27	17	3.2	207	197
To R. #50 Fl. R.	39-20.8	76-10.2	58	48	3.0	238	228
To R. #54 Fl. R.	39-22.1	76-07.7	65	55	2.5	245	235
To G. #1 ER Fl. G. 2.5 sec.	39-23.9	76-03.3	76	66	4.0	256	246
To R. #8 Fl. R.	39-27.9	75-58.2	53	43	5.5	233	223
To G. #19 Qk. G. (start of Canal)	39-30.3	75-55.2	56	46	3.4	236	226

* **Note:** Vessels entering Deep Channel at R #2 should announce their intentions on VHF Ch. 16. Vessels drawing less than 40' should use inbound channel ***north*** of Deep Channel (keeping red buoys on port hand) and use outbound channel ***south*** of Deep Channel (keeping green buoys on port hand).

CHESAPEAKE LT. STATION TO NORFOLK & NEWPORT NEWS
Charts 12221, 12245, 12253, 12254

COURSES AS GIVEN ARE FOR VARIATION FOR CURRENT YEAR	Lat.	Long.	Courses °M.	Courses °T.	Naut. Miles	Reverse of Courses °M.	Reverse of Courses °T.
* Chesapeake Lt. Sta., Racon (N)	36-54.3	75-42.8					
To R. #2 Fl. R. 2.5 sec.	36-49.7	75-46.8	225	215	5.7	45	35
To R. #10 Qk. R.	36-52.4	75-52.1	311	301	5.1	131	121
To R. #12 Fl. R. 2.5 sec.	36-53.3	75-53.1	334	324	1.1	154	144
To R. #16 Qk. R.	36-55.1	75-55.2	328	318	2.7	148	138
To GR "CBJ" 3 Fl. G. 6 sec. Gong	36-56.1	75-57.5	312	302	2.0	132	122
To G. #1 TS Fl. G. 2.5 sec. Bell (Thimble Sh. Entr.)	36-57.1	76-01.4	298	288	3.4	118	108
To R. #18 Qk. R.	37-00.7	76-15.0	298	288	11.5	118	108
To G. #21 Fl. G. 4 sec. Bell	37-00.1	76-17.1	260	250	1.8	80	70
To G. #1ER Qk. G. Horn	36-59.3	76-18.7	250	240	1.2	70	60
To G. #3 Fl. G. 2.5 sec.			235	225	1.2	55	45
Thence to Norfolk/Newport News by marked ch.							

DELAWARE RIVER COURSES, RANGES & DISTANCES
Charts 12304, 12214, 12311, 12312

	Lat.	Long.	Courses °M.	Courses °T.	Naut. Miles	Reverse °M.	Reverse °T.
Brown Shoal Lt. Fl. W. 2.5 sec., Racon (B)	38-55.3	75-06.0					
To G. #19 Qk. G.	39-05.0	75-11.3	347	337	10.5	167	157
To Elbow of Cr. Ledge L.H. abm. Iso. W.	39-10.9	75-16.1	335	325	7.0	155	145
To R. #32 Qk. R., Liston Rge.	39-14.0	75-18.0	346	336	3.5	166	156
To G. #1B Qk. G., Baker Rge.	39-27.3	75-33.7	328	318	18.0	148	138
To R. #2R Qk. R. Bell, Reedy Rge.	39-29.0	75-33.7	5	355	1.7	185	175
To G. #1N Qk. G. Gong, N. Castle Rge.	39-33.1	75-32.6	25	15	4.3	205	195
▶ **Into Ches. & Del. Canal**							
G. #1N Qk. G. Gong	39-33.1	75-32.6					
To RG "CD" 3 Fl. R. 6 sec. Bell	39-33.8	75-33.2	327	317	.9	147	137
Thence to Reedy Point			269	259	.7	89	79
G. #1N Qk. G. Gong	39-33.1	75-32.6					
To R. #2B Qk. R. Bell, Blkhd. Bar Rge.	39-36.8	75-34.6	345	334	4.3	165	154
To G. #1D Fl. G. 2.5 sec., Dpwtr. Rge.	39-37.7	75-34.7	19	8	.7	199	188
To G. #1C Qk. G. Bell, Cherry Is. Rge.	39-40.4	75-31.6	53	42	3.6	233	222
To R. #2B Qk. R, Bellevue Rge.			28	17	4.3	208	197
To R. #2M Qk. R. Bell, Marcus Hk. Rge.	39-46.9	75-27.5	47	36	3.0	227	216
To G. #1C Qk. G. Gong, Chester Rge.	39-49.4	75-23.0	69	58	4.3	249	238
To R. #6C Qk. R.	39-50.5	75-20.9	61	50	1.9	241	230
To R. #2T Qk. R., Tinicum Rge.	39-50.8	75-19.9	76	65	.9	256	245
To G. #5T Qk. G.	39-51.0	75-15.9	103	92	3.1	283	272
To G. #1F Qk. G. Pt. Mifflin Bar Rge.	39-51.4	75-14.5	82	71	1.3	262	251
To R. #44 Qk. R. Bell, Horseshoe Rge.	39-52.7	75-11.8	65	54	2.4	245	234
To G. #35 Qk. G.	39-52.9	75-04.3	105	94	1.8	285	274
To G. #37 Qk. G.	39-53.0	75-08.8	81	70	.5	261	250
To various courses to Philadelphia							

CHESAPEAKE BAY ENTR. NORTH TO NEW YORK
Charts 12221, 12210, 12211, 12214, 12318, 12323, 12326

	Lat.	Long.	Courses °M.	Courses °T.	Naut. Miles	Reverse °M.	Reverse °T.
GR "CBJ" 3 Fl. G. 6 sec. Gong	36-56.1	75-57.5					
To R. #14 Fl. R. 2.5 sec. Bell	37-07.4	75-41.0	60	50	17.3	240	230
To R. #12 Fl. R. 2.5 sec. Bell	37-17.7	75-34.7	37	27	11.5	217	207
To R. #10P Fl. R. 4 sec. Gong	37-32.1	75-25.9	36	26	16.1	216	206
To R. #8 Fl. R. 2.5 sec. Gong	37-50.6	75-12.1	40	30	21.5	220	210
To R. "6WQS" Fl. R. 6 sec. Bell	37-59.7	75-01.4	50	40	8.3	230	220
To R. #4 Fl. R. 4 sec. Wh.	38-16.4	75-00.4	22	12	19.7	202	192
To G. "1 FS" Fl. G. 4 sec. Gong	38-26.8	74-57.4	22	12	10.7	202	192
To Y. "DB" Fl. Y. 4 sec.	38-38.2	74-52.2	29	19	12.5	209	199
To Y. "FA" Fl. Y. 4 sec.	38-46.8	74-44.2	47	38	10.2	227	218
To G. "3FB" Fl. G. 4 sec. Gong	38-58.2	74-40.5	23	13	12.9	203	193
To RG Brig. Wreck By: "WR2" Qk. R.	39-24.8	74-13.8	50	38	33.7	230	218
To RW #2 Fl. R. 6 sec. Wh.	39-45.5	73-59.5	40	28	23.5	220	208
To Ambrose Lt. Tower (NY Hbr. Entr.)	40-27.6	73-49.9	22	10	42.5	202	190

LOOKING BACK AT LEADLINES:

While it is not known when leadlines were first used, it seems apparent that it was long before Columbus's time. They are certainly one of the earliest navigation instruments. As recently as 50 years ago, the sounding lead was regarded as invaluable for coastwise piloting. In his book ***The Coastwise Navigator,*** Captain Warwick Tompkins says of the sounding lead: "Today it is second only to the magnetic compass in the list of the ship's essential equipment." That was 1938.

The sounding lead of many years ago was most often attached to a very long line, as there were no electronic sounders to probe great depths. Instead of a knot every fathom, more elaborate markings were used to indicate unambiguously a variety of depths:

2 fathoms - a 2-ended scrap of leather
3 fathoms - a 3-ended scrap of leather
5 fathoms - a scrap of white calico
7 fathoms - a strip of red wool bunting
10 fathoms - leather with a round hole in it
13 fathoms - a piece of thick blue serge
15 fathoms - a piece of white calico
17 fathoms - a piece of red wool bunting
20 fathoms - a cord with 2 knots
30 fathoms - a cord with 3 knots

Why did sailors of old need to know about such great depths? For one, fishing schooners needed to know when they were over the relatively shallow banks offshore where great schools of fish congregated. (Sailors would sometimes "arm the lead" by putting tallow into the star chamber to bring up a bottom sample.) Also, charts were often not so reliable in places, and it was prudent to check depths. Consider the recent groundings of ocean-going ships, incidents which underline the prudence every skipper should exercise when proceeding in "thin" waters.

NEW YORK HARBOR COURSES & DISTANCES - To Verrazano Bridge by Ambrose Channel, 40 feet. Charts 12326, 12327 - Mid Channel Courses

COURSES AS GIVEN ARE FOR VARIATION FOR CURRENT YEAR	Lat.	Long.	Courses °M.	Courses °T.	Naut. Miles	Reverse of Courses °M.	Reverse of Courses °T.
Ambrose Lt. Tower, Racon (N) Fl. W. 5 sec.	40-27-35	73-49-49					
To RW Fairway "A" Mo (A) Wh.	40-28-47	73-53-37	306	294	3.1	126	114
To R. #2 Fl. R. 2.5 sec. Bell (Ch. Entr.)	40-29-49	73-55-45	309	297	2.0	129	117
To R. #10 Qk. R. Bell	40-31-40	74-00-23	309	297	4.1	129	117
To R. #14 Qk. R. Bell	40-32-19	71-01-12	335	323	.9	155	143
To R. #20 Fl. R. 4 sec. Bell	40-34-01	74-01-42	0	348	1.7	180	168
To Ctr. of Verrazano Bridge, lvg. G. #21 Qk. G. Gong abm. .2 mi.	40-35-08	74-02-29	353	340	2.5	173	160

To Verrazano Bridge by Sandy Hook & Chapel Hill Channels, 30 feet. Charts 12326, 12327

	Lat.	Long.	Courses °M.	Courses °T.	Naut. Miles	Reverse °M.	Reverse °T.
Ambrose Lt. Tower, Racon (N)	40-27-35	73-49-49					
To RW Scotl'd Ltd. By. "S" Mo (A) Horn	40-26-33	73-55-00	268	256	4.1	88	76
To R. #2 Fl. R. 2.5 sec. Bell	40-27-18	73-56-11	320	308	1.2	140	128
To R. Nun #10	40-29-13	73-59-22	320	308	3.1	140	128
To R. #14 Qk. R. Bell	40-28-57	74-00-51	269	257	1.1	89	77
To R. #18 Fl. R. 4 sec. Bell	40-28-34	74-02-02	259	247	1.0	79	67
To R. #20 Fl. R. 2.5 sec.	40-28-34	74-02-18	281	269	.2	101	89
To intersection of Chapel Hill S. Ch. Rge., near "SP" 3 Fl. R. 6 sec. Gong	40-28-42	74-02-58	296	284	.5	116	104
To G. #15 Fl. G. 2.5 sec.	40-31-59	74-02-27	19	7	3.4	199	187
To G. #17 Fl. G. 4 sec. Bell	40-32-48	74-02-20	21	8	1.6	201	188
To center of Verrazano Bridge			6	354	3.0	186	174

Ambrose Light Tower to Perth Amboy, Chart 12327

	Lat.	Long.	Courses °M.	Courses °T.	Naut. Miles	Reverse °M.	Reverse °T.
Ambrose Lt. Tower, Racon (N)	40-27-35	73-49-49					
To RW Scotl'd Ltd. By."S" Mo(A) Horn, Racon (M)	40-26-33	73-55-00	268	256	4.1	88	76
To R. #8 Fl. R. 4 sec. Bell	40-28-51	73-58-46	323	310	3.7	43	30
To R. #12 Fl. R. 2.5 sec. Bell	40-29-06	74-00-05	298	285	1.0	118	105
To .1 mi. S. of R. #20 Fl. R. 2.5 sec.	40-28-28	74-02-18	264	250	1.8	84	70
Various courses to Ward Pt. Fl. R.							

Verrazano Bridge to Yonkers, North River, Charts 12327, 12341, 12345

	Lat.	Long.	Courses °M.	Courses °T.	Naut. Miles	Reverse °M.	Reverse °T.
Verrazano Bridge							
To R. #22 Fl. R. 2.5 sec. Bell	40-38.0	74-03.0	5	353	1.7	185	173
To R. #24 Qk. R. Bell	40-38-39	74-03-14	7	355	.9	187	175
To Battery (abm. to E. lvg. R. #26 + R. #28 to E.)			34	22	3.4	214	202
To Pier 64 abm. to E.			21	9	3.0	201	189
To Pier 99 abm. to E.			37	25	1.5	217	205
To 135th St. abm. to E.			42	30	3.3	222	210
To Ft. Washington			34	22	1.8	214	202
To Spuyten Duyvil Creek			39	27	2.0	219	207
Thence to Yonkers, City Hall abm.			30	18	3.3	210	198

Verr. Br. to Brooklyn Br. & Exec. Rk. Charts 12327, 12334, 12335, 12339, 12366

	Lat.	Long.	Courses °M.	Courses °T.	Naut. Miles	Reverse °M.	Reverse °T.
Verrazano Bridge							
To G. #1 Fl. G. 4 sec.	40-39-28	74-03-41	10	358	1.7	190	178
To G. Can #7 (abm. to W. .1 mi.)	40-39-31	74-04-18	48	36	2.0	228	216
To Buttermilk Ch. Rge. Lt. Qk. G.	40-40-52	74-01-01	7	355	1.0	187	175
To Can #7 (abm. to W. .1 mi.)			66	54	.7	246	234
To Brooklyn Bridge			41	29	1.3	221	209
To Corlears Hook			91	79	1.0	271	259
To Poorhouse Flats Rge. Front Lt. Qk. R.	40-43-28	73-57-46	28	16	1.4	208	196
To Belmont Is. Lt. Fl. G. 4 sec. (abm. to E.)			12	0	.9	192	180

COURSES AS GIVEN ARE FOR VARIATION FOR CURRENT YEAR	Lat.	Long.	Courses °M.	Courses °T.	Naut. Miles	Reverse of Courses °M.	Reverse of Courses °T.
To Horns Hook thru Western Ch.			46	34	2.1	226	214
To Hallet's Pt.			58	46	.3	238	226
To Ctr. of Triborough Bridge			97	85	.4	277	265
To Ctr. of Hell Gate Bridge			62	50	.3	242	230
To Lawrence Pt. Ledge Fl. W. 4 sec. (abm.)	40-47-36	73-54-14	54	42	1.0	234	222
To S. Bro. Is. G. #1 Fl. G. 2.5 sec. (abm.)	40-47-59	73-53-29	81	69	.7	261	249
To R. #6 Fl. R. (abm. to N. .1 mi.)	40-47-56	73-52-12	111	99	1.0	291	279
To Whitestone Pt. abm. Fl. G.	40-48-06	73-49-10	94	82	2.4	274	262
To Throgs Neck Lt. F.R.	40-48-16	73-47-26	109	97	1.5	289	277
To Stepping Stones Lt. Occ. G. 4 sec. (.1 mi. abm.)	40-49-30	73-46-28	29	17	1.5	209	197
To Execution Rk. (abm. to W. .2 mi.)	40-52-41	73-44-16	44	32	3.7	224	212

EXECUTION ROCK TO NARRAG. BAY ENTR. "NB," outside Fisher's Is., Charts 12363, 12354, 13218

	Lat.	Long.	°M.	°T.	Naut. Miles	°M.	°T.
Exec. Rk. Lt. Fl. W. 10 sec. Horn Racon (B)	40-52-14	73-44-16					
To Eaton's Pt. G. #11B Fl. G. 4 sec. Gong	41-00-01	73-23-42	76	64	17.0	256	244
To Stratford Sh. R. #2 Fl. R. 4 sec. Bell	41-03-04	73-06-12	90	77	13.5	270	257
To RW "PI" Mo (A) Wh.	41-13-17	72-10-48	90	76	43.1	270	256
To RW "WH" Mo (A) Wh.	41-15-50	71-50-55	95	81	15.3	275	261
To Pt. Judith R. #2 Fl. R. 2.5 sec. Wh.	41-18-28	71-28-21	97	82	17.4	277	262
To Entr. "NB" Fl. W. 4 sec. Hn. Racon (B)	41-23-00	71-23-18	55	40	6.0	235	220

EATON'S PT. TO NARRAG. BAY ENTR. "NB" thru Fisher's Is. Sd.; Entr. to Bridgeport, New Haven, Essex, New London, Mystic. Charts 13214, 13212, 12375, 12371, 12369, 12363

	Lat.	Long.	°M.	°T.	Naut. Miles	°M.	°T.
Eaton's Pt. G. #11B Fl. G. 4 sec. Gong	41-00-01	73-23-42					
To Bridgeport R. #2 Fl. R. 2.5 sec Gong	41-07-17	73-11-19	65	52	11.8	245	232
▶ **Thence to Tongue Pt. and Bridgeport Hbr.: Chart 12369**			26	13	2.8	206	193
Bridgeport R. #2 Fl. R. 2.5 sec. Gong	41-07-17	73-11-19					
To Stratford Pt. R. #18 Fl. R. 6 sec. Bell	41-06-49	73-08-09	114	101	2.5	294	281
To RW "NH" Mo (A) Wh.	41-12-07	72-53-47	77	66	12.0	257	244
▶ **Into New Haven: Chart 12371**							
"NH" to G. #7 Fl. G. 2.5 sec.	41-14-05	72-55-06	348	334	2.1	168	154
To G. #11 Fl. G. 4 sec.	41-15-17	72-54-52	21	7	1.2	201	187
To G. #15 Fl. G. 2.5 sec.	41-16-18	72-54-45	19	5	1.0	199	185
To New Haven Reach			15	1	.6	195	181
"NH" Mo (A) Wh.	41-12-07	72-53-47					
To Falkner Is. Rf. G. #15 Fl. G. 4 sec. Gong	41-13-23	72-39-15	97	83	11.0	277	263
To Long Sand Sh. RG "W" 3 Fl. R. 6 sec. Horn	41-13-35	72-27-36	103	89	8.8	283	269
To Cornfield Pt. Sh. #2 Fl. R. 4 sec. Bell	41-15-01	72-23-03	81	67	3.8	261	247
To 200 yds. S. mag. of Saybrook Lt. Fl. G. 6 sec. (outer bar. Ch. Entr.)	41-15-48	72-20-34	85	71	2.0	265	251
To G. #5 Fl. G. 4 sec.	41-16-18	72-20-30	17	3	.6	197	183
▶ **Thence to Essex: Chart 12375**							
Saybrook Breakwater Lt. Fl. G. 6 sec. Horn	41-15-48	72-20-34					
To Saybrook Bar R. #8 Fl. R. 4 sec. Bell	41-14-51	72-18-51	137	123	1.5	317	303
To Bartlett Rf. R. #4 Fl. R. 4 sec. Bell	41-15-34	72-08-21	100	85	8.0	280	265
To R. #2 Fl. R. 2.5 sec. Bell	41-17-36	72-04-36	69	56	3.6	249	236
▶ **Thence to New London by marked channel: Chart 13212**							
R. #2 Fl. R. 2.5 sec. Bell	41-17-36	72-04-36					
To Intrepid Rk. "DIR" Can R. & B. bands	41-17-56	72-00-15	97	83	3.3	277	263
▶ **Thence to Mystic (inadvisable to enter Ch. without Chart 13214)**							
Intrepid Rk. "DIR" Can R. & B. bands	41-17-56	72-00-15					
To Ram Is. Rf. R. #20 Fl. R. 4 sec. Bell	41-18-11	71-58-24	95	80	1.4	275	260
To Napatree Pt. R. #6 Fl. R. 4 sec. Bell	41-18-01	71-53-20	108	93	3.8	288	273
To Gangway Rk. R. #2 Fl. R. 6 sec. Bell	41-18-00	71-51-32	106	91	1.4	286	271
To Pt. Judith R. #2 Fl. R. 4 sec. Wh.	41-19-16	71-28-31	101	86	17.6	281	266
To Narrag. Bay Entr."NB"Fl. W. 4 sec. Horn Racon(B)	41-23-00	71-23-20	46	31	7.5	226	211

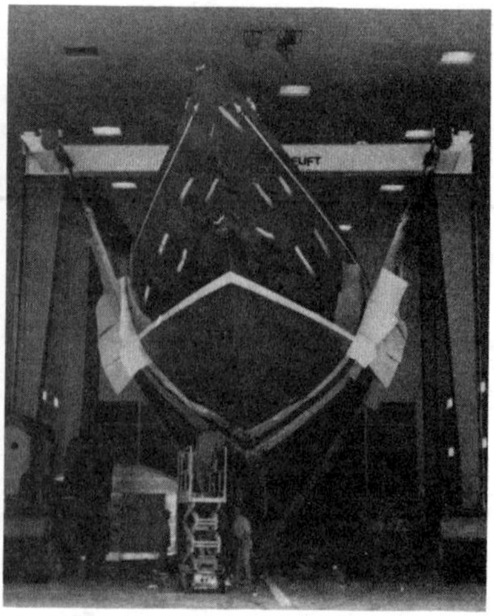

NARRAGANSETT BAY COURSES & DISTANCES, Narrag. "NB" to Providence and Fall River. Chart 13221

COURSES AS GIVEN ARE FOR VARIATION FOR CURRENT YEAR	Lat.	Long.	Courses °M.	Courses °T.	Naut. Miles	Reverse of Courses °M.	Reverse of Courses °T.
Entr. "NB" Fl. W. 4 sec. Hn. Racon (B)	41-23-00	71-23-20					
To Castle Hill Lt. Iso. R. (abm. 1/4 mi. to E.)	41-27-42	71-21-46	38	23	2.4	218	203
To G. #11 Qk. G. Bell	41-29-00	71-21-02	56	41	1.3	236	221
To #13 Can (lvg. R. #12 to E.)	41-30-34	71-20-55	16	1	1.8	196	181
To R. #24 Fl. R. 2.5 sec. Bell	41-35-30	71-18-14	38	23	5.4	218	203
To RG Sandy Pt. Jct. "SP" 3 Fl. R. 6 sec. Bell	41-36-55	71-17-36	35	20	1.6	215	200
To Poppasquash G. #5 Qk. G.	41-39-05	71-18-43	350	335	2.1	170	155
To Ch. Entr. R. #10 Qk. R. Bell (passing R. #8 Fl. R. and G. #7 Fl. G.)	41-41-21	71-18-44	14	359	2.4	194	179
To Conimicut Lt. Fl. W. 2.5 sec.	41-43-01	71-20-42	337	322	2.3	157	142
To R. #26 Fl. R. 4 sec.	41-44-48	71-22-10	320	305	1.1	140	125
To Sabin Pt. Daybeacon "SP"	41-45.7	71-22.5	358	343	2.1	178	163
To Popham Lt. F. R.	41-46-39	71-22-10	28	13	1.0	208	193
▶ **Thence to Ship Anchorage, Providence**							
Sandy Pt. Jct. Fl. R. Bell	41-36-55	71-17-36					
To G. #3 Fl. G. 4 sec. Bell	41-37-50	71-16-15	64	49	1.2	244	229
To Mt. Hope Br. (lvg. Mt. Hope Lt. to S.)			60	45	1.0	240	225
To RG "MH" Fl. R. 6 sec. Gong (Jct. Lt.)	41-39-32	71-14-03	59	44	1.5	239	224
Then up dredged ch. 59° Mag., .3 mi., Borden Flats Lt. ahead (41-42.3 71-10.5)							
To #10 Fl. R. Bell lvg. Borden Flats Lt. to N.	41-41-39	71-11-16					
▶ **Thence to Anchorage, Fall River**							

BUZZARDS BAY AND CANAL, Charts 13218, 13230, 13236

	Lat.	Long.	Courses °M.	Courses °T.	Naut. Miles	Reverse °M.	Reverse °T.
Pt. Judith R. #2 Fl. R. 2.5 sec. Wh.	41-18-28	71-28-21					
To Buzz. Bay Entr. Lt. Hn. Fl. W. 2.5 sec. Racon (B)	41-23-48	71-02-00	89	74	20.5	269	254
To R. #8 Fl. R. 2.5 sec. Gong	41-28-58	70-53-34	66	51	8.2	246	231
To R. #10 Fl. R. 4 sec. Gong	41-33-04	70-46-39	66	51	6.6	246	231
To R. #12 Qk. R.	41-37-22	70-42-19	52	37	5.4	232	217
To Cleveland Ledge Entr. Buoys	41-41-33	70-40-27	51	36	.7	231	216
To G. #1 Qk. G. (abm. to W.)	41-37-50	70-41-56	31	16	3.2	211	196
Cleveland Ledge #14 Fl. R. 2.5 sec.:	41-37-34	70-42-00					
To Woods Hole G. #13 Fl. G. 4 sec. Bell	41-31-41	70-41-52	194	179	5.9	14	359
To Robinson's Hole G. #1 Buoy	41-26-46	70-48-14	222	207	11.2	42	27
To Quick's Hole Entr. Ltd. Bell #1	41-25-48	70-50-22	228	213	12.4	48	33

TO NEW BEDFORD AND MARION VIA MAIN CHANNEL, Charts 13218, 13230

	Lat.	Long.	Courses °M.	Courses °T.	Naut. Miles	Reverse °M.	Reverse °T.
Buzz. Bay Entr. Lt. Hn. Fl. W. 2.5 sec. Racon (B)	41-23-48	71-02-00					
To R. #4 Qk. R. Bell	41-27-02	70-59-14	60	45	5.8	240	225
To RW "BB" Mo (A) Bell	41-30-47	70-50-05	75	60	5.7	255	240
To Ch. Entr. G. #9 Fl. G. 6 sec. (lvg. R. #4 Bell to E.; G. #1, G. #5, GR Bell to W.)	41-35.2	70-53.0	348	333	5.1	168	153
Then up ch. to New Bedford			351	336	2.5	171	156
▶ **Via West Channel (Small Craft only):**							
Buzz. Bay Entr. Lt. Hn. Fl. W. 2.5 sec. Racon (B)	41-23-48	71-02-00					
To Coxens Ledge R. #4 Qk. R. Bell	41-27-02	70-59-14	60	45	5.8	240	225
To Mishaum Ledge G. #5 Fl. G. 4 sec. Gong	41-28-58	70-57-22	52	37	2.4	232	217
To G. Can #1 (SW Sh.)	41-35.5	70-55.2	46	31	2.5	226	211
To "AB" 3 W. Fl. Gong	41-33.3	70-54.4	35	20	.6	215	200
To Ch. Entr. #9 (lvg. #11 G. Can Inez Rock to port)	41-35.2	70-53.0	44	28	.2	224	208
Thence to New Bedford			351	336	2.5	171	156
▶ **To Marion (chart 13230):**							
Buzz. Bay Entr. Lt. Hn. Fl. W. 2.5 sec. Racon (B)	41-23-48	71-02-00					
To R. #8 Fl. R. 2.5 sec. Gong	41-28-58	70-53-34	66	51	8.2	246	231
To R. #10 Fl. R. 4 sec. Gong	41-33-04	70-46-39	67	52	6.7	247	232
To Centerboard Sh. R. #2 Fl. R. 4 sec.	41-39-44	70-43-34	34	19	7.1	214	199
To G. #5 Can	41-40-50	70-44-24	345	330	1.3	165	150
To R. #6 Nun	41-41-37	70-45-00	345	330	.9	165	150
To Ram Is. R. #8 Fl. R. 4 sec.	41-41-54	70-45-13	348	333	.3	168	153

MARION TO CAPE COD CANAL, Charts 13218, 13230

COURSES AS GIVEN ARE FOR VARIATION FOR CURRENT YEAR	Lat.	Long.	Courses °M.	Courses °T.	Naut. Miles	Reverse of Courses °M.	Reverse of Courses °T.
Centerboard Sh. R. #2 Fl. R. 4 sec.	41-39-34	70-43-34					
To Bird Is. G. #17 Bell	41-39-59	70-42-23	87	72	.9	267	252
To G. #1 Qk. G. (Canal Entr.)	41-37-50	70-41-56	68	53	1.5	248	233

PT. JUDITH THRU VINEYARD & NANTUCKET SOUNDS, Charts 13218, 13237

	Lat.	Long.	°M.	°T.	Naut. Miles	°M.	°T.
Pt. Judith R. #2 Fl. R. 2.5 sec. Wh.	41-18-28	71-28-21					
To R. #32 Fl. R. 2.5 sec. Wh.	41-23-48	70-59-04	95	80	23.6	275	260
To R. #30 Fl. R. 4 sec. Bell	41-24-00	70-50-12	86	71	5.8	266	251
To Nobska R. #26 Fl. R. 4 sec. Bell	41-30-19	70-38-43	68	54	10.8	248	234
To Hedge Fence R. #22 Fl. R. 4 sec. Gong	41-28-21	70-29-00	121	105	7.6	301	285
To .1 mi. N. of G. #21 Fl. G. 2.5 sec. Gong	41-26-51	70-17-29	116	100	8.7	296	280
To Tuckernuck Sh. G. #1 Fl. G. 4 sec. Bell	41-24-19	70-09-48	128	112	6.5	308	292
To Nantucket Bar "NB" Mo (A) Bell	41-19-01	70-06-13	170	155	5.9	350	334

EASTWARD THRU GREAT ROUND SHOAL, Chart 13237

	Lat.	Long.	°M.	°T.	Naut. Miles	°M.	°T.
Cross Rip G. #21 Fl. G. 2.5 sec. Gong	41-26-51	70-17-29					
To G. #17 Fl. G. 6 sec. Gong	41-26-35	70-11-12	110	94	4.7	290	274
To G. #13 Fl. G. 2.5 sec. Bell	41-25-55	69-58-54	110	94	9.2	290	274
To RW "GRC" Mo (A) Wh.	41-24-21	69-54-55	134	118	3.4	314	298
To RW "GRS" Mo (A) Wh. Racon "G"	41-26-06	69-43-19	95	79	8.8	275	259
To RW "PR" Mo (A) Wh.	41-33-54	69-50-44	341	325	9.5	161	145

EASTWARD THRU POLLOCK RIP CHANNEL, Chart 13237

	Lat.	Long.	°M.	°T.	Naut. Miles	°M.	°T.
Cross Rip G. #21 Fl. G. 2.5 sec. Gong	41-26-51	70-17-29					
To R. #18 Fl. R. 4 sec. Bell	41-27-14	70-14-04	100	84	2.6	280	264
To Handkerchief Sh. #14 Nun	41-29-10	70-05-00	91	75	7.0	271	255
To Handkerchief Sh. #12 Nun	41-30-05	70-03-14	71	55	1.7	251	235
To R. #10 Fl. R. 4 sec. Bell	41-31-51	70-00-31	65	49	2.7	245	229
To R. #8 Fl. R. 6 sec.	41-32-43	69-58-55	70	54	1.5	250	234
To R. #4 Fl. R. 4 sec. Gong	41-33-23	69-54-33	95	79	3.5	275	259
To RW "PR" Mo (A) Wh.	41-33-54	69-50-44	341	325	9.5	161	145
Thence to Boston via Y. "BD," (see below).							

TO BOSTON, OUTSIDE CAPE COD, Charts 13246, 13267

	Lat.	Long.	°M.	°T.	Naut. Miles	°M.	°T.
RW "PR" Mo (A) Wh.	41-33-54	69-50-44					
To Y. "BD" 4 Fl. Y. 20 sec.	42-08-15	69-53-35	13	357	34.5	193	177
To Boston RW Ltd. By. "B" (Mo) A Horn	42-22-42	70-46-58	314	298	33.8	134	118
To Graves G. #5 Fl. G. 4 sec. Wh.	42-22-33	70-51-28	284	268	3.3	104	88
To Finn's Ledge R. #2 Qk. R. Bell	42-22-11	70-55-09	278	262	2.8	98	82
Thence down Boston North Channel							

CAPE COD CANAL TO BOSTON, Charts 13246, 13267

	Lat.	Long.	°M.	°T.	Naut. Miles	°M.	°T.
▶ **Alongshore:**							
G. #3 Qk. G. Bell (Canal E. Entr.)	41-46-48	70-28-55					
To Mary Ann Rks. R. #12 Fl. R. 2.5 sec. Wh.	41-55-06	70-30-22	13	357	8.5	193	177
To Farnham Rk. #6 Fl. R. 4 sec. Bell	42-05-36	70-36-26	349	333	11.5	169	153
To Jason Sh. G. #21 Fl. G. 4 sec. Wh.	42-16-35	70-42-22	354	338	11.8	174	158
To Graves G. #5 Fl. G. 4 sec. Wh.	42-22-33	70-51-28	327	311	9.0	147	131
To Finn's Ledge R. #2 Qk. R. Bell	42-22-11	70-55-09	278	262	2.8	98	82
Thence down Boston North Channel							
▶ **Offshore:**							
G. #3 Qk. G. Bell (Canal E. Entr.)	41-46-48	70-28-55					
To RW "CC" Mo (A) Bell	41-48-53	70-27-39	40	24	2.3	220	204
To Humarock RW "H" Mo (A) Wh.	42-09-49	70-31-09	9	353	21.1	189	173
To Boston RW Ltd. By. "B" (Mo) A Horn	42-22-42	70-46-58	334	318	17.4	154	138
Thence to Graves #5, to Finn's Ledge #2 and down Boston North Channel (see above).							

PILOTING ALONGSHORE

CROSS BEARINGS
by Compass (or RDF) **Notes A, B, C**
(Accuracy: fair-good)

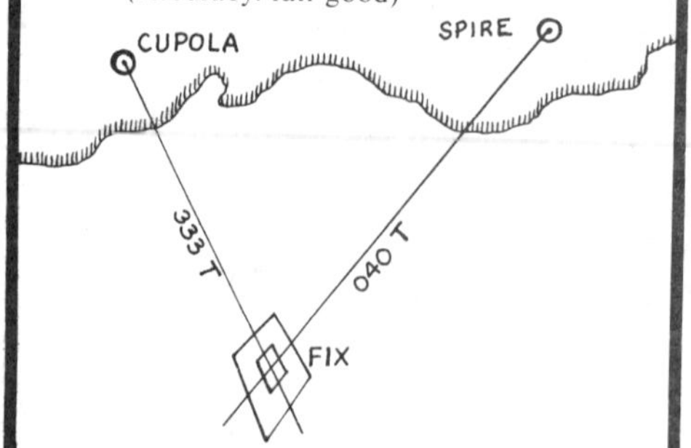

1. Cupola (or RBN) bears 333°T; construct LOP.
2. Spire bears 040°T; construct LOP.
3. Where LOP's cross is your position (FIX).

RANGE & SOUNDING
by Visual Range and Sounder **Note D**
(Accuracy: good-excellent)

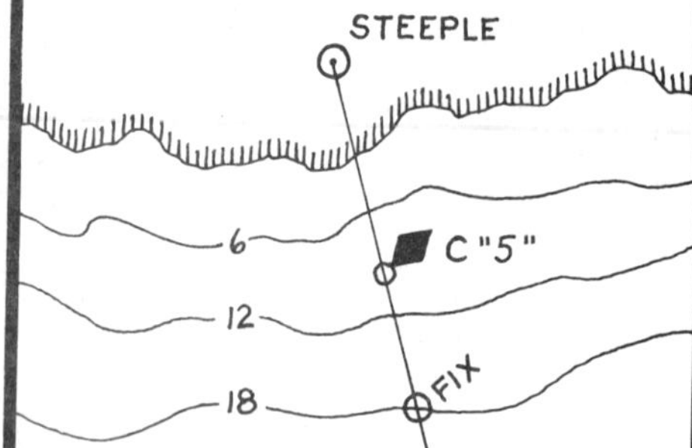

1. Steeple appears over Can "5", giving LOP.
2. Depth sounder indicates 17′. Subtract for ht. of tide above datum level (say 1′). Add for depth of transducer below surface (say 2′), for adjusted water depth of 18′.
3. FIX at 18′ depth along LOP.

RUNNING FIX
by Pelorus/Compass, & Log Notes E, F
(Accuracy: fair)

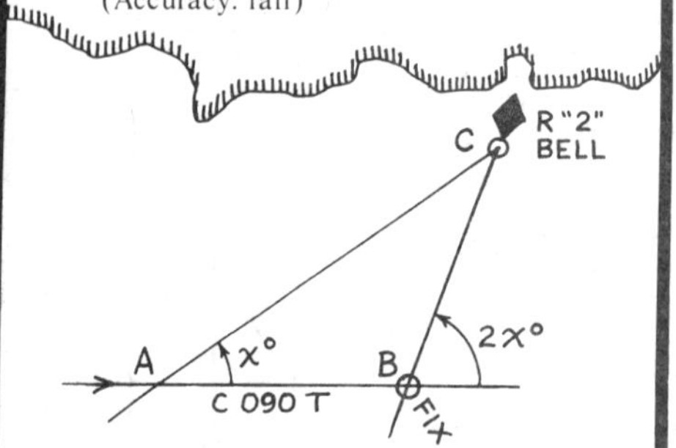

1. On CMG of 090°T, Bell "2" appears at X° off bow. Note Log. Construct LOP "CA".
2. Later, Bell "2" bears 2X° off bow; Note Log and plot LOP "CB".
3. Distance AB from Log = Distance CB for FIX.

FORE & AFT BEARINGS
by Compass and Log **Note F**
(Accuracy: fair)

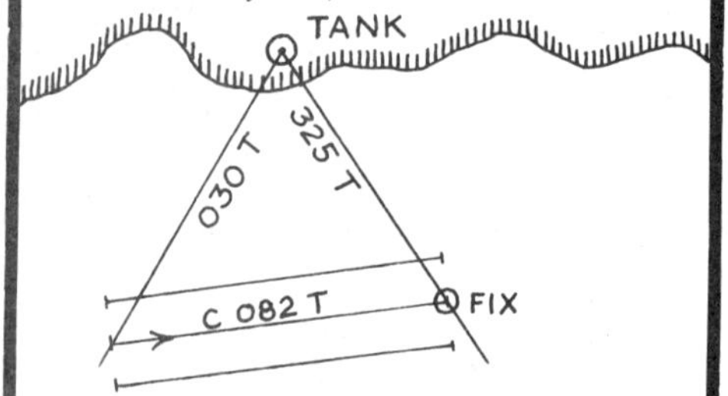

1. On CMG of 082°T, Tank bears 030°T. Note Log. Construct LOP.
2. Later, Tank bears 325°T. Note Log. LOP.
3. Set dividers to logged distance, find where points touch LOP's parallel to CMG, and plot FIX.

Caution: The prudent navigator will assume some error in these and similar situations, unless data are irrefutable. See Note A below.

NOTES:

A) Large box surrounding FIX represents Area of Uncertainty with ±5° error in instruments or procedure. Small box shows Area of Uncertainty with ±2° error.

B) When possible, LOP's should cross at more than 60° (or less than 120°) to keep Area of Uncertainty to a minimum.

C) If there is no handbearing compass aboard, aim your boat at the object and read bearing from the steering compass.

D) Although this LOP is likely to be quite accurate, buoys can move. Two land-based objects give the best range. Depth readings must be adjusted to give accurate positions, especially with widely spaced bottom contours. Sounder dead? Use lead.

E) First bearing must be no greater than 45°, second no greater than 90°. For other useful bearing angles, consult the Table of Equivalents near the back of ELDRIDGE.

F) Course-Made-Good (CMG) is required here (not just compass course converted to True), since current will skew the results. Distance, too, must be over the bottom, not just through water.

BOSTON TO THE EASTWARD, Chart 13267

COURSES AS GIVEN ARE FOR VARIATION FOR CURRENT YEAR	Lat.	Long.	Courses °M.	Courses °T.	Naut. Miles	Reverse of Courses °M.	Reverse of Courses °T.
Out Broad Sound to Finn's Ledge #2			44	28		224	208
From Finn's Ledge #2:	42-22-11	70-55-09					
To Pig Rks. R. #4 Fl. R. 4 sec. Bell	42-27.6	70-50.2	50	34	6.6	230	214
To Marblehd RG 3 Fl. R. 6 sec. Bell	42-30.2	70-49.1	35	19	2.8	215	199
To Cape Ann R. #2 Fl. R. 6 sec. Wh.	42-37-56	70-31-09	65	49	23.8	245	229
From Peaked Hill Bar "2PH" Fl. R. Wh.:	42-06-59	70-06-13					
To Boston Lt. Fl. W. 10 sec. Horn	42-19-41	70-53-24	306	290	37.2	126	110
To Boston RW Ltd. By. "B" (Mo) A Horn	42-22-42	70-46-58	314	298	33.8	134	118
To Marblehead Rk. Beacon			322	306	40.0	142	126
To Newcomb Ledge G. #1 Fl. G. 4 sec. Wh.	42-30-28	70-44-23	325	309	37.0	145	129
To Eastern Pt. R. #2 Fl. R. Wh.	42-34-14	70-39-50	334	318	37.2	154	138
To Cape Ann R. #2 Fl. R. 6 sec. Wh	42-37-56	70-31-09	345	329	36.3	165	149
From Boston Ltd. By. "B":	42-22-42	70-46-58					
To Marblehead Rk. Beacon			2	346	8.0	182	166
To Eastern Pt. R. #2 Fl. R. Wh.	42-34-14	70-39-50	39	23	13.1	219	203
To Cape Ann R. #2 Fl. R. 6 sec. Wh	42-37-56	70-31-09	53	37	19.4	233	217

CAPE ANN TO MOOSE PEAK, Charts 13278, 13286, 13288, 13302, 13312, 13325

	Lat.	Long.	Courses °M.	Courses °T.	Naut. Miles	Reverse of Courses °M.	Reverse of Courses °T.
To Cape Ann R. #2 Fl. R. 6 sec. Wh	42-37-56	70-31-09					
To Dry Salvages G. #1 Fl. G. 2.5 sec. Bell	42-40-37	70-33-31	344	328	3.3	164	148
To Isles of Sh. Lt. (abm. 1 mi. to E.)	42-58-02	70-37-24	3	347	17.9	183	167
(To Whaleback Lt. 2 Fl. W. 10 sec.)	43-03-32	70-41-47	354	338	6.1	174	158
To R. "24YL" Fl. R. 4 sec. Wh.	43-04-25	70-34-27	43	27	7.2	223	207
To RW "CP" Mo (A) Wh.	43-20-16	70-23-37	42	25	17.8	222	205
To RW "WI" Mo (A) Wh.	43-27-42	70-17-23	48	31	8.8	228	211
To Portland Ltd. By. "P" Fl. W. Hn. Racon "M"	43-31-36	70-05-28	82	65	9.5	262	245
To W. Cod Ledge Rk. #4 Fl. R. 4 sec. Wh.	43-34-02	70-07-47	341	324	2.9	161	144
To Pine Tree Ledge R. #8 Qk. R. Wh.	43-36-36	70-10-51	337	320	3.4	157	140
To Cushing Is. R. #12 Fl. R. 4 sec. Bell	43-37-57	70-12-31	336	319	1.8	156	139
To Spring Pt. Lt. (abm.) Fl. W. 6 sec.	43-39-08	70-13-26	359	342	1.3	179	162
Portland Ltd. By. "P" Racon "M"	43-31-36	70-05-28					
To Seguin R. "18SI" Wh.	43-40-11	69-46-56	75	58	16.0	255	238
To Cuckolds Lt. Fl. W. (abm. .5 mi. to N.)	43-46-47	69-38-59	62	44	8.8	242	224
(Then to #4 Fl. R. 2.5 sec., Boothbay)	43-48-57	69-38-14	30	13	2.5	210	193
Seguin "18SI" Wh.	43-40-11	69-46-56					
To Manana R. "14M" Fl. R 4 sec. Wh.	43-45.3	69-22.5	92	74	18.5	272	254
To Marshall Pt. "MP" Mo (A) Wh	43-55-18	69-10-52	58	40	13.0	238	220
To Two Bush Is. RW "TBI" Mo (A) Wh.	43-58-17	69-00-16	86	69	8.3	266	249
To Curtis Is. Lt. F. G. (abm. passing	44-12-05	69-02-55	20	2	13.8	200	182
close to "PA" Mo (A) Gong)	44-01.1	69-00.3					
(Then to Searsport #5 Fl. G. Bell)	44-26-35	68-54-07	34	16	15.1	214	196
RW "TBI" Mo (A) Wh.	43-58-17	69-00-16					
To R. "2RB" Fl. R. 6 sec. Bell	43-59-22	68-37-56	112	94	16.4	292	274
To Gr. Duck Is. Lt. Fl. R. 5 sec.	44-08-31	68-14-44	77	58	20.4	257	238
(1 mi. abm. to N.)							
(Thence to Baker Is. R. "8BI" Wh.)	44-13-38	68-10-50	40	21	6.5	220	201
To R. #2 Fl. R. 4 sec. Gong (to Seal Hbr.)	44-16-32	68-13-02	355	336	3.2	175	156
Baker Is. R. "8BI" Wh.	44-13-38	68-10-50					
To G. #3 Fl. G. 4 sec. Gong	44-20-06	68-10-03	24	5	6.5	204	185
(Then to Bar Harbor)			18	359	3.6	198	179
Baker Is. R. "8BI" Wh.	44-13-38	68-10-50					
To Southeast Rk. R. #6A Fl. R. 4 sec. Wh.	44-19-46	67-48-37	87	68	17.4	267	248
To Moose Pk. (abm. 1 mi. to N.) Fl. W. Horn	44-28-28	67-31-55	76	57	14.8	256	237

While all possible care has been taken in compiling these courses and distances the publishers are not responsible for any errors that may be found.

VICE ADMIRAL ROBERT FITZROY, RN

(See pp. 246 & 247 for FitzRoy's Barometer Instructions)

One Sunday morning at 8 o'clock, Robert FitzRoy shut himself in his dressing room and cut his throat with a razor. He was 59 years old. *The Times* reported that "a painful feeling of regret agitated the members of the Board of Trade...when the suicide became known."

Robert FitzRoy was born in 1805 in Suffolk, a son of the second son of the third Duke of Grafton. The first Duke (Robert's great-great grandfather) was the illegitimate son of Charles II and the Duchess of Cleveland. Young FitzRoy's mother was the sister of the Marquis of Londonderry; perhaps from this side of the family Robert inherited his highly-strung and hard-working temperament.

At age 13 Robert FitzRoy entered the Royal Naval College at Portsmouth where he studied mechanics, hydrostatics, physics, astronomy and much more. In 1824, as a lieutenant, he served on the *Thetis*, off the South American coast, and later, at age 23 was promoted to the command of the *Beagle*. The *Beagle*, a ten-gun brig (known as a "coffin" because of its tendency to turn turtle) was engaged together with the *Adventure* in surveying the coasts of Patagonia, Tierra del Fuego and the Straits of Magellan. On FitzRoy's second voyage (to South America and the Galapagos) the *Beagle* also carried a young and unknown Cambridge graduate, Charles Darwin, on his first job as a naturalist.

The 5 years of this second voyage were the most important years of FitzRoy's life. The Straits of Magellan and a large part of the mainland coast were surveyed and a chronometric line was run around the world, fixing the longitudes of many second meridians. But the relationship between FitzRoy and Darwin which had at first been an enthusiastic one, altered drastically. FitzRoy had hoped that Darwin's naturalist findings would confirm his own fundamentalist beliefs about the creation of the universe (the Biblical version). Later, in his *Narrative of the Voyages of HMS's Adventure and Beagle,* he indulges this belief, allying his scientific observations of cliffs, lava beds, marine fossils, etc. with the Biblical Deluge. These same observations are the ones that led Darwin to formulate his theory of evolution! (Also the failure of FitzRoy, a zealous humanitarian, to "civilize" three Fuegian natives added another disappointment). This mismatch of the two men engendered frustration, mitigating FitzRoy's personal sense of achievement in what was a remarkable surveying expedition.

When the *Beagle* returned to England in 1836, the Admiral was acclaimed as a navigator, surveyor and chart maker. He was awarded the gold medal to the Royal Geographic Society and two years later he published his famous surveying *Narrative*. From 1841 to 1845 he served as MP for Durham and later as Governor and Commander-in Chief of New Zealand. Here, in defending the rights of the aborigines over those of the white settlers, his relationship with the Colonial Secretary deteriorated and in 1845, FitzRoy was dismissed. The blow to his reputation and self-esteem was severe.

Over the next 9 years he was Superintendent of the naval dockyard at Woolwich, was elected a Fellow of the Royal Society and in 1854 became a Chief of the new meteorological department of the Board of Trade. This post established his enduring reputation as a practical meteorologist. His "wind star" charts, his system of storm warnings and barometer readings to "forecast" (his own coinage) the weather are familiar to sailors everywhere. He had already devised the simple barometer bearing his name, set up weather stations, produced "synoptic charts" (another term he coined) and produced weather forecasts which were later published by the newspapers. He also devised the system of weather cones still used at ports and fishing villages.

Already the author of the *Narrative* and a basic *Barometer Manual* (containing weather rhymes composed on the *Beagle*: "... soon you may set sail again"), FitzRoy in 1862 published his *Weather Book*, an overwhelmingly comprehensive guide to all things relating to weather, with emphasis on barometer readings telling future rather than present weather conditions. The book's success was a high point in his career.

In 1859 Darwin published his *Origin of Species* and FitzRoy began desperately defending his own fundamentalist beliefs, in person and at meetings. Leading articles in *The Times* began to cast derisory remarks on his forecasts and the famous American meteorologist, Lt. M. F. Maury, showed himself less impressed by his English counterpart's achievements.

FitzRoy's reaction to all this was predictable and inevitable. He forced himself to work harder until the strain of "proving" himself became intolerable. Thus on April 30, 1865, Vice Admiral Robert FitzRoy, RN, ended his life.

Reprinted from the 1980 ***Eldridge Tide and Pilot Book***

Approximate

Bar Harbor to
Halifax, N.S. 259
Yarmouth, N.S. 101
Saint John, N.B. 122
Machiasport. 49
Rockland 60
Boothbay Hbr. 85
Portland. 114
Marblehead 178

Rockland to
Boothbay Harbor via
Muscle Ridge Ch. 38
Belfast 22
Bucksport 33

Boothbay Harbor to
Kennebec River 11
Monhegan 15
Portland. 34

Portland Ltd. Horn By. "P" to
Biddeford 17
Portsmouth 54
Cape Cod Light 99
Cape Cod Canal, E. Entr. 118
Pollack Rip Slue 141

Portsmouth (Whaleback) to
York River 7
Biddeford Pool 30
Newburyport Entr. 15
Gloucester
Via Annisquam. 28

Gloucester to
Boston 26
Scituate 25
Plymouth 40
Cape Cod Canal, E. Entr. .50
Cape Cod Light 44
Provincetown. 47

Marblehead to
Portsmouth 43
Biddeford Pool 68
Portland. 87
Boothbay Harbor. 104
Rockland 133
Plymouth 38
Cape Cod Canal, E. Entr. .47

Boston (Comm. Pier) to
Marblehead 17
Isles of Shoals 52
Portsmouth 58
Portland. 95
Kennebec River 107
Boothbay Harbor. 116
Rockland 149
North Haven 148
Bangor 194
St. John, N.B.. 286
Halifax, N.S. 380
Cohasset 14
Cape Cod Canal, E. Entr. .50
Provincetown. 50
Vineyard Haven 77
New Bedford 81
Newport. 107
Fall River. 122
New London. 140
New York 234

***W. Entr. Cape Cod Canal to**
E. Entr. 8
Woods Hole. 15
Quicks Hole 20
New Bedford 24
Newport. 50
New London. 83

Woods Hole to
Hyannis 19
Chatham 32
Cuttyhunk 14
Marion 11

Vineyard Haven to
Edgartown 9
Marblehead (around
Cape. 114
Canal (via Woods Hole) . . .20
Newport. 43
New London. 75
New Haven 115
South Norwalk 132
City Island. 153

******Western entr.*, the beginning of the "land cut" at Bourne Neck, 7.3 naut. mi. up the channel from Cleveland Ledge Lt.**

How Current Affects Fishing - *by Mark Sosin*

Where They'll Be

Current is essential to the feeding strategies of many species, whether it be on the inshore grounds or in blue water. The swirling turmoil of a rip gives striped bass and bluefish an advantage over hapless prey struggling in the flow. Snook linger at the mouth of a tidal creek waiting for the last trickle of water to flush frightened minnows from their quickly shrinking sanctuary. Even the wind can create a current, such as when it pushes water down a canal, causing predators to stack up near ambush points along the sides.

Any fish trying to remain stationary in a current faces into the flow of water and expects food to come toward it. By facing into the flow, a fish gains stability much the way an airplane does when taking off into the wind. A bait or lure can be swept across the nose of the fish and appear natural, but anything that swims upcurrent usually isn't recognized as food.

Fish on the move tend to work into the current or tack across it, expecting to intercept their prey. And moving water carries odor molecules, enabling a predator to locate prey with its keen sense of smell.

When fishing a current, one must account not only for its direction, but its velocity as well. Holding against a strong current saps a considerable amount of energy from a fish. Therefore, they instinctively seek out sanctuaries from the main force of the water, while still being able to dart out and nab a snack that's swimming or drifting by.

Working the Bottom

Current velocity usually diminishes near the bottom. Fish often hold deep during the period of maximum flow, particularly where there's a depression in the bottom. This way they can rise up quickly, grab something scooting past, and return to their lie. If there's no visible cover above the surface, always fish near the bottom. To produce consistently, your bait or lure must cover the lowest five or ten percent of the water column. However, remember that you must take into account the force of the current, which can sweep the line and terminal gear above the feeding window. In this case you may have to cast farther upcurrent, or use additional weight, a heavier lure, or a deeper-diving plug to reach the payoff zone.

Rocks, wrecks, bridge pilings, and any other type of obstruction create dead zones where the current velocity is decreased. Fish hold in these spots, conserving energy while waiting for food to sweep past. As the direction of the current reverses with the change of tide, the fish swap sides and face the other way. A dead zone obviously occurs behind a solid object that blocks the current, but there is also a dead spot directly in front of the object, caused by the water piling up against the structure and countering the current. These are the places you want to fish. Work both the front and back of the structure, and experiment to find the right depth.

Take Notes

Once the current starts to moderate, fish begin moving around more. Some species cruise only during the short time on either side of slack water. It depends on the species and the habitat. Tidal currents may get off to a slow start, build rapidly to a peak, rip through an area for a while, and then taper off. A bit of observation and perhaps a few written notes will help you determine the most productive current stages for the places you fish.

Many species like to hold in "funnel necks" - places where the water is forced through a narrow opening, creating a powerful current. An inlet during an outgoing tide ranks as one of the most common and most productive types of necks. Countless creatures find themselves carried along by the increased flow, and then have to run a gauntlet of predators waiting around the neck. Look for a similar situation to occur where extensive shallows drop into a narrow channel.

continued on page 200

Approximate

Nantucket Entr. Bell NB to
Boston (around Cape) . . .105
(via Canal). 94
Chatham 23
Edgartown. 23
Hyannis 21
Woods Hole. 30
Cape Cod Canal, W. Entr. .45
Newport 66

New Bedford (State Pier) to
Woods Hole. 14
Newport 35
New London. 70
New York (Gov. I.) 163

Newport to
Providence 21
Stonington 31
New London. 45
New Haven 81
City Island. 122

Block Island (FR Horn) to
Nantucket 79
Vineyard Haven 52
Cleveland Ledge Lt. 50
New Bedford 44
Newport 22
Race Pt. Lt. 21
New London. 27

New London to
Greenport 25
New Haven 46
Bridgeport 57
City Island. 86

Port Jefferson to
Larchmont 30
So. Norwalk 15
Milford. 14
Old Saybrook 43
New London. 53

City Island to
Governors Island 17
Execution Rocks 3

Execution Rocks to
Port Chester 8
Stamford 12
Oyster Bay Harbor 14
So. Norwalk 19
Bridgeport 29
Port Jefferson 30
Milford. 37
New Haven 49
Conn. River 69
Mystic 84
Montauk Point. 87

New York (Battery) to
Jones Inlet 34
Fire Island Inlet 47
Moriches Inlet 74
Shinnecock Inlet 88
Montauk Point. 117
Keyport 22
Asbury Park 35
Manasquan 41
Little Egg Inlet 81
Atlantic City 90
Philadelphia 235
Chesapeake L. Station . .247
Cape Henry Light 262
Norfolk. 288
Baltimore. 418

Brielle-Manasquan to
E. Rockaway Inlet 32
Jones Inlet 35
Fire Island Inlet 45
Montauk Point. 117
Barnegat Inlet. 21
Atlantic City 51

Delaware Breakwater to
Reedy Pt. Entr.
C&D Canal 53
Annapolis (via Canal) 97
Norfolk. 167
New York 150
New London. 242
Providence 275
New Bedford 278
Boston (outside) 399
Portland (outside) 443

Old Point Comfort to
Baltimore. 163
Philadelphia 240
New York 276
New London. 363
Providence 392
New Bedford 397
Boston (outside) 512
Portland. 553

Work the sides and drop-offs, and pay particular attention to any points and pockets.

Chumming in a Current

Effective chumming of any type demands that the current carry the goodies and the scent to the fish. As the flow falls off, so will the action. If you are trying to tempt fish away from some form of structure, it's essential to position the boat with great care so that the chum reaches the target. When the current changes, you have to anchor on the new upcurrent side.

Velocity of flow becomes a factor in how far from the structure you have to be. In fast-moving water, it takes more distance for the chum to get down to the fish. Skilled skippers continuously adjust the scope of their anchor line according to current speed in order to get the chum to the right spot.

Fishing the Shallows

Tide is defined as the vertical rise and fall of water, while current refers to horizontal movement. Fish that prowl shallow flats are much more affected by the depth of the water than by the current, so tide becomes the essential factor when fishing for them.

The greater the range between high and low water, the stronger the current will be during both flood and ebb. Keep in mind that the change in current flow in some areas may follow the tide change by as much as several hours. That means the tide may actually be rising while the current is flowing out, and vice versa.

As a general rule, fish roam more on the incoming tide, especially in shallow water. On exceptionally high tides, they may enter and exit the shallows in different locations, or move back into shoreline cover where you can't reach them. As the water level falls, the fish drop back to deeper water. It's important to learn which way they travel, particularly if there is deeper water on more that one side of the shallow area.

Always check the tides before setting the dates for a fishing trip to an exotic locale. You can tell if you will have spring or neap tides simply by knowing the moon phase. But you also have to know which set of tides are better for the species you seek and the area you intend to fish.

Offshore Currents

Moving water affects fishing in the ocean as well as in an estuary. Major ocean currents, such as the Gulf Stream or the Loop Current, take on aspects of a river. If you're trolling in this type of water, work across the flow, with it, and finally against it. Work along the edges of upwellings, a type of vertical current that brings nutrients to the surface, creating plankton blooms and drawing in baitfish and their predators.

Coastal currents and gyres become important, even when you are not fishing a named current like the Gulf Stream. In the Atlantic, for example, most of the minor currents move north or south, and the direction should be noted. It takes local knowledge to discover the better spots when the water heads in one direction or other. Don't be afraid to experiment, but always do so in relation to the current.

Moving water within a few miles of the beach frequently flows with more force near the surface, but this may occur within a narrow vertical band. Get a bait or lure just below the fast-moving water and you have an excellent chance of finding fish.

Marine species live with currents every day and use them to their advantage. Once you recognize how important current is to fish, it's easier to narrow your search area. Learn to "go with the flow" and the results should be rewarding.

Mark Sosin teaches outdoor techniques through clinics and seminars, sharing his expertise with conservation groups and government agencies. An award-winning writer, photographer and TV producer, he has been voted into the International Fishing Hall of Fame. He broadcasts to all 50 states on "TNN Outdoors" as on-camera host and producer. (This article originally published by Salt Water Sportsman.)

MEASURED MILE SPEED TABLES

Secs.	1 Min.	2 Min.	3 Min.	4 Min.	5 Min.	6 Min.	7 Min.	8 Min.	9 Min.	10 Min.	11 Min.
0	60.00	30.00	20.00	15.00	12.00	10.00	8.57	7.50	6.66	6.00	5.45
2	58.06	29.50	19.78	14.87	11.92	9.94	8.53	7.46	6.64	5.98	5.43
4	56.25	29.03	19.56	14.75	11.84	9.89	8.49	7.43	6.61	5.96	5.42
6	54.54	28.57	19.35	14.63	11.76	9.83	8.45	7.40	6.59	5.94	5.40
8	52.94	28.12	19.14	14.51	11.68	9.78	8.41	7.37	6.56	5.92	5.38
10	51.42	27.69	18.94	14.40	11.61	8.73	8.37	7.34	6.54	5.90	5.37
12	50.00	27.27	18.75	14.28	11.53	9.67	8.33	7.31	6.52	5.88	5.35
14	48.64	26.86	18.55	14.17	11.46	9.62	8.29	7.28	6.49	5.86	5.34
16	47.36	26.47	18.36	14.06	11.39	9.57	8.26	7.25	6.47	5.84	5.32
18	46.15	26.08	18.18	13.95	11.32	9.52	8.22	7.22	6.45	5.82	5.30
20	45.00	25.71	18.00	13.84	11.25	9.47	8.18	7.20	6.42	5.80	5.29
22	43.90	25.35	17.82	13.74	11.18	9.42	8.14	7.17	6.40	5.78	5.27
24	42.85	25.00	17.64	13.63	11.11	9.37	8.11	7.14	6.38	5.76	5.26
26	41.86	24.65	17.47	13.53	11.04	9.32	8.07	7.11	6.36	5.75	5.24
28	40.90	24.32	17.30	13.43	10.97	9.27	8.03	7.09	6.33	5.73	5.23
30	40.00	24.00	17.14	13.33	10.90	9.23	8.00	7.05	6.31	5.71	5.21
32	39.13	23.68	16.98	13.23	10.84	9.18	7.96	7.03	6.29	5.69	5.20
34	38.29	23.37	16.82	13.13	10.77	9.13	7.93	7.00	6.27	5.67	5.18
36	37.50	23.07	16.66	13.04	10.71	9.09	7.89	6.97	6.25	5.66	5.17
38	36.73	22.78	16.51	12.95	10.65	9.04	7.86	6.95	6.22	5.64	5.15
40	36.00	22.50	16.36	12.85	10.58	9.00	7.83	6.92	6.20	5.62	5.14
42	35.29	22.22	16.21	12.76	10.52	8.95	7.79	6.89	6.18	5.60	5.12
44	34.61	21.95	16.07	12.67	10.46	8.91	7.76	6.87	6.16	5.59	5.11
46	33.96	21.68	15.92	12.58	10.40	8.86	7.72	6.84	6.14	5.57	5.09
48	33.33	21.42	15.78	12.50	10.34	8.82	7.69	6.81	6.12	5.55	5.08
50	32.72	21.17	15.65	12.41	10.28	8.78	7.66	6.79	6.10	5.53	5.07
52	32.14	20.93	15.51	12.32	10.22	8.73	7.63	6.76	6.08	5.51	5.05
54	31.57	20.69	15.38	12.24	10.16	8.69	7.59	6.74	6.06	5.50	5.04
56	31.03	20.45	15.25	12.16	10.11	8.65	7.56	6.71	6.04	5.48	5.02
58	30.50	20.22	15.12	12.08	10.05	8.61	7.53	6.69	6.02	5.47	5.01
	30.00	20.00	15.00	12.00	10.00	8.57	7.50	6.66	6.00	5.45	5.00
		3 Min.	4 Min.	5 Min.	6 Min.	7 Min.	8 Min.	9 Min.	10 Min.	11 Min.	12 Min.

The blank table below should be used for making a permanent record of the speed obtained at any given number of R.P.M. on this yacht.

RPM	SPEED

Table applies to either nautical or land miles; answer in knots in former case, miles-per-hour in latter. **Example:** if time elapsed to run measured mile is 4 min. and 2 sec., the 4 min. column opp. 2 sec. gives 14.87, meaning knots or miles-per-hour, depending on type of mile run.
1 nautical mile 6080 ft.
1 land mile 5280 ft.

If a current is involved, run the distance up and downstream, noting the time for each run. Your true boat speed will be found as follows:

$$\frac{60 \times (\text{Distance}) \times (\text{time up} + \text{time downstream})}{2 \times (\text{time up} \times \text{time downstream})}$$

Example: $\frac{60 \times 1.0\ (1\ \text{mi.}) \times (10 + 8)}{2 \times (10 \times 8)}$ $\frac{1080}{160} = 6.75 = \text{speed}$

THE VERSATILE HAND BEARING COMPASS
and Finding a Magic Spot

Not many boats have a steering compass from which it is easy to take bearings and these boats, without a Hand Bearing Compass, must rely blindly on the accuracy of the steering compass. A Hand Bearing Compass can change all that and give the boat owner a continuous checking device for all conditions. There are many types of Hand Bearing Compasses and it is generally true that the more expensive ones are the better ones. They vary widely in size, portability and legibility and the buyer should select the one that fits his own purposes best.

Checking for Sensitivity: With the Compass at rest, well away from anything magnetic, deflect the compass card precisely 3°, using a knife or screwdriver or other magnetic object. When you remove your magnet the compass card should return without delay to its at-rest position, proving the Compass is without friction or stickiness.

Checking for Built-in Error: Take your Compass to any identifiable spot ashore, free from magnetic materials. Sight very carefully at any distant mark having a known magnetic bearing. (Use a local chart.) If the magnetic bearing by the Compass is the same as the known bearing, the Compass has no built-in error. If it reads 2° high for example, then you must *always* subtract 2° from any bearing taken. If it reads 2° low, *always* add 2°. This is a permanent error.

Finding the "Magic Spot": Find a place on your boat on the center line and preferably as far aft as possible where it is likely that there are no magnetic effects from the boat. On sailing vessels this may be on the fantail or on top of the aftermost end of the cabin. Powerboats vary more widely and finding a spot may be more difficult.

In any case, on a clear, quiet day, with a friend at the wheel and the controls, go to this spot and take a bearing with your Hand Bearing Compass on the most distant, stationary object. Remember this bearing. Then with your rudder hard over, give the boat a kick ahead and make a short turn of 50 or 60°. Standing in the same spot, take another bearing of the same object. Continue this procedure until you have completed a circle. If all the bearings taken have been the same, no matter what the boat's heading, you have found a spot free of the boat's magnetism. Mark this "magic spot" carefully for future reference. If the bearings have not been the same, try other locations on the boat till you find the right one. The reason for standing as far aft as possible is to permit the greatest accuracy later on in checking the boat's exact magnetic course. When sighting from this "magic spot" towards the mast or jackstaff, the greater the distance is, the greater the accuracy will be.

Checking Your Steering Compass for Deviation: Now, having found your "magic spot" on the boat, you can check the deviation of your steering compass at any time of day, in thick fog or darkest night. All you have to do is go to your spot and sight with your Hand Bearing Compass forward against the mast or jackstaff and that is what your steering compass should read. You will then have confidence in the integrity of your course.

Sometimes it is impossible to find a "magic spot" well aft. In this case, you can put your back against the mast and sight aft at the sternstaff or mizzen or any other center-line object but you must add or subtract 180° in order to correlate with the heading of the steering compass.

Since you can establish the deviation of your steering compass on a *single* heading, you can also find it on *all* headings and if your steering compass has a built-in corrector system you can adjust it to agree with the headings observed from your Hand Bearing Compass.

Finally, from this "magic spot" you can take all your bearings for piloting purposes. The Hand Bearing Compass is much more useful than a pelorus. A pelorus can be used only during the daytime under conditions of good visibility and requires considerable manipulations and calculations. If you would cruise with absolute assurance of your course, the Hand Bearing Compass is the tool you need.

GPS and DGPS

GPS and DGPS (Global Positioning System and Differential GPS) are satellite-based navigational technologies developed by the U.S. Department of Defense. Now made available for commercial and private use, the systems promise to revolutionize the marine navigation industry, allowing mariners to know their position with unprecedented accuracy anywhere on earth. Because of both accuracy and global scope, GPS and DGPS are generally regarded as superior to LORAN and to conventional satellite navigational systems now in place.

GPS is a system that positions any object on earth -- or above the earth -- through the simple principle of triangulation. Extremely accurate atomic clocks on board GPS satellites, perfectly synchronized, are the basis for measuring the time it takes an encoded radio signal to travel from the satellite to the GPS receiver. When signal-travel times are calculated from four well-spaced orbiting satellites whose positions are precisely known, the position of the receiver becomes known.

There are, of course, a number of variables and sources of distortion. Atmospheric conditions can affect signal travel time. The quality of the on-board receiver will affect the quality of the information it processes. For security reasons, commercial GPS signals are purposefully distorted by the government to generate a margin of error (the government's military system is said to be accurate within a fraction of an inch). But as a rule, GPS has the capacity to locate a vessel within 100 meters or better of its true position anywhere at sea. Under optimum conditions, accuracy can be much greater.

The constellation of GPS satellites (3 of which are "spares") are spaced roughly equidistant from each other around the earth. As of this writing, 26 satellites are now in orbit. The Coast Guard cautions that although the system has reached operational capability it is usable only for mariners who have the right equipment, and who are aware of its limitations.

Typically, GPS receivers have small antennae, some only a few inches long. GPS satellites transmit identical "pseudo-random" codes along an extremely narrow band-width, requiring very little power for reception. Although the signal is so weak it is barely discernible from the earth's natural background noise, the receiver "recognizes" the GPS code as a distinct pattern and processes it accordingly. Land-based receivers and transmitters are located at Brunswick, ME, Portsmouth, NH, Chatham, MA, Montauk Pt., NY, Sandy Hook, NJ, Cape Henlopen, DE, Cape Henry, VA, Ft. Macon, NC, Charleston, SC, Cape Canaveral, Miami and Key West, FL.

The Coast Guard has now developed "Differential GPS" as an improvement on the basic system, to provide mariners with computerized corrections to their receivers, further refining the accuracy of their readouts to within 10 meters or better. Marine Radiobeacons now broadcast correction factors to be used by on-board receivers to offset errors in satellite signals.

GPS and Differential GPS are fast emerging as the navigation systems of choice for both coastal and ocean-going vessels, but mariners are advised to be aware of new develop-ments as the technology changes.

The status of Coast Guard DGPS broadcasts is available from the C.G. Navigation System Center (NAVCEN) at 703-313-5900. For Electronic Bulletin Board call 703-313-5910, http://www.navcen.uscg.mil. WWV (Ft. Collins, CO) broadcasts GPS information at 14-15 minutes past the hour.

We cover the Waterfront and each of the Seven Seas
From DAWN,
when the mist is still rising off the sea, to mid-day when the glare of a thousand suns reflect in your face.
The right time and place for the Swift 825R Eaglet. Multi-coated lenses cut through the haze and cut down the reflection and guarantee a bright image in all weather. Waterproof and armored the Swift 825R Eaglet weighs only 22 ounces but brings distant objects 7 times closer.
825R EAGLET
To DUSK,
when channel markers begin to merge with the lengthening shadows and distant land-marks fade along with the light.
Time to reach for your Swift 716 Armored Sea King. Waterproof and nitrogen filled the Sea King is equipped with a Suunto compass and vertical and horizontal reticle. With a light efficiency of 84.2 this glass was designed as the perfect night glass for mariners.
716 SEA KING
Discover the entire line of Swift Marine Binoculars at better specialty stores including the 7x, 50 rubber armored, immersion tested Sea Wolf.
SWIFT 713 SEA WOLF - Armored - Waterproof
7x, 50 BCF - (372 ft.) - 33 oz. - R.L.E. 76.5
SWIFT
Swift Instruments, Inc.
952 Dorchester Ave., Boston, MA 02125
In Canada: Vision Canada LTD., Pickering, Ontario L1N 3SI
Some binoculars take you to the visual edge.
Swift binoculars take you a step beyond.
For the name of your closest Swift retailer call !-617-436-2960

YACHT FLAGS AND HOW FLOWN

The Ensign: 8 AM to Sundown only.
At Stern Staff of all vessels at anchor, or under way under power or sail.
At the leech of the aftermost sail, approximately 2/3 of the leech above the clew. When the aftermost sail is gaff-rigged, the Ensign is flown immediately below the peak of the gaff.

The Club Burgee:
At Bow Staff of Launches and Power Yachts with one mast.
At Main Peak of Yawls, Ketches, Catboats, Sloops and Cutters.
At Fore Peak of Schooners and Power Yachts with two masts.

The Private Signal: 8 AM to Sundown only, when owner is in command.
At Masthead of Catboats, Sloops, Cutters and Power Yachts with one mast.
At Mizzen Peak of Yawls and Ketches.
At Main Peak of Schooners and Power Yachts with two masts.

Flag Officer's Flags: Commodore (blue), Vice Commodore (red) and Rear Commodore (white) flags are flown in place of the private signal on all rigs except single-masted sailboats when it is flown on the masthead replacing the club burgee.

The Union Jack: 8 AM to Sundown, only at anchor on Sundays, holidays or occasions for "dressing-ship," flown from Bowstaff. Sailing vessels without Bowstaff may lash the Jack to the Forestay a few feet above the Stem Head.

The U.S. Power Squadron Ensign: 8 AM to Sundown. May be flown in place of Yacht Ensign, or flown additionally at the Starboard Yardarm or Main Spreader, but only when Squadron Member is in command.

How Do You Define A Shipyard?

D.N. KELLEY & SON, INC. SHIPYARD

Since 1864 Quality Yacht Repairs

Kelley Power

- Fiberglass, structural & cosmetic
- Painting, Awlgrip, Imron, Sterling & other systems
- Machining • Shafting & Propellor repair • Welding • Rigging
- Engine & Generator Repairs
- Antonio Marques Marine Carpentry
- Piers handling 50'-150', 15' draft

800 & 650 Ton Railways
160 Ton Travelift
Deep Water inside the Hurricane Barrier

D. N. KELLEY & SON, INC. SHIPYARD

32 Water St. Fairhaven, MA 02719
(508) 999-6266 Fax: (508) 999-2513
We monitor Channels 9 & 68

Perkins

GANNON AND BENJAMIN MARINE RAILWAY, INC.

Box 1095, Beach Road, Vineyard Haven — Martha's Vineyard, Mass. 02568
Phone: 508-693-4658 — FAX: 508-693-1818
Next to the Black Dog

Wooden Boat Design and Construction

Repairs, Traditional Rigging, 8' Draft
50 Ton Railway

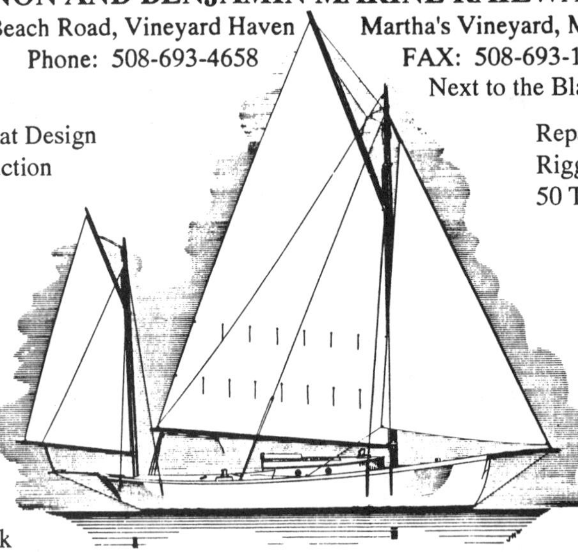

The Loft
Sails: New & Repairs
Canvas Work
508-693-3312

Charters on Classic Boats 40'-72' by the day, weekend, term

"Candle in the Wind"
Launched 1992

Moorings • Launch Service: Call Vineyard Haven Launch on Channel 72

THE INTERNATIONAL CODE OF SIGNALS

The Code comprises 40 flags: 1 Code Flag; 26 letters; 10 numerals; 3 repeaters. With this Code it is possible to converse freely at sea with ships of different countries. (For further information, see H.O. Vol. 102)

SINGLE FLAG SIGNALS

A : I have a diver down; keep well clear at slow speed.
B : I am taking in, or discharging, or carrying dangerous goods.
C : Yes
D : Keep clear of me; I am maneuvering with difficulty.
E : I am altering my course to starboard.
F : I am disabled; communicate with me.
G : I require a pilot. (When made by fishing vessels operating in close proximity on the fishing grounds it means: "I am hauling nets.")
H : I have a pilot on board.
I : I am altering my course to port.
J : I am on fire and have dangerous cargo on board: keep well clear of me.
K : I wish to communicate with you.
L : You should stop your vessel instantly.
M : My vessel is stopped and making no way through water.
N : No
O : Man overboard.
P : *In Harbor: All* persons should report on board as the vessel is about to proceed to sea.
At Sea: It may be used by fishing vessels to mean "My nets have come fast upon an obstruction."
Q : My vessel is "healthy" and I request free pratique.
S : My engines are going astern.
T : Keep clear of me; I am engaged in pair trawling.
U : You are running into danger.
V : I require assistance.
W : I require medical assistance.
X : Stop carrying out your intentions and watch for my signals.
Y : I am dragging my anchor.
Z : I require a tug. (When made by fishing vessels operating in close proximity on the fishing grounds it means: "I am shooting nets."

TWO FLAG SIGNALS – URGENT AND IMPORTANT

N C : International Distress
A C : I am abandoning vessel.
A N : I need a doctor.
C B : I require immediate assistance.
D V : I am drifting.
D X : I am sinking.
G M : I cannot save my vessel.
I M : I request to be escorted until further notice.
I N : I require a diver.
I T : I am on fire.
J F : I am in distress and need immediate attention.
J G : I am aground; I am in dangerous situation.
J W : I have sprung a leak.
K F : I require a tug.
Y G : You do not appear to be complying with traffic separation scheme.

INTERNATIONAL MORSE CODE

A	· —	**J**	· — — —	**S**	· · ·	**2**	· · — — —
B	— · · ·	**K**	— · —	**T**	—	**3**	· · · — —
C	— · — ·	**L**	· — · ·	**U**	· · —	**4**	· · · · —
D	— · ·	**M**	— —	**V**	· · · —	**5**	· · · · ·
E	·	**N**	— ·	**W**	· — —	**6**	— · · · ·
F	· · — ·	**O**	— — —	**X**	— · · —	**7**	— — · · ·
G	— — ·	**P**	· — — ·	**Y**	— · — —	**8**	— — — · ·
H	· · · ·	**Q**	— — · —	**Z**	— — · ·	**9**	— — — — ·
I	· ·	**R**	· — ·	**1**	· — — — —	**0**	— — — — —

International Code of Signals

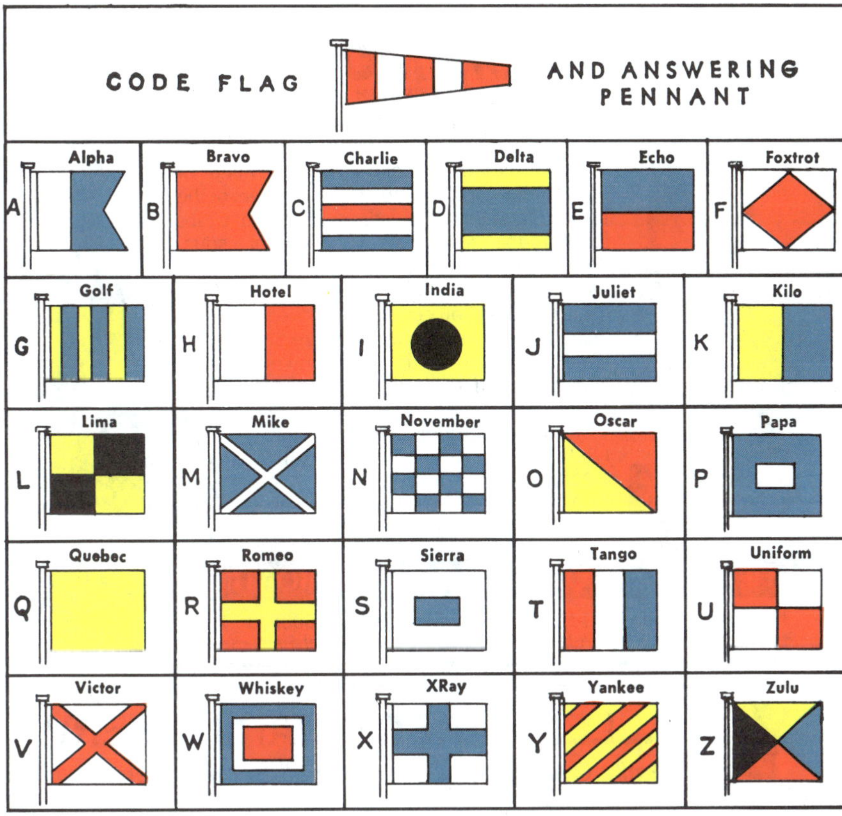

NUMERAL PENNANTS

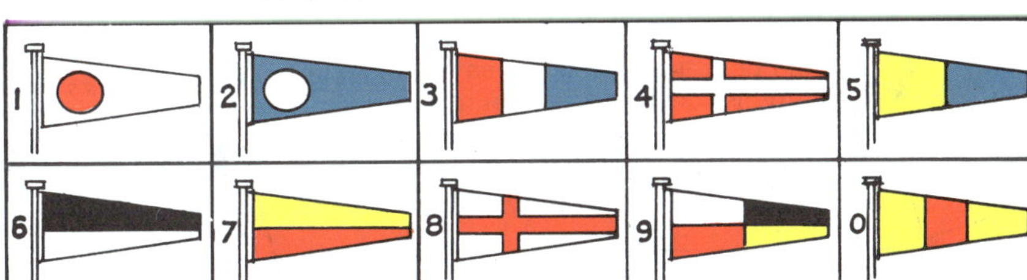

REPEATERS

U. S. STORM SIGNALS

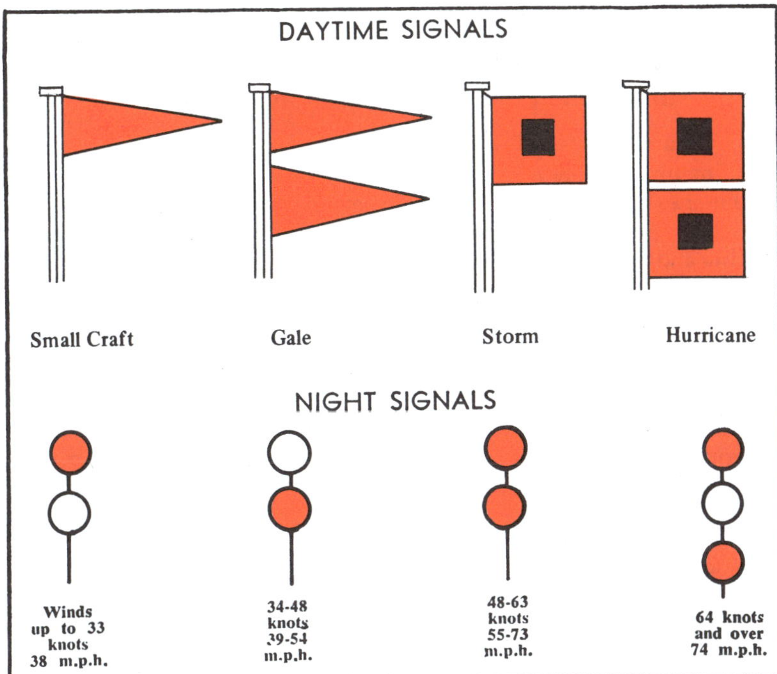

The above signals are displayed regularly on Light Vessels, at Coast Guard shore stations, and at many principal lighthouses. Each Coast and Geodetic Survey Chart lists those locations which appear within the area covered by that chart.

When the Rain before the Wind,
Topsail Sheets and Halliards Mind,
But when the Wind before the Rain,
Then you may set Sail again.

Rain before 7

Clear before 11

DISTANCE OF VISIBILITY

(Approximate ranges with observer's eye at sea level.Add to any distance the distance shown for observer's ht. of eye above sea level. Ex: 30 ft. object, Ht. of eye 10 ft. Range of visibility 9.9 mi.) Hts. of Objects are in feet; distances given are in nautical miles.

Ht.	Dist.	Ht,	Dist.	Ht.	Dist.	Ht.	Dist.	Ht.	Dist.
4	2.3	30	6.3	80	10.3	340	21.1	860	33.6
6	2.8	32	6.5	90	10.9	380	22.3	900	34.4
8	3.2	34	6.7	100	11.5	420	23.5	1000	36.2
10	3.6	36	6.9	120	12.6	460	24.6	1400	42.9
12	4.0	38	7.1	140	13.6	500	25.7	1800	48.6
14	4.3	40	7.3	160	14.5	540	26.7	2200	53.8
16	4.6	42	7.4	180	15.4	580	27.6	2600	58.5
18	4.9	44	7.6	200	16.2	620	28.6	3000	62.8
20	5.1	46	7.8	220	17.0	660	29.4	3400	66.9
22	5.4	48	8.0	240	17.8	700	30.4	3800	70.7
24	5.6	50	8.1	260	18.5	740	31.1	4200	74.3
26	5.9	60	8.9	280	19.2	780	32.0	4600	77.7
28	6.1	70	9.6	300	19.9	820	32.8	5000	81.0

IALA BUOYAGE SYSTEM - also see p. 8.

Lateral Aids marking the sides of channels seen when entering from Seaward.

Port Side - Odd Numbers:

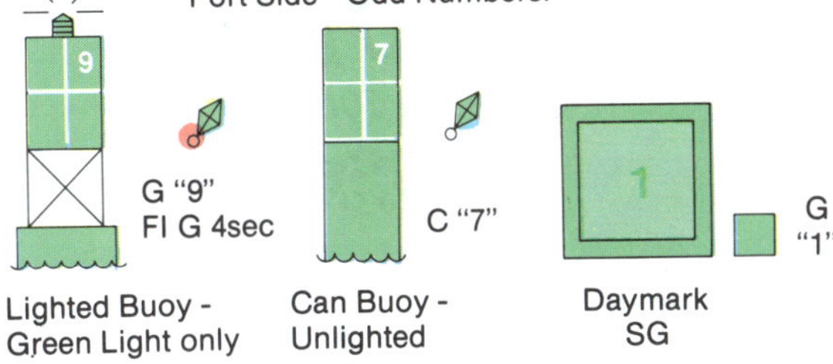

Port-hand aids are Green, some with Flashing Green Lights.
Daymarks:
1st letter "S" = Square
2nd letter "G" = color Green

Starboard Side - Even Numbers:

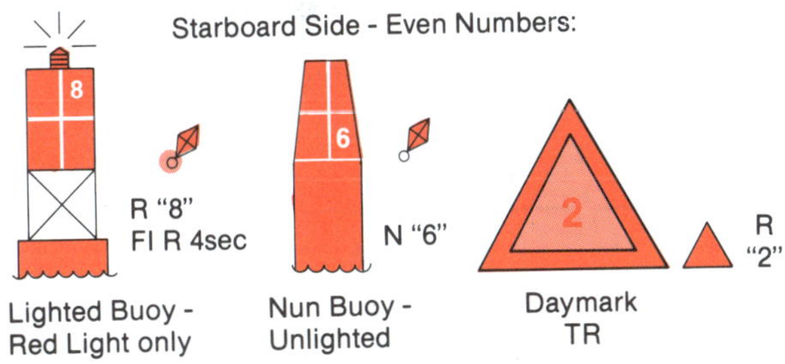

Starboard-hand aids remain Red, some with Flashing Red Lights.
Daymarks:
1st letter "T" = Triangle
2nd letter "R" = color Red

Safe Water Aids Marking Mid-Channels & Fairways - No Numbers - May Be Lettered:

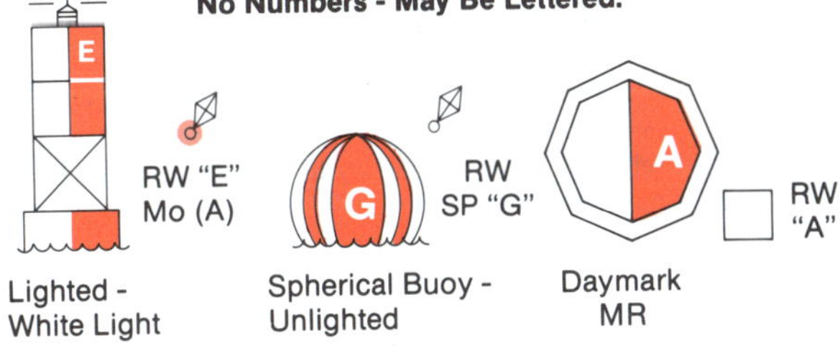

Red and White replaces Black and White. Buoys are spherical; or have a Red spherical topmark. Flashing White Light only: Mo (A).
Daymarks:
1st letter "M" = Octagon
2nd letter "R" = color Red

Preferred Channel Aids - Mark Bifurcations - No Numbers -

Preferred Ch. to Starboard (Aid to Port):

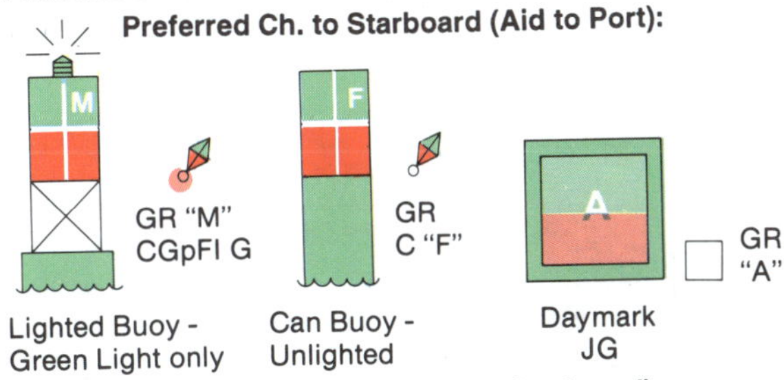

Green replaces Black. Flashing Light (Red or Green) is Composite Gp. Fl. (2 + 1).
Daymarks:
1st letter "J" = Square or Triangle
2nd letter "R" or "G" is color of top band.

Preferred Ch. to Port (Aid to Starboard):

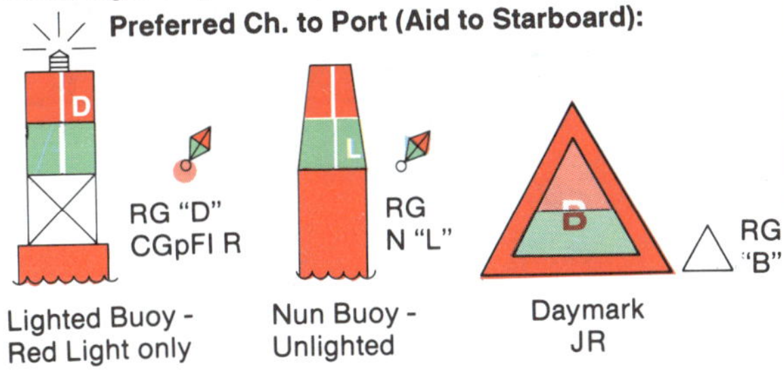

Note: ISOLATED DANGER BUOYS, Black and Red with two Black spherical topmarks - no numbers, may be lettered (if lighted, white light only, Fl (2) 5s). Stay Clear. SPECIAL AIDS BUOYS will be all YELLOW (if lighted, with yellow light only, Fixed Flashing): Anchorage Areas, Fish Net Areas, Spoil Grounds, Military Exercise Zones, Dredging Buoys (where conventional markers would be confusing), Ocean Data Systems, some Traffic Separation Zone Mid-Channel Buoys.

YACHT CLUB FLAGS

Frequently seen when cruising

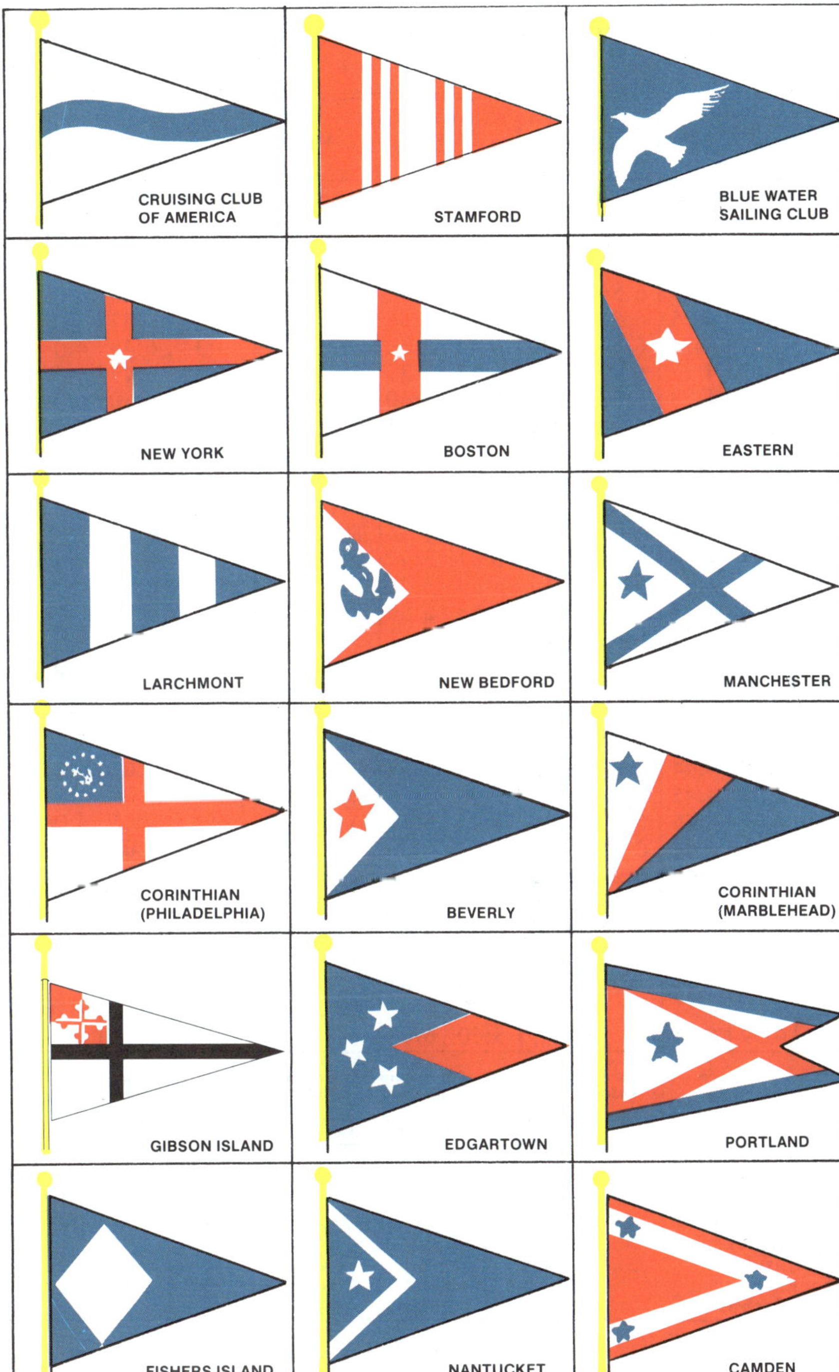

1999 SUN'S RISING AND SETTING AT BOSTON

42° 20'N 71°W

Add one hour for Daylight Saving Time, covering summer months

Vernal Equinox: March 20th, 8:46 p.m. Summer Solstice: June 21st, 2:49 p.m.

Day of Month	Jan.		Feb.		Mar.		April		May		June	
	Rises H. M.	Sets H. M.	Rises H. M.	Sets H. M.	Rises H. M.	Sets H. M.	Rises H. M.	Sets H. M.	Rises H. M.	Sets H. M.	Rises H. M.	Sets H. M.
1	7:13	4:21	6:58	4:57	6:21	5:33	5:28	6:09	4:40	6:43	4:11	7:14
2	7:13	4:22	6:57	4:58	6:19	5:34	5:26	6:10	4:39	6:44	4:10	7:15
3	7:13	4:23	6:56	5:00	6:17	5:35	5:25	6:11	4:37	6:45	4:09	7:15
4	7:13	4:25	6:55	5:01	6:16	5:36	5:23	6:13	4:36	6:47	4:09	7:16
5	7:13	4:26	6:54	5:02	6:14	5:38	5:22	6:14	4:34	6:48	4:08	7:17
6	7:13	4:27	6:53	5:03	6:13	5:39	5:20	6:15	4:33	6:49	4:08	7:18
7	7:13	4:28	6:52	5:05	6:11	5:40	5:18	6:16	4:32	6:50	4:08	7:18
8	7:13	4:29	6:50	5:06	6:09	5:41	5:16	6:17	4:31	6:51	4:08	7:19
9	7:12	4:30	6:49	5:08	6:08	5:42	5:15	6:18	4:29	6:52	4:07	7:19
10	7:12	4:31	6:48	5:09	6:06	5:44	5:13	6:19	4:28	6:53	4:07	7:20
11	7:12	4:32	6:47	5:10	6:05	5:45	5:11	6:20	4:27	6:54	4:07	7:20
12	7:12	4:33	6:45	5:12	6:03	5:46	5:09	6:21	4:26	6:55	4:07	7:21
13	7:11	4:34	6:44	5:13	6:01	5:47	5:08	6:22	4:25	6:56	4:07	7:21
14	7:11	4:36	6:42	5:15	5:59	5:48	5:06	6:24	4:24	6:57	4:07	7:22
15	7:10	4:37	6:41	5:16	5:58	5:50	5:05	6:25	4:23	6:58	4:07	7:22
16	7:10	4:38	6:40	5:17	5:56	5:51	5:03	6:26	4:22	6:59	4:07	7:22
17	7:09	4:39	6:38	5:18	5:54	5:52	5:01	6:27	4:21	7:00	4:07	7:23
18	7:09	4:40	6:37	5:20	5:52	5:53	5:00	6:28	4:20	7:01	4:07	7:23
19	7:08	4:42	6:35	5:21	5:51	5:54	4:58	6:30	4:19	7:02	4:07	7:24
20	7:08	4:43	6:34	5:22	5:49	5:56	4:57	6:31	4:18	7:03	4:07	7:24
21	7:07	4:44	6:33	5:23	5:48	5:57	4:55	6:32	4:17	7:04	4:07	7:24
22	7:07	4:45	6:31	5:24	5:46	5:58	4:53	6:33	4:17	7:05	4:07	7:24
23	7:06	4:46	6:30	5:26	5:44	5:59	4:52	6:34	4:16	7:06	4:08	7:25
24	7:05	4:48	6:28	5:27	5:42	6:00	4:50	6:36	4:15	7:07	4:08	7:25
25	7:05	4:49	6:27	5:28	5:41	6:01	4:49	6:37	4:15	7:08	4:08	7:25
26	7:04	4:50	6:25	5:29	5:39	6:02	4:47	6:38	4:14	7:09	4:08	7:25
27	7:03	4:51	6:24	5:30	5:37	6:03	4:46	6:39	4:14	7:10	4:09	7:25
28	7:02	4:52	6:22	5:32	5:35	6:04	4:44	6:40	4:13	7:11	4:09	7:25
29	7:01	4:54			5:33	6:05	4:43	6:41	4:12	7:11	4:10	7:25
30	7:00	4:55			5:32	6:07	4:41	6:42	4:12	7:12	4:10	7:25
31	6:59	4:56			5:30	6:08			4:11	7:13		

For correct SETTING of Sun any day of the year at places specified below, FOR FLAG USE, add or subtract from above table.

	Jan. 15	Feb. 15	Mar. 15	Apr. 15	May 15	June 15
New London	+7	+6	+4	+2	0	−1
Newport	+4	+3	+1	−1	−2	−3
New Bedford	+3	+2	0	−1	−2	−3
Vineyard Haven	+1	−1	−2	−4	−5	−6
Nantucket	−1	−2	−4	−6	−7	−8
Portland	−8	−6	−3	−1	+1	+2
Rockland	−14	−12	−8	−6	−4	−2
Bar Harbor	−18	−15	−11	−8	−5	−3

Note: Times shown in table are first tip of Sun at Sunrise, last tip at Sunset.

1999 SUN'S RISING AND SETTING AT BOSTON

42° 20'N 71°W

Add one hour for Daylight Saving Time, covering summer months

Autumnal Equinox: Sept. 23rd, 6:31 a.m. Winter Solstice: Dec. 22nd, 2:44 a.m.

Day of Month	July		Aug.		Sept.		Oct.		Nov.		Dec.	
	Rises H. M.	Sets H. M.	Rises H. M.	Sets H. M.	Rises H. M.	Sets H. M.	Rises H. M.	Sets H. M.	Rises H. M.	Sets H. M.	Rises H. M.	Sets H. M.
1	4:11	7:25	4:36	7:04	5:09	6:19	5:40	5:28	6:17	4:38	6:53	4:12
2	4:11	7:25	4:37	7:02	5:10	6:18	5:41	5:26	6:18	4:36	6:54	4:12
3	4:12	7:24	4:38	7:01	5:11	6:16	5:42	5:24	6:19	4:35	6:55	4:12
4	4:12	7:24	4:39	7:00	5:12	6:14	5:43	5:22	6:21	4:34	6:56	4:12
5	4:13	7:24	4:40	6:59	5:13	6:12	5:44	5:20	6:22	4:32	6:57	4:12
6	4:13	7:24	4:41	6:58	5:14	6:11	5:46	5:19	6:24	4:31	6:58	4:12
7	4:14	7:23	4:42	6:56	5:15	6:09	5:47	5:17	6:25	4:30	6:59	4:12
8	4:15	7:23	4:43	6:55	5:16	6:07	5:48	5:15	6:26	4:29	7:00	4:12
9	4:15	7:22	4:44	6:54	5:17	6:05	5:49	5:13	6:27	4:28	7:01	4:12
10	4:16	7:22	4:45	6:53	5:18	6:03	5:50	5:12	6:29	4:27	7:02	4:12
11	4:17	7:21	4:46	6:51	5:19	6:02	5:52	5:10	6:30	4:26	7:03	4:12
12	4:18	7:21	4:47	6:50	5:20	6:00	5:53	5:09	6:31	4:25	7:04	4:12
13	4:18	7:20	4:48	6:48	5:21	5:58	5:54	5:07	6:32	4:24	7:05	4:12
14	4:19	7:20	4:49	6:47	5:22	5:56	5:55	5:05	6:33	4:23	7:05	4:12
15	4:20	7:19	4:50	6:46	5:23	5:55	5:56	5:04	6:35	4:23	7:06	4:13
16	4:20	7:18	4:51	6:44	5:25	5:53	5:57	5:02	6:36	4:22	7:06	4:13
17	4:21	7:18	4:52	6:43	5:26	5:52	5:58	5:01	6:37	4:21	7:07	4:13
18	4:22	7:17	4:53	6:41	5:27	5:50	5:59	4:59	6:38	4:20	7:08	4:13
19	4:23	7:17	4:54	6:40	5:28	5:48	6:00	4:58	6:39	4:19	7:08	4:14
20	4:24	7:16	4:55	6:38	5:29	5:46	6:01	4:56	6:41	4:19	7:09	4:14
21	4:25	7:15	4:56	6:37	5:30	5:45	6:03	4:55	6:42	4:18	7:09	4:15
22	4:26	7:14	4:58	6:35	5:31	5:43	6:04	4:53	6:43	4:17	7:10	4:15
23	4:27	7:13	4:59	6:34	5:32	5:41	6:05	4:52	6:44	4:16	7:10	4:16
24	4:28	7:12	5:00	6:32	5:33	5:39	6:06	4:50	6:45	4:16	7:11	4:16
25	4:29	7:11	5:01	6:30	5:34	5:38	6:07	4:49	6:47	4:15	7:11	4:17
26	4:30	7:10	5:02	6:29	5:35	5:36	6:09	4:47	6:48	4:15	7:12	4:17
27	4:31	7:09	5:04	6:27	5:36	5:35	6:10	4:46	6:49	4:14	7:12	4:18
28	4:32	7:08	5:05	6:26	5:37	5:33	6:11	4:44	6:50	4:14	7:12	4:19
29	4:33	7:07	5:06	6:24	5:38	5:31	6:12	4:42	6:51	4:13	7:13	4:20
30	4:34	7:06	5:07	6:22	5:39	5:29	6:14	4:41	6:52	4:13	7:13	4:20
31	4:35	7:05	5:08	6:21			6:15	4:39			7:13	4:21

For correct SETTING of Sun any day of the year at places specified below, FOR FLAG USE, add or subtract from above table.

	July 15	Aug. 15	Sept. 15	Oct. 15	Nov. 15	Dec. 15
New London	0	+1	+2	+5	+6	+7
Newport	−2	−1	0	+2	+4	+5
New Bedford	−2	−1	0	+1	+2	+3
Vineyard Haven	−5	−4	−3	−2	0	+1
Nantucket	−7	−6	−5	−3	−2	−1
Portland	+1	0	−2	−4	−6	−7
Rockland	−4	−5	−7	−10	−12	−14
Bar Harbor	−5	−7	−9	−13	−17	−18

Note: Times shown in table are first tip of Sun at Sunrise, last tip at Sunset.

1999 SUN'S RISING AND SETTING AT NEW YORK (THE BATTERY)

40° 42'N 74°W

Add one hour for Daylight Saving Time, covering summer months

Vernal Equinox: March 20th, 8:46 p.m. Summer Solstice: June 21st, 2:49 p.m.

Day of Month	Jan.		Feb.		Mar.		April		May		June	
	Rises H. M.	Sets H. M.	Rises H. M.	Sets H. M.	Rises H. M.	Sets H. M.	Rises H. M.	Sets H. M.	Rises H. M.	Sets H. M.	Rises H. M.	Sets H. M.
1	7:20	**4:38**	7:06	**5:13**	6:30	**5:46**	5:41	**6:20**	4:56	**6:51**	4:28	**7:20**
2	7:20	**4:39**	7:05	**5:14**	6:29	**5:47**	5:39	**6:21**	4:55	**6:52**	4:27	**7:21**
3	7:20	**4:40**	7:04	**5:16**	6:28	**5:48**	5:38	**6:22**	4:54	**6:53**	4:27	**7:21**
4	7:20	**4:41**	7:03	**5:17**	6:26	**5:49**	5:36	**6:24**	4:52	**6:55**	4:26	**7:22**
5	7:20	**4:42**	7:02	**5:18**	6:25	**5:51**	5:35	**6:25**	4:51	**6:56**	4:26	**7:23**
6	7:20	**4:43**	7:01	**5:19**	6:23	**5:52**	5:33	**6:26**	4:50	**6:57**	4:26	**7:24**
7	7:20	**4:44**	7:00	**5:20**	6:22	**5:53**	5:31	**6:27**	4:49	**6:58**	4:26	**7:24**
8	7:20	**4:45**	6:59	**5:22**	6:20	**5:54**	5:30	**6:28**	4:48	**6:59**	4:26	**7:25**
9	7:20	**4:46**	6:58	**5:23**	6:19	**5:55**	5:28	**6:29**	4:46	**7:00**	4:25	**7:25**
10	7:20	**4:47**	6:57	**5:24**	6:17	**5:57**	5:27	**6:30**	4:45	**7:01**	4:25	**7:26**
11	7:20	**4:48**	6:56	**5:25**	6:16	**5:58**	5:25	**6:31**	4:44	**7:02**	4:25	**7:26**
12	7:20	**4:49**	6:55	**5:26**	6:14	**5:59**	5:23	**6:32**	4:43	**7:03**	4:25	**7:27**
13	7:19	**4:50**	6:53	**5:28**	6:12	**6:00**	5:22	**6:33**	4:42	**7:04**	4:25	**7:27**
14	7:19	**4:52**	6:52	**5:29**	6:11	**6:01**	5:20	**6:34**	4:41	**7:05**	4:25	**7:28**
15	7:19	**4:53**	6:51	**5:30**	6:09	**6:02**	5:19	**6:35**	4:40	**7:06**	4:25	**7:28**
16	7:18	**4:54**	6:50	**5:31**	6:08	**6:03**	5:17	**6:36**	4:39	**7:07**	4:25	**7:28**
17	7:18	**4:55**	6:48	**5:32**	6:06	**6:04**	5:16	**6:37**	4:38	**7:08**	4:25	**7:29**
18	7:17	**4:56**	6:47	**5:34**	6:04	**6:05**	5:14	**6:38**	4:37	**7:09**	4:25	**7:29**
19	7:17	**4:58**	6:45	**5:35**	6:03	**6:06**	5:13	**6:39**	4:36	**7:09**	4:25	**7:30**
20	7:16	**4:59**	6:44	**5:36**	6:01	**6:07**	5:11	**6:40**	4:35	**7:10**	4:25	**7:30**
21	7:15	**5:00**	6:43	**5:37**	6:00	**6:08**	5:10	**6:41**	4:34	**7:11**	4:25	**7:30**
22	7:14	**5:01**	6:41	**5:38**	5:58	**6:09**	5:09	**6:42**	4:33	**7:12**	4:26	**7:30**
23	7:13	**5:02**	6:40	**5:40**	5:56	**6:10**	5:07	**6:43**	4:33	**7:13**	4:26	**7:31**
24	7:13	**5:04**	6:38	**5:41**	5:54	**6:11**	5:06	**6:44**	4:32	**7:13**	4:26	**7:31**
25	7:12	**5:05**	6:37	**5:42**	5:53	**6:12**	5:04	**6:45**	4:32	**7:14**	4:26	**7:31**
26	7:11	**5:06**	6:35	**5:43**	5:51	**6:13**	5:03	**6:46**	4:31	**7:15**	4:27	**7:31**
27	7:10	**5:07**	6:34	**5:44**	5:49	**6:14**	5:02	**6:47**	4:30	**7:16**	4:27	**7:31**
28	7:09	**5:08**	6:32	**5:45**	5:47	**6:15**	5:00	**6:48**	4:30	**7:17**	4:27	**7:31**
29	7:09	**5:10**			5:46	**6:16**	4:59	**6:49**	4:29	**7:17**	4:28	**7:31**
30	7:08	**5:11**			5:44	**6:18**	4 57	**6:50**	4 29	**7:18**	4 28	**7:31**
31	7:07	**5:12**			5:43	**6:19**			4:28	**7:19**		

For correct SETTING of Sun any day of the year at places specified below, FOR FLAG USE, add or subtract from above table.

	Jan. 15	Feb. 15	Mar. 15	Apr. 15	May 15	June 15
Hampton Rds., Va.	+18	+15	+9	+2	0	−1
Oxford, Md.	+14	+12	+9	+5	+4	+3
Annapolis, Md.	+14	+13	+10	+7	+6	+5
Cape May, N. J.	+8	+7	+4	+1	0	−1
Atlantic City	+5	+4	+2	0	−1	−2
Manasquan, N. J.	+2	+1	0	−1	−2	−2
Port Jefferson, N. Y.	−5	−4	−4	−3	−3	−3
Bridgeport, Ct.	−4	−4	−3	−2	−2	−1
New Haven	−7	−6	−4	−3	−3	−3

Note: Times shown in table are first tip of Sun at Sunrise, last tip at Sunset.

1999 SUN'S RISING AND SETTING AT NEW YORK (THE BATTERY)

40° 42'N 74°W

Add one hour for Daylight Saving Time, covering summer months

Autumnal Equinox: Sept. 23rd, 6:31 a.m. Winter Solstice: Dec. 22nd, 2:44 a.m.

Day of Month	July		Aug.		Sept.		Oct.		Nov.		Dec.	
	Rises H. M.	Sets H. M.	Rises H. M.	Sets H. M.	Rises H. M.	Sets H. M.	Rises H. M.	Sets H. M.	Rises H. M.	Sets H. M.	Rises H. M.	Sets H. M.
1	4:28	**7:31**	4:52	**7:12**	5:22	**6:30**	5:52	**5:40**	6:25	**4:53**	7:00	**4:29**
2	4:29	**7:31**	4:53	**7:10**	5:23	**6:28**	5:53	**5:38**	6:26	**4:52**	7:01	**4:29**
3	4:29	**7:31**	4:54	**7:09**	5:24	**6:27**	5:54	**5:36**	6:27	**4:51**	7:02	**4:29**
4	4:30	**7:30**	4:55	**7:08**	5:25	**6:25**	5:55	**5:34**	6:28	**4:50**	7:03	**4:29**
5	4:30	**7:30**	4:56	**7:07**	5:26	**6:23**	5:56	**5:33**	6:30	**4:48**	7:04	**4:29**
6	4:31	**7:30**	4:57	**7:06**	5:27	**6:22**	5:57	**5:31**	6:31	**4:47**	7:05	**4:29**
7	4:31	**7:30**	4:58	**7:04**	5:28	**6:20**	5:58	**5:30**	6:32	**4:46**	7:06	**4:29**
8	4:32	**7:29**	4:59	**7:03**	5:29	**6:18**	5:59	**5:28**	6:33	**4:45**	7:07	**4:29**
9	4:32	**7:29**	5:00	**7:02**	5:30	**6:16**	6:00	**5:26**	6:34	**4:44**	7:08	**4:29**
10	4:33	**7:29**	5:01	**7:01**	5:31	**6:14**	6:01	**5:25**	6:36	**4:43**	7:08	**4:29**
11	4:34	**7:28**	5:02	**7:00**	5:32	**6:13**	6:03	**5:23**	6:37	**4:42**	7:09	**4:29**
12	4:35	**7:28**	5:03	**6:58**	5:33	**6:11**	6:04	**5:22**	6:38	**4:41**	7:10	**4:29**
13	4:35	**7:27**	5:04	**6:57**	5:34	**6:09**	6:05	**5:20**	6:39	**4:40**	7:11	**4:29**
14	4:36	**7:27**	5:05	**6:56**	5:35	**6:07**	6:06	**5:18**	6:40	**4:39**	7:11	**4:30**
15	4:37	**7:26**	5:06	**6:55**	5:36	**6:06**	6:07	**5:17**	6:42	**4:39**	7:12	**4:30**
16	4:38	**7:25**	5:06	**6:53**	5:37	**6:04**	6:08	**5:15**	6:43	**4:38**	7:12	**4:30**
17	4:39	**7:25**	5:07	**6:52**	5:38	**6:03**	6:09	**5:14**	6:44	**4:37**	7:13	**4:30**
18	4:39	**7:24**	5:08	**6:50**	5:39	**6:01**	6:10	**5:12**	6:45	**4:36**	7:14	**4:31**
19	4:40	**7:24**	5:09	**6:49**	5:40	**5:59**	6:11	**5:11**	6:46	**4:35**	7:14	**4:31**
20	4:41	**7:23**	5:10	**6:48**	5:41	**5:58**	6:12	**5:09**	6:48	**4:35**	7:15	**4:31**
21	4:42	**7:22**	5:11	**6:46**	5:42	**5:56**	6:13	**5:08**	6:49	**4:34**	7:15	**4:32**
22	4:43	**7:21**	5:12	**6:45**	5:43	**5:55**	6:14	**5:06**	6:50	**4:33**	7:16	**4:32**
23	4:43	**7:21**	5:13	**6:43**	5:44	**5:53**	6:15	**5:05**	6:51	**4:33**	7:16	**4:33**
24	4:44	**7:20**	5:14	**6:42**	5:45	**5:51**	6:16	**5:04**	6:52	**4:32**	7:17	**4:33**
25	4:45	**7:19**	5:15	**6:40**	5:46	**5:50**	6:17	**5:03**	6:54	**4:32**	7:17	**4:34**
26	4:46	**7:18**	5:16	**6:39**	5:47	**5:48**	6:19	**5:01**	6:55	**4:31**	7:18	**4:34**
27	4:47	**7:17**	5:17	**6:37**	5:48	**5:47**	6:20	**5:00**	6:56	**4:31**	7:18	**4:35**
28	4:48	**7:16**	5:18	**6:36**	5:49	**5:45**	6:21	**4:59**	6:57	**4:30**	7:18	**4:36**
29	4:49	**7:15**	5:19	**6:34**	5:50	**5:43**	6:22	**4:58**	6:58	**4:30**	7:19	**4:36**
30	4:50	**7:14**	5:20	**6:33**	5:51	**5:41**	6:23	**4:56**	6:59	**4:30**	7:19	**4:37**
31	4:51	**7:13**	5:21	**6:31**			6:24	**4:55**			7:20	**4:37**

For correct SETTING of Sun any day of the year at places specified below, FOR FLAG USE, add or subtract from above table.

	July 15	Aug. 15	Sept. 15	Oct. 15	Nov. 15	Dec. 15
Hampton Rds., Va.	−1	+2	+7	+13	+18	+20
Oxford, Md.	+3	+5	+8	+11	+14	+15
Annapolis, Md.	+5	+7	+9	+12	+14	+15
Cape May, N. J.	−1	+1	+3	+6	+8	+9
Atlantic City	−2	0	+2	+4	+6	+6
Mannasquan, N. J.	−2	−1	0	+1	+2	+2
Port Jefferson, N. Y.	−3	−3	−4	−4	−5	−5
Bridgeport, Ct.	−1	−2	−3	−4	−4	−5
New Haven	−3	−3	−4	−5	−6	−7

Note: Times shown in table are first tip of Sun at Sunrise, last tip at Sunset.

USING OUR SUN TABLES FOR CHECKING YOUR COMPASS

You have chartered a boat at Cape May, NJ on August 10 and you are not sure of the accuracy of the compass and there is no hand-bearing compass aboard to check against the main compass. What you need to know, to establish a proper deviation table, is the Sun's *Magnetic* Bearing over a period of an hour or so for your area. Here's what you do:

In the hour before Sunset (or after Sunrise), the bearing of the Sun changes only about 1° each 6 minutes or 10° in 1 hour. It could be 9° or 11° but using 10° is close enough for this purpose, and Sunset is more convenient than Sunrise. From the Sun's Declination Table, you find the Declination to be 15° 22′ N. Your latitude from your chart is about 39° N. From the Sun's True Bearing Table for 15° N you find the Sunset Bearing at 40° N. to be 289.7°, and at 38° N. to be 289.2°. You're halfway between, so the Bearing would be about 289.5° at 39° N. Lat.

But, the Declination is 15° 22′ N., about ⅓ between 15° N. and 16° N. At 15° N. and 16° N., the difference in the Sun's True Bearing is 1.4°, divided by 3, which equals about .5°, so the True Bearing at Sunset would be 290°. Your local westerly variation from your chart is about 10° W. so the *Magnetic* Bearing at Sunset would be 300°. One hour before Sunset the *Magnetic* Bearing would be 290°.

Now, get the approximate time of Sunset at Cape May for August 10 (see New York Sunrise and Sunset Table). It appears to be about the same as New York, 8 p.m. EDT. Make your table accordingly:

1 hour before Sunset,	7 p.m. EDT	Mag. Bearing	290°
	7:06		291°
	7:12		292°
	7:18		293°
	7:24, etc. ending at:		294°
	8:00 Sunset		300°

Using the above system and interpolating as necessary, you are in good shape if your compass is out in the sun, as on a steering pedestal, where the pivot of the compass will give a good shadow on the card. If your compass is not out in the sun, and you have no bearing instrument, you'll need to improvise a flat compass card. You could cut one out carefully from an old chart or other nautical publication. Mount it on any flat horizontal surface where the sun can get at it. Vertically, through its center, stick a pin or needle or anything else sharp and slender that will cast a fine shadow. On this horizontal surface, draw 2 short lines out from the compass rose, 1 dead forward and 1 dead aft, to serve as lubber lines. If heading E. by compass, set E. point of compass rose at forward line. At 7:18 p.m. EDT, the shadow of the pin should be at the reciprocal of 293° or at 113°. If shadow of pin is at 110°, you have 3° E. deviation on E. (See table above).

You should check every 45° around the compass. Each time you come to a new heading, rotate the compass rose to your new heading and check deviation as before. By following this procedure, you should end up with a pretty good idea of the accuracy of the compass aboard and by making a list of the errors on each heading, you can be sure of steering the correct course.

1999 SUN'S RISING AND SETTING AT JACKSONVILLE, FL

30° 20'N 81° 37'W

Add one hour for Daylight Saving Time, covering summer months

Vernal Equinox: March 20th, 8:46 p.m. Summer Solstice: June 21st, 2:49 p.m.
Autumnal Equinox: Sept. 23rd, 6:31 a.m. Winter Solstice: Dec. 22nd, 2:44 a.m.

	Jan.		Feb.		Mar.		April		May		June	
	Rises	Sets	Rises	Sets	Rises	Sets	Rises	Sets	Rises	Sets	Rises	Sets
5	7:24	**5:39**	7:14	**6:06**	6:49	**6:27**	6:11	**6:46**	5:40	**7:06**	5:24	**7:25**
10	7:24	**5:43**	7:11	**6:10**	6:43	**6:31**	6:06	**6:49**	5:36	**7:09**	5:23	**7:27**
15	7:24	**5:47**	7:07	**6:14**	6:36	**6:34**	6:00	**6:53**	5:32	**7:12**	5:23	**7:29**
20	7:22	**5:52**	7:02	**6:18**	6:30	**6:37**	5:55	**6:56**	5:29	**7:15**	5:24	**7:31**
25	7:20	**5:56**	6:57	**6:22**	6:24	**6:40**	5:49	**6:59**	5:27	**7:19**	5:25	**7:32**
30	7:18	**6:01**	28th:		6:18	**6:43**	5:44	**7:02**	5:26	**7:22**	5:27	**7:32**
			6:54	**6:24**								

	July		Aug.		Sept.		Oct.		Nov.		Dec.	
	Rises	Sets	Rises	Sets	Rises	Sets	Rises	Sets	Rises	Sets	Rises	Sets
5	5:29	**7:32**	5:47	**7:17**	6:05	**6:45**	6:21	**6:08**	6:43	**5:37**	7:07	**5:25**
10	5:31	**7:31**	5:50	**7:12**	6:07	**6:39**	6:24	**6:02**	6:47	**5:33**	7:11	**5:26**
15	5:34	**7:30**	5:53	**7:08**	6:10	**6:33**	6:27	**5:56**	6:51	**5:30**	7:14	**5:27**
20	5:36	**7:28**	5:56	**7:03**	6:13	**6:26**	6:30	**5:51**	6:54	**5:27**	7:18	**5:29**
25	5:39	**7:25**	5:59	**6:58**	6:16	**6:19**	6:34	**5:46**	6:58	**5:26**	7:20	**5:32**
30	5:43	**7:22**	6:02	**6:52**	6:19	**6:13**	6:38	**5:41**	7:03	**5:25**	7:22	**5:35**

FOR FLAG USE, add or subtract from above table:

		1/15	2/15	3/15	4/15	5/15	6/15	7/15	8/15	9/15	10/15	11/15	12/15
Morehead City —	Rises.	−9	−14	10	25	27	−30	−29	−25	−21	−14	−11	−7
	Sets:	**−28**	**−24**	**−20**	**−14**	**−10**	**−7**	**−9**	**−13**	**−19**	**−24**	**−28**	**−29**
Wilmington —	Rises:	−6	−10	−14	−19	−22	−24	−23	−20	−15	−10	−7	−3
	Sets:	**−22**	**−18**	**−14**	**−10**	**−5**	**−3**	**−5**	**−8**	**−14**	**−18**	**−22**	**−23**
Myrtle Beach —	Rises:	−4	−7	−9	−15	−16	−18	−17	−15	−11	−8	−4	−1
	Sets:	**−16**	**−14**	**−10**	**−6**	**−3**	**−1**	**−3**	**−5**	**−7**	**−12**	**−17**	**−17**
Charleston —	Rises:	−1	−3	−6	−10	−11	−12	−11	−9	−6	−4	−3	+1
	Sets:	**−11**	**−9**	**−6**	**−3**	**+1**	**0**	**−1**	**−3**	**−7**	**−9**	**−11**	**−12**
Savannah —	Rises:	+5	+4	+2	+1	−1	−2	−2	0	+2	+4	+5	+7
	Sets:	**−1**	**0**	**+2**	**+4**	**+6**	**+6**	**+5**	**+4**	**+3**	**0**	**−2**	**−2**
Brunswick —	Rises:	+1	+1	0	−1	−2	−2	−2	−1	0	+1	+1	+2
	Sets:	**+1**	**−1**	**0**	**+1**	**+2**	**+2**	**+1**	**+1**	**−1**	**+1**	**+2**	**+2**
Ponce Inlet —	Rises:	−5	−4	−2	−1	+1	+1	+1	−1	−2	−3	− 5	−5
	Sets:	**+1**	**−1**	**−1**	**−2**	**−4**	**−5**	**−5**	**−4**	**−2**	**0**	**+1**	**+1**
Melbourne —	Rises:	−9	−7	−3	−2	+1	+2	+1	−2	−3	−5	−8	−9
	Sets:	**+1**	**−2**	**−4**	**−5**	**−7**	**−9**	**−9**	**−8**	**−5**	**−1**	**0**	**+1**
N.Palm Beach —	Rises:	−14	−10	−6	−3	+1	+3	+2	−2	−4	−8	−12	−13
	Sets:	**+2**	**−2**	**−6**	**−8**	**−11**	**−14**	**−14**	**−11**	**−7**	**−4**	**0**	**+2**
Miami —	Rises:	−15	−10	−4	−1	+3	+6	+5	0	−3	−8	−13	−14
	Sets:	**+6**	**0**	**−3**	**−8**	**−13**	**−15**	**−15**	**−11**	**−7**	**−2**	**+3**	**+5**
Key West —	Rises:	−11	−5	+1	+5	+12	+15	+13	+8	+3	−2	−9	−12
	Sets:	**+13**	**+7**	**+2**	**−3**	**−8**	**−12**	**−12**	**−6**	**−1**	**+6**	**+11**	**+14**

Note: Times shown in table are first tip of Sun at Sunrise, last tip at Sunset.

SUN'S TRUE BEARING AT RISING AND SETTING

(figures are correct for all longitudes)

Sun's Decl.	38° N Rise	38° N Set	40° N Rise	40° N Set	42° N Rise	42° N Set	44° N Rise	44° N Set	Sun's Decl.
N 23°	60.3°	299.7°	59.3°	300.7°	58.3°	301.7°	57.1°	302.9°	N 23°
22	61.6	298.4	60.7	299.3	59.7	300.3	58.6	301.4	22
21	63.0	297.0	62.1	297.9	61.2	298.8	60.1	299.9	21
20	64.3	295.7	63.5	296.5	62.6	297.4	61.6	298.4	20
19	65.6	294.4	64.9	295.1	64.0	296.0	63.1	296.9	19
18	66.9	293.1	66.2	293.8	65.4	294.6	64.6	295.4	18
17	68.2	291.8	67.6	292.4	66.8	293.2	66.0	294.0	17
16	69.5	290.5	68.9	291.1	68.2	291.8	67.5	292.5	16
15	70.8	289.2	70.3	289.7	69.6	290.4	68.9	291.1	15
14	72.1	287.9	71.6	288.4	71.0	289.0	70.4	289.6	14
13	73.4	286.6	72.9	287.1	72.4	287.6	71.8	288.2	13
12	74.7	285.3	74.3	285.7	73.8	286.2	73.2	286.8	12
11	76.0	284.0	75.6	284.4	75.1	284.9	74.6	285.4	11
10	77.3	282.7	76.9	283.1	76.5	283.5	76.0	284.0	10
9	78.6	281.4	78.2	281.8	77.9	282.1	77.4	282.6	9
8	79.8	280.2	79.5	280.5	79.2	280.8	78.9	281.1	8
7	81.1	278.9	80.9	279.1	80.6	279.4	80.3	279.7	7
6	82.4	277.6	82.2	277.7	81.9	278.1	81.7	278.3	6
5	83.7	276.3	83.5	276.5	83.3	276.7	83.0	277.0	5
4	84.9	275.1	84.8	275.2	84.6	275.4	84.4	275.6	4
3	86.2	273.8	86.1	273.9	86.0	274.0	85.8	274.2	3
2	87.5	272.5	87.4	272.6	87.3	272.7	87.2	272.8	2
N 1°	88.7	271.3	88.7	271.3	88.7	271.3	88.6	271.4	N 1°
0	90.0	270.0	90.0	270.0	90.0	270.0	90.0	270.0	0
S 1°	91.3	268.7	91.3	268.7	91.3	268.7	91.4	268.6	S 1°
2	92.5	267.5	92.6	267.4	92.7	267.3	92.8	267.2	2
3	93.8	266.2	93.9	266.1	94.0	266.0	94.2	265.8	3
4	95.1	264.9	95.2	264.8	95.4	264.6	95.6	264.4	4
5	96.3	263.7	96.5	263.5	96.7	263.3	97.0	263.0	5
6	97.6	262.4	97.8	262.2	98.1	261.9	98.3	261.7	6
7	98.9	261.1	99.1	260.9	99.4	260.6	99.7	260.3	7
8	100.2	259.8	100.5	259.5	100.8	259.2	101.1	258.9	8
9	101.4	258.6	101.8	258.2	102.1	257.9	102.6	257.4	9
10	102.7	257.3	103.1	256.9	103.5	256.5	104.0	256.0	10
11	104.0	256.0	104.4	255.6	104.9	255.1	105.4	254.6	11
12	105.3	254.7	105.7	254.3	106.2	253.8	106.8	253.2	12
13	106.6	253.4	107.1	252.9	107.6	252.4	108.2	251.8	13
14	107.9	252.1	108.4	251.6	109.0	251.0	109.6	250.4	14
15	109.2	250.8	109.7	250.3	110.4	249.6	111.1	248.9	15
16	110.5	249.5	111.1	248.9	111.8	248.2	112.5	247.5	16
17	111.8	248.2	112.4	247.6	113.2	246.8	114.0	246.0	17
18	113.1	246.9	113.8	246.2	114.6	245.4	115.4	244.6	18
19	114.4	245.6	115.1	244.9	116.0	244.0	116.9	243.1	19
20	115.7	244.3	116.5	243.5	117.4	242.6	118.4	241.6	20
21	117.0	243.0	117.9	242.1	118.8	241.2	119.9	240.1	21
22	118.4	241.6	119.3	240.7	120.3	239.7	121.4	238.6	22
S 23°	119.7	240.3	120.7	239.3	121.7	238.3	122.9	237.1	S 23°

For maximum use of this table, it is convenient to know that the sun's bearing changes about 1° each 6 minutes during the first hour after sunrise or before sunset. (See tables for times of Sunrise and Sunset and Sun's Declination.)

To obtain the Magnetic Bearing or Correct Compass Bearing, add the Local Westerly Variation to the True Bearing tabulated above. The variation can be taken from the nearest chart rose.

The difference between the Sun's Magnetic Bearing and the Compass Bearing will be the deviation on the boat's heading at that time. Easterly deviation if the Compass Bearing is less than the Sun's, westerly if more.

Example: Your Latitude, 42° N; Sun's Declination for the date, 19° N; Sun's True Bearing, 296° plus local Westerly variation, 15° = 311° magnetic

Bearing by Compass, ship headed N: 308°

Your compass has: 3° E. Dev. on N

TIME OF LOCAL APPARENT NOON (L.A.N.) 1999
FOR THE CENTRAL MERIDIAN OF ANY TIME ZONE

	JAN.	FEB.	MAR.	APR.	MAY	JUN.		JUL.	AUG.	SEP.	OCT.	NOV.	DEC.
	H M S	H M S	H M S	H M S	H M S	H M S		H M S	H M S	H M S	H M S	H M S	H M S
1	12:03:30	12:13:33	12:12:22	12:03:56	11:57:06	11:57:46	1	12:03:48	12:06:19	12:00:02	11:49:43	11:43:36	11:48:58
2	12:03:58	12:13:42	12:12:10	12:03:38	11:56:59	11:57:56	2	12:03:59	12:06:15	11:59:43	11:49:24	11:43:34	11:49:20
3	12:04:25	12:13:47	12:11:58	12:03:20	11:56:53	11:58:05	3	12:04:10	12:06:10	11:59:24	11:49:05	11:43:34	11:49:44
4	12:04:52	12:13:53	12:11:45	12:03:02	11:56:47	11:58:15	4	12:04:21	12:06:05	11:59:04	11:48:46	11:43:34	11:50:08
5	12:05:19	12:13:58	12:11:32	12:02:45	11:56:42	11:58:26	5	12:04:32	12:06:00	11:58:44	11:48:28	11:43:36	11:50:32
6	12:05:46	12:14:03	12:11:18	12:02:28	11:56:37	11:58:36	6	12:04:42	12:05:53	11:58:24	11:48:10	11:43:38	11:50:57
7	12:06:12	12:14:06	12:11:04	12:02:11	11:56:32	11:58:48	7	12:04:52	12:05:46	11:58:04	11:47:53	11:43:41	11:51:23
8	12:06:38	12:14:09	12:10:49	12:01:54	11:56:29	11:58:59	8	12:05:01	12:05:39	11:57:44	11:47:36	11:43:44	11:51:49
9	12:07:02	12:14:11	12:10:34	12:01:38	11:56:26	11:59:10	9	12:05:10	12:05:31	11:57:23	11:47:19	11:43:49	11:52:16
10	12:07:27	12:14:12	12:10:19	12:01:21	11:56:23	11:59:22	10	12:05:19	12:05:22	11:57:02	11:47:03	11:43:55	11:52:43
11	12:07:51	12:14:13	12:10:04	12:01:06	11:56:21	11:59:35	11	12:05:28	12:05:13	11:56:41	11:46:48	11:44:01	11:53:10
12	12:08:14	12:14:13	12:09:48	12:00:50	11:56:20	11:59:47	12	12:05:36	12:05:03	11:56:20	11:46:32	11:44:08	11:53:38
13	12:08:37	12:14:12	12:09:32	12:00:34	11:56:19	12:00:00	13	12:05:43	12:04:53	11:55:59	11:46:18	11:44:16	11:54:06
14	12:09:00	12:14:10	12:09:14	12:00:20	11:56:19	12:00:12	14	12:05:50	12:04:42	11:55:38	11:46:04	11:44:25	11:54:35
15	12:09:21	12:14:08	12:08:58	12:00:05	11:56:20	12:00:25	15	12:05:56	12:04:30	11:55:16	11:45:50	11:44:34	11:55:03
16	12:09:42	12:14:04	12:08:42	11:59:51	11:56:21	12:00:38	16	12:06:02	12:04:18	11:54:55	11:45:37	11:44:45	11:55:32
17	12:10:02	12:14:00	12:08:25	11:59:37	11:56:22	12:00:51	17	12:06:08	12:04:06	11:54:34	11:45:24	11:44:56	11:56:02
18	12:10:22	12:13:56	12:08:07	11:59:23	11:56:24	12:01:04	18	12:06:12	12:03:53	11:54:12	11:45:12	11:45:08	11:56:31
19	12:10:41	12:13:50	12:07:50	11:59:10	11:56:27	12:01:17	19	12:06:17	12:03:39	11:53:51	11:45:01	11:45:22	11:57:00
20	12:10:59	12:13:45	12:07:32	11:58:57	11:56:30	12:01:30	20	12:06:20	12:03:25	11:53:30	11:44:50	11:45:35	11:57:30
21	12:11:16	12:13:38	12:07:15	11:58:45	11:56:34	12:01:43	21	12:06:24	12:03:11	11:53:08	11:44:40	11:45:50	11:58:00
22	12:11:32	12:13:31	12:06:57	11:58:33	11:56:38	12:01:56	22	12:06:26	12:02:56	11:52:47	11:44:30	11:46:05	11:58:30
23	12:11:48	12:13:23	12:06:39	11:58:22	11:56:43	12:02:09	23	12:06:28	12:02:40	11:52:26	11:44:22	11:46:21	11:58:59
24	12:12:03	12:13:14	12:06:21	11:58:11	11:56:48	12:02:22	24	12:06:30	12:02:24	11:52:05	11:44:14	11:46:38	11:59:29
25	12:12:17	12:13:05	12:06:02	11:58:00	11:56:54	12:02:35	25	12:06:30	12:02:08	11:51:44	11:44:06	11:46:56	11:59:59
26	12:12:30	12:12:55	12:05:44	11:57:50	11:57:00	12:02:48	26	12:06:30	12:01:51	11:51:23	11:44:00	11:47:14	12:00:28
27	12:12:43	12:12:45	12:05:26	11:57:40	11:57:06	12:03:00	27	12:06:30	12:01:34	11:51:02	11:43:54	11:47:33	12:00:58
28	12:12:55	12:12:34	12:05:08	11:57:31	11:57:14	12:03:12	28	12:06:29	12:01:16	11:50:42	11:43:48	11:47:53	12:01:27
29	12:13:06		12:04:50	11:57:22	11:57:21	12:03:24	29	12:06:28	12:00:58	11:50:22	11:43:44	11:48:14	12:01:56
30	12:13:16		12:04:32	11:57:14	11:57:29	12:03:36	30	12:06:25	12:00:40	11:50:02	11:43:40	11:48:35	12:02:26
31	12:13:25		12:04:14		11:57:38		31	12:06:22	12:00:21		11:43:38		12:02:54

Explanatory Notes: The noon sight and the Sun's Declination (p. 229) results in the vessel's parallel of latitude. It is taken at the time of the sun's meridian passage, when the sun is at maximum altitude.

The moment of meridian passage is called Local Apparent Noon (L.A.N.), and only rarely is it the same time as noon Standard Time or Local Mean Time. Instead, as this Table shows, the sun is either ahead of or behind its theoretical schedule.

Two corrections are involved. 1) To correct for your difference in longitude from the central meridian of your time zone (i.e. 75° for U.S. Atlantic Coast), either a) add 4 minutes of time for each degree West or b) subtract 4 minutes of time for each degree East. 2) If necessary, convert from Daylight Savings Time to Standard Time by subtracting 1 hour from your watch.

Thus for Boston, at 71° West longitude (or 4° East of 75°), L.A.N. occurs 16 minutes before the times listed in the Table.

For New York, at 74° West (1° East of 75°), L.A.N. occurs 4 minutes earlier than times shown.

Converting arc to time:

360°	=	24 hours
15°	=	1 hour
1°	=	4 minutes
15'	=	1 minute
1'	=	4 seconds

THE SUN'S DECLINATION 1999

THE SUN'S DECLINATION 1999

MEAN NOON – 75° MERIDIAN (1700 G.M.T.)

	JAN.	FEB.	MAR.	APR.	MAY	JUN.	JUL.	AUG.	SEPT.	OCT.	NOV.	DEC.	
	South ° '	South ° '	South ° '	North ° '	North ° '	North ° '	North ° '	North ° '	North ° '	South ° '	South ° '	South ° '	
1	-23 00	-17 05	-7 34	+4 33	+15 05	+22 04	–23 06	+18 00	+8 16	-3 12	-14 26	-21 48	1
2	22 55	16 48	7 11	4 57	15 23	22 11	23 02	17 45	7 54	3 35	14 45	21 57	2
3	22 49	16 30	6 48	5 20	15 41	22 19	22 57	17 29	7 32	3 59	15 04	22 06	3
4	22 43	16 13	6 25	5 42	15 59	22 26	22 52	17 14	7 10	4 21	15 23	22 14	4
5	22 36	15 55	6 02	6 05	16 16	22 33	22 47	16 58	6 48	4 45	15 41	22 22	5
6	22 29	15 36	5 39	6 28	16 33	22 39	22 41	16 41	6 26	5 08	15 59	22 30	6
7	22 22	15 18	5 16	6 51	16 50	22 45	22 35	16 25	6 03	5 31	16 17	22 37	7
8	22 14	14 59	4 52	7 13	17 06	22 51	22 28	16 08	5 41	5 54	16 35	22 43	8
9	22 06	14 40	4 29	7 35	17 22	22 56	22 21	15 50	5 18	6 17	16 52	22 49	9
10	21 57	14 20	4 05	7 58	17 38	23 01	22 14	15 33	4 55	6 39	17 09	22 55	10
11	21 48	14 01	3 42	8 20	17 53	23 05	22 06	15 15	4 33	7 02	17 26	23 00	11
12	21 38	13 41	3 18	8 42	18 09	23 09	21 58	14 57	4 10	7 25	17 42	23 05	12
13	21 28	13 21	2 55	9 04	18 24	23 13	21 49	14 39	3 47	7 47	17 58	23 09	13
14	21 18	13 00	2 31	9 25	18 38	23 16	21 40	14 21	3 24	8 10	18 14	23 13	14
15	21 07	12 40	2 07	9 47	18 53	23 19	21 31	14 02	3 01	8 32	18 30	23 16	15
16	20 56	12 19	1 44	10 08	19 07	23 21	21 21	13 43	2 38	8 54	18 45	23 19	16
17	20 44	11 58	1 20	10 29	19 20	23 23	21 11	13 24	2 14	9 16	18 59	23 21	17
18	20 32	11 37	0 56	10 50	19 34	23 24	21 01	13 05	1 51	9 38	19 14	23 23	18
19	20 19	11 16	0 32	11 11	19 47	23 25	20 50	12 46	1 28	10 00	19 28	23 25	19
20	20 07	10 54	-0 09	11 32	19 59	23 26	20 39	12 26	1 05	10 21	19 42	23 26	20
21	19 53	10 33	+0 15	11 52	20 12	23 26	20 28	12 06	0 41	10 43	19 55	23 26	21
22	19 40	10 11	0 39	12 13	20 24	23 26	20 16	11 46	+0 18	11 04	20 08	23 26	22
23	19 26	9 49	1 02	12 33	20 35	23 26	20 04	11 26	-0 05	11 25	20 10	23 26	23
24	19 12	9 27	1 26	12 53	20 46	23 25	19 51	11 05	0 29	11 46	20 33	23 25	24
25	18 57	9 04	1 50	13 12	20 57	23 23	19 39	10 45	0 52	12 07	20 45	23 24	25
26	18 42	8 42	2 13	13 32	21 08	23 21	19 25	10 24	1 15	12 27	20 57	23 22	26
27	18 27	8 20	2 37	13 51	21 18	23 19	19 12	10 03	1 39	12 48	21 08	23 19	27
28	18 11	-7 57	3 00	14 10	21 28	23 17	18 58	9 42	2 02	13 08	21 19	23 17	28
29	17 55		3 24	14 29	21 37	23 14	18 44	9 21	2 25	13 28	21 29	23 13	29
30	17 39		3 47	+14 47	21 47	+23 10	18 30	8 59	-2 49	13 47	-21 39	23 10	30
31	-17 22		+4 10		+21 55		+18 15	+8 38		-14 07		-23 06	31

Vernal Equinox: March 20th, 8:46 p.m E.S.T.
Summer Solstice: June 21st, 2:49 p.m. E.S.T.

Autumnal Equinox: September 23rd, 6:31 a.m. E.S.T.
Winter Solstice: December 22nd, 2:44 a.m. E.S.T.

To find Sun's Declination in the Atlantic Time Zone (1 hour earlier than E.S.T.), take 1/24 of the difference between Day 1 and Day 2. Add or subtract this figure from Day 2 to find the Declination for Day 2.
If Declination is increasing (N. or S.), *subtract*.
If Declination is decreasing (N. or S.), *add*.

1999 MOONRISE AND MOONSET

BOSTON, MA

Add one hour for Daylight Saving Time, covering summer months

Day	JANUARY		FEBRUARY		MARCH		APRIL		MAY		JUNE		Day
	Rise h m	Set h m	Rise h m	Set h m	Rise h m	Set h m	Rise h m	Set h m	Rise h m	Set h m	Rise h m	Set h m	
1	1619	0624	1815	0731	1704	0604	1857	0605	1944	0534	2114	0607	1
2	1720	0723	1919	0807	1807	0637	1955	0633	2040	0607	2200	0656	2
3	1825	0815	2022	0839	1908	0707	2053	0702	2135	0643	2242	0750	3
4	1930	0859	2122	0908	2008	0736	2150	0733	2227	0724	2320	0849	4
5	2035	0937	2221	0936	2107	0804	2246	0807	2316	0810	2355	0950	5
6	2137	1010	2319	1004	2205	0832	2340	0845	....	0901	....	1054	6
7	2237	1040	....	1032	2302	0902	....	0928	0001	0957	0028	1200	7
8	2336	1108	0016	1103	2359	0934	0031	1016	0041	1057	0100	1308	8
9	....	1135	0113	1137	....	1010	0119	1110	0119	1201	0132	1419	9
10	0034	1203	0210	1215	0054	1050	0203	1209	0154	1307	0205	1532	10
11	0131	1232	0305	1257	0148	1136	0244	1312	0227	1416	0242	1646	11
12	0228	1304	0358	1346	0238	1227	0321	1419	0300	1528	0323	1801	12
13	0325	1340	0448	1441	0326	1324	0357	1528	0334	1642	0411	1912	13
14	0422	1420	0535	1542	0409	1426	0431	1640	0410	1758	0506	2018	14
15	0517	1506	0617	1646	0450	1533	0505	1754	0450	1914	0608	2115	15
16	0609	1558	0656	1754	0527	1642	0541	1910	0536	2027	0714	2203	16
17	0657	1655	0732	1903	0602	1753	0619	2025	0627	2135	0821	2244	17
18	0742	1757	0806	2013	0636	1906	0701	2138	0725	2234	0928	2318	18
19	0822	1902	0839	2124	0711	2019	0749	2247	0828	2325	1033	2349	19
20	0858	2008	0913	2235	0747	2132	0842	2348	0933	....	1135	....	20
21	0932	2116	0949	2345	0826	2244	0940	....	1038	0008	1235	0018	21
22	1005	2224	1028	....	0909	2353	1041	0042	1142	0045	1335	0045	22
23	1037	2334	1111	0054	0957	....	1145	0128	1244	0117	1433	0112	23
24	1111	....	1200	0200	1050	0057	1248	0208	1344	0146	1531	0140	24
25	1147	0043	1255	0301	1148	0154	1350	0242	1443	0214	1628	0211	25
26	1228	0153	1354	0356	1249	0244	1451	0313	1542	0241	1724	0244	26
27	1313	0302	1457	0445	1352	0327	1551	0341	1640	0308	1819	0322	27
28	1405	0408	1600	0527	1454	0405	1650	0409	1738	0337	1911	0404	28
29	1503	0509	....	....	1557	0438	1748	0436	1835	0409	1959	0452	29
30	1606	0603	....	....	1658	0509	1846	0504	1930	0443	2043	0545	30
31	1710	0650	....	....	1758	0537	....	....	2024	0523	....	....	31

Day	JULY		AUGUST		SEPTEMBER		OCTOBER		NOVEMBER		DECEMBER		Day
	Rise h m	Set h m	Rise h m	Set h m	Rise h m	Set h m	Rise h m	Set h m	Rise h m	Set h m	Rise h m	Set h m	
1	2122	0642	2137	0849	2155	1116	2219	1232	....	1346	0027	1324	1
2	2158	0743	2208	0957	2239	1226	2318	1330	0026	1420	0129	1352	2
3	2231	0846	2241	1105	2328	1333	....	1421	0130	1451	0230	1419	3
4	2303	0951	2316	1214	....	1436	0022	1505	0234	1519	0330	1446	4
5	2333	1057	2356	1325	0024	1533	0128	1543	0336	1547	0429	1515	5
6	....	1205	....	1435	0126	1623	0233	1617	0437	1614	0528	1547	6
7	0005	1314	0042	1542	0231	1705	0338	1647	0537	1642	0625	1623	7
8	0039	1426	0134	1645	0338	1743	0442	1716	0636	1713	0721	1703	8
9	0117	1538	0234	1740	0445	1816	0544	1743	0735	1746	0813	1748	9
10	0200	1650	0339	1828	0551	1846	0645	1812	0831	1824	0902	1838	10
11	0250	1758	0447	1910	0655	1915	0746	1841	0926	1906	0946	1932	11
12	0348	1859	0555	1946	0757	1943	0845	1913	1017	1952	1026	2030	12
13	0452	1952	0702	2018	0857	2012	0943	1948	1104	2044	1101	2130	13
14	0600	2037	0807	2047	0957	2042	1038	2027	1146	2140	1134	2232	14
15	0708	2115	0910	2115	1055	2115	1131	2111	1225	2239	1204	2337	15
16	0815	2149	1011	2143	1152	2152	1221	2200	1300	2341	1233	....	16
17	0921	2219	1110	2212	1247	2233	1307	2254	1332	....	1303	0043	17
18	1023	2247	1209	2243	1339	2319	1348	2352	1403	0046	1335	0152	18
19	1124	2315	1306	2318	1427	....	1426	....	1434	0153	1410	0303	19
20	1223	2343	1402	2356	1512	0011	1501	0054	1506	0303	1450	0417	20
21	1322	....	1456	....	1553	0108	1534	0159	1540	0416	1538	0532	21
22	1419	0012	1547	0040	1630	0209	1606	0307	1619	0531	1633	0645	22
23	1516	0044	1635	0129	1705	0314	1638	0418	1704	0647	1737	0753	23
24	1611	0120	1718	0224	1738	0421	1712	0530	1756	0802	1846	0851	24
25	1705	0201	1758	0323	1810	0530	1749	0644	1855	0911	1957	0940	25
26	1755	0247	1834	0426	1842	0641	1830	0759	2000	1012	2107	1022	26
27	1840	0338	1907	0531	1917	0753	1917	0913	2108	1104	2215	1057	27
28	1922	0435	1939	0639	1954	0905	2011	1022	2216	1147	2319	1127	28
29	2000	0535	2011	0747	2036	1017	2110	1125	2322	1224	....	1156	29
30	2034	0638	2043	0856	2124	1127	2214	1220	....	1256	0022	1223	30
31	2106	0743	2117	1006	....	....	2320	1306	....	....	0123	1250	31

Time meridian 75° W. 0000 is midnight. 1200 is noon.

1999 MOONRISE AND MOONSET

NEW YORK, NY

Add one hour for Daylight Saving Time, covering summer months

Day	JANUARY Rise	JANUARY Set	FEBRUARY Rise	FEBRUARY Set	MARCH Rise	MARCH Set	APRIL Rise	APRIL Set	MAY Rise	MAY Set	JUNE Rise	JUNE Set	Day
	h m	h m	h m	h m	h m	h m	h m	h m	h m	h m	h m	h m	
1	1636	0632	1831	0740	1719	0614	1908	0618	1953	0549	2122	0624	1
2	1737	0731	1934	0817	1821	0648	2006	0647	2049	0623	2208	0713	2
3	1841	0823	2035	0849	1921	0719	2103	0717	2143	0700	2250	0807	3
4	1946	0908	2135	0920	2020	0748	2159	0749	2235	0741	2329	0905	4
5	2049	0946	2233	0949	2118	0817	2254	0823	2323	0827	....	1006	5
6	2151	1020	2330	1017	2215	0846	2348	0902	....	0918	0005	1109	6
7	2250	1051	....	1047	2312	0917	....	0945	0009	1014	0039	1214	7
8	2348	1120	0026	1118	....	0950	0039	1033	0050	1113	0111	1321	8
9	....	1148	0123	1153	0008	1026	0127	1127	0128	1216	0144	1430	9
10	0045	1217	0218	1231	0102	1107	0211	1225	0204	1321	0219	1542	10
11	0141	1247	0313	1314	0155	1153	0253	1328	0238	1429	0257	1656	11
12	0238	1320	0406	1403	0246	1244	0331	1433	0312	1540	0339	1810	12
13	0334	1356	0456	1458	0334	1341	0407	1542	0347	1653	0428	1920	13
14	0430	1437	0543	1558	0418	1442	0442	1653	0425	1808	0523	2026	14
15	0525	1523	0626	1702	0459	1548	0518	1806	0506	1923	0625	2123	15
16	0617	1615	0706	1809	0537	1656	0554	1920	0552	2035	0731	2211	16
17	0705	1712	0742	1917	0613	1807	0634	2035	0644	2142	0838	2253	17
18	0750	1813	0817	2026	0649	1918	0717	2147	0742	2242	0943	2329	18
19	0831	1917	0852	2136	0724	2030	0805	2255	0845	2333	1047	....	19
20	0908	2023	0927	2245	0801	2143	0859	2356	0949	....	1148	0000	20
21	0943	2129	1003	2355	0841	2253	0957	....	1054	0017	1248	0030	21
22	1017	2237	1043	....	0925	....	1058	0050	1157	0055	1346	0058	22
23	1050	2345	1128	0103	1013	0001	1201	0137	1258	0128	1443	0126	23
24	1125	....	1217	0208	1107	0105	1303	0217	1357	0158	1540	0155	24
25	1202	0054	1312	0309	1205	0202	1404	0252	1456	0226	1637	0226	25
26	1243	0203	1411	0404	1305	0252	1504	0324	1553	0254	1733	0300	26
27	1330	0311	1513	0453	1407	0336	1603	0353	1650	0322	1827	0338	27
28	1422	0416	1616	0536	1510	0414	1701	0421	1747	0352	1918	0421	28
29	1520	0517	....	....	1611	0449	1759	0450	1843	0424	2007	0509	29
30	1622	0611	....	....	1711	0520	1856	0519	1938	0500	2051	0602	30
31	1726	0659	....	....	1810	0549	....	....	2032	0540	....	....	31

Day	JULY Rise	JULY Set	AUGUST Rise	AUGUST Set	SEPTEMBER Rise	SEPTEMBER Set	OCTOBER Rise	OCTOBER Set	NOVEMBER Rise	NOVEMBER Set	DECEMBER Rise	DECEMBER Set	Day
	h m	h m	h m	h m	h m	h m	h m	h m	h m	h m	h m	h m	
1	2131	0659	2149	0903	2211	1126	2236	1239	....	1355	0040	1336	1
2	2208	0759	2221	1009	2255	1235	2336	1338	0041	1430	0142	1404	2
3	2242	0901	2255	1116	2345	1342	....	1429	0145	1502	0242	1432	3
4	2314	1005	2331	1225	....	1444	0039	1514	0247	1531	0341	1500	4
5	2346	1110	....	1334	0041	1541	0144	1552	0348	1559	0439	1530	5
6	....	1217	0012	1443	0143	1631	0249	1627	0448	1628	0537	1603	6
7	0018	1325	0058	1550	0248	1714	0352	1658	0547	1657	0634	1640	7
8	0053	1436	0151	1652	0354	1752	0455	1728	0646	1728	0729	1720	8
9	0132	1547	0251	1748	0500	1826	0556	1756	0743	1803	0821	1806	9
10	0216	1658	0356	1837	0604	1858	0657	1826	0840	1841	0909	1856	10
11	0307	1806	0503	1919	0707	1927	0756	1856	0933	1923	0954	1949	11
12	0405	1907	0611	1956	0808	1956	0854	1929	1024	2010	1034	2046	12
13	0509	2000	0717	2029	0908	2026	0951	2004	1111	2101	1110	2146	13
14	0616	2046	0820	2059	1007	2057	1046	2044	1154	2157	1143	2247	14
15	0724	2125	0922	2128	1104	2131	1139	2128	1233	2255	1215	2350	15
16	0830	2159	1022	2157	1200	2208	1229	2217	1309	2357	1245	....	16
17	0934	2231	1121	2227	1254	2250	1314	2311	1342	....	1316	0055	17
18	1036	2300	1218	2259	1346	2336	1356	....	1414	0100	1349	0203	18
19	1136	2328	1315	2334	1435	....	1435	0009	1446	0207	1425	0314	19
20	1234	2357	1410	....	1520	0028	1511	0110	1519	0315	1506	0427	20
21	1332	....	1504	0013	1601	0124	1545	0214	1555	0427	1555	0541	21
22	1428	0027	1555	0057	1639	0225	1618	0321	1635	0541	1651	0653	22
23	1525	0100	1642	0146	1715	0329	1651	0430	1720	0656	1754	0800	23
24	1619	0137	1726	0241	1749	0435	1726	0542	1813	0810	1903	0859	24
25	1712	0218	1807	0339	1822	0543	1804	0655	1912	0919	2013	0949	25
26	1802	0304	1843	0442	1855	0653	1846	0809	2017	1019	2122	1031	26
27	1848	0355	1918	0546	1931	0804	1934	0921	2124	1112	2229	1107	27
28	1931	0451	1951	0652	2010	0915	2028	1030	2232	1156	2333	1139	28
29	2009	0551	2023	0800	2053	1026	2128	1133	2337	1233	....	1208	29
30	2044	0654	2056	0908	2141	1135	2231	1228	....	1306	0034	1236	30
31	2117	0758	2132	1017	....	....	2336	1315	....	....	0134	1304	31

Time meridian 75° W. 0000 is midnight. 1200 is noon.

Visibility of The Planets, 1999

On August 11 there will be a Total Eclipse of the Sun, visible from N.E. USA and E. Canada.

VENUS is a brilliant object in the evening sky from the beginning of the year until mid-August when it becomes too close to the Sun for observation. During the last week of August it reappears in the morning sky where it stays until the end of the year. Venus is in conjunction with Jupiter on February 23, with Saturn on March 20 and with Mercury on August 26. Venus is brighter on all occasions.

MARS rises around midnight at the beginning of the year and is at opposition on April 24 when it is visible throughout the night. Its eastward elongation gradually decreases passing 3° N of Antares on September 17.

JUPITER can be seen at the beginning of the year in the evening sky after which it becomes too close to the Sun for observation. It reappears in the morning sky in mid-April. Its westward elongation gradually increases and from late July it can be seen for more than half the night. It is at opposition on October 23 when it is visible throughout the night. Its eastward elongation gradually decreases and after mid-October it passes into Pisces, in which constellation it remains for the rest of the year. Jupiter is in conjunction with Venus on February 23 (Jupiter is fainter) and with Mercury on May 1 (Jupiter is brighter).

SATURN can be seen in the evening sky in mid-January. It moves into Cetus in late March and into Aries from early April, in which constellation it remains for the rest of the year. It becomes too close to the Sun for observation during the second week of April, reappearing in the morning sky from mid-May. Its westward elongation gradually increases and is at opposition on November 6 when it is visible throughout the night. Its eastward elongation then gradually decreases and for the rest of the year can be seen for more than half the night. Saturn is in conjunction with Venus on March 20 and with Mercury on May 13. Saturn is fainter on both occasions.

MERCURY can only be seen low in the east before sunrise, or low in the west after sunset (about the time of the beginning or end of civil twilight). It is visible in the mornings between the following approximate dates: January 1 to January 21, March 27 to May 18, August 4 to August 31, and November 22 to December 30. The planet is brighter at the end of each period. It is visible in the evenings between the following approximate dates: February 16 to March 12, June 2 to July 18 and September 20 to November 10. The planet is brighter at the beginning of each period.

Visibility of Planets in Morning and Evening Twilight

	Morning	Evening
VENUS	August 25 – December 31	January 1 – August 16
MARS	January 1 – April 24	April 24 – December 31
JUPITER	April 15 – October 23	January 1 – March 18 October 23 – December 31
SATURN	May 16 – November 6	January 1 – April 10 November 6 – December 31

THE TIDES AND THE MOON

The principal forces creating our ocean tides are the gravitational pulls of the moon and sun acting on the earth. While the sun's mass is 26 million times the moon's, it is 390 times as far away. To find the relative gravitational forces, we cube 390 (59,000,000), and find that the effect of the moon is about 2.17 times that of the sun.

High tide occurs on the side of the earth nearest the moon because of the gravitational force. High tide also occurs simultaneously on the opposite side of the earth; this arises from the fact that the rotational axis of the combined mass of the earth and the moon is not at the center of the earth, but located close to the earth's surface nearest the moon. This means that the side of the earth opposite the moon is subject to considerable centrifugal force, and bulges outward, causing high tide. Simultaneously, the oceans are depressed (low water) halfway between these two bulges. Since the earth's rotation and the moon's orbit make it appear to travel around the earth every 24 hours and 50 minutes on the average, we have the following usual daily sequence: high tide; 6 hrs. 12 min. later a low tide; a second high tide 12 hrs. 25 min. after the first high tide; and 6 hrs. 12 min. later a second low tide.

How much the tide rises and falls depends basically on three things.

First, when the sun and moon are in line with the earth, their gravitational forces work together to produce a greater range of tide than usual. This occurs at full moon (moon opposite the earth from the sun), and new moon (moon on same side as sun). Note, in the Boston High and Low Water Tables, that the dates for Full Moon and New Moon bring the highest tides of the month. These tides are called "spring tides". When the moon is at right angles to the sun (moon at first quarter and last quarter), their forces are working against each other and the result is a lower range of tide than usual. These are called "neap tides". There are about 13 "lunar" months in a year, so we have 26 spring tides and 26 neap tides in the year.

Secondly, the moon's orbit around the earth is not circular but elliptical, ranging from 252,000 miles at Apogee (A), down to 221,000 miles at Perigee (P), and its effect on our tides at "P" is much greater than at "A". Note, in Boston High and Low Water Tables, how much higher the tide is when the Full Moon is at "P" than when the Full Moon is at "A". The position of the moon in its elliptical orbit is thus very important to the height of the tides.

Thirdly, both the sun and moon pass through an apparent curve called the "ecliptic". The sun's ecliptic takes a full year (see "Sun's Declination," p. 227). The moon goes through a similar ecliptic once a month. When the moon is on the Equator twice each month, see "E", the day's two high tides will be about the same. But during the northernmost declination (N) or southernmost (S), the two high waters will be markedly different in height.

This effect is known as "semidiurnal inequality". It occurs because the gravitational force of the moon when it has, say, northerly declination, is creating the largest bulge in our northern hemisphere while it is passing above us, but on the other side of the world, at the same time, its principal bulge is directly opposite us, or in the southern hemisphere. Thus, 12½ hrs. later when the moon is over China, its greatest effect on our side is in the S. Atlantic, with a lesser effect in the N. Atlantic. The above principles are based on the "equilibrium tide" theory, which presupposes a spherical earth which has no land masses and is entirely covered to a depth of 14 miles, by water with zero viscosity. Because of the existence of land masses with irregular shorelines, and widely varying ocean depths, the actual tides may frequently not be in accord with the equilibrium tide theory.

In summary, the moon's phase is most important, with highest tides at the full and new moon (at new moon, the mid-day tide will be higher than the night tide and at full moon, the night tide will be higher than the mid-day tide); secondly is the moon's position in its elliptical orbit, tides being highest when the moon is at Perigee; and lastly is the effect of the moon's ecliptic in creating tides of different heights on the same day.

PHASES OF THE MOON 1999 E.S.T.

● New Moon, ◐ 1st Quarter, ○ Full Moon, ◑ Last Quarter, A in Apogee
P in Perigee, N, S Moon farthest North or South of Equator, E on Equator

January

N	1	6pm
○	1	10pm
E	8	1pm
◑	9	9am
A	11	7am
S	16	1am
●	17	11am
E	22	7pm
◐	24	2pm
P	26	4pm
N	29	3am
○	31	11am

February

E	4	11pm
A	8	4am
◑	8	7am
S	12	10am
●	16	2am
E	19	1am
P	20	10am
◐	22	10pm
N	25	10am

March

○	2	2am
A	8	midn
E	4	7am
◑	10	4am
S	11	7pm
●	17	2pm
E	18	10am
P	19	7pm
◐	24	5am
N	24	3pm
E	31	2pm
○	31	6pm

April

A	4	5pm
S	8	3am
◑	9	1am
E	14	8pm
●	15	11pm
P	17	midn
N	20	10pm
◐	22	2pm
E	27	8pm
○	30	10am

May

A	2	1am
S	5	10am
◑	8	noon
E	12	7am
●	15	7am
P	15	10am
N	18	7am
◐	22	1am
E	25	3am
A	29	3am
○	30	2am

June

S	1	5pm
◑	6	11pm
E	8	5pm
P	12	8pm
●	13	2pm
N	14	6pm
◐	20	1pm
E	21	10am
A	25	11am
○	28	5pm
S	29	midn

July

E	6	midn
◑	6	7am
P	11	1am
N	12	5am
●	12	9pm
E	18	7pm
◐	20	4am
A	23	1am
S	26	7am
○	28	6am

August

E	2	6am
◑	4	noon
P	7	7pm
N	8	2pm
●	11	6am
E	15	4am
◐	18	9pm
A	19	6pm
S	22	3pm
○	26	7pm
E	29	noon

September

P	2	1pm
◑	2	5pm
N	4	9pm
●	9	5pm
E	11	noon
A	16	2pm
◐	17	3pm
S	18	11pm
○	25	6am
E	25	8pm
P	28	noon

October

◑	1	11pm
N	2	2am
E	8	8pm
●	9	7am
A	14	9am
S	16	7am
◐	17	10am
E	23	6am
○	24	4pm
P	26	8am
N	29	9am
◑	31	7am

November

E	5	3am
●	7	11pm
A	11	1am
S	12	3pm
◐	16	4am
E	19	5pm
○	23	2am
P	23	5pm
N	25	6pm
◑	29	6pm

December

E	2	9am
●	7	6pm
A	8	6am
S	9	10pm
◐	15	8pm
E	17	3am
P	22	6am
○	22	1pm
N	23	6am
◑	29	9am
E	29	4pm

Midnight is the *beginning* of the day.

Herreshoff 12½

LOA 15′8½″
LWL 12′6″
Beam 5′10″
Draft 2′5″
Displ 1350 lbs.

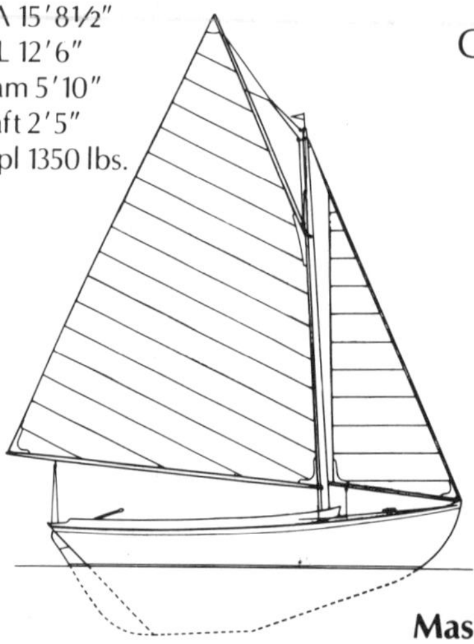

Capt Nat's time-honored design. Large, comfortable open cockpit, deep lead keel, self-tending jib. Available as original gaff rig or marconi. Combine low maintenance with history-making tradition. Deal direct with builder. Call or write: Cape Cod Shipbuilding Co., Wareham MA 02571 Telephone 508 295 3550

Cape Cod Shipbuilding Co.

Master Builders of Fine Sailboats since 1899

ANDREWS COMPASS SERVICE

Professional Compass Adjuster

Serving The New Bedford Fleet Since 1983

FACTORY AUTHORIZED SALES/SERVICE FOR:

DANFORTH/WHITE RITCHIE DIRIGO PLASTIMO SUUNTO C-PLATH

FULL SERVICE ADJUSTER: SUN AZIMUTH AND GYRO

Serving Government, Commercial and Pleasure Craft
In Southern New England, Cape Cod and Islands

800-566-7277 (Kompass) Beeper: (978) 547-3405

Local / Fax: (508) 758-3001

Ray Andrews, 11 Baptist Street., Mattapoisett, MA 02739

ATLANTIC COAST RADIOBEACON STATIONS

Revised as of September, 1998

	kHz	Signal	Range	Lat.N.	Lon.W.
CANADA					
Fox Island, NS	307	cont.	162	43-19	61-04
Partridge Is., NB	295	cont.	162	45-14	66-03
UNITED STATES					
ME. and N.H.					
*Brunswick	316	cont.	115	43-53-42	69-56-19
*Portsmouth	288	cont.	70	43-04-16	70-42-36
MA.					
*Chatham	325	cont.	95	41-40-17	69-57-02
N.Y.					
*Montauk Point	293	cont.	125	41-04-02	71-51-25
N.J.					
*Sandy Hook	286	cont.	100	40-28-18	74-00-42
Del. Md. Va.					
*Cape Henlopen	298	cont.	180	38-46-34	75-05-16
*Cape Henry	289	cont.	130	36-55-35	76-00-27
N.C.					
*Fort Macon	294	cont.	130	34-41-53	76-41-00
S.C.					
*Charleston	298	cont.	150	32-45-28	79-50-36
Florida					
*Cape Canaveral	289	cont.	250	28-27-36	80-32-36
*Miami	322	cont.	75	25-43-56	80-09-38
*Key West	286	cont.	110	24-34-54	81-39-06
Bermuda					
St. Davids Head	323	cont.	173	32-22-00	64-38-54

* These stations are land-based receivers and transmitters for Differential GPS (see page 205). The carrier of these Radiobeacons is modulated with a GPS correction (differential) signal, which may be used to greatly improve the accuracy of GPS. Mariners should see no degradation in the usability of the radiobeacon signal for direction finding although a warbling of the identification signal may be noticed. Correction broadcasts are changing from Type 1 format to Type 9-3. Your equipment may need upgrading. For status of GPS broadcasts: call NAVCEN (703) 313-5900; Electronic Bulletin Board: (703) 313-5910. Please see our May 1999 Free Supplement for further updates.

WARNING: *GPS system time will roll over at midnight August 21-22, 1999, 132 days before the year 2000. On August 22, 1999, unless repaired, many GPS receivers will claim that it is January 6, 1980 and August 23rd as January 7 and so on. Some manufacturers have solved the problem, but some have not. Users are encouraged to contact their manufacturer to determine if their receiver will be affected. For further information please visit U.S.C.G. Navigation Centers website at http://www.navcen.uscg.mil.*

RACONS

RACONS are Radar Beacons operating in the marine radar frequency bands, 2900-3000 mHz and 9300-9500 mHz. When triggered by a vessel's radar signal they provide a bearing by sending a coded reply (e.g. "T": –). This signal received takes the form of a single line or narrow sector extending radially towards the circumference of the radarscope from a point slightly beyond the spot formed by the echo from the lighthouse, buoy, etc. at the Racon site. Thus "distance off" may be measured to the point at which the Racon coded flash begins. (The figure obtained will be a few hundred feet greater than the actual distance of the ship from the Racon due to the slight response delay in the Racon apparatus.) Hours of transmission are continuous and coverage is all around the horizon unless otherwise stated. Their ranges depend on the effective range of the ship's radar and on the power and elevation of the Racon apparatus. Under conditions of abnormal radio activity, reliance should only be put on a Racon flash that is consistent and when the ship is believed to be within the area of the Racon's quoted range. Mariners are advised to turn off the interference controls of their radar when wishing to receive a Racon signal or else the signal may not come through to the ship. See p. 237 for list of Racons.

ATLANTIC COAST RACONS

RACON SITE	SIGNAL	RANGE	LAT. N	LONG. W
Cranberry Island	– • • • (B)	8 mi.	45-19-29	60-55-41
Halifax Alpha Lt. & Wh. By., NS	– • • • (B)	8 mi.	44-21-45	63-24-15
Chebucto Head, NS	– – • • (Z)	10 mi.	44-30-26	63-31-24
Cape Sable, NS	– • – • (C)	10 mi.	43-23-24	65-37-17
Lurcher Shoal West Lt. & Wh. By. "NK"	– • – (K)	8 mi.	43-47-30	66-37-00
Fundy Entr. So. Lt. & Wh. By. "M," NS	– • • • (B)	10 mi.	44-13-36	66-50-12
Black Pt. Lt. & Wh. By., NB	– • (N)	8 mi.	45-12-55	66-02-39
Gannet Rock, NB	– – • (G)	10 mi.	44-30-37	66-46-56
Portland Ltd. Hn. By., "P," ME	– – (M)	under 16 mi.	43-31-36	70-05-30
Frenchman Bay Ltd. Wh. By. "FB," ME	– • • • (B)	" "	44-19-21	68-07-26
Boston Ltd. Hn. By. "B"	– • • • (B)	" "	42-22-42	70-46-58
Boston Entr. Ltd. Wh. By. "NC"	– • (N)	" "	42-22-30	70-54-18
Cleveland Ledge E. Lt.	– • – • (C)	" "	41-37-52	70-41-39
Great Round Shoal Ltd. Whistle By. "GRS," MA	– – • (G)	" "	41-26-06	69-43-19
Nantucket Shoals Ltd. Horn By. "N," MA	– • (N)	" "	40-30-00	69-25-30
Nantucket Traffic Lane Ltd. Whistle By. "NA," MA	– • (N)	" "	40-25-42	73-11-30
Stratford Shoal	– • – – (Y)	" "	41-03-36	73-06-05
Execution Rocks	– • • – (X)	" "	40-52-41	73-44-16
Buzzards Bay Entr. Lt. "B," Horn	– • • • (B)	" "	41-23-48	71-02-00
Narrag. Bay E. Entr. Ltd. Hn. By. "NB"	– • • • (B)	" "	41-23-00	71-23-24
Narrag.-Buzz. Bay Appr. Ltd. Wh. By. "A," RI	– • (N)	" "	41-06-00	71-23-24
Twenty-Eight Ft. Sh. Ltd. By. "TE"	– (T)	" "	41-09-16	72-30-25
Ambrose Lt., NY	– • (N)	" "	40-27-35	73-49-51
Montauk Pt. Ltd. Wh. By. "M"	– – (M)	" "	41-01-48	71-45-42
Valiant Rock Bell By. #11	– • • • (B)	" "	41-13-46	72-04-00
Southwest Ledge By. #2	– • • • (B)	" "	41-06-44	71-40-16
Hudson Canyon Traffic Ltd. By. "HA"	– • – • (C)	" "	40-07-36	73-21-24
Scotland Ltd. Hn. By. "S," NJ	– – (M)	" "	40-26-30	73-55-00
Sandy H. Ch. Rge. Front Lt., NJ	– • – • (C)	" "	40-29-15	73-59-36
Barnegat Ltd. Horn By., NJ	– • • • (B)	" "	39-45-48	73-46-06
Del. Bay Appr. Ltd. Whistle By. "CH"	– • – (K)	" "	38-46-19	75-01-19
Del. Ltd. By. "D," DE	– • – (K)	" "	38-27-18	74-41-46
Del. R., Pea Patch Is.	– • • – (X)	" "	39-36-42	75-34-54
Brown Shoal Lt.	– • • • (B)	" "	33-55-20	75-06-03
Miah Maul Sh. Lt.	– – (M)	" "	39-07-36	75-12-30
Ship John Sh. Lt.	– – – (O)	" "	39-18-19	75-22-36
Chesapeake Light Tower, VA	– • (N)	" "	36-54-18	75-42-48
Ches. Ch. Ltd. Bell By. #62	– • – – (Y)	" "	37-46-29	76-10-16
Ches. Ch. Ltd. Bell By. #78	– – • – (Q)	" "	38-33-18	76-24-42
Ches. Bay Bridge (on main chann. span)	– • – (K)	" "	38-59-36	76-23-00
Cape Fear R. Ch. LWB CF, SC	– • – • (C)	" "	38-48-18	78-05-11
Charleston Ltd. Wh. By. "2C," SC	– • – (K)	" "	32-40-30	79-42-48
St. Simons Ltd. Wh. By. "STS," GA	– • • • (B)	" "	31-03-12	81-15-06
St. John's R. Ltd. By. "STJ," FL	– – (M)	" "	30-23-36	81-19-06
Miami Ltd. By. "M," FL	– – (M)	" "	25-46-06	80-05-00
North East Breaker Buoy - Bermuda	– • (N)	" "	32-28-42	64-40-54

For more on RACONS see pg. 235

MARINE EMERGENCY AND DISTRESS CALLS

Speak Slowly and Clearly. Use: Ch. 16 (156.8 mHz) or 2182 kHz or CB Radio Ch. 9

I. **Distress Signal** (top priority)
If you are in DISTRESS (i.e. when threatened by grave and imminent danger) transmit the International DISTRESS call on either 2182 kHz or 156.8 MHz (Channel 16) – "MAYDAY MAYDAY MAYDAY" "THIS IS (Your vessel's name and call sign repeated three times)"

IF CALLING FROM A VESSEL IN TROUBLE – give:

1. WHO you are (Your vessel's call and name).
2. WHERE you are (Your vessel's position in latitude/longitude or true bearing and distance in nautical miles from a known geographical point. Local names known only in the immediate vicinity are confusing.)
3. WHAT is wrong (Nature of Distress or Difficulty, if not in Distress).
4. Kind of assistance desired.
5. Number of persons aboard and condition of any injured.
6. Present seaworthiness of your vessel.
7. Description of your vessel (length, type cabin, masts, power, color of hull, superstructure and trim).
8. Your listening frequency and schedule.

IF CALLING WHILE OBSERVING ANOTHER VESSEL IN DIFFICULTY – give:

1. Your position and the bearing and distance of the vessel in difficulty.
2. Nature of Distress or Difficulty.
3. Description of the vessel in Distress or Difficulty, (see item 7 above).
4. Your intentions, course and speed, etc.
5. Your radio call sign, name of your vessel, listening frequency and schedule.

If there is no immediate response, repeat appropriate messages above; if still no response, you may send on any other available frequency until you make contact.

IF YOU HEAR A MAYDAY CALL – Immediately discontinue any transmission. Note details in your radio log right away. Do not make any transmission of this DISTRESS channel until MAYDAY condition is lifted by the Coast Guard, *unless* you are in a position to be of assistance.

II. **URGENCY SIGNAL** (second in priority)
If you have an URGENT message to send (threat to a vessel's safety or to someone on board, overboard or within sight), use the same procedure as above but say the word "PAN" three times. "PAN" (pronounced "PAWN") is also used as a warning signal that a Distress Signal may be sent out at a later stage.

III. **SAFETY SIGNAL** (third in priority)
If you wish to report navigation or weather warnings (ice, derelicts, tropical storms, etc.), use the same procedure as above, but say the word "SECURITY" three times.

OTHER DISTRESS SIGNALS – Flag hoist "NC"; Flying ensign upside down; Hoisting any square flag with a ball above or below it; Flames in a bucket; Red flare or rocket; Orange smoke; Dye marker; Flashlight "SOS" (• • • — — — • • •); Firing a gun repeatedly; Fog horn sounded continuously; Raising and lowering arms slowly. **NOTE**: 3 aerial or hand-held flares now mandatory for night use for all vessels over 16′

EPIRBS (Emergency Position Indicating Radio Beacons) when used for distress on the high seas, have a built in frequency of 121.5 and 243 MHz and are monitored by transoceanic aircraft who relay the distress message to the Coast Guard. The 406.025 MHz system, now available, is vastly more effective. See p. 241.

U.S. COAST GUARD STATIONS – MAINE TO KEY WEST
(monitoring Ch. 16, 2182 kHz, 156.8 MHz)

1st District – Boston – (800) 848-3942
Jonesport, ME (207) 497-2134
Southwest Harbor, ME (207) 244-4250
Rockland, ME (207) 596-6667
Boothbay Hbr., ME (207) 633-2644
S. Portland, ME (207) 767-0360
Portsmouth, NH (603) 436-4415
Merrimac-Newburyport, MA (978) 462-3498
Gloucester, MA (978) 283-0704
Boston, MA (617) 223-3224
Point Allerton-Hull, MA (617) 925-0166
Scituate, MA (781) 545-3801
Cape Cod Canal-E. Entr. (508) 888-0020
Provincetown, MA (508) 487-0077
Chatham, MA (508) 945-3830
Brant Pt.-Nantucket, MA (508) 228-0388
Woods Hole, MA (508) 457-3250
Menemsha, MA (508) 645-2661
Castle Hill, Newport, RI (401) 846-3676
Point Judith, RI (401) 789-0444
New London, CT (860) 442-4471
New Haven, CT (203) 468-4497
Fire Island, NY (516) 661-9101
New York-Staten Island, NY (718) 727-9188
Eatons Neck, NY (516) 261-6959
Fort Totten, NY (718) 352-4422
Rockaway, NY (718) 634-2848
Jones Beach, NY (516) 785-2921
Shinnecock, NY (516) 728-0078
Montauk, NY (516) 668-2773
Sandy Hook, NJ (908) 872-0326
Shark River, NJ (908) 776-6730

5th District – Portsmouth, VA (757) 398-6486
Manasquan Inl., NJ (908) 899-0130
Barnegat, NJ (609) 494-2661
Beach Haven, NJ (609) 494-2661
Atlantic City, NJ (609) 344-6594
Great Egg Hbr., NJ (609) 399-0144
Cape May, NJ (609) 898-6995
Indian River Inl., DE (302) 227-2440
Ocean City, MD (410) 289-7457
Chincoteague, VA (757) 336-2874

5th District, cont.
Little Creek-Norfolk, VA (757) 464-9371
Parramore Beach, VA (757) 787-9526
Portsmouth, VA (757) 483-8527
Taylors Island, MD (410) 397-3103
Crisfield, MD (410) 968-0323
Annapolis, MD (410) 267-8108
Curtis Bay-Baltimore, MD (410) 576-2625
Still Pond, MD (410) 778-2201
Cape Charles, VA (757) 331-2000
Milford Haven, VA (804) 725-3732
Oregon Inlet, NC (919) 441-6260
Hatteras Inlet, NC (919) 986-2175
Hobucken, NC (919) 745-3132
Ocracoke, NC (919) 928-4731
Fort Macon, NC (919)247-4509
Elizabeth City, NC (919) 335-6086
Wrightsville Beach, NC (919) 256-4224
Swansboro, NC (919) 354-2719
Oak Island, NC (919) 278-5592

7th District – Miami, FL (305) 536-5621
Georgetown, SC (803) 546-2052
Charleston, SC (803) 724-7629
Tybee, GA (912) 786-5440
Brunswick, GA (912) 267-7999
Mayport, FL (904) 247-7350
Ponce de Leon Inl., FL (904) 428-9085
Cape Canaveral, FL (407) 868-4200
Fort Pierce Inl., FL (561) 464-6100
Fort Meyers Beach, FL (813) 463-5754
Ft. Lauderdale, FL (954) 927-1611
Miami Beach, FL (305) 535-4368
Islamorada, FL (305) 664-4404
Marathon, FL (305) 743-1945
Key West, FL (305) 292-8862

Canada-Nova Scotia & New Brunswick
Department of Fisheries and Oceans
344 Slater St.
Ottawa, Ontario, Canada K1A 0N7
(613) 998-0215
Maritimes - Dartmouth, N.S. (902) 426-3907

The Coast Guard Navigation Information Service (NIS)
All U.S.C.G. Districts: Local Notices to Mariners (without enclosures). Electronic Bulletin Board Service - (703) 313-5910
300-28,800 bps, 8 data bits, no parity, 1 stop bit
NIS Watchstander, 24 hrs. a day - (703) 313-5900
Internet - HTTP://www.notmar.com
or
FTP://ftp.navcen.uscg.mil
Canada: Biweekly information available on line - (613) 990-3022.

EPIRBS

EPIRBS - Emergency Position Indicating RadioBeacons are designed to save your life by alerting rescue authorities and indicating your position. *Keep it above decks at all times, when at sea.*

Class A	121.5/243 MHz	Float free, automatically activated, detectable by aircraft and satellite. Coverage limited.
Class B	121.5/243 MHz	Manually activated version of Class A.
Class C	VHF Ch. 15/16	Manually activated, operates on marine channels only. Not detectable by satellite.
Class S	121.5/243 MHz	Similar to Class B, except it floats, or is an integral part of a survival craft.
Cat. I	406/121.5 MHz	Float-free, automatically activated. Detectable by satellite anywhere in the world.
Cat. II	406/121.5 MHz	Similar to Category I, except is manually activated.

Classes A, B and S are the most common and least expensive type of EPIRB, designed to be detected by overflying aircraft. Satellites were also designed to detect these EPIRBS, but are limited because the satellite must be within line of sight of both the EPIRB and a ground terminal for detection to occur.

Design of these EPIRBS and frequency congestion cause these devices to be subject to a very high false alert/false alarm rate. Consequently confirmation is required before search and rescue forces can be deployed.

EPIRBS manufactured before October, 1989 may have design or construction problems (some models may leak and cease operating when immersed in water) and may not be detectable by satellite.

Class C EPIRBS are manually activated and intended for pleasure craft who do not venture far offshore and for vessels in the Great Lakes. They transmit a short burst on Ch. 16 and a longer homing signal on Ch. 15. Their usefulness depends upon a Coast Guard Station or other vessel guarding Ch. 16 and recognizing the brief recurring tone of the EPIRB. Class C EPIRBS are not recognized outside the U.S.

Categories I and II are 406 MHz EPIRBS designed to operate with satellites. Their signals allow a satellite local user terminal to locate accurately the EPIRB (much more accurately than the 121.5/243 MHz devices) and identify the vessel (the signal is encoded with the vessel's identity) anywhere in the world. There is no range limitation. They also include a 121.5 MHz homing signal, allowing aircraft and rescue craft to find the vessel in distress very quickly. These are the only types which must be certified by Coast Guard-approved independent laboratories before they can be sold in the U.S.

The automatically-activated version will be required on Safety of Life at Sea Convention vessels (passenger ships and ships over 300 tons, on international voyages) of any nationality by Aug. 1, 1993. The Coast Guard requires U.S. commercial fishing vessels to carry this device (by August 1, 1991), and will require the same for other U.S. commercial uninspected vessels which travel more than 3 miles offshore.

A word about the Satellite System: COSPAS (a Russian acronym for Space System for Search of Distress Vessels) and SARSAT (Search and Rescue Satellite-Aided Tracking) is an international satellite-based search and rescue system established by the U.S., U.S.S.R., Canada and France to locate emergency radio beacons transmitting on frequencies 121.5, 243 and 406 MHz. The U.S. Coast Guard operates 2 local user terminals (satellite earth stations designed to receive EPIRB Distress Calls forwarded from COSPAS-SARSAT satellites, located in Kodiak, Alaska and Point Reys, California. The Air Force operates a third terminal at Scott Air Force Base, Illinois.

The Coast Guard urges all EPIRB owners to examine them periodically for water tightness, battery expiration date and signal presence. FCC rules allow Classes A, B and S EPRIBS to be turned on briefly for 1 second only during the first 5 minutes of each hour. Signal presence can be detected by an FM radio tuned to 99.5 MHz or an AM radio tuned to any vacant frequency and located close to the EPIRB. Class C EPIRBS can be detected by a marine radio tuned to Ch. 15 or 16. 406 MHz EPIRBS can be tested through their self-test function which is an integral part of the device.

RADIO TELEPHONE INFORMATION — VHF SYSTEM

GENERAL INFORMATION ON THE 156-162 MHz VHF-FM MARINE RADIO

Calling Guidelines: Avoid excessive calling. Make calls as brief as possible. Give name of called vessel first, then "This is (name of your vessel)", your call sign, and the word "Over." If station does not answer, delay your repeat call for 2 minutes. At the end of your message, sign off with "This is (your vessel's name)," your call sign, and the word "Out."

Range and Power: Operation is essentially line-of-sight. Since the elevation of antennas at both communications points extends the "horizon," range may be 20 to 50 miles on a 24 hour basis between a boat and a land station. Effective range between boats will be less because of lower antenna heights. 25 watts is the maximum output permitted.

Interference Factor: Most VHF-FM equipment has 6 or more channels, so it is possible to shift to a clear channel. Like the FM in your home radio, the system is practically immune to interference from ignition noise, static etc. except under unusual conditions.

Channelization: A minimum of 3 channels is required by the FCC. Two are mandatory: Channel 16 (156.80 MHz), the National Distress frequency; and Channel 06 (156.30), the Intership Safety Frequency. The Coast Guard *strongly recommends* that you have Channel 22 as your third channel.

Channel	Purpose and Comments
16 156.80 MHz	**DISTRESS and SAFETY:** Ship to Shore and Intership. Guarded 24 hours by the Coast Guard. No routine messages allowed other than to establish use of a working channel. See page 241 for Distress calling procedure. **CALLING:** Ship to Shore and Intership. Use Channel 16 to establish contact, then switch to the appropriate working channel (see list below). **Calling Channel:** New England waters. Commercial and Pleasure Craft.
09 156.450 MHz	**Calling Channel:** New England waters. Commercial and Pleasure Crafts.
06 156.30 MHz	**INTERSHIP SAFETY:** No routine messages allowed. This channel is limited to: talking with the Coast Guard and others at the scene of an emergency, and to information on the movement of vessels.
22 157.10 MHz (**21** in Canada) 161.65 MHz	**COAST GUARD COMMUNICATION** Channel. Not guarded by the C.G., but after vessel makes contact with the C.G. for non-distress calls on Channel 16, the C. G. will tell you to switch to and use *only* 22 for communicating. C.G. Weather Advisories and Notices to Mariners are also broadcast on Channel 22. C.G. announces *times* on these broadcasts on Channel 16.
12, 14, 20, 65, 66, 73, 74, 77	**Ship to Shore and Intership:** Port operations, harbormasters, etc. (Your electronics dealer should have local frequencies.)
8, 67, 88	**Commercial (Intership only).** For ocean vessels, dredges, tugs, etc.
7, 10, 11, 18 **19, 79, 80**	**Commercial Only.**
09, 13, 67	**Bridge Information.** Monitored 24 hours by Commercial craft.
68, 69, 71, 72, 78	**Ship to Shore and Intership, Pleasure Craft** only. Shore stations, marinas, etc. For pleasure craft these are the best channels for general communication.
70	**Digital Selective Calling (DSC).** Ships must have special equipment for DSC.
24, 25, 26, 27, 28 **84, 85, 86 and 87**	***Marine Telephone*** Operators (see p. 243)
15	***NOAA Weather*** Reports - primarily weather channels. (see p. 243) **WX-1, WX-2, WX-3** (receive only)

VHF MARINE TELEPHONE INFORMATION

by MariTEL VHF Marine Telecommunications Service

Operator's Channels from Maine to Key West (updated 8/98):

Portland, ME	24, 28, 87	Atlantic City, NJ	26	Wilmington, NC	26
Gloucester, MA	25	Cape May, NJ	24	Georgetown, SC	24
Boston, MA	26	Philadelphia, PA	26	Charleston, SC	26
Nantucket, MA	85	Odessa, DE	28	Savannah, GA	27
Hyannisport, MA	28	Dover, DE	84	Jacksonville, FL	26
New Bedford, MA	24, 26	Lewes, DE	27	Daytona Bch., FL	28
Providence, RI	27	Bethany Beach, DE	86	Cocoa Bch., FL	26
Narragansett, RI	84	Ocean City, MD	26	Vero Bch., FL	27
New London, CT	26, 86	Cambridge, MD	28	St. Lucie, FL	26
Bridgeport, CT	27	North Bay, MD	24	W. Palm Beach, FL	28
Staten Island, NY	28	Baltimore, MD	25, 26	Ft. Lauderdale, FL	84
Sandy Hook, NJ	24	Pt. Lookout Ridge, MD	26	Miami, FL	24, 25
Toms River, NJ	27	Belle Haven, VA	25, 84	Key Largo, FL	27, 28
Ship Bottom	28	Virginia Beach, VA	26, 27	Marathon, FL	27
Beach Haven, NJ	25	Morehead City, NC	28	Key West, FL	26, 84

Calling from a Vessel: (VHF Radio) Select the working channel for your area. If channel is clear key your VHF microphone button and hold down for 6 seconds . After the "Welcome Message" has been activated, Key again for 6 seconds, the MariTEL operator will answer and ask for your vessel name, the telephone number to call and your subscriber number or method of billing.

Calling to a Vessel: (telephone) Dial 1-888-627-4632 and tell the MariTEL operator the name and location of the vessel, your subscriber number or method of billing and your return telephone number. The MariTEL operator will hail the vessel on Channel 16 and connect your call. If contact is not made you may leave a message and MariTEL will continue to call the vessel at regular intervals for 24 hours. *To subscribe call 1-888-MariTEL (627-4835).*

MARINE WEATHER FORECASTS

VHF-FM NOAA WEATHER REPORTS

	MHz		
WX-1	*162.55	NOAA Weather	Primary weather
WX-2	**162.40	" "	channel. Ship
WX-3	***162.475	" "	receive only

Continuous broadcasts are provided by the National Weather Service with taped messages repeated every 4-6 minutes. These are up-dated every 3-6 hours and include weather and radar summaries, wind observations, visibility, sea conditions and detailed local forecasts. NOAA VHF-FM broadcasts can usually be received 20-40 miles from transmitting site.

**Ellsworth, ME	**Meriden, CT	*Norfolk, VA	***Beaufort, SC
***Dresden, ME	*New York, NY	**Heathsville, VA	**Savannah, GA
*Portland, ME	***Riverhead, NY	***Richmond, VA	*Jacksonville, FL
**Concord, NH	**Atlantic City, NJ	***Cape Hatteras, NC	**Daytona Bch., FL
**Burlington, VT	*Lewes, DE	**New Bern, NC	*Melbourne, FL
***Boston, MA	***Philadelphia, PA	*Buxton, NC	***W. Palm Bch., FL
*Hyannis, MA	*Washington, DC	*Wilmington, NC	*Miami, FL
**Providence, RI	***Salisbury, MD	**Myrtle Bch., SC	*Key West, FL
*New London, CT	**Baltmore, MD	*Charleston, SC	

TIME SIGNALS

Bureau of Standards Time Signals: WWV, Ft. Collins, Col., every min. on 2500, 5000, 10000, 15000, 20000, 25000 kHz. **Canadian Time Signals:** CHU, (frequently easier to get than WWV) 45° 17' 47" N, 75° 45' 22" W. Continuous transmission on 3330 kHz, 7335 kHz and 14670 kHz.

Omission of a tone indicates the 29th second of each minute. The new minute is marked by the full tone *immediately* following the voice announcement. Five sets of two short tones mark the first five seconds of the next minute. The hour is identified by a pulse of one full second followed by 12 seconds of silence.

COAST GUARD RADIO BROADCASTS – WEATHER FORECASTS AND NAVIGATIONAL WARNINGS

Halifax to Key West

Station		Channel:	MHz	kHz	
		Atlantic Standard Time			
Sydney, NS	**VCO**	**21:**	161.65 —		Cont. reports
			–	2749	0118, 0618, 1018, 1248, 1718, 1820
Halifax, NS	**VCS**	**21:**	161.65 —		Cont. reports
			–	2749	0203, 0803, 0918, 1403, 1635, 1910
Grindstone, MI	**VCN**	**21:**	161.65 —		Cont. reports
Western Hd., NS		**21:**	161.65 —		Cont. reports
Yarmouth, NS	**VAU**	**21:**	161.65 —		Cont. reports
			–	2749	0140, 0640, 0848, 1225, 1330, 1548, 1948
		Eastern Standard Time			
SW Harbor, ME	**NMF**	**22:**	157.1	—	0635, 1835
Portland, ME	**NMF**	**22:**	157.1	—	0605, 1805
York, ME	**NMF**	**22:**	157.1	—	0635, 1835
Gloucester, MA	**NMF**	**22:**	157.1	—	0535, 1735
Boston, MA	**NMF**	**22:**	157.1	—	0535, 1735
		—	—	2670	1140, 2340
Provincetown, MA	**NMF**	**22:**	157.1	—	0505, 1705
Brant Pt., MA	**NMF**	**22:**	157.1	—	0535, 1735
Woods Hole, MA	**NMF**	**22:**	157.1	—	0535, 1735
Long Is. Sound		**22:**	157.1	—	0620, 1820
NY	**NMY**	**22:**	157.1	2670	0550, 1750
Moriches, NY		**22:**	157.1	2670	0720, 1920
Sandy Hook, NJ		**22:**	157.1	2670	0520, 1720
Atlantic City, NJ	**NMK**	**22:**	157.1	2670	0603, 1803
Philadelphia, PA		**22:**	157.1	—	0735, 1935
Baltimore, MD	**NMX**	**22:**	157.1	2670	0705, 2030
Chincoteague, VA	**NMN**	**22:**	157.1	—	0645, 2100
		—	—	2670	0903, 2133
Hampton Roads, VA	**NMN**	**22:**	157.1	—	0620, 2130
		—	—	2670	0833, 2103
Ft. Macon, NC	**NMN**	**22:**	157.1	—	0530, 2030
		—	—	2670	0733, 2003
C. Hatteras, NC	**NMN**	**22:**	157.1	—	0555, 2000
		—	—	2670	0803, 2033
Charleston, SC	**NMB**	**22:**	157.1	—	0700, 1700
		—	—	2670	0920, 2120
Mayport, FL	**NMA**	**22:**	157.1	—	0715, 1715
		—	—	2670	1120, 2320
Miami Bch., FL	**NCF**	**22:**	157.1	—	0730, 1730
		—	—	2670	1050, 2250
Key West, FL	**NOK**	**22:**	157.1	—	0700, 1700
Bermuda Hbr. Radio	**ZMB**	**27:**	—	2582	0035, 0435, 0835, 1635 2035
		38:	162.4		Cont. reports (Berm. St. Time)

NATIONAL WEATHER SERVICE

National Weather Service provides a Radiofacsimile service for weather and charts prepared by the U.S. Coast Guard Communication Station, Boston (NMF). Continuous Broadcasts are on 6340.5 and 12750 kHz for East Coast and Gulf Coast waters at 1403 and 2143 E.S.T. For more specific information, contact the **Marine Prediction Ctr.** 301-763-8000, ext. 7401 or FAX 301-763-8085. Internet http://www.ncep.noaa.gov/MPC/ for information on receiving radiofax charts via E-mail.

1999 COMMERCIAL RADIO WEATHER BROADCASTS

Listed below are forecasts valuable to the mariner. In addition, all stations have weather reports in their regular newscasts.

Ellsworth, ME	WDEA	1370	–	Daily: 33 past, 5 a.m.-11 p.m.. Weatherphone 207-667-8910.
	WWMJ	95.7	–	Mon.-Fri.: 01 and 31 past, 6-10 a.m. Mon.-Sat. 18 and 48 past, 9 p.m. - 5 a.m.
Rockland, ME	WRKD	1450	–	Daily: 02, 50, 32 min. past. WMCM FM 103.3 – 35 and 57 min. past.
Portsmouth, NH	WHEB	750	–	Daily: 5:55, 6:55, 7:55, 8:55, 11:55 a.m.; 3:55, 4:55,5:55p.m.;
	& FM	100.3	–	Sat.: 6:55, 7:55, 8:55 a.m.; Sun.: varies.
Boston, MA	WBZ	1030	–	Daily: ev. 10 min.
W. Yarmouth, MA	WXTK	94.9	–	Daily: hourly, twice an hour each a.m., Fishing Reports on weekend.
Hyannis, MA	WOCN FM	103.9	–	Daily: 06 and 29 past the hour; 15 & 45 past the hour, 5-9a.m.
	WQRC FM	99.9	–	Daily: detailed reports & tides, 5:20, 5:50, 6:20, 6:50, 7:20, 7:50, 8:20, 8:50 a.m.; 12:10, 4:06, 4:36, 5:06, 5:36 p.m.
New Bedford, MA	WBSM	1420	–	Marine forecasts for Georges Bank, Nantucket Shoals: Mon-Fri: 6:35, 7:35, 8:35 a.m.; 12:20, 3:10, 5:20, 11:15 p.m. Commercial & Pleasure Craft: Sat: 6:35, 7:30, 8:35 a.m. Pleasure Craft, summer only: Sat: 6:15, 7:15 a.m.; 12:15 p.m.; Sun: 6:15, 7:15, 8:15 a.m.; 12:15, 6:15 p.m.
Providence, RI	WPRO	630	–	Daily: on the hour.
New London, CT	WNLC FM	98.7	–	Daily: 2 times an hour. 20 & 50 min. past the hour.
	WTYD	100.9	–	Daily: 2 times an hour.
Old Saybrook, CT	WLIS	1420	–	Daily: ev. 30 min. Marine Weather, Tides: 30 min. past.
New Haven, CT	WELI	960	–	Daily: on the hr. & half hr. Sat & Sun: 15 & 45 min. past the hr. (live from boat).
Long Island, NY	WGSM	740	–	Daily: at :07 and :37 past the hour.
New Brunswick, NJ	WCTC	1450	–	Daily: 05, 25, 35, 45, 55 past the hour 5 a.m. to 8 p.m.; 05 past 8 p.m. to 5 a.m.
Wilmington, DE	WDEL	1150	–	Daily: 20 and 50 past the hour. Also at: 6:12, 6:42, 7:12, 7:42, 8:12, 8:42 a.m., 12:10, 4:12, 4:42, 5:12, 5:42, 11:05 p.m.
Annapolis, MD	WNAV	1430	–	Daily: 22 past the hour.
Baltimore, MD	WBAL	1090	–	Daily: once an hour, 5 a.m.-midnight.
Wilmington, NC	WMFD	630	–	Daily: 18 & 40 min. past the hour.
Morehead City, NC	WMBL	740	–	Daily: at :20 and :50 past the hour.
Jacksonville, FL	WJAX	930	–	Daily: 20 and 50 min. past the hour.
Daytona Beach, FL	WROD	1340	–	Daily: 10, 20, 40, 50 past. 24 hrs. a day. Marine & Radar report.
Ft. Pierce, FL	WIRA	1400	–	Daily: 05, 35 past 6 a.m.-7 p.m.; 05 past 7 p.m.-6 a.m. Marine forecast & tides.
W. Palm Beach, FL	WEAT	850	–	Daily: Marine Weather,7:20 a.m., hourly updates include tides.
Coral Gables, FL	WVCG	1080	–	Mon-Sat: 7-7:05 a.m. for S.W.N. Atlantic, Nassau tides, Tropical Storms.
Key West, FL	WKIZ	1500	&	WEOW FM 92.5 – Daily: on the hour and half hour.

(All times are Local times, standard or daylight whichever applies.)

While all possible care has been taken in collecting these schedules, changes may occur during the year.

FITZROY'S BAROMETER INSTRUCTIONS

The words on scales of barometers should not be so much regarded for weather indications as the *rising* or *falling* of the mercury; for if it stand at *changeable* (29.50), and then rise towards *fair* (30.00), it presages a change of wind or weather, though not so great as if the mercury had risen higher; and, on the contrary, if the mercury stand above *fair* and then fall, it presages a change, though not to so great a degree as if it had stood lower: beside which, the direction and force of wind are not in any way noticed.

It is not from the point at which the mercury may stand that we are alone to form a judgment of the state of the weather, but from its *rising* or *falling* and from the movements of immediately *preceding* days as well as hours, keeping in mind effects of change of *direction*, and dryness, or moisture, as well as alteration of force or strength of wind.

It should always be remembered that the state of the air *foretells coming* weather, rather than shows the weather that is *present*—(an invaluable fact too often overlooked)—that the longer the time between the signs and the change foretold by them, the longer such altered weather will last; and, on the contrary, the less the time between a warning and a change, the shorter will be the continuance of such foretold weather.

If the barometer has been about its ordinary height, say near thirty inches at the sea-level, and is steady or rising, while the thermometer falls, and dampness becomes less—north-westerly, northerly, or north-easterly wind, or less wind, less rain or snow may be expected.

On the contrary, if a fall takes place with a rising thermometer and increased dampness, wind and rain may be expected from the south-eastward, southward, or south-westward.

A fall with low thermometer foretells snow.

When the barometer is rather below its ordinary height, say down to near twenty-nine inches and a half (at sea-level), a rise foretells less wind, or a change in its direction towards the northward—or less wet: but when it has been very low, about twenty-nine inches, the first rising usually precedes or indicates strong wind—at times heavy squalls—from the north-westward, northward or north-eastward; *after* which violence a gradually rising glass foretells improving weather, if the thermometer falls; but if the warmth continue, probably the wind will back (shift against the sun's course), and more southerly or southwesterly wind will follow, especially if the barometer rise is sudden.

The most dangerous shifts of wind, or the *heaviest* northerly gales, happen *soon* after the barometer *first* rises from a very low point; or, if the wind veers *gradually*, at some time afterwards.

Indications of approaching change of weather, and the directions and force of winds, are shown less by the height of the barometer than by its falling or rising. Nevertheless, a height of more than thirty (30.0) inches (at the level of the sea) is indicative of fine weather and *moderate* winds, except from east to north *occasionally*.

A rapid rise of the barometer indicates unsettled weather; a slow movement the contrary; as, likewise, a *steady* barometer, which, when continued, and with dryness, foretells very fine weather.

A rapid and considerable fall is a sign of stormy weather, and rain or snow. Alternate rising and sinking indicates unsettled and threatening weather.

The greatest depressions of the barometer are with gales from S.E., S. or S.W.; the greatest elevations with wind from N.W., N. or N.E., or with calm.

A sudden fall of the barometer, with a westerly wind, is sometimes followed by a violent storm from N.W., or N. or N.E.

If a gale sets in from the E. or S.E. and the wind veers by the South, the barometer will continue falling until the wind is near a marked change, when a lull *may* occur; after which the gale will soon be renewed perhaps suddenly and violently, and the veering of the wind towards the N.W., N., or N.E., will be indicated by a rising of the barometer, with a fall of the thermometer.

After very warm and calm weather a storm or squall, with rain, may follow; likewise at any time when the atmosphere is *heated* much above the *usual* temperature of the season.

To know the state of the air; not only the barometer and *thermometer*, but appearance of the sky should be vigilantly watched.

SIGNS OF WEATHER

Whether clear or cloudy, a rosy sky at sunset presages fine weather; and a red sky in the morning, bad weather, or much wind, perhaps rain; a grey sky in the morning, fine weather; high dawn, wind; low dawn, fair weather.

A high dawn is when the first indications of daylight are seen above a bank of clouds. A low dawn is when the day breaks on or near the horizon, the first streaks of light being very low down.

Soft-looking or delicate clouds foretell fine weather, with moderate or light breezes; hard-edged, oily-looking clouds, wind. A dark, gloomy, blue sky is windy; but a light, bright-blue sky indicates fine weather. Generally, the *softer* the clouds look, the less wind (but perhaps more rain) may be expected; and the harder, more "greasy," rolled, tufted, or ragged, the stronger the coming wind will prove. Also, a bright yellow sky at sunset presages wind; a pale yellow, wet; and thus, by the prevalence of red, yellow, or grey tints, the coming weather may be foretold very nearly—indeed, if aided by instruments, almost exactly.

Small inky-looking clouds foretell rain; light scud clouds driving across heavy masses show wind and rain; but if alone may indicate wind only.

High upper clouds crossing the sun, moon, or stars in a direction different from that of the lower clouds, or the wind then felt below, foretell a change of wind.

After fine, clear weather, the first signs in the sky of a coming change are usually light streaks, curls, wisps, or mottled patches of white distant clouds, which increase, and are followed by an overcasting of murky vapour that grows into cloudiness. This appearance, more or less oily, or watery, as wind or rain will prevail is an infallible sign.

Light, delicate, quiet tints or colours with soft, undefined forms of clouds, indicate and accompany fine weather; but gaudy or unusual hues, with hard, definitely outlined clouds, foretell rain, and probably strong wind.

When seabirds fly out early and far to seaward, moderate wind and fair weather may be expected. When they hang about the land, or over it, sometimes flying inward, expect a strong wind, with stormy weather. As many creatures beside birds are affected by the approach of rain or wind, such indications should not be slighted by an observer who wishes to foresee weather.

Remarkable clearness of atmosphere near the horizon, distant objects, such as hills, unusually visible, or raised (by refraction),* and what is called "a good *hearing* day," may be mentioned among signs of wet, if not wind, to be expected.

More than usual twinkling of the stars, indistinctness or apparent multiplication of the moon's horns, haloes, "wind-dogs" (fragments or pieces of rainbows, sometimes called "wind-galls") seen on detached clouds, and the rainbow, are more or less significant of increasing wind, if not approaching rain, with, or without wind.

Lastly, the dryness or dampness of the air, and its temperature (for the season) should *always* be considered *with other* indications of change, or continuance of wind and weather.

On Barometer scales, the following contractions may be useful:

RISE for N.E.ly (N.W.-N.-E.) DRY or LESS WIND—except wet from N.E. d.

FALL for S.W.ly (S.-E.S.-W.) WET or MORE WIND—except wet from N.E.d.

When the wind shifts against the sun,
Trust it not, for back it will run.

First rise after very low,
Indicates a stronger blow.
Long foretold—long past.
Short notice—soon past.

*Much refraction is a sign of Easterly wind.

BAROMETER RISING

28.8 to 29.2 inches.	Clearing, high winds and cool wave.
29.2 to 29.6 inches.	High winds, cool wave, preceded by squalls.
29.6 to 29.9 inches.	Fair weather, fresh winds during next 24 hours.
29.9 to 30.2 inches.	Fair weather and brisk winds, diminishing.
30.2 to 30.5 inches.	Fair weather, cooler, variable winds.
30.5 to 30.8 inches.	Continued cool, clear weather, light winds.
30.8 to 31.0 inches.	High winds, southeast with rain.

BAROMETER FALLING

30.8 to 30.5 inches.	Fair, warmer, followed by rain and wind.
30.5 to 30.2 inches.	Approaching storm.
30.2 to 29.9 inches.	Cloudy, warmer, unsettled weather.
29.9 to 29.6 inches.	Unsettled weather, warmer with increasing wind.
29.6 to 29.3 inches.	Squally, clearing, fair and cooler weather.
29.3 to 29.0 inches.	Clearing, high winds with squalls and cooler weather.
29.0 to 28.7 inches.	Stormy weather.

Forecasting from Wind Direction and Barometric Pressure - in inches
(General Rule of Thumb for Sailors)

Wind direction is indicated in sectors, i.e. blowing from the SW - NW sector:

SW-NW
(30.10-30.20) Steady
Fair-little temperature change

(30.10-30.20) Rising rapidly
Fair, perhaps followed by warmer weather and rain

(+30.20) Steady
Fair, no temperature change

(+30.20) Falling slowly
Fair, gradual rising temperature

S-SE
(30.20) Falling rapidly
Increasing wind, rain indicated later

SE-NE
(30.10-30.20) Falling slowly
Rain indicated

(30.10-30.20) Falling rapidly
Increasing wind and rain indicated

(-30.00) Falling slowly
Rain continuing

SE-NE
(-30.00) Falling rapidly
Rain and high winds followed by clearing and cooler

E-NE
(+30.10) Falling slowly
Rain with light winds

(+30.10) Falling rapidly
Rain or snow with increased winds

S-SW
(-30.00) Rising slowly
Clearing followed by fair

S-E
(-29.80) Falling rapidly
Severe NE gale with heavy rain/snow

Shifting W
(-29.80) Rising rapidly
Clearing and cooler

BEAUFORT SCALE

Wind Speed in knots (33′ above sea level)	Wind Condition	Sea Conditions	Approx. Wave ht. (in ft.)	Force
0-1	**Calm**	Smooth, mirror-like sea.	-	**0**
1-3	**Light Air**	Scale-like ripples; no foam crests.	¼	**1**
4-6	**Light Breeze**	Short wavelets; glassy crests; non-breaking.	½	**2**
7-10	**Gentle Breeze**	Large wavelets; glassy crests, some breaking; occasional white foam.	2	**3**
11-16	**Moderate Breeze**	Small waves becoming longer; frequent white foam crests.	3½	**4**
17-21	**Fresh Breeze**	Moderate waves with more pronounced long form; many white foam crests; some spray.	6	**5**
22-27	**Strong Breeze**	Large waves form; white foam crests everywhere; probably more spray.	9½	**6**
28-33	**Near Gale**	Sea heaps up; white foam from breaking waves is blown in streaks with the wind.	13½	**7**
34-40	**Gale**	Moderately high waves with greater length; edges of crests break into spindrift; foam is blown in well-defined streaks with the wind.	18	**8**
41-47	**Strong Gale**	High waves; crests of waves start to topple and roll over; spray may affect visibility.	23	**9**
48-55	**Storm**	Very high waves; overhanging crests; resulting foam is blown in dense white patches with the wind; sea surface takes on a whiter look; the tumbling of the sea becomes heavy and shock-like; visibility affected.	29	**10**
56-63	**Violent Storm**	Exceptionally high waves; sea covered with long white patches of foam blown in direction of wind; all wave crests are blown into froth; visibility affected.	37	**11**
64+	**Hurricane**	Air is filled with spray and foam; sea is completely white with driving spray; visibility seriously affected.	45	**12**

Note: In enclosed waters or near land with an offshore wind, wave heights will be less, the waves steeper and not so long. In many tidal waters wave heights are apt to increase considerably in a very short time and conditions can be more dangerous near land than in the open sea.

WIND CHILL FACTOR

Wind Speed (mph)	Temperature in Fahrenheit +30	+20	+10	0	-10	-20
10	+16	+ 2	- 9	-22	-31	-45
20	+ 3	- 9	-24	-40	-52	-68
30	- 2	-18	-33	-49	-63	-78
40	- 4	-22	-36	-54	-69	-87
50	- 7	-24	-38	-56	-70	-88

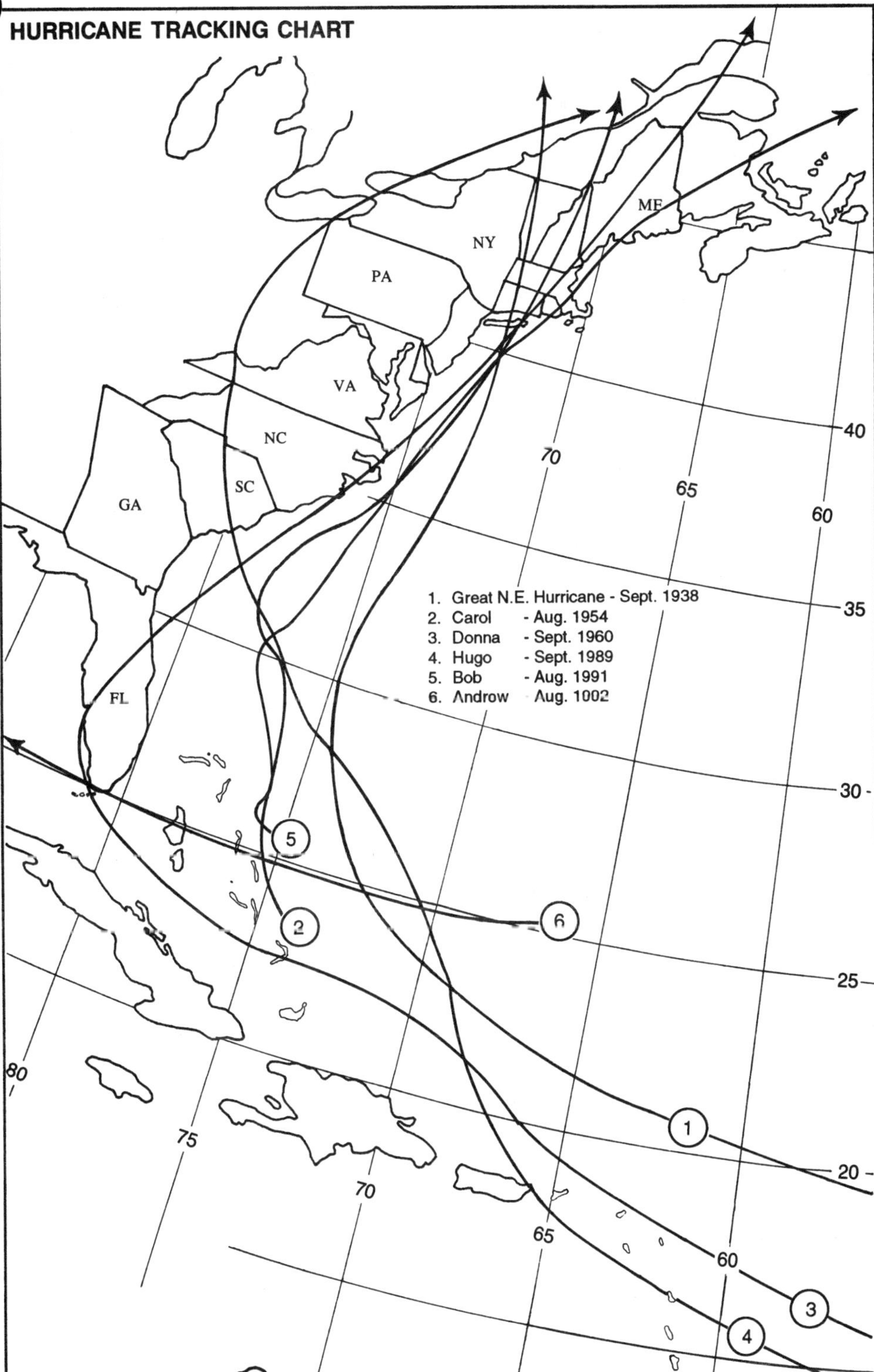

Hurricane Hot Line: 1-900-410-NOAA

Hurricanes

In recent years, hurricanes hitting the Atlantic coast have taken a great toll in lives and property. We present these two pages in the interest of promoting greater safety along the coast.

Hurricanes affecting the N. Atlantic usually move W. or N.W. from the Eastern Caribbean, veering to the N. and E., and losing force as they spin out to sea. They are areas of low barometric pressure; the sharper and deeper the drop in pressure, the more violent the winds will generally be. The "eye", or storm center, may travel forward from 5 to 50 m.p.h. The winds around it travel COUNTER-CLOCKWISE, spiraling inwards and ***accelerating*** as they approach the eye. See diagram below.

In the right or dangerous semicircle, where the spiral winds are heading in the same general direction as the eye, the two velocities are combined and therefore maximum. In the left or navigable semicircle they are subtractive and lesser velocities will be met. If the eye is moving directly at you, the wind direction will remain constant and the velocity increases until the eye arrives. Then the wind will drop, perhaps to a dead calm, and the sky will suddenly clear. When the eye has passed, the wind will suddenly rise, stronger than ever, but from the opposite direction. This treacherous shift plus the extreme velocity make the vicinity of the eye most dangerous.

In the diagram, vessels are at positions A1, B1 and C1 as the storm approaches them. By the time the storm has passed, the vessels will be at positions A2, B2 and C2 respectively. Each vessel will have experienced an entirely different series of wind directions and velocities. See illustration below.

If the storm in diagram is moving to the N:

Vessel at A1 has wind from NE, ***backing*** to N, NW and finally, at A2, from W. (Least wind from N.)

Vessel at B1 has ***constant*** wind from ENE, increasing until eye arrives. After calm of the eye, wind rises violently from WSW and stays in that quarter to position B2.

Vessel at C1 has wind from E, ***veering*** to SE, S, and finally, at C2 to SW. (Maximum wind from S). Note that winds in "LEAST VELOCITY" area are moving in opposite direction—S—to eye—N. Also, that winds in "MOST VELOCITY" area are moving in identical direction—N—with the eye.

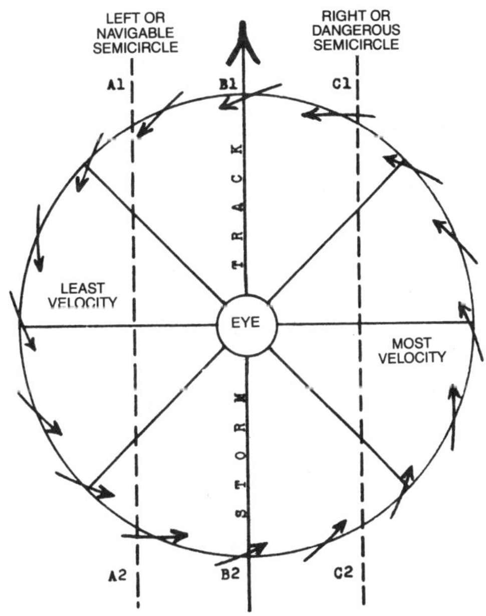

The aim, if underway, and if time permits, is to run at right angles to the storm track. A vessel at A1 should put wind just abaft her starboard beam. A vessel at B1 should put wind on starboard quarter and run for the navigable semicircle. Vessel C1 should put wind on starboard bow.

PRECAUTIONS ALONG SHORE

YOUR PRECAUTIONS ALONG SHORE should include the following: PULL SMALL CRAFT well above high water mark; dismast and strip, remove outboard motors to safety. SEEK MOST PROTECTED ANCHORAGE POSSIBLE, considering probable wind reversal and extreme tides. If on mooring or anchor, use maximum scope, but keep room to swing clear of other

boats as wind shifts. (You can easily slack off, but rarely shorten scope in a real blow.) Use of chain next to anchor will tend to keep anchor stock flat and flukes well-dug in. Another piece of chain or weight made fast to mooring or anchor rode, halfway along it, will tend to prevent sudden jerking strain. USE CHAFING GEAR (rags, canvas, etc.) liberally at bitts and bow chocks to minimize fraying of lines. SHUT GAS AND STOVE TANKS OFF. Douse any fires in wood or coal stoves. In crowded harbor, rig fenders. SECURE all portholes, skylights, ventilators, hatch covers, companionways, etc. Pump boat dry. Do not wait till the last moment to get ashore. Boats are replaceable.

IF TYING UP TO A DOCK: rig one or more anchors out abreast of boat, to keep off dock, if tide rises above spiles. Use fenders liberally and observe all applicable precautions above.

Extremely high tides also accompany the eye; if it arrives anywhere near the usual high water time, low-lying areas may be flooded. Especially high tides will occur in all bays or V-mouthed harbors (as in the Chesapeake, Naragansett Bays, coast of Maine, etc.), and high water will remain a much longer time.

PRECAUTIONS OFF-SHORE

The U.S. Coast Guard warns all vessels off-shore to seek shelter AT LEAST 72 HOURS AHEAD of hurricane. But IF CAUGHT OFF-SHORE with no chance to get to shelter: watch wind most carefully. First, note that if you face the wind, the eye is about 10 points or 112° to your right. If the wind "backs" you are already on the LESS dangerous side of the track. If it "veers" you are on the right, and MORE dangerous side. IF THE WIND REMAINS CONSTANT you know the eye is headed directly at you, so make every effort to maneuver to the left side of track, by putting the wind on your starboard quarter, going with top speed. Use radio-telephone to advise Coast Guard and other ships of your position. Have life-boat ready and equipped with flares, flashlights, oars, buckets, etc. Don life-jackets. If impossible to hold course, head directly into wind and sea, using only enough power to keep certain steerageway. If power fails, rig sea anchor or drag to keep bow to wind.

Sailing vessels heaving to should do so on the starboard tack in the dangerous semicircle, or on port tack in navigable semicircle. Thus the wind will always be drawing aft and the danger of being taken back averted.

A final note: This information is of the most general nature, the diagram over-simplified, and the suggestions based on a straight storm track. If the storm track curves to the right, A1 and B1 will have an easier time, but C1 may wind up directly on the track. As science cannot yet predict the hurricane path accurately, seeking shelter well ahead of time remains the very best course of action.

National Weather Service - http://www.nws.noaa.gov.
National Hurricane Ctr. - http://www.nhc.noaa.gov.
NOAA Hurricane Research - http://www.aoml.noaa.gov/hrd.
Southern New England Forecast - http://bristol.ll.mit.edu.

WEATHER NOTES
from Eastport to the Chesapeake

Sea Fog

There is always invisible moisture in the air, and the warmer it is, the more moisture the air can contain invisibly. But when such a mass of moist air is cooled off, as by passing over a cool body of water, the moisture condenses into a visible vapor, fog. The fog clears up when it is dried up, by heat from the sun or from stored-up heat in a land mass nearby, or by a warm dry wind.

To predict fog accurately, one should use a wet and dry bulb thermometer (or sling psychrometer). One of the thermometers has its bulb covered with a wet fabric, the other bulb is exposed to air. The evaporation of the wet fabric results in lowering the temperature of the wet bulb. Knowing the dry bulb reading (air temperature) and the difference between this and the wet bulb reading, Dew Point (temperature to which air must be cooled for condensation to take place) is determined.

Eastport to Cape Cod

The cold water (48-55°) off the northern New England Coast causes heavy fog conditions in the spring and summer when a warm moist southerly or southwesterly (prevailing summer wind) flow of air passes over it. East of Portland to the Bay of Fundy fog is not apt to occur when the dew point is under 55° unless there is a very warm moist wind. The effect of the cold water on the warm air will be reduced, however, if the winds become very brisk, as they are apt to do in the afternoon. Visibility will then improve.

From Portland south to Cape Cod, if the dew point is under 60° and fog *does* develop, it will generally clear up by mid-day. Visibility will also improve in the coastal bays where the water is a bit warmer. Sometimes in the summer, after the wind shifts from its customary southwest direction to northeast, the humid air that has been blown out to sea will return and a dense fog will result until a dry westerly or northwesterly clears it up.

Cape Cod to Eastern Long Island

Fog is not apt to develop here if the dew point is under 65°: When fog does develop, it may take a long time to burn off because the small land areas of Nantucket, Martha's Vineyard and Block Island do not produce enough heat to dry up the moisture and fog will continue until at least mid-day along the coast, into Buzzards Bay and up the Cape Cod Canal.

Prevailing winds in this area are southwest during the spring and summer, freshening in the afternoon, sometimes to 25–30 knots. Buzzards Bay has a few quirks worth mentioning. It can be smooth as a millpond but if the prevailing southwest wind freshens as usual in the afternoon and if the tide is ebbing at the same time from the relatively shallow Bay (opposite direction to the wind), short steep seas as high as 4 to 5 feet result, making it an uncomfortable passage for small craft. In the fall, three-day northeasterlies with driving rain, strong winds, cold temperatures and rough seas are not uncommon.

Long Island Sound

Here also, there is small chance of fog if the dew point is less than 65°. Any fog that does develop usually disperses by mid-morning because of the large amount of heat in the big land masses surrounding the Sound.

Typical spring and summer weather is light breezes in the morning and brisker winds (15 to 20 knots) in the afternoon usually diminishing at dusk. As the day progresses, the prevailing southwest wind (or an occasional northeasterly) becomes stronger as it travels over the water and will build up uncomfortable seas at the leeward end of the Sound. Occasional brief thunderstorms in summer can bring winds gusting to 50 knots. Here also, in the fall, one can expect three-day northeast storms with high winds, cold rain and short steep seas.

DEW POINT

Dry-bulb temp. F	Difference between dry-bulb and wet-bulb temperatures														Dry-bulb temp. F
	1°	2°	3°	4°	5°	6°	7°	8°	9°	10°	11°	12°	13°	14°	
+50	+48	+46	+44	+42	+40	+37	+35	+32	+29	+25	+21	+17	+12	+5	+50
52	50	48	46	44	42	40	37	35	32	29	25	21	17	11	52
54	52	50	49	47	44	42	40	37	35	32	28	25	21	16	54
56	54	53	51	49	47	45	42	40	37	35	32	28	25	21	56
58	56	55	53	51	49	47	45	43	40	38	35	32	28	25	58
+60	+58	+57	+55	+53	+51	+49	+47	+45	+43	+40	+38	+35	+32	+28	+60
62	60	59	57	55	54	52	50	48	45	43	41	38	35	32	62
64	62	61	59	57	56	54	52	50	48	46	43	41	38	35	64
66	64	63	61	60	58	56	54	52	50	48	46	44	41	39	66
68	67	65	63	62	60	58	57	55	53	51	49	46	44	42	68
+70	+69	+67	+66	+64	+62	+61	+59	+57	+55	+53	+51	+49	+47	+45	+70
72	71	69	68	66	64	63	61	59	58	56	54	52	50	47	72
74	73	71	70	68	67	65	63	62	60	58	56	54	52	50	74
76	75	73	72	70	69	67	66	64	62	61	59	57	55	53	76
78	77	75	74	72	71	69	68	66	65	63	61	59	57	55	78
+80	+79	+77	+76	+74	+73	+72	+70	+68	+67	+65	+64	+62	+60	+58	+80
82	81	79	78	77	75	74	72	71	69	67	66	64	62	61	82
84	83	81	80	79	77	76	74	73	71	70	68	67	65	63	84
86	85	83	82	81	79	78	76	75	74	72	70	69	67	66	86
88	87	85	84	83	81	80	79	77	76	74	73	71	70	68	88
+90	+89	+87	+86	+85	+84	+82	+81	+79	+78	+76	+75	+73	+72	+70	+90
92	91	89	88	87	86	84	83	82	80	79	77	76	74	73	92
94	93	92	90	89	88	86	85	84	82	81	79	78	76	75	94
96	95	94	92	91	90	88	87	86	84	83	82	80	79	77	96
98	97	96	94	93	92	91	89	88	87	85	84	82	81	80	98
+100	+99	+98	+96	+95	+94	+93	+91	+90	+89	+87	+86	+85	+83	+82	+100

Note: South Long Island inlets can become dangerous when breezes blowing toward shore are at the same time as an ebbing tide and produce short steep seas. Offshore swells close to the inlet mouths can also become very high.

New Jersey Coast
Summertime warm water (in the 70's) in this area rarely cools down enough any warm moist air mass to produce fog. (Not so, farther off the coast where water temperature is no higher than the 50's.)

Prevailing winds in summer are southerly, increasing in mid-morning to rarely more than 20 knots and usually dying down at dusk. Occasional summer thunderstorms can be expected. Any brisk winds from the east, northeast or southeast (and New Jersey has its share of stormy northeasterlies in fall and winter) can produce dangerous conditions along this lee shore and at mouths of inlets. (When the wind is from offshore like this, inlets should be entered on a *flood* tide). A very strong southeast wind with a simultaneous ebbing current at the mouth of Delaware Bay can cause seas to build up to a hazardous degree.

Chesapeake Bay
There is little chance of fog in this area because of the warmth of the water. This Bay also has quirks of its own in weather and sea conditions. The Bay is a narrow and fairly shallow body of water and winds tend to blow up or down it. Sharp seas result depending on the direction of the tide and wind. Opposing forces make for rough water. Prevailing winds in spring and summer are southerly, freshening in the afternoon after a morning of calm. Summer thunderstorms occur frequently in afternoon and early evening, usually from the west. In the fall, after a cold front passes through, the winds will shift to the north or northeast, usually for three days, and increase in velocity causing seas to build up. Calm follows for a day or so until the wind shifts to the southwest.

COAST GUARD POLLUTION REGULATIONS FOR COMMERCIAL & RECREATIONAL VESSELS

As of July 31, 1990, the Coast Guard's garbage pollution regulations have been amended by adding requirements for certain U.S. documented or state registered vessels to have on board waste management plans and/or discharge warning placards. The following rules apply:

• **Placards**. All documented or registered vessels 26 feet or more in length must display a placard in prominent locations to warn persons of the penalties for illegally dumping garbage. The placard must:

(1) be at least nine inches wide by four inches high, made of durable material with lettering at least 1/8 inch high; and
(2) Notify the reader of the following:
   - (i) The discharge of plastic or garbage mixed with plastic into any waters is prohibited.
   - (ii) The discharge of all garbage is prohibited in the navigable waters of the United States and, in all other waters within three nautical miles of the nearest land.
   - (iii) The discharge of dunnage, lining, and packing materials that float is prohibited within 25 nautical miles of the nearest land.
   - (iv) Other unground garbage may be discharged beyond 12 nautical miles from the nearest land.
   - (v) Other garbage ground to less than one inch may be discharged beyond three nautical miles of the nearest land.
   - (vi) A person who violates the above requirements is liable for a civil penalty of up to $25,000, a fine of up to $50,000, and imprisonment for up to five years for each violation.
   - (vii) Regional, State, and local restrictions on garbage discharges may apply.

• **Waste Management Plans**. Documented or registered oceangoing vessels 40 feet or more in length that either are engaged in commerce **or** have a galley and berthing must have waste management plans. Each plan must be in writing and:

(1) Provide for the discharge of garbage by means that meet the requirement of law.
(2) Describe procedures for collecting, processing, storing and discharging garbage; and
(3) Designate the person who is in charge of carrying out the plan.

Note: Marinas will now be required to have facilities to receive refuse.

DOCKSIDE SEWAGE PUMPING STATIONS

Please be sure to call or radio in advance for rates and availability. While we have taken all possible care in compiling this list, changes may have occurred and we cannot guarantee accuracy. (Stations with * are seasonal) Additions to Conn. and N.Y. State courtesy of New York Sea Grant Program.

MAINE & NEW HAMPSHIRE

Bangor
Town Dock 207-947-5251
Bar Harbor
* Harbor Place 207-288-3346
Boothbay Harbor
Carousel Marina 207-633-2922
Brunswick
Paul's Marina 207-729-3067
Camden
Frenchmans Bay Boat Co. 207-288-8941
Wayfarer Marine Corp. 207-236-4378
Medomak
Broad Cove Marina 207-529-5427
Northeast Harbor
Northeast Harbor Town Pier 207-276-5737
N. Edgecomb
Eddy Yt. Sales & Marina 207-882-7776
Robinhood
Robinhood Marine Ctr. 207-371-2525
South Freeport
Brewers S. Freeport Marina 207-865-3181
South Portland
* Marineast 207-799-8191
Spring Point Marina 207-767-3213
Southwest Hbr. - Manset
Town Pier 207-244-5221
Thomaston
* Thomaston Marine 207-354-2200
Winter Harbor
Winter Harbor Marina 207-963-7449
Portsmouth, NH
Wentworth-by-the-Sea Marina 603-433-5050

MASSACHUSETTS

Boston
* Rowe's Wharf Marina 617-330-7667
Bourne
Town Marina 508-759-2512
Cataumet
Parker's Boat Yard 508-563-9366
Charlestown
Constitution Marina 617-241-9640
Shipyard Quarters Marina 617-242-2020
Chilmark
Chilmark Twn. Dk., Menem. 508-645-2846
Menemsha Texaco Service 508-645-2641
Duxbury
Snug Harbor Town Pier 617-934-2866
East Boston
Boston Boat Yard 617-561-0348
East Falmouth
Edward's Boat Yard 508-548-2216
East Orleans
Nauset Marine East 508-255-1777
Edgartown
Edgartown Marine 508-627-4388
Memorial Pier 508-627-9794
Town Dock Bulkhead 508-627-4746
Fairhaven
Acushnet R. Boating Club 508-993-9667
Brightman's Marina 508-997-2444
D.N. Kelley & Sons 508-999-6266
* Earl's Marina 508-993-8600
Fairhaven Shipyard 508-996-8591
Seaport Inn Marina 508-992-7985
Falmouth
Town Pier 508-457-2550
Fiddler's Cove Marina (private only)
Haverhill - Merrimack R.
Slavit's Shipyard 508-372-3420
Hingham
Hewitts Cove Marina 617-749-6647
Hull
Sunset Marina 617-925-2653
Hyannis
Hyannis Marina 508-775-5662
Lewis Bay Marina 508-775-6633
Lynn
Seaport Landing Marina 617-592-5821
Marion
Burr Brothers Boats 508-748-0541
Town Pier 508-748-3535
Marshfield
Town Pier (Aug. '92) 617-834-6741
Marston Mills
Prince's Cove Marina 508-428-5885
Mattapoisett
Ringuette's Marina 508-758-9511
Nantucket
Nantucket Boat Basin 508-228-1333
North Falmouth
Fiddler's Cove Marina 508-564-6326
Quincy
Bay Pointe Marine 617-471-1777
* Captain's Cove Marina 617-770-0573
Clipper Marina 617-479-1449
Marina Bay 617-847-1800
Salisbury
* Cove Marina 508-462-4998
Scituate
James Landing Marina 617-545-3000
South Dartmouth
Padanarum Pump-out Boat 508-999-0759
Vineyard Haven
Maciel Marina 508-693-4174
Martha's Vineyard Shipyard 508-693-0400
Wareham
Bevans Marine 508-759-5451
Onset Bay Marina 508-295-0338
Pt. Independence Yt. Club 508-295-3972
Town Pier 508-295-8160
Wareham Boat Yard 508-748-1472
Warr's Marina 508-295-0022
Wellfleet
Town Marina 508-349-0320
* Pump-out Boat
Westport Harbor
* Pump-out Boat 508-636-1105

RHODE ISLAND

Barrington
Stanley's Boat Yard 401-245-5090
Block Island (New Hbr.)
Pump-out Boat 401-466-3204
E. Greenwich
Norton's Shipyard & Marina 401-884-8828
Jamestown
Conanicut Marina 401-423-1556
Newport
Goat Is. Marina 401-849-5655
Portsmouth
Brewer's Sakonnet Marina 401-683-3551
East Passage Yachting Center 401-683-4000
Little Harbor Marine 401-683-5600
Providence
Shooter's Marina 401-273-6433
Wakefield
Ram Point Marina 401-783-4535
Warren
Blount Marine Corp. 401-245-8300

CONNECTICUT

Branford
Bruce & Johnson's Marina 203-488-8329
Goodsell Pt. Marina 203-488-5292
Chester
Chrisholm Marina 203-526-5147
Clinton
Cedar Island Marina 203-669-8681
* Clinton Yacht Haven 203-669-7254
Cos Cob
Greenwich Harbor Club 203-869-9567

Deep River
Chrisholm Marina 203-526-5147
Deep River Marina 203-526-5560
Essex
Brewer's Dauntless Shipyard 203-767-2483
Five Mile River (to Rowayton)
* Sound Disposal 203-853-7247
Greenwich
Mianus R. Yacht Club 203-869-4689
* Town Fuel Dock 203-869-9689
Guilford (West River)
Guilford Yt. Club Marina 203-453-8746
Lyme
Cove Landing Marina 203-434-5240
Milford
Milford Boat Works 203-877-1475
Mystic River
Brewer Yacht Yard 203-536-2293
* Gwenmoor Marina 203-536-3123
* Harbor Marine Service 203-572-8939
Mystic Cove Yacht Club 203-572-8939
New London
* American Wharf Marina 203-886-1511
Ferry Slip Dockominium 203-442-9038
Niantic
Boats, Inc. 203-739-6251
Noank
* Noank Shipyard 203-536-9436
Noank Village Boatyard 203-536-1770
Spicer's Marina 203-536-4978
Norwalk
Norwalk Cove Marina 203-838-2326
Norwich
American Wharf Marina 203-886-1511
Old Saybrook
River Landing Marina 203-388-1431
Pawcatuck
Norwest Marine 203-599-2442
Stamford Hbr. E. Branch
* Cove Municipal Marina 203-977-4687
Czescik Marina 203-977-4687
Harbor Square Marina 203-324-3331
Stamford Harbor Club 203-348-7700
Stamford Hbr. W. Branch
* Cummings Marina 203-977-4687
Stamford Landing Marina 203-965-0065
Stratford
Marina at the Dock 203-378-9300
Stonington
* Dodson Boat Yard 203-535-1507
Westbrook
Harry's Marine Repair 203-399-6165
Pilot's Pt. Marina 203-399-7906

NEW YORK

(in geographical order)
Catskill
Riverview Marine Services 518-943-5311
Rye-Milton Harbor
Boat Basin 914-967-2011
City Island Harbor
* Maritime Landings, Inc. 212-885-1800
Manhassett Bay
North Bay Marina 516-767-0110
* Pt. Washington Town Dock 516-328-8500
Tom's Point Marina 516-883-6630
Glen Cove
* Glen Cove Yt. Serv. & Repair 516-676-0777
Hempstead Harbor
* Brewer Yacht Yard 516-671-5563
Harry Tappen Marina 516-671-0484
Oyster Bay
Roosevelt Town Bt. Basin 516-795-1000
Cold Spring Harbor
* Town Dock 516-351-3255
Huntington Harbor
* Castle Cove Marina 516-673-3299
West Shore Marina 516-427-3444
* Mill Dam Marina 516-351-3089
* Halesite Town Marina 516-351-3255
Northport Harbor
* Seymour's Marina 516-261-6574
* Woodbine Marina 516-351-3255
Brittannia Yacht Club 516-261-5600
Nissequogue River
* Old Dock Mooring Area 516-862-6663
Stony Brook Harbor
* Long Beach Town Marina 516-862-6663
Port Jefferson
* Town Marina 516-331-3567
* Bayle's Dock, Inc. 516-928-5200
Mt. Sinai Harbor
Old Man's Boat Yard 516-473-7330
Mt. Sinai Yacht Club 516-473-2993
Town Marina 516-928-0199
Mattituck Creek
* Matt-A-Mar Marina 516-298-4739
Shelter Island
Coecles Harbor Marina 516-749-0700
* Dering Harbor Marina 516-749-0045
* Island Bt. Yd. W. Neck Hbr. 516-749-3333
Southold Area
* Brick Cove Mar., Sage Cr. 516-477-0830
* Port of Egypt Mar., Budds Pd. 516-765-2445
Riverhead Area
Dreamers Cove Motel 516-722-3212
* Larry's Lighthouse Marina 516-722-3400
Peconic River Yacht Basin 516-727-8386
Shinnecock Canal
* County Boat Basin 516-852-8291
North Sea Harbor
Conscience Pt. Marina 516-283-8295
Sag Harbor
* Municipal Dock 516-725-2368
Three Mile Harbor
* Harbor Marina 516-324-5666
E. Hampton Town Bulkhead 516-329-3078
* Maidstone Harbor Marina 516-324-2651
Gardiner's Marina 516-324-9894
Lake Montauk
Captain's Cove Marina 516-668-5995
* Montauk Marine Basin 516-668-5900
* Montauk Yacht Club 516-668-3100
Gone Fishing Marina 516-668-3232
Old Fort Pond
* Corr's Best Boat Works 516-283-7359
Quantuck Canal
* Bath & Tennis Yacht Club 516-288-0380
Seatuck Cove
* Seatuck Cove Marina 516-325-0367
Moriches Bay
* Cerullo Bros. Marina 516-878-1180
Forge River
* Brookhaven Town Dock 516-451-6100
Baypoint Marina 516-878-0767
Swan River
* Morgan's Marina 516-758-3524
Great South Bay
Watch Hill 516-597-6644
Sailors Heaven Marina 516-597-6183
* Atlantique Marina 516-583-8610
Patchogue Bay
* Leeward Cove Marina 516-758-2550
* Mascot Dock 516-475-4302
* Sandspit Town Marina 516-475-1592
Wagner's River Pt. Marina 516-475-0512
* Pier 66 516-654-3106
Corey Creek
* Park and Marina 516-363-6045
Connetquot River
* Suffolk County Park Marina 516-581-9299
Quintuck Creek & Penataquit Creek
* East Islip Marina 516-224-5648
* Bay Shore Marina 516-224-5648
Nequntatogue Creek
* Surfside 3 Marina 516-957-5900
* The Anchorage 516-957-7300
Fire Island Inlet
* Moses State Park Boat Basin 516-669-0449
State Boat Channel
* Captree St. Park Marina 516-669-0449
* Cedar Beach Boat Basin 516-669-5949
* Tobay Town Boat Basin 516-795-1000
* West End Boat Basin 516-785-1600
Babylon Cove
* Tanner Park Marina 516-893-2100
Island Creek
Treasure Is. Marine Basin 516-221-7156
Bellmore Creek
* Wantagh Pk. Cty. Marina 516-785-7777

Hudson Canal
* Edgo Marina 516-623-8115
Reynolds Channel
* Hempstead Marina West 516-431-9200

NEW JERSEY

Atlantic City
* The Harbour at Harrah's 609-441-5315
Senator Frank S. Farley State Marina 609-441-8482
Atlantic Highlands
Atlantic Highlands Marina 908-291-1670
Bayville
River Bank Marina 201-244-2106
Forked River
The Berth Place Marina 609-693-9423
Jersey City
Newport Marina 201-626-5550
Keyport
Hans Pedersen & Sons 908-264-0971
Leonardo
Leonardo State Marina 908-291-1333
Longport
Seaview Harbor Marina 609-823-2626
Mantoloking
Winter Yacht Basin 908-477-6700
Margate
* Bay Club Marina 609-487-1000
Marmora
All Seasons Marina 609-390-1852
Neptune
* Sea View Marina 908-775-5522
Pt. Pleasant
Johnson Bros. Boat Works 908-892-9000
Ship Bottom
Shelter Harbor 609-492-8645
Somers Pt.
Harbour Cove Marina 609-927-9600
Tuckertown
American Boat Exchange 609-296-6900
Waretown
Holiday Harbor Marina 609-693-2217
Weehawken
Port Imperial Marina 201-902-8787

PENNSYLVANIA & WASHINGTON, D.C.

Philadelphia
Philadelphia Marine Center 215-931-1000
Washington
Gangplank Marina 202-554-5000

MARYLAND

Annapolis
Annapolis Landing Marina 410-263-0090
Sandy Point Marina 410-974-0772
Baltimore
Inner Harbor Marina 410-837-5339
Middle Branch Moorings 410-539-2628
Berlin
* Ocean Pines Yacht Club 410-641-7447
Bladensburg
Safford Marine & Port of Bladensburg 301-799-4133
Cambridge
Yacht Maintenance Co. 410-228-8878
Chesapeake City
Bohemia Bay Yacht Harbour 410-885-2601
Chestertown
Worton Creek Marina 410-778-3282
Drayden
Dennis Point Marina 410-994-2288
Earleville
Long Point Marina 410-275-8181
Edgewater
Bay View Marina 410-798-6060
Fort Washington
Fort Washington Marina 301-792-7700
Galena
Gregg Neck Boat Yard 410-648-5360
Georgetown
Georgetown Yacht Basin 410-648-5112
Sassafras Boat Club 410-648-5355

Skipjack Cove Yachting Center 410-275-2122
Grasonville
Mears Point Marina 410-827-8888
Scott Marine Sales & Service 410-827-8150
Joppa
Gunpowder Cove Marina 410-679-5454
Mechanicsville
Cape St. Mary's Marina 410-373-2001
North East Heights
McDaniel Yacht Basin 410-287-8121
Ocean City
* Ocean City Yacht Club 410-289-3474
Perryman
Bush River Boat Works 410-272-1882
Port Tobacco
Harbor at Port Tobacco 301-932-1407
Ridge
Point Lookout Marina 410-872-5145
Salisbury
Port of Salisbury 410-548-3176
Scotland
Point Lookout State Park 410-872-5688
Snow Hill
Shad Landing Marina, Pocomoke R. 410-632-2566
Solomons
Zahniser's Sailing Center 410-326-2166
Worton
Green Point Marina 410-778-1615
Wharf at Handy's Point 410-778-4363

VIRGINIA

Alexandria
Alexandria City Marina 703-838-4265
Belle Haven Marina 703-768-0018
Washington Sailing Marina 703-548-9027
Battery Park
Pagan River Marina 804-357-7405
Coles Point
Ragged Point Harbor 804-472-3955
Deltaville
Chesapeake Cove Marina 804-776-6855
J&M Marina 804-776-9860
Porpoise Cove Marina 804-776-7606
Waldens Marina 804-776-9440
Farnham
Whelen's Marina 804-394-9500
Fredericksburg
Fairview Beach Yacht Club 703-775-5971
Waugh Point Marina 703-775-7121
Gloucester Point
York River Yacht Haven 804-642-2156
Grafton
Chisman Creek Marina 804-898-3000
Grimstead
The Narrows Marina 804-725-2151
Hampton
Bluewater Yacht Sales 804-723-0795
Hampton Roads Marina 804-723-6774
Joy's Marina 804-851-1550
Hayes
Cook's Landing 804-642-6177
Irvington
* The Tides Lodge 804-438-6000
Kinsale
Kinsale Harbour Marina 804-472-2514
Lancaster
Greenvale Creek Marina 804-462-7350
Yankee Point Marina 804-462-7018
Lorton
* Capt. John S. Beach Marina 703-339-9650
Newport News
Menchville Marine Supply 804-877-0207
Norfolk
City Marina 804-441-2222
Cobb's Marina 804-588-5401
Little Creek Marina 804-583-3600
Oak Grove
Outerworld Harborview Mar. 804-224-8164
Poquoson
Poquoson Marina 804-868-6171
Portsmouth
Lee's Yachting Center 804-484-2652
Tidewater Yacht Agency 804-393-2525

Reedville
Leroy's Marina 804-453-6806
Stafford
Willow Landing Marina 703-659-2653
Suffolk
Brady's Marina 804-539-8221
Topping
Regent Point Marina 804-758-4457
Urbanna
* Urbanna Bridge Marina 804-758-5124
Urbanna Yacht Sales 804-758-2342
Virginia Beach
Lynnhaven Yacht Marina 804-481-6909
Westmoreland
Sandy Point Marina 804-472-3237
Williamsburg
Jamestown Yacht Basin 804-229-8366
Kingsmill Marina 804-253-3919
Windmill Point
Windmill Point Marine 804-435-1166
Woodbridge
* Tyme 'n Tyde 703-491-5116
Yorktown
Wormley Creek Marina 804-898-5060

NORTH CAROLINA WATERWAY

Arapahoe
Minnesott Beach Yacht Club 919-249-1424
Beaufort
Beaufort Docks 919-728-2503
Belhaven
River Forest Marina 919-943-2151
Callabash
Marsh Harbor Marina (Little R. Inlet) 919-579-3500
Manteo
Salty Dawg Marina (Roanoke Sound) 919-473-3405
New Bern
Ramada Hotel & Marina (Trent River) 919-636-2888
Sheraton Hotel & Marina (Trent River) 919-638-3585
Southport
Southport Marina 919-457-5261
Wilmington
Masonboro Boatyard & Marina 919-791-1893

SOUTH CAROLINA WATERWAY

Charleston
Ashley Marina (Ashley River) 803-722-1996
Buzzard's Roost Marina (Stono River) 803-559-5516
Charleston City Marina (Ashley River) 803-724-7357
Dolphin Cove Marina (Ashley River) 803-744-2562
Ripley Light Marina (Ashley River) 803-766-2100
Georgetown
Belle Isle Marina 803-546-8491
Georgetown Landing 803-546-1776
Hilton Head Island
Harbour Town Yacht Basin 803-671-2704
Palmetto Bay Marina 803-785-3910
Shelter Cove-Palmetto Dunes 803-842-7001
Skull Creek Marina 803-681-4234
Windmill Harbor 803-681-9235
Isle of Palms
Caroline Marine Service 803-886-4213
Wild Dunes Yacht Harbor 803-886-5100
Johns Island
Bohicket Marina 803-768-1280
Stono Marina 803-559-2307
Little River
Lightkeepers Village (Coquina Hbr.) 803-344-1604
Mt. Pleasant
Toler's Cove Marina 803-881-0325
Murrells Inlet
Marlin Quay Marina 803-651-4444
N. Myrtle Beach
Coquina Harbor, Inc. (Little River) 803-249-9333
Myrtle Beach Yacht Club (Coquina Hbr.) 803-249-5376
Port Royal
Port Royal Landing 803-525-6664

GEORGIA WATERWAY

Jekyll Island
Jekyll Island Marina 912-635-2891
Skidaway Island
Delegal Creek Marina (Savannah Area) 912-598-0023
St. Simon's Island
Hampton River Club Marina 912-265-1010
Thunderbolt
Savannah Bend Marina (Savannah Area) 912-897-3625

FLORIDA

Betw. St. Aug. & Dayton
Palm Coast Marina 904-445-4141
Daytona Beach
Halifax Harbor Marina 904-253-0575
Ft. Lauderdale Beach
Hall of Fame Marina 305-764-3975
Lauderdale Marina 305-523-8507
Pier 66 305-525-6666
Fort Pierce
Fort Pierce City Marina 407-464-1245
Harbortown Marina 407-466-7300
Riverside Marina 407-464-5720
Indian Harbor Beach
Indian Harbour Marina 407-773-2468
Islamorada
Matecumbe Marina 305-664-2402
Jacksonville
Beach Marine 904-249-8200
Mandarin Holiday Marina 904-268-1036
Jupiter
Admirals Cove Marina 407-744-1700
Jonathan's Landing 407-747-8980
Key Biscayne
Crandon Marina 305-361-1281
Key West
Galleon Marina 305-292-1292
Marathon
Boot Key Marina 305-743-4200
Faro Blanco Marina 305-743-9018
Melbourne
Anchorage Yacht Basin 407-773-3620
Melbourne Harbor Marina 407-725-9054
Miami
Biscayne Bay Marriott 305-374-4900
Miami Beach
Sunset Harbour Marina 305-673-0044
Miami River
Jones Boat Yard 305-635-0891
Palm Beach Gardens
Frenchman's Creek Marina 407-627-6358
Harbour Point Marina 407-622-6890
Soverel Harbour 407-622-6910
Port Canaveral
Cape Marina 407-783-8410
Riviera Beach
Municipal Marina 407-842-7806
St. Johns River
Seafarer's Marina 904-765-8152
Sanford
Monroe Harbor Marina 305-322-2910
Sanford Boat Works 305-322-6613
S. Titusville
Kennedy Point Yt. Club 407-383-0280
Stuart
Indian River Plantation 407-225-3700
Mariner Cay Marina 407-287-2900
Stuart Yacht Club 407-692-4000
Titusville
Municipal Marina 407-269-7255
Vero Beach
Municipal Marina 407-231-2819
Williams Island
Williams Island 305-935-5555

CRUISING CHARTS

DESIRABLE CRUISING CHARTS FROM CAPE BRETON I. TO KEY WEST, FL. Numbers listed to the left are general coastal charts. Indented numbers refer to harbor charts. (C) indicates Loran-C lines of position on the chart. (E) indicates New Edition - prior edition **Obsolete**. 1:80(000), 1:40(000), etc. indicates scale.

Canada

4013 (C) Halifax to Sydney 1:350
4279 Bras d'Or Lake 1:60 metric
4321 (C) Cape Canso to Liscomb Is. 1:108.8
4317 (C) Liscomb Is. to Egg Is. 1:109
4320 (C) Egg Is. to W. Ironbound Is. 1:145
4012 (C) Yarmouth to Halifax 1:300
4385 (C) Osborne Hd. to Betty Is. 1:36.5
4386 St. Margaret's Bay 1:39.4
4381 Mahone Bay 1:38.9
4384 Pearl Is. to Cape LaHave 1:39
4211 Cape LaHave to Liverpool Bay 1:37.5
4212 Western Hd. to Little Hope Is. 1:37.5
4213 Little Hope Is. to Lockeport Hbr. 1:37.5
4214 Lockeport Hbr. to Cape Negro 1:37.5
4215 Cape Negro to Cape Sable 1:37.5
4242 (C) Cape Sable, Tusket Is., Pubnico 1:60 metric
4243 (C) to Cape St. Mary's 1:60 metric
4010 (C) Bay of Fundy (inner portion) 1:200
4011 (C) Appr. to Bay of Fundy 1:300
4118 Briar Is., St. Mary's 1:60 metric
4396 Annapolis Basin 1:24
4141 St. John to Evandale 1:30
4128 (C) Appr. to St. John Hbr., Pt. Lepreau to Cape Spencer 1:50
4331 Passamaquoddy Bay to St. Croix R. 1:40.6
4340 Grand Manan 1:60

U.S. East Coast

13325 (C) Quoddy Narrows to Petit Manan Is. 1:80
13312 (C) Frenchman & Blue Hill Bays & apprs. 1:80
13302 (C) Penobscot Bay and apprs. 1:80
- **13315 Deer Is. Thoro. and Casco Pass. 1:20**
- **13308 Fox Islands Thorofare 1:15**

13288 (C) Monhegan Is. to Cape Elizabeth 1:80
- **13290 Casco Bay 1:40**

13286 (C) Cape Elizabeth to Portsmouth 1:80
- **13283 Cape Neddick Hbr. to Isles of Shoals 1:20, Portsmouth Hbr. 1:10**

13278 (C) Portsmouth to Cape Ann 1:80, Hampton Harbor 1:30
- **13281 Gloucester Hbr. and Annisquam R. 1:10**

13267 (C) Massachusetts Bay 1:80
- **13275 Salem and Lynn Harbors 1:25, Manchester Harbor 1:10**
- **13276 (E) Salem, Marblehead & Beverly Hbrs. 1:10**

13270 Boston Harbor 1:25
13246 (C,E) Cape Cod Bay 1:80
- **13253 Plymouth, Kingston and Duxbury Hbrs. 1:20, Greens Hbr. 1:10**
- **13236 Cape Cod Canal and approaches 1:20**

13237 (C) Nantucket Sound and approaches 1:80
- **13241 Nantucket Island 1:40**
- **13242 Nantucket Harbor 1:10**

13218 (C,E) Martha's Vineyard to Block Island 1:80
- **13230 (E) Buzzards Bay 1:40, Quicks Hole 1:20**
- **13233 Martha's Vineyard 1:40, Menemsha Pond 1:20**
- **13221 (E) Narragansett Bay 1:40**

13205 (C) Block Island Sound and apprs. 1:80
- **13214 Fishers Island Sound 1:20**
- **13212 Approaches to New London Hbr. 1:20**
- **13219 Point Judith Harbor 1:15**
- **13217 Block Island 1:15**
- **13211 North Shore of Long Is. Sd.-Niantic Bay & Vicinity 1:20**
- **13213 (E) New London Harbor and Vicinity 1:10, Bailey Point to Smith Cove 1:05**
- **13209 (E) Block Is. Sd. & Gardiners Bay, Long Is., 1:40**

12354 (C) Long Island Sound - eastern part 1:80
- **12375 Connecticut R. -Long Is. Sd. to Deep R. 1:20**
- **12374 Duck Island to Madison Reef 1:20**
- **12373 Guilford Hbr to Farm R. 1:20**
- **12371 New Haven Harbor 1:20**
- **12370 Housatonic R. and Milford Hbr. 1:20**
- **12362 Port Jefferson & Mt. Sinai Hbrs. 1:10**

12363 (C,E) Long Island Sound - western part 1:80
- **12369 Stratford to Sherwood Pt. 1:20**
- **12368 Sherwood Pt. to Stamford Hbr. 1:20**
- **12367 Greenwich Pt. to New Rochelle 1:20**
- **12366 L.I. Sd. and East R., Hempstead Hbr. to Tallman Is. 1:20**
- **12365 L.I. Sd. S. Shore, Oyster and Huntington Bays 1:20**

CRUISING CHARTS

12353 (C) Shinnecock Light to Fire Island Light 1:80
- 12352 Shinnecock B. to E. Rockaway In. 1:20; 1:40
- 12339 East R. - Tallman I. to Queensboro Br. 1:10
- 12331 Raritan Bay and Southern Part of Arthur Kill 1:15
- 12327 New York Harbor 1:40
- 12335 Hudson & E. Rs. -Governors I. to 67 St. 1:10

12326 (C) Appr. to N.Y., Fire I. to Sea Girt 1:80
- 12350 Jamaica Bay and Rockaway In. 1:20

12323 (C) Sea Girt to Little Egg In. 1:80
- 12324 Sandy Hook to Little Egg Harbor 1:40

12318 (C) Little Egg In. to Hereford In. 1:80, Absecon In. 1:20
- 12316 Little Egg Harbor to Cape May 1:40

12304 (C) Delaware Bay 1:80
- 12311 Delaware R.- Smyrna R. to Wilmington 1:40
- 12312 Wilmington to Philadelphia 1:40
- 12277 Chesapeake and Delaware Canal, Salem R. Ext. 1:20

12214 (C) Cape May to Fenwick I. 1:80

12211 (C) Fenwick I. to Chincoteague In.1:80, Ocean City In. 1:20

12210 (C) Chincoteague In. to Great Machipongo In. 1:80, Chincoteague In. 1:20

12221 (C,E) Chesapeake Bay Entrance 1:80
- 12222 Cape Charles to Norfolk Hbr. 1:40
- 12224 Cape Charles to Wolf Trap 1:40
- 12228 Pocomoke and Tangier Sds. 1:40
- 12231 Tangier Sd.-northern part 1:40

12225 (C) Wolf Trap to Smith Point 1:80

12230 (C,E) Smith Point to Cove Point 1:80

12263 (C,E) Cove Point to Sandy Point 1:80

12273 (C,E) Sandy Point to Susquehanna River 1:80
- 12233 Chesapeake Bay to Piney Pt. 1:40
- 12286 Piney Pt. to Lower Cedar Pt. 1:40
- 12288 Lower Cedar Pt. to Mattawoman Cr. 1:40
- 12289 Mattawoman Cr. to Georgetown 1:40; Washington Hbr. 1:20
- 12282 (E) Severn and Magothy Rs. 1:25
- 12278 (E) Baltimore Hbr. Appr. 1:40
- 12253 Norfolk Hbr. and Elizabeth R. 1:20
- 12254 (E) Cape Henry to Thimble Shoal Lt. 1:20
- 12256 Chesapeake Bay-Thimble Shoal Channel 1:20
- 12245 (E) Hampton Roads 1:20

12207 (C) Cape Henry to Currituck Bch. Lt. 1:80
- 12205 Cape Henry to Pamlico Sd. incl. Albemarle Sd. 1:40; 1:80

12204 (C) Currituck Beach Lt. to Wimble Shoals 1:80

11555 (C) Cape Hatteras-Wimble Shoals to Ocracoke In. 1:80

11548 (C) Pamlico Sd.-western part 1:80
- 11550 Ocracoke In. and N. Core Sd. 1:40
- 11545 Beaufort In. and S. Core Sd. 1:40, Lookout Bight 1:20

11544 (C) Portsmouth I. to Beaufort incl. Cape Lookout Shoals 1:80

11543 (C) Cape Lookout to New R. 1:80

11539 (C) New R. In. to Cape Fear 1:80

11536 (C) Appr. to Cape Fear R. 1:80

11535 (C) Little R. In. to Winyah Bay Entr. 1:80

11531 (C) Winyah B. Entr to Isle of Palms 1:80
- 11532 Winyah Bay 1:40

11521 (C) Charleston Hbr. & Appr. 1:80

11513 (C) St. Helena Sd. to Savannah R. 1:80

11509 (C) Tybee I. to Doboy Sd. 1:80

11502 (C) Doboy Sd. to Fernandina 1:80

11488 (C) Amelia I. to St. Augustine 1:80

11486 (C,E) St. Augustine Lt. to Ponce de Leon In. 1:80

11484 (C,E) Ponce de Leon In. to Cape Canaveral 1:80

11476 (C) Cape Canaveral to Bethel Shoal 1:80

11474 (C) Bethel Shoal to Jupiter In. 1:80

11466 (C) Jupiter In. to Fowey Rocks 1:80, Lake Worth In. 1:10

11462 (C) Fowey Rocks to Alligator Reef 1:80

11452 (C) Alligator Reef to Sombrero Key 1:80

11442 (C) Sombrero Key to Sand Key 1:80

11434 (C) Florida Keys-Sombrero Key to Dry Tortugas 1:180

TABLE OF EQUIVALENTS
(and useful nautical information)

LENGTH EQUIVALENTS:

1 meter = 39.37 inches
= 3.28 feet
= .55 fathoms
1 kilometer = .62 statute miles
= .54 nautical miles
or: 1 inch = 2.54 centimeters
1 foot = .30 meters
1 fathom = 1.83 meters
1 stat. mi. = 1.61 kilometers
1 naut. mi. = 1.85 kilometers

METRIC UNITS:

10 millimeters = 1 centimeter
100 centimeters = 1 meter
1000 meters = 1 kilometer

(Note: 1 fathom = 6 feet
1 cable = 608 feet
(approx. 100 fathoms)
1 naut. mi. = 6080 feet
= 1.15 stat. mi.
1 knot is 1 naut. mi. per hour)

CAPACITY EQUIVALENTS:

1 liter = 1.06 quarts
= .26 U.S. gallons
or: 1 qt. = .95 liters
1 gal. = 3.78 liters

METRIC UNITS:

10 milliliters = 1 centiliter
100 centiliters = 1 liter
(1000 cu. cm.)
1000 liters = 1 kiloliter

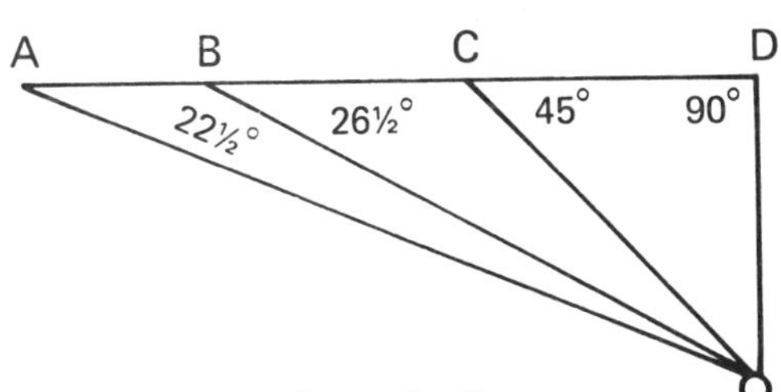

Doubling the Angle on the Bow

1. Angle DCO = 45°; Angle CDO = 90°; True distance run (CD) = distance DO.
2. Angle DAO = 22½°; Angle DCO = 45°; True distance run (AC) = Distance CO.

Other Useful Bow Bearings

3. Angle DBO = 26½°; Angle DCO = 45°; True distance run (BC) = distance DO.
4. Example 3 also works with angles of 25° and 41°; 32° and 59°; 35° and 67°; 37° and 72° when distance run will be distance DO.

WEIGHT EQUIVALENTS:

1 gram = .035 ounces
1 kilogram = 2.20 lbs.
1 metric ton = 2204.6 lbs
= .98 U.S. tons
or: 1 ounce = 28.35 grams
1 lb. = .45 kilograms
1 U.S. ton = .907 metric tons
= .893 long tons (Brit.)

METRIC UNITS

10 milligrams = 1 centigram
100 centigrams = 1 gram
1000 grams = 1 kilogram
1000 kilograms = 1 metric ton

OTHER EQUIVALENTS

1 sq. centimeter = .16 sq. inches
1 sq. meter = 10.76 sq. ft. or 1.20 sq. yds.
1 cu. meter = 35.31 cu. ft.
1 displacement ton = 35 cu. ft. salt water (36 cu. ft. fresh)
1 cu. ft. salt water = 64 lbs. (fresh water 62.4 lbs.)
1 Solar Day = 24 hours
1 Sidereal Day = 23 hrs., 56 min., 4.1 sec.
1 Lunar Day = 24 hrs., 50 min., average

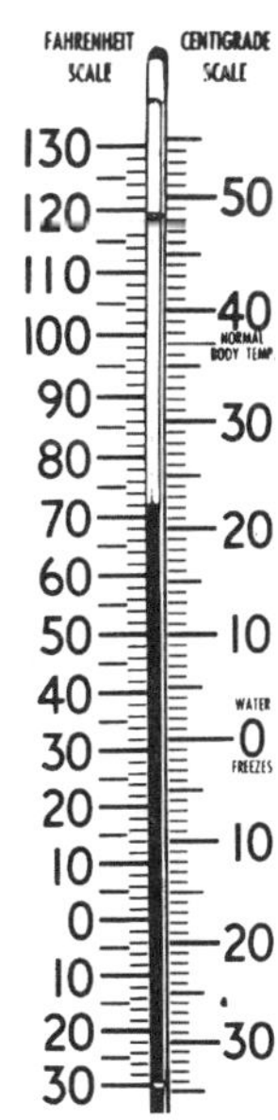

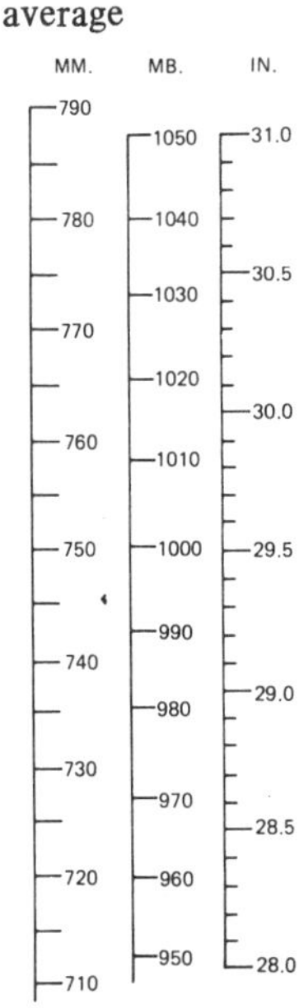

Temperature:
$C° = (F° - 32) \times 5/9$
$F° = C° \times 9/5 + 32$

Barometric Pressure: millimeter millibars, inches

FERRY SERVICE INFORMATION

Maine
*Bar Harbor–Yarmouth, N.S., "The Bluenose," 6 hours Marine Atlantic, 1-800-341-7981
*Bass Harbor–Frenchboro, "Everett Libby," 50 minutes, all year
Bass Harbor–Swans Island, 40 minutes all year
Maine Ferry Service, Bass Harbor 04653, 207-244-3254
*Islesboro–Lincolnville, "Margaret Chase Smith," 20 minutes, all year
Maine Ferry Service, Lincolnville 04849, 207-789-5611, 734-6935
*North Haven–Rockland, "North Haven," 1 hour 10 minutes, all year
Maine Ferry Service, North Haven 04853, 207-867-4441
*Vinalhaven–Rockland, "Governor Curtis," 1 hour 15 minutes, all year
Maine Ferry Service, Rockland 04841, 207-594-5543
Rockland–Matinicus Island, "William S. Silsby," 2 hours 15 minutes, all year once a month
Maine Ferry Service, Rockland Terminal 04841, 207-594-5543
Port Clyde–Monhegan Island, May–October
Monhegan–Thomaston Boat Line, Box 238, Port Clyde 04855, 207-372-8848

Massachusetts
*Hyannis-Nantucket, 2 hours 15 minutes, all year
Steamship Authority, Falmouth 02543, 508-477-8600
Hyannis—Oak Bluffs, Martha's Vineyard, May-October
Hyannis—Nantucket, May-October
Hyline, Ocean St. Dock, Hyannis 02601, 508-778-2600
Falmouth—Oak Bluffs, Martha's Vineyard, "Island Queen," 40 minutes, May-October
Island Commuter Corp., Falmouth 02540, 508-548-4800
*Woods Hole—Vineyard Haven/Oak Bluffs, 45 minutes, all year
Woods Hole Steamship Authority, Falmouth 02543, 508-477-8600
New Bedford—Vineyard Haven, "The Schamonchi," May-October
Cape-Island Express Line, Box J4095, Billy Woods Wharf, New Bedford 02741, 508-997-1688
New Bedford—Cuttyhunk Island, "Alert II," May-October
Cuttyhunk Boat Line, 508-992-1432, passenger reservations required

Rhode Island
*Newport–Block Island, 2 hours
*Jerusalem (Pt. Judith)–Block Island, 1 hour 20 minutes
Interstate Navigation, Galilee State Pier, Pt. Judith 02882, 401-789-3502, 421-4050
Bristol–Prudence Island and Hog Islands, Church St. Wharf, 401-245-7411

Connecticut
*New London–Orient Point, L.I., NY
Cross Sound Ferry, Box 33, New London 06320, 203-443-5281, 516-323-2415
*New London–Block Island, RI, 2 hours 30 minutes
Nelseco Navigation Co., Box 482, New London 06320, 203-442-7891
*New London–Fishers Island, NY, 1 hour, 203-442-0165, car res. 516-788-7463
*Bridgeport–Port Jefferson, L.I., NY, 203-367-3043

New Jersey
*Cape May–Lewes, Del., Lewes Terminal, Lewes 19958, 302-645-6345

North Carolina
*Ocracoke–Hatteras, free ferry
Ocracoke–Cedar Island–Swan Quarter, Ocracoke 919-928-3841, Swan Quarter 919-926-1111

*Will take cars; reservations required.

Where To Buy The Eldridge Tide and Pilot Book

NOVA SCOTIA
Halifax
Binnacle Yachting Equip.
ONTARIO
Toronto
Nautical Mind Bookstore
QUEBEC
Montreal
McGill Maritime

MAINE
Augusta
Kennebec Books
Bar Harbor
Harbor Place
Sherman's Book Store
Belfast
Fertile Mind Bookshop
Biddeford
Rumerys Boatyard
Blue Hill
Blue Hill Books
McVays Hardware Inc.
North Light Books
Boothbay
Carousel Marine
Pod of Edgecombe
Samples Shipyard
Sherman's Book Store
Brunswick
Bookland
Camden
Owl and the Turtle
Wayfarer Marine Corp.
Castine
Four Flags
Damariscotta
Maine Coast Book Shop
Eastport
Moose Island Marine
Falmouth Foreside
HandyBoat Serv.
Kennebunkport
KennebunkBookPort
Port Fly Shop
Standish Marine Books
Kittery
*Dion's Yacht Yard
Jackson Hardware
Little Deer Isle
Buck's Harbor Marine
Northeast Harbor
F.T. Brown Co.
Portland
Chase Leavitt
West Marine
Robinhood
*Robinhood Marine Ctr.
Rockland
Reading Corner
Rockland Boat
Saco
Saco Bay Tackle
Searsport
Hamilton Marine
South Freeport
*Brewer's
Ring's Marine Service
South Portland
Bookland
Southwest Harbor
Henry Hinckley Co.
Stonington
Billings Diesel & Marine
Thomaston
Lyman Yacht Services
West Southport
Boothbay Region Boatyard
Yarmouth
Harbour Books
Landing Boat Supply
Yankee Marina
York
York Harbor Mar. Serv.

VT & NH
Waitsfield
Tempest Bookshop
Exeter
Water Street Bookstore
Hudson
Aqua Venture
Manchester
Goodman's Meteorological
Newington
Great Bay Marina
Portsmouth
Little Professor Book Ctr.
West Marine
Salem
American Angling Supplies

MASS.
Attleboro
L.H. Cooper, Inc.
Bass River
Ship Shops, Inc.
Beverly Farms
The Bookshop
Boston
Boat/US (Waltham)
*Boston Boatworks
Boston Hbr. Sailing Club
Boston Yacht Supplies
*Boxell's Chandlery
C.G. Edwards & Co.
Globe Corner Bookstore
Mallard Discount Marine
*Robert E. White Instr.
Braintree
West Marine
Buzzards Bay
Red Top Sporting Goods
Cambridge
Can Am Sailcraft
Cataumet
*Kingman Marine Inc.
*Parker's Boat Yard
Charlestown
Constitution Marina
*Shipyard Quarters Marina
Chatham
Cabbages & Kings
Cape Fisherman's Supply
Chatham Hardware
Mayflower Shop
Stage Harbor Marine
Yellow Umbrella Books
Chilmark
Menemsha Texaco
Cohasset
Buttonwood Books
Fleming's
Concord
Concord Book Shop
Cotuit
Peck's Boats
Danvers
Baert Marine
Dedham
E&B Marine
Duxbury
Bayside Marine Corp.
East Bridgewater
Sub Sea Divers
East Falmouth
Booksmith
Edwards Boatyard
East Wareham
Ken-Mark Bookstore
Edgartown
Bickerton & Ripley
Edgartown Marine
Everett
Book Stall
Fairhaven
Euro Ship Store
*Fairhaven Shipyard
*D.N. Kelley Yachtyard
Nautical Nonsense
Falmouth
East Marine
Eastman's Sport & Tackle
*Falmouth Marine
*MacDougall's
Market Bookshop
West Marine
Gloucester
The Bookstore
Building Center
Enos Marine
Three Lanterns Ship Supp.
Gosnold
Island Market
Hanover
Paperback Booksmith
Sylvester Co.
Harwich Port
*Allen Harbor Mar. Serv.
Harwich Port Boat Works
Hingham
Hewitt's Cove Marina
Landfall Marine
Old Salt Outfitters
Hull
Sunset Marina
Hyannis
E&B Marine
Hyannis Marina
Sports Port
West Marine
Lincoln
Three S Pharmacy
Malden
Ultimate Marine Supply
Manchester
Manchester Marine Corp.
Marblehead
Lynn Marine Supply Co.
Marblehead Marine Outfitters
*Marblehead Trading Co.
Spirit of '76 Bookstore
*Fred L. Woods
Marion
Book Stall
*Burr Bros. Boats
Mashpee
Bosun's Marine
Mattapoisett
Seaport Pharmacy
Medfield
Lords
Nantucket
*Nantucket Ship Chandlery
Newburyport
Book Rack
Hudson Tackle
New Bedford
Bay Colony News
C.E. Beckman Co.
The Gear Locker
New Bedford Ship Supply
Skip's Marine Supply
West Marine
North Andover
Woodscrew Sporting Goods
North Chatham
Ryder's Cove Boat Yard
North Dartmouth
Baker Books
North Falmouth
Fiddlers Cove Marina
North Weymouth
Tern Harbor Marina
Norwell
Armstrong Boats
Norwood
Green Pond Marina

Onset
Onset Bay Marina
Orleans
Arey's Pond Boatyard
Booksmith
Compass Rose Bookshop
Goose Hummock
Nauset Marina
Osterville
Crosby Yachts
*Nauticus Marina
Oyster Harbors Marine
Sign of the Owl
Peabody
Marine Exchange
West Marine
Plymouth
*Plymouth Marine
Provincetown
Land's End Mar. Supply
Marine Specialties
Quincy
*Marina Bay
Raynham
Slip's Capeway
Reading
Recreational Equipment
Rockport
Toad Hall Bookstore
Salem
Derby Sq. Bookstore
Nautical Traders
Sandwich
Sandwich Ship Supply
Scituate
Front St. Bookshop
*James Landing Marina
Seekonk
E&B Marine
South Dartmouth
Navigator Shop
South Weymouth
Viking Inst. & Photo
Taunton
Book City
Vineyard Haven
Bunch of Grapes Book Store
*Coastwise Packet
Island Marine
Maciel Marine
Martha's Vineyard Fuel & Ice
*Martha's Vineyard Shipyard
Ship's Store of the Vineyard
Watertown
Assoc. Yacht
Webster
Webster Lake Sporting
Wellfleet
Bay Sails Marine
West Dennis
Bass River Marina
Westport
The Village Store
Weymouth
Monahan's Marina
Whitman
Chart House Books
Winthrop
Ward Marine
Woburn
E&B Marine

R.I.
Barrington
*Cove Haven Marina
Little Professor Book Ctr.
Block Island
Ned Phillips Co.
Bristol
Good Books
East Greenwich
Norton's Shipyard
West Marine
Jamestown
*Conanicut Marine
*Dutch Harbor Marina
*Jamestown Boatyard
Narragansett
Narr. Lobster Trap Supply
R.I. Engine Co.
Wilcox Marine Supply
Newport
Armchair Sailor Bookstore
The Book Bay
J.T.'s Chandlery
Newport Nautical
Team One
West Marine
Portsmouth
*Brewer's Sakonnet Mar.
*Little Harbor Marine
Ship's Store & Rigging
Providence
E&B Marine
New England Marine Supply
The Map Center
Riverside
Bullocks Pt. Bt. Yd.
Wakefield
Ram Point Marina
Snug Harbor Marina
Warwick
*Brewer's Cowesett Yt. Yd.
Masthead Marina
Salk's Hardware
Westerly
Frank Hall Boat Yard
Wickford
Kayak Center
*Wickford Cove Marina

CONN.
Branford
Birbarie Marine Sales
*Bruce & Johnson's Marina
E&B Marine
Bridgeport
Hitchcock Marine
Orbit Marine
Clinton
Cedar Is. Marina
The Needle Loft
Riverside Basin Marina
West Marine
Cos Cob
Beacon Point Marine
Coastal Marine Ent.
Darien
West Marine
Deep River
Deep River Marina
East Norwalk
*Norwalk Cove Marina
Essex
*Brewer's Chandlery East
Clipper Ship Book Store
*Dauntless Shipyard
Essex Boat Works
Fairfield
E&B Marine
Greenwich
Landfall Navigation
Outdoor Traders
Guilford
Breakwater Books
Milford
Milford Harbor Chandlery
West Marine
Mystic
Bank Sq. Books
*Brewer's Yacht Yard
Mystic Seaport Stores
Seaport Marine Inc.
Wilcox Marine Supply
West Marine
New Canaan
Great Explorations
New London
Burr's Yacht Haven
Crocker's Boatyard
Hellier Yacht Sales
Niantic
Boats Inc.
Noank
Noank Ship Yard
Sail the Sounds
Spicer's Marinas
Norwalk
Boat/US
*Norwalk Cove Marina
Small Boat Shop
Norwich
Thayer's Marine
Old Lyme
Old Lyme Marina
Old Saybrook
Emerson Book Co.
Kellogg Marine
Rivers End Tackle
Saybrook Point Marina
Sedgwick Marine
Pawcatuck
Connors & O'Brien Mar.
Portland
Wm. J. Petzold Inc.
Portland Boat Works
South Norwak
Rex Marine Ctr.
South Windsor
Mellen's Marine Inc.
Stamford
Basset Book Shop
*Brewer's Yacht Haven
Hathaway Reiser & Raymond
West Marine
Stonington
Book Mart
Dodson's Boat Yard
Don's Dock
Stratford
Marina at the Dock
Waterford
Defender Industries
Electronautical Equip.
Hillyer's Tackle Shop
Westbrook
*Pilots Point Marinas
West Hartford
Norman Marine
West Mystic
Mystic Shipyard
Westport
Coastwise Service
Wilton
Outdoor Sports
Woodbridge
Woodbridge Fishing & Tackle

N.Y.
Albany
The Book House
Amityville
Yacht Service
Aquebogue
Larry's Lighthouse
Babylon
West Marine
Bay Shore
Gibson & Cushman
Brooklyn
Bernie's Tackle
Catskill
Hop-o-Nose Marine
Centereach
Diesel Marine Service
City Island
J.J. Burck
Cutchogue
Cutchogue Marine
Easthampton
East Hampton Pt. Marina
Three Mile Harbor Boat Yard
Fisher's Island
Pirate's Cove Marine
Freeport
Fred Chall Marine
Freeport Mar. Supply
Garden City
E&B Marine

Glen Cove
*Brewer Yacht Yard
Glen Cove Marina
Greenport
*Brewer Yacht Yard
*S.T. Preston
Washington White's Sons
Hampton Bays
Modern Yachts
Huntington
Book Revue
Coney's Marine
E&B Marine
Knutson Marine Supply
Island Park
Allstate Marine
Mamaroneck
R.G. Brewer Inc.
*Post Rd. Boat Yard
Montauk
Gone Fishing Marine
Montauk Marine
Star Island Yacht Club
New Rochelle
Post Marine Supply
West Harbor Yacht Serv.
New York City
E&B Marine
Hagstrom Map
Nautical Charts
N.Y. Kayak
*N.Y. Naut. Instr. & Serv.
Seaport Museum Shops
Northport
Northport Marine Ctr.
Oyster Bay
Nobman's Marine Hdwre.
Oyster Bay Marine Supply
Seawanhaka Boat Yard
Patchoque
B. Sack
Port Jefferson
E&B Marine
Port Washington
Capri Marina
Dock Shop
Northstar Marine & Chandlery
Riverhead
West Marine
Sag Harbor
Ace Hardware
Baron's Marina
The Emporium, Inc.
Henry Persan & Sons
Saugerties
Atlantic Kayak
Shelter Island
Coecles Harbor Marina
Stony Point
Willow Cove Marina
Syosset
West Marine
Upper Nyack
*Julius Peterson, Inc.
Westhampton Beach
Chesterfield Assoc.
West Islip
E&B Marine

NJ
Atlantic Highlands
The Skipper's Shop
West Marine
Beachwood
Barnegat Bay Sailing Ctr.
Belfort
Mariners Mart
Berlin
Goldberg's
Bricktown
Green Cove Marina
West Marine
Brielle
Brielle Marine Basin
Cape May
Sea Gear Marine
Tony's Marine
West Marine
Collingwood
M&E Marine
Eatontown
E&B Marine
Hillside
Sailors Bookshelf
Homdel
Boat/US
Lodi
E&B Marine
Mantoloking
Winter Yacht Basin
Mt. Laurel
E&B Marine
Paramus
Ramsey Outdoor
Perthy Amboy
E&B Marine
Ridgefield
International Map Co.
Somerspoint
Brennan Marine Supply
E&B Marine
Somerville
Baker & Taylor
South Amboy
Lockwood Boat Works
Morgan Marina
Toms River
E&B Marine
Weehawken
Port Imperial Marina

PA
Paoli
Sail Gear
Phildelphia
E&B Marine
Victor Auguste Gustin
The Pilot House
West Conshohocken
Barnes-Wright Boats

DC
Washington Marina Co.

DE
Newark
Eastern Marine
Wilmington
First State Map Co.

MD
Annapolis
Boat/US
*Fawcett Boat Supplies
C. Plath
West Marine
Baltimore
E&B Marine
Maryland Nautical Sales
Vane Bros. Marine
West Marine
Beltsville
Boater's World
Cambridge
*Yacht Maintenance
Chesapeake City
Schaefer's Marina
Chestertown
Worton Creek Marina
Easton
West Marine
Georgetown
Georgetown Yacht Basin
Glen Burnie
E&B Marine
Kent Island
West Marine
Lanham
E&B Marine
Oxford
Crockett Bros. Boatyard
Pasadena
West Marine
Rosedale
E&B Marine
Solomon's Island
Zahniser's

VA
Alexandria
Boat/US
West Marine
Deltaville
West Marine
Glen Allen
E&B Marine
Hampton
West Marine
Norfolk
Boat/US
W.T. Brownley Co.
E&B Marine
Portsmouth
Holiday Harbor Marina
Tidewater Yacht
Virginia Beach
Lynnhaven Yt. Marina
West Marine
Woodbridge
West Marine

NC
Beaufort
N.C. Maritime Museum
Scuttlebutt
Morehead City
Dee Gee's
West Marine
New Bern
E&B Marine
Wilmington
West Marine

SC
Charleston
E&B Marine

GA
Savannah
E&B Marine

FLA
Daytona
West Marine
Ft. Lauderdale
Bahia Mar Marine Store
Bluewater Books
Boat/US
West Marine
Fort Pierce
West Marine
Hollywood
E&B Marine
Jacksonville
E&B Marine
Pier "17" Marina
West Marine
Key Largo
West Marine
Key West
West Marine
Marathon
West Marine
Miami
E&B Marine
Hopkins-Carter Hdwre.
West Marine
North Palm Beach
West Marine
Pompano Beach
E&B Marine
Stuart
West Marine
West Palm Beach
Boat/US
Hopkins Mar. Hdwre.
Spencer Boat Co.

Advertisers in book.
Refer to page 270.

INDEX TO ADVERTISEMENTS

ELDRIDGE TIDEMARK

A Cruising Aid

To save time when cruising, list ports of call below, with time differences from nearest port of your choice. (See section on Times of High Water.)

Harbor	Time Diff. Hrs.	Min.	Before or after (Circle One)
			Before After
			Before After
			Before After
			Before After
			Before After
			Before After
			Before After
			Before After
			Before After

1999

JAN.

S	M	T	W	T	F	S
—	—	—	—	—	1	2
3	4	5	6	7	8	9
10	11	12	13	14	15	16
17	18	19	20	21	22	23
24	25	26	27	28	29	30
31	—	—	—	—	—	—

APRIL

S	M	T	W	T	F	S
—	—	—	—	1	2	3
4	5	6	7	8	9	10
11	12	13	14	15	16	17
18	19	20	21	22	23	24
25	26	27	28	29	30	—

JULY

S	M	T	W	T	F	S
—	—	—	—	1	2	3
4	5	6	7	8	9	10
11	12	13	14	15	16	17
18	19	20	21	22	23	24
25	26	27	28	29	30	31

OCT.

S	M	T	W	T	F	S
—	—	—	—	—	1	2
3	4	5	6	7	8	9
10	11	12	13	14	15	16
17	18	19	20	21	22	23
24	25	26	27	28	29	30
31	—	—	—	—	—	—

FEB.

S	M	T	W	T	F	S
—	1	2	3	4	5	6
7	8	9	10	11	12	13
14	15	16	17	18	19	20
21	22	23	24	25	26	27
28	—	—	—	—	—	—

MAY

S	M	T	W	T	F	S
—	—	—	—	—	—	1
2	3	4	5	6	7	8
9	10	11	12	13	14	15
16	17	18	19	20	21	22
23	24	25	26	27	28	29
30	31	—	—	—	—	—

AUG.

S	M	T	W	T	F	S
1	2	3	4	5	6	7
8	9	10	11	12	13	14
15	16	17	18	19	20	21
22	23	24	25	26	27	28
29	30	31	—	—	—	—

NOV.

S	M	T	W	T	F	S
—	1	2	3	4	5	6
7	8	9	10	11	12	13
14	15	16	17	18	19	20
21	22	23	24	25	26	27
28	29	30	—	—	—	—

MAR.

S	M	T	W	T	F	S
—	1	2	3	4	5	6
7	8	9	10	11	12	13
14	15	16	17	18	19	20
21	22	23	24	25	26	27
28	29	30	31	—	—	—

JUNE

S	M	T	W	T	F	S
—	—	1	2	3	4	5
6	7	8	9	10	11	12
13	14	15	16	17	18	19
20	21	22	23	24	25	26
27	28	29	30	—	—	—

SEPT.

S	M	T	W	T	F	S
—	—	—	1	2	3	4
5	6	7	8	9	10	11
12	13	14	15	16	17	18
19	20	21	22	23	24	25
26	27	28	29	30	—	—

DEC.

S	M	T	W	T	F	S
—	—	—	1	2	3	4
5	6	7	8	9	10	11
12	13	14	15	16	17	18
19	20	21	22	23	24	25
26	27	28	29	30	31	—

LOGARITHMIC TIME/SPEED/DISTANCE SCALE

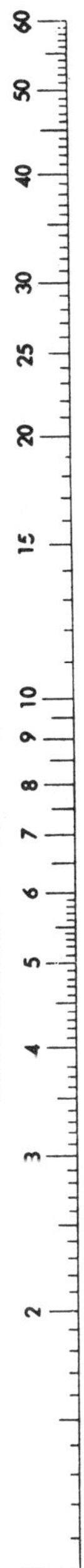

To find SPEED, place one point of dividers on nautical miles run and the other on minutes run. Without changing divider spread, place right point on 60 and left point will then indicate speed in knots. Example: with 4.0 miles run in 15 minutes, the speed is 16.0 knots.

To find TIME to run a given distance, place right point of dividers on 60 and left point on boat speed. Without changing divider spread, put left point on distance to be run and right point will show number of minutes it will take. Example: If speed is 10 knots and distance to be run is 5 miles, it will take 30 minutes.

ELDRIDGE TideWare™

Tradition Meets Technology

Real-time, Tide, Current and Coastal Map Software

"The world's most comprehensive, reliable and versatile software program for tides and currents has been launched"

Of course, our great-grand father Captain George W. Eldridge never did own a laptop computer. However, if he were around today he would certainly have one, loaded with all of the latest software. After all, nothing is more fundamental to the study of tides and currents than change.

In response to the opportunities of new technology, we felt you would want to have the benefit of our tidal observations fully integrated with **50 vector shoreline charts** for your computer. We make this announcement with full assurance that this new software meets our most rigorous standards for accuracy, comprehensiveness and reliability.

What's more, we have added features which you will find helpful, perhaps even invaluable. As a navigational aid, Eldridge TideWare's advanced design easily takes you from simple - accessing **NOAA East Coast and Gulf of Mexico** tide and current data - to the sophisticated. Our Custom Ports feature for zeroing in on your own dock, inlet or fishing spot. Eldridge Tideware's Real-Time Window provides a 24-hour display of tides and currents for any time, any place in the past, present or future. We trust with this marriage of tradition and technology, Eldridge TideWare will end your tide and current guesswork forever.

We think Captain George W. Eldridge would be proud!

Yours for a fair tide,

The Publishers of Eldridge

To purchase Eldridge TideWare visit your favorite boating or marine outlet, or call us toll free at 1-877-TIDEWARE (843-3927), or contact us on the internet at www.tideware.com